METHODS FOR STUDYING THE GENETICS, MOLECULAR BIOLOGY, PHYSIOLOGY, AND PATHOGENESIS OF THE STREPTOCOCCI

Methods for studying the genetics, molecular biology, physiology, and pathogenesis of the streptococci

Edited by

PAULA M. FIVES-TAYLOR and DONALD J. LEBLANC

Reprinted from *Methods in Cell Science*, Volume 20 (1–4), 1998

KLUWER ACADEMIC PUBLISHERS

DORDRECHT / BOSTON / LONDON

Library of Congress Cataloging-in-Publication Data

ISBN 0-7923-5815-5

Published by Kluwer Academic Publishers,
P.O. Box 17, 3300 AA Dordrecht, The Netherlands

Sold and distributed in North, Central and South America
by Kluwer Academic Publishers,
101 Philip Drive, Norwell, MA 02061, U.S.A.

In all other countries, sold and distributed
by Kluwer Academic Publishers,
P.O. Box 322, 3300 AH Dordrecht, The Netherlands

Printed on acid-free paper

Printed in the Netherlands

Methods in Cell Science **20**: vii–ix (1998).

EDITORIAL

Methods for studying the genetics, molecular biology, physiology, and pathogenesis of the streptococci

Paula Fives-Taylor[1] & Donald J. LeBlanc[2]
[1] *Department of Microbiology and Molecular Genetics, University of Vermont, Burlington, Vermont, USA*
[2] *Lilly Research Laboratories, Eli Lilly and Company, Indianapolis, Indiana, USA*

This issue of Methods in Cell Science is devoted entirely to methods, or new research tools, developed in and for studies of members of the bacterial family, *Streptococcaceae.* The methods included were specifically designed for analysis of different species belonging to what was once a single genus, *Streptococcus*, but which now includes three genera of *Streptococcaceae*: the original genus, *Streptococcus*, in which are retained the β-hemolytic species, such as *S. pyogenes* and *S. agalactiae*, the oral streptococci, e.g., *S. mutans* and *S. parasanguis*, and *S. pneumoniae*; *Enterococcus*, formerly represented among the Lancefield Group D streptococci, and including such species as *E. faecalis* and *E. faecium*; and *Lactococcus*, formerly referred to as the Group N, or dairy, streptococci, most notably *L. lactis*. Many investigators conducting research on the so-called 'pathogenic streptococci' seldom interact with those who study dairy organisms, such as the lactococci, and many of the former also fail to recognize the pathogenic potential of the enterococci. Yet, it has become clear in recent years that procedures developed specifically for members of one genus are very often applicable to investigations of all three genera. Thus, it is hoped that those working with any one streptococcal genus will find valuable information in articles written by investigators involved with species classified in one of the other two genera.

Certainly, many of the studies that have been conducted on the *Streptococcaceae* were initiated because of the diseases they cause, or to enhance their utility from an industrial perspective. However, the results of many of these investigations have demonstrated a complexity among some members of the family that would seem to warrant an interest in them in their own right, apart from, or in addition to, any biomedical or industrial considerations. For example, many of the streptococci and pneumococci have as many as 30 multifunctional surface proteins. Some of these prokaryotic organisms are capable of glycosylating proteins in a manner similar to eukaryotes, a concept that just a few years ago would have been thought to be heresy. Clearly, there is much to be learned from these complex organisms.

Although many of the papers appearing in this issue describe highly sophisticated genetic and molecular techniques, the genetics and molecular biology of the Gram-positive cocci lagged behind that of the *Enterobacteriaceae* for many years. This seems somewhat ironic, since the origins of bacterial genetics and molecular biology can be traced to two seminal papers that describe the results of studies on *S. pneumoniae*. The first of these papers, of course, represented the beginning of bacterial genetics, i.e., the transformation experiements of Griffith, in 1928 [5]. The second, and a major boost to both bacterial genetics and molecular biology, was the identification of DNA as Griffith's transforming principle, by Avery, MacLeod, and McCarty, in 1944 [1]. However, very little work in streptococcal genetics was published until 30 years later, when Jacob and Hobbs reported the first definitive demonstration of conjugation between streptococcal (enterococcal) strains [6]. Despite the pioneering work presented in those three papers on members of the *Streptococcaceae*, four years after the appearance of the latter paper, the ever widening gap between the streptococci and other bacterial groups, relative to advances in genetics, molecular biology, and by that time recombinant DNA technology, as well, was pointed out in the inaugural address at the VIIth International Symposium on Streptococci and Streptococcal Diseases, by Sir Robert Williams, President of the Symposium [8]. He noted that although there were a large number of papers to be presented on 'the fine details of the structure and behaviour of the antigens of *Streptococcus pyogenes*, but in comparison with most other areas of bacteriology, rather little on genetics and antibiotic susceptibility'. In fact, of the 156 presentations at that meeting, only seven were concerned with genetic transfer mechanisms, mostly on the conjugative transfer of antibiotic resistance plasmids. Sir Robert jokingly referred to a stage play in London called 'No sex please, we're British . . .', and to the fact that for most symposia on streptococci prior to the present one, one might give credence to the comparable assertion, 'no sex please, we're streptococci'. However, that symposium, with its seven papers on genetic transfer presented in a single poorly attended session, served as a new beginning for streptococcal genetics, a field which continues to grow to this day.

The enthusiastic and optimistic speakers at that

session met at dinner in the evening to discuss the possibility of an international conference on streptococcal genetics, sometime in the future. Three years later, in November of 1981, the first ASM International Conference on Streptococcal was held in Sarasota, FL, and was attended by 140 scientists from 14 countries, who had come to listen to 33 oral presentations, and read and discuss 44 posters devoted exclusively to streptococcal genetics [7]. Of the many highlights of that meeting, two stand out in particular, the keynote address by Maclyn McCarty, titled 'Streptococci and the Birth of Molecular Genetics', and the session titled 'Development and Use of Recombinant DNA Technology'. There have now been a total of four International Streptococcal Genetics meetings since then, the latest in 1999 in Vichy, France, and attendance has exceeded 300 scientists from more than 20 different countries. Many new genetic and molecular tools have been introduced at these meetings, and presented in publications stemming from them [2–4]. However, the streptococci are often difficult to work with, and there are often minute, yet extremely critical details that are omitted from the proceedings of scientific conferences, primarily because of space constraints. In the articles appearing in this issue of Methods in Cell Science devoted to the *Streptococcaceae*, the experimental methods have been written in such a manner as to permit someone unfamiliar with the field to repeat the experiments presented without the need of additional sources.

Two additional occurrences were noted in the inaugural address of the VIIth International Symposium on Streptococci and Streptococcal Diseases. These included the resuscitation of *S. mutans* as a cause of dental caries, and an upsurge of interest in Lancefield group B streptococci and in group B infections. Recognition of these two species as the causes of important diseases has resulted in a large number of studies on their virulence over the past 20 years, and consequently, in the development of a significant amount of the new methodology, much of which will be applicable to other streptococcal species. In fact, strains of *S. mutans*, or one or more other oral streptococcal species, and the group B streptococci, were employed in the development of more than half of the methods to be described in the articles appearing in this issue.

The 27 papers appearing in this issue are grouped into seven sections. The first includes methods of mutagenesis, which is accomplished by the use of transposons (Cvitkovitch et al.), suicide vectors (Yim and Rubens), or a combination of both (Qin et al.). The second section contains several different genetic transfer, and/or host-vector systems. Three of the papers in this section describe new integration vectors (Leenhouts et al., McShan et al., and Tao), which can be used for multiple purposes, including mutagenesis, while the remaining four provide protocols for gene transfer between bacterial strains via transformation (Gaustad and Morrison), conjugation (Mills et al.), electroporation (Hirt et al.), and by mobilization or electroporation (LeBlanc et al.). The papers in section 3 describe methods for gene isolation (Xu et al., and Nizet et al.), gene mapping (Lee et al.), and for the study of gene expression (Djordejvic and Klaenhammer, Shiroza and Kuramitsu, and Grey et al.). Section 4 concerns studies on the regulation of gene expression at the transcriptional (Froeliger and Fives-Taylor, and Peruzzi et al.) and translational (Quivey et al.) levels, or by cells in an adherent versus planktonic state (Burne and Chen). The fifth section includes methods for the study of parasite-host interactions in vitro (Winram et al.) and in vivo (Munro), or by analyses of bacterial cell surface proteins (McNab and Jenkinson). Section 6 contains papers that provide tools for epidemiologic analysis, including species-specific isolation media (Spatafora and Moore), and methods of species identification (Paster et al.), and strain differentiation (Patterson and Kelly). Finally, section 7 contains a single paper on the in vitro production of antibodies (Stephenson et al.).

Most of the papers in this issue deal with microorganisms that are pathogenic for humans. Such a host:parasite relationship involves a complex array of highly specific molecular and cellular interactions. A great deal of progress has been made in the identification of the causes of epidemic diseases such as tuberculosis and plague. However, many of the diseases associated with organisms like the streptococci and enterococci, that can coexist with their host indefinitely, and in some instances may even be considered normal members of the human flora, are more intractable. Koch's postulates cannot be fulfilled for the most part. The real challenge is to understand the mechanisms of the disease processes of these bacteria that have evolved together with their human hosts, resulting in the elaboration of well developed and subtle strategies for coexistence. Some properties of these bacteria, such as the ability to usurp host functions, or to invade host cells and tissues, may facilitate peaceful coexistence with the host, or contribute to pathogenesis. Other properties, such as the ability to secrete toxins and other factors, and to cause tissue damage through stimulation of the host inflammatory response, would seem to be overt virulence traits. Clearly, whether coexistence or pathogenesis prevails must depend on changes in the regulation of many of these traits. How, then, does one decipher the mechanisms and conditions associated with such regulation? Perhaps the most important area in need of study relates to the types of signaling involved in cell-cell communication; i.e., bacteria to bacteria, bacteria to host, host to bacteria, bacteria in mixed communities (biofilms), bacteria in their environment. Significant insights into the evolution of microbial pathogenicity have been gained from studies of such properties at the molecular level, but much remains to be learned if the information is to provide a basis for the development of new therapeutic agents to prevent or interrupt microbial infections. Furthermore, a greater understand of the role of the immune system in health and disease will be required. While advances have been made in understanding the immune response to specific pathogens, what is still needed is an understanding of how the immune system influences all the phases of disease progression. It is expected that the advanced methods contained in this issue of Methods in Cell Science will provide many of the tools required to obtain a better understanding of how microorganisms interact with each other, their environment, and their hosts.

Acknowledgment

The editors wish to express their thanks to Dr Eunice Froeliger for all her help in receiving and mailing papers while one of the editors was on sabbatic leave.

References

1. Avery OT, Macleod CM, McCarty M (1944). Studies on the chemical nature of the substance inducing transformation of pneumococcal types. Induction of transformation by a desoxyribonucleic fraction isolated from pneumococcus type III. J Exp Med 79: 137–158.
2. Dunny GM, Cleary PP, McKay LL (eds) (1991). Genetics and molecular biology of streptococci, lactococci, and enterococci. Washington, DC: American Society for Microbiology.
3. Ferretti JJ, Curtiss R III (eds) (1987). Streptococcal genetics. Washington, DC: American Society for Microbiology.
4. Ferretti JJ, Gilmore MS, Klaenhammer TR, Brown F, (eds) (1995). Developments in Biological Standardization, Vol. 85. Genetics of Streptococci, enterococci and lactococci. New York: Karger.
5. Griffith F (1928). The significance of pneumococcal types. J Hygiene 27: 113–159.
6. Jacob AE, Hobbs SJ (1974). Conjugal transfer of plasmid-borne multiple antibiotic resistance in *Streptococcus faecalis* var. *zymogenes*. J Bacteriol 117: 360–372.
7. Schlessinger D (ed) (1982). Microbiology – 1982. Washington, DC: American Society for Microbiology.
8. Williams REO, In: Parker MT (ed) (1979). Pathogenic Streptococci. Surrey: Reedbooks, Ltd.

Dedication

The editors have agreed to dedicate this special issue on methods for the study of the streptococci to the memory of the late Dr. Ronald Gibbons. The pioneering studies of Gibbons and coworkers with the oral viridans group of streptococci were among the first to establish the critical initial role of microbial adherence in the colonization of host mucosal surfaces. In the late sixties and early seventies very few papers were published that documented the ability of microorganisms to attach to human tissues in a specific fashion. Dr. Gibbons first papers on the topic appeared in the early seventies. He clearly showed that microorganisms must have specific adhesins for receptors on oral tissues and that this specificity was responsible for their ecological niche. His work so dramatically influenced the field that the number of papers on bacterial adhesion increased dramatically (Figure 1). In 1980, Drs. Ofek and Beachey, well-known for their work on the adhesion of Gram-negative organisms, recognized the contributions made by Gibbons and coworkers when they stated: 'Although the adhesiveness of bacteria for some mammalian cells was recognized as early as 1908 and selectivity was demonstrated by Duguid and Old (1980), an understanding of the role of microbial adhesion in the initiation of the infectious processes obtained a major stimulus through the studies of the attachment of oral bacteria to the surfaces of the mouth'. While insights into the molecular basis of streptococcal adhesion have continued with the studies performed over the past 25 years, much still remains to be discovered.

References

1. Gibbons RJ (1989). Bacterial adhesion to oral tissues: A model for infectious diseases. J Dent Res 68: 750–760.
2. Ofek I, Beachey EH (1980). Bacterial adherence. Adv Intern Med 25: 503–532.

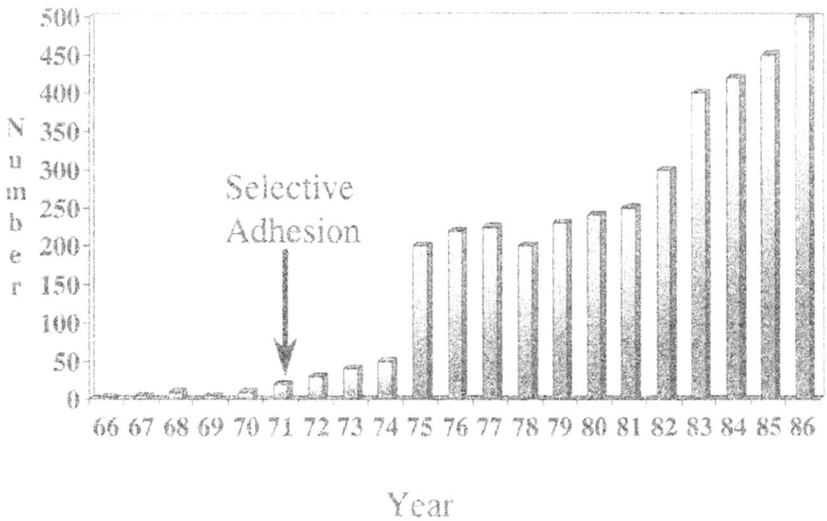

Figure 1. Papers published on bacterial adhesion.

List of contributors

Kathleen A. Baldwin
University of Minnesota
Department of Food Science and Nutrition
1334 Eckles Avenue
St. Paul, MN 55108, USA

Irena M. Bartoszyk
Forsyth Dental Center
Department of Molecular Genetics
140 Fenway
Boston, MA 02115, USA

Arnold S. Bleiweis*
University of Florida
Department of Oral Biology
Gainesville, FL 32610, USA

Nicole D. Buckley
Department de Biochimie (Sciences)
Universite Laval, Quebec
Canada

Robert A. Burne*
University of Rochester
School of Medicine and Dentistry
Departments of Dental Research, Microbiology and Immunology
601 Elmwood Avenue
Rochester, NY 14642, USA

Yi Chen
University of Minnesota
Medical School
Department of Microbiology
Minneapolis, MN 55455-0312, USA

Yi-Ywan M. Chen
University of Rochester
School of Medicine and Dentistry
Department of Dental Research
601 Elmwood Avenue
Rochester, NY 14642, USA

Paula J. Crowley
University of Florida
Department of Oral Biology
Gainesville, FL 32610, USA

Roy Curtiss III
Washington University
Department of Biology
St. Louis, MO 63130, USA

Dennis G. Cvitkovitch
University of Toronto
Dental Research Institute
124 Edward Street
Toronto, Ontario,
Canada M5G 1G6

Lolita Daneo-Moore*
Temple University School of Medicine
Department of Microbiology and Immunology
Philadelphia, PA 19140, USA

Floyd E. Dewhirst
Forsyth Dental Center
Department of Molecular Genetics
140 Fenway
Boston, MA 02115, USA

G. M. Djordjevic
North Carolina State University
Department of Microbiology
Raleigh, NC 27695-7624, USA

Gary M. Dunny*
University of Minnesota
Medical School
Department of Microbiology
Minneapolis, MN 55455-0312, USA

Roberta C. Faustoferri
University of Rochester
School of Medicine and Dentistry
Department of Dental Research
Box 611, 601 Elmwood Avenue
Rochester, NY 14642, USA

Joseph J. Ferretti*
University of Oklahoma Health Sciences Center
Department of Microbiology and Immunology
Oklahoma City, OK 73190, USA

Paula M. Fives-Taylor*
University of Vermont
Departments of Microbiology and Molecular Genetics
Stafford Building
Burlington, VT 05405, USA

Eunice H. Froeliger
University of Vermont
Departments of Microbiology and Molecular Genetics
Stafford Building
Burlington, VT 05405, USA

Peter Gaustad*
National Hospital
Microbiological Institute
Oslo, Norway

W. Todd Grey
The University of North Carolina at Charlotte
Department of Biology
Charlotte, NC 28223, USA

Juan A. Gutierrez
Millenium Pharmaceuticals, Inc.
640 Memorial Drive
Cambridge, MA 02137, USA

Jeffrey D. Hillman
University of Florida
Department of Oral Biology
Gainesville, FL 32610, USA

Helmut Hirt
University of Minnesota
Medical School
Department of Microbiology
Minneapolis, MN 55455-0312, USA

Michael C. Hudson*
The University of North Carolina at Charlotte
Department of Biology
Charlotte, NC 28223, USA

Howard F. Jenkinson*
University of Bristol
Departments of Oral and Dental Science
Dental Hospital and School
Lower Maudlin Street
Bristol BS1 2LY, UK

Lingxia Jiang
University of Texas Medical School
Departments of Biochemistry and Molecular Biology
Houston, Texas 70030
Current address: The Institute for Genomic Research
Rockville, MD 20850, USA

Xiaomei Jin
University of Texas Medical School
Department of Medical
Division of Infectious Diseases
Houston, TX 77030, USA

Cindy C. Kelly
University of Texas Health Science Center at San Antonio
Department of Pathology
San Antonio, TX, USA

T. R. Klaenhammer*
North Carolina State University
Department of Food Science
Raleigh, NC 27695-7624, USA

Jan Kok*
University of Groningen
Biomolecular Sciences and Biotechnology Institute
Department of Genetics
Haren, The Netherlands

Wendi L. Kuhnert
University of Rochester
School of Medicine and Dentistry
Department of Microbiology and Immunology
Box 611, 601 Elmwood Avenue
Rochester, NY 14642, USA

Howard Kuramitsu*
State University of New York
Department of Oral Biology
3435 Main Street
Buffalo, NY 14214, USA

Joshua D. Lasker
The University of North Carolina at Charlotte
Department of Biology
Charlotte, NC 28223, USA

Donald J. LeBlanc*
Department of Oral Biology
Indiana University School of Dentistry
Indianapolis, IN, USA

Linda N. Lee
Department of Medicine
Infectious Diseases
University of Texas
Health Science Center at San Antonio
San Antonio, TX, USA

Martin H. Lee
University of Washington
Children's Hospital Medical Center CH-32
Department of Pediatrics
Seattle, WA 98105, USA

Kees Leenhouts
University of Groningen
Biomolecular Sciences and Biotechnology Institute
Department of Genetics
Haren, The Netherlands

Robert J. Melamede
University of Vermont
Departments of Microbiology and Molecular Genetics
Burlington, VT, USA

Larry L. McKay
University of Minnesota
Department of Food Science and Nutrition
1334 Eckles Avenue
St Paul, MN 55108, USA

Robert E. McLaughlin
University of Oklahoma Health Sciences Center
Departments of Microbiology and Immunology
Oklahoma City, OK 73190, USA

Roderick McNab
University of Bristol
Department of Oral and Dental Science
Dental Hospital and School
Lower Maudlin Street
Bristol BS1 2LY, UK

W. Michael McShan
University of Oklahoma Health Sciences Center
Department of Microbiology and Immunology
Oklahoma City, OK 73190, USA

David A. Mills
North Carolina State University
Box 7624
Raleigh, NC 27695-7624, USA

Meagan W. Moore
Middlebury College
Department of Biology
Middlebury, VT 05753, USA

Donald A. Morrison
University of Illinois at Chicago
Department of Biological Sciences
Laboratory for Molecular Biology
Chicago, IL, USA

Cindy L. Munro*
Virginia Commonwealth University
Box 980567
Richmond, VA 23298-0567, USA

Barbara E. Murray*
University of Texas Medical School
Departments of Microbiology and Molecular Genetics
Medicine and Center for the Study of Emerging and Re-emerging Pathogens
Houston, TX 77030, USA

Aphakorn Nittayajarn
University of Washington
Children's Hospital Medical Center CH-32
Department of Pediatrics
Seattle, WA 98105, USA

Victor Nizet*
University of California-San Diego
Department of Pediatrics
La Jolla, CA, USA

Annika Nordstrand
Umeå University
Department of Clinical Microbiology
Umeå, Sweden

Bruce J. Paster*
Forsyth Dental Center
Department of Molecular Genetics
140 Fenway
Boston, MA 02115, USA

Jan E. Patterson*
University of Texas Health Science Center at San Antonio
Department of Medicine/Infectious Diseases
San Antonio, TX 78284-7881, USA

Francesca Peruzzi
Temple University of School of Medicine
Departments of Microbiology and Immunology
Philadelphia, PA 19140, USA

Trevor G. Phister
University of Minnesota
Department of Food Science and Nutrition
1334 Eckles Avenue
St Paul, MN 55108, USA

Patrick J. Piggot
Temple University School of Medicine
Department of Microbiology and Immunology
Philadelphia, PA 19140, USA

Xiang Qin
University of Texas Medical School at Houston
Division of Infectious Diseases
Department of Medicine
Departments of Microbiology and Molecular Genetics
Center for the Study of Emerging and Re-emerging Pathogens
Houston, TX 77030, USA

Robert G. Quivey, Jr*
University of Rochester
School of Medicine and Dentistry
Department of Dental Research
Box 611, 601 Elmwood Avenue
Rochester, NY 14642, USA

Craig E. Rubens*
University of Washington
Children's Hospital and Medical Center
Department of Pediatrics
Division of Infectious Diseases
Seattle WA 98105, USA

Patrick M. Schlievert
University of Minnesota
Medical School
Department of Microbiology
Minneapolis, MN, USA

Teruaki Shiroza
Nihon University
School of Dentistry
Department of Biochemistry
Matsudo, Japan

Kavindra V. Singh
University of Texas Medical School of Houston
Division of Infectious Diseases
Department of Medicine
Center for the Study of Emerging and Re-emerging Pathogens
Houston, TX, USA

Arnold L. Smith
University of Missouri-Columbia Medical School
Departments of Microbiology and Molecular Genetics
Burlington, VT, USA

Grace Spatafora*
Middlebury College
Department of Biology
Middlebury, VT 05753, USA

Aimee E. Stephenson
University of Vermont
Departments of Microbiology and Molecular Genetics
Burlington, VT, USA

Paul M. Sullam
University of California-San Francisco
Veteran Affairs Medical Center
Department of Medicine
San Francisco, CA, USA

Glen S. Tamura
University of Washington
Children's Hospital and Medical Center
Department of Pediatrics
Division of Infectious Diseases
Seattle, WA 98105, USA

Lin Tao*
University of Missouri-Kansas City
School of Dentistry
Department of Oral Biology
Kansas City, MO 64108, USA

Fang Teng
University of Texas Medical School at Houston
Departments of Biochemistry and Molecular Biology
Center for the Study of Emerging and Re-emerging Pathogens
Houston, TX 77030, USA

Gerard Venema
University of Groningen
Biomolecular Sciences and Biotechnology Institute
Department of Genetics
Haren, The Netherlands

George M. Weinstock*
University of Texas Medical School
Departments of Biochemistry and Molecular Biology
Microbiology and Molecular Genetics and Center for the Study of Emerging and Re-emerging Pathogens
Houston, TX 77030, USA

Scott B. Winram
University of Washington
Children's Hospital and Medical Center
Department of Pediatrics
Division of Infectious Diseases
Seattle, WA 98105, USA

Laura Wojciechowski
University of Florida
Department of Oral Biology
Gainesville, FL 32610, USA

Yi Xu
University of Texas Medical School at Houston
Departments of Biochemistry and Molecular Biology
Center for the Study of Emerging and Re-emerging Pathogens
Houston, TX 77030, USA

Harry H. Yim
University of Washington
Children's Hospital and Medicine Center
Department of Pediatrics
Division of Infectious Diseases
Seattle, WA 98105, USA

* Corresponding Author

Methods in Cell Science **20**: 1–12 (1998).

Tn*917* transposon mutagenesis and marker rescue of interrupted genes of *Streptococcus mutans*

Dennis G. Cvitkovitch[1], Juan A. Gutierrez[2], Paula J. Crowley, Laura Wojciechowski, Jeffrey D. Hillman & Arnold S. Bleiweis

Department of Oral Biology, University of Florida, Gainesville, Florida, USA; [1] Present address: University of Toronto Dental Research Institute, Toronto, Ontario, Canada; [2] Present address: Millenium Pharmaceuticals Inc., Cambridge, Massachusetts, USA

Abstract. In order to study virulence factors of the opportunistic oral pathogen *Streptococcus mutans* we have used Tn*917* mutagenesis to generate mutants defective in specific phenotypic properties believed to be associated with virulence. This work describes the procedures required to use the temperature-sensitive transposon Tn*917* delivery vector pTV1-OK, along with two techniques to recover inactivated genes. This vector has proven useful in generating transposon mutants of a poorly transformable *S. mutans* strain. We have successfully isolated mutants with diminished growth at pH 5.0, with defects in production of a mutacin, formerly known as BLIS or bacteriocin-like inhibitory substance and with nutritional requirements in minimal media including adenine, glutamate and arginine auxotrophy. The transposon has been used successfully in two strains of *S. mutans*, JH1005 and NG8, where transposition frequencies of 10^{-5} and 10^{-4} were observed, respectively. Rescue of inactivated genes was achieved using a recovery vector, pTV21Δ2TetM, that generates a replicative *E. coli* plasmid by reciprocal recombination with the mutant *S. mutans* chromosomal transposon site, and shot-gun cloning chromosomal DNA of mutants into pUC18, followed by selection of ampicillin and erythromycin resistant transformants in *E. coli*. The techniques described here can be adapted for generating mutants and recovering genes from a number of other Gram-positive and possibly Gram-negative bacteria.

1. Introduction

Streptococcus mutans is an organism critical to the onset and progression of dental caries, a disease characterized by the dissolution of tooth enamel and dentin by acid end-products of carbohydrate metabolism by this bacterium. In its role as an opportunistic pathogen, *S. mutans* predominates in plaque biofilm communities located at sites that develop caries [16]. Important virulence factors include its ability to withstand a rapidly changing external pH. In fact, it has been shown that the direct application of sucrose to human plaque *in situ* can result in a pH drop from 7.0 to 3.0 within 3 minutes [23].

Initial methodologies utilizing transposons to study the genetics of virulence factors of *S. mutans*, including acid tolerance, have used transposon Tn*916* or Tn*4001* on suicide vectors [3, 19, 24]. The main limitation of these delivery systems is the fact that the introduction of this transposon to the *S. mutans* chromosome requires both transformation and transposition, two low frequency events. In addition, once a mutant of interest has been isolated, it is difficulty to recover Tn*916*, a large (18 kb), relatively unstable, element.

Work in our laboratory has utilized transposon Tn*917*, a small (5.2 kb) mobile genetic element that caries a selectable erythromycin resistance gene harbored on a replication-conditional (temperature-sensitive) vector (pTV1-OK). Transposon Tn*917* and its derivatives have been successfully used in other Gram-positive organisms including species in the genera *Lactococcus* [14], *Clostridium* [2], Listeria [4, 8], *Staphylococcus* [6, 21], and *Enterococcus* [7], where it was first discovered [22].

pTV1-OK (Figure 1) is a temperature-sensitive deriative of pWVO1, a broad-host-range *Lactococcus lactis* plasmid [17]. Among the key features of pTV1-OK include *repAts*, an origin of replication that allows the plasmid to replicate at a permissive temperature of 28–30 °C, but not at 37–42 °C. pTV1-OK, like pWVO1 is also capable of replication in both Gram-negative and Gram-positive bacteria [17]. The plasmids also harbor a selectable kanamycin resistance gene, *aphA3* which is expressed in both *E. coli* and Gram-positive hosts [17].

Initial studies in our laboratory using Tn*917* mutagenesis resulted in the isolation of a variety of mutants including those with an acid-sensitive phenotype [13]. In order to recover transposon-mutagenized genes two different techniques were utilized: allelic exchange in *S. mutans* between the inserted transposon and a second vector pTV21Δ2TetM (Figure 2) followed by marker rescue in *E. coli* and; shot-gun cloning of the transposon and flanking regions of the chromosome into pUC18 followed by

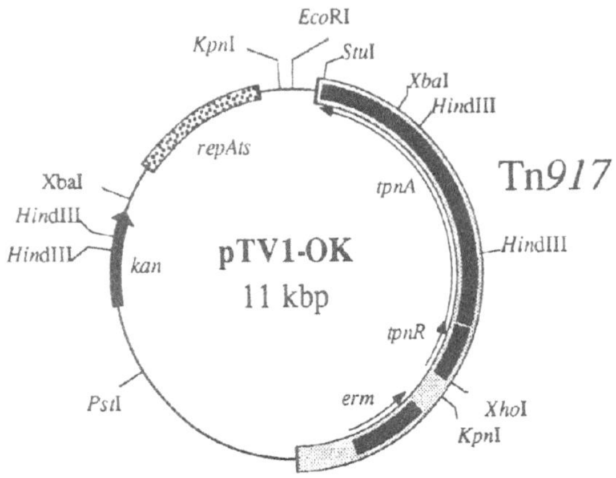

Figure 1. Plasmid pTV1-OK harbors the transposon Tn*917* and contains the temperature-sensitive origin of replication from *Lactococcus lactis* plasmid pWVO1 (*repAts*-pWVO1). It also carries the kanamycin resistance gene *aphA3* that is expressed in *E. coli* and *S. mutans*.

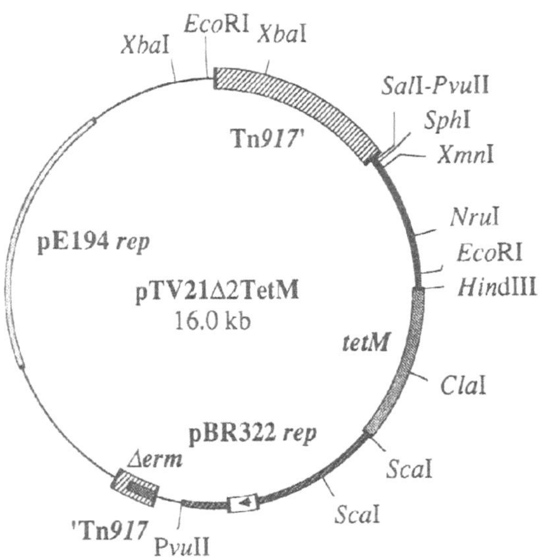

Figure 2. Recovery vector pTV21Δ2TetM contains replicons *rep*-pE194 (*Bacillus subtilis*) and *rep*-pBR322, that functions in *E. coli*, but not in *S. mutans*, where it is introduced into strains harboring Tn*917* insertions after being linearized by *Xba*I (see Figure 3). 'Tn*917* and Tn*917*' are Erm-proximal and Erm-distal ends of transposon Tn*917*, respectively, Δerm is a deletion of the erythromycin resistance gene that inactivates Erm-resistance upon recombination, *tetM* confers tetracycline resistance in the *E. coli* replicative form and in the chromosomally integrated form in *S. mutans*.

selection of *E. coli* transformants for ampicillin and erythromycin resistance. The mutagenesis technique and the use of complementary rescue and recovery techniques described here in detail have allowed us to isolate a variety of mutants and recover the relevant genetic loci. This paper will use the example of Tn*917* mutagenesis followed by a genetic screen for mutants with an acid-sensitive (AS) phenotype, followed by confirmation of the association of the phenotype to the mutation and recovery of the inactivated genetic loci. The selection techniques used to isolate auxotrophic (AX) and mutacin defective (MUT$^-$ or BLIS$^-$) strains are described elsewhere (13).

2. Materials

A. Equipment

1. Water bath adjustable from 28 °C–45 °C, Cat. No. 6651-25.[7]
2. Circulating cooled water bath Model D.[8]
3. Shaking incubator adjustable from 28 °C–45 °C.
4. UV-Visible recording spectrophotometer UV-160.[9]
5. Bench-top centrifuge Cat. No. 67390.[7]
6. GasPak Anaerobic culture jar Cat. No. 11-814-21.[5]
7. Bench-top microfuge and tubes (1.5 ml).

B. Chemicals and media

1. Tetracycline Cat. No. T3258.[1]
2. Erythromycin Cat. No. E6376.[1]
3. Kanamycin monosulfate Cat. No. K4378.[1]
4. Ampicillin Sodium Crystalline Cat. No. A9518.[1]
5. Tryptone Cat. No. 0123-07-5 Lot No. 108093J1.[10]
6. Yeast extract Cat. No. 0127-17-9 Lot No. 100392JB.[10]
7. Todd- Hewitt broth Cat. No. 11736 H8292-181202 Lot No. J5DMUS.[6]
8. Bacto agar Cat No. 0140-01 Lot No. 97810JA.[10]
9. Sodium Chloride Cat. No. 22351-4.[1]
10. ddH$_2$O.
11. Tris crystallized free base Cat. No. BP152-1.[5]
12. EDTA Cat. No. ED255.[1]
13. Heat inactivated horse serum Cat. No. H1138.[1]
14. Gycerol Cat. No. G5516.[1]
15. 5N HCl Cat. No. A144^{SI}-212.[5]
16. *Eco*RI Cat. No. R6011.[3]
17. *Hin*dIII Cat. No. R6041.[3]
18. *Xba*I Cat. No. R6181.[3]
19. *Nru*I Cat. No. R7091.[3]
20. *Bgl*II Cat. No. R6081.[3]
21. Agarose Cat. No. BP1356-100 Lot No. 965988.[5]
22. λ-*Hin*dIII biotinylated standards Cat. No. 15617-012.[4]
23. Isoamyl alcohol Cat No. A393-4.[5]
24. Chloroform Cat. No. C298-4.[5]
25. Ethanol, Absolute 200 proof Cat. No. 94L20QA.[11]

26. Transfer RNA from Baker's Yeast Cat. No. 109495.[12]
27. T4 DNA ligase and buffer Cat. No. M1801.[3]
28. DL-threonine Cat. No.T1520.[1]
29. Glycine Cat. No. 332-9.[1]
30. Mutanolysin Cat. No. M9901.[1]
31. RNase Cat. No. R6501.[1]
32. Lysozyme Cat. No. L7651[C].
33. Sodium dodecyl sulfate (SDS) Cat. No. L3771.[1]
34. Ammonium acetate Cat. No. A2706.[1]

C. Utensils and kits
1. BRL Photogene nucleic acid detection system Cat. No. 8192SA.[4]
2. BRL Bionick labeling system Cat. No. 8247SA.[4]
3. Photogene nylon membrane Cat. No. 8194SA.[4]
4. Disposable gas generator envelopes Cat No. 43-7140.[2]
5. 15 ml 125 × 16 mm Falcon Polystyrene, round-bottom tubes Cat. No. 2307.[6]
6. 50 ml Falcon Blue Max 30 × 115 ml polypropylene conical tubes Cat. No. 2098.[6]
7. 500 ml Centrifuge bottles Cat. No. B0658.[1]
8. 2 l Srew cap flasks Kimax Cat. No. 10-26505-1000.[5]
9. 1 l culture bottles Kimax Cat. No. 10-26505-2000.[5]
10. Petri dishes 100 × 15 mm Cat. No. 08-757-12.[5]
11. Micropipetors and tips.
12. Milk dilution bottles Cat. No. 03-422-13.[5]
13. Dialysis tubing Cat. No. D9277.[1]

D. Plasmids
1. pTV1-OK.[13]
2. pTV21Δ2TetM.[13]
3. pUC18 *Eco*RI/BAP.[13]

E. Bacteria
1. *Escherichia coli* MC1061.[13]
2. *E. coli* DH5α.[4]
3. *Streptococcus mutans* JH1005.[12]
4. *S. mutans* NG8.[13]

3. Procedures

A. Preparation of pTV1-OK from an *E. coli* host

Important note: All growth and maintenance of *E. coli* strains harboring pTV1-OK must be performed at 28 °C. Growth or incubation of the cultures at a higher temperature at any stage can result in loss of the plasmid.

Prepare stock solutions of antibiotics as follows: erythromycin (Erm) or tetracycline (Tet) – dissolve 250 mg of the antibiotic in 10 ml of 95% ethanol, filter sterilize through a 0.45 μm syringe filter, aliquot into 1 ml volumes, and store at –20 °C; for kanamycin (Kan) or ampicillin (Amp) – dissolve 250 mg antibiotic in 10 ml of H_2O, filter sterilize through a 0.45 μm syringe filter and aliquot into 1 ml volumes. Prepare LB agar plates containing 50 μg/ml kanamycin by dissolving 10 g tryptone, 5 g yeast extract and 5 g NaCl in 1 l of ddH_2O in a 2 l bottle. Add 17 g of agar and autoclave for 20 min at 121 °C. When the medium has cooled to 55–58 °C add 2 ml of Kan stock and dispense medium into 100 mm × 15 mm petri plates, cool at room temperature until solid, and store inverted in a sealed bag at 4 °C. Inoculate an LB agar plate with a sterile loop from a glycerol or stab culture of *E. coli* harboring pTV1-OK. Incubate the plate for 2 days at 28 °C. Make 1 l of LB broth medium by dissolving 10 g tryptone, 5 g yeast extract and 5 g NaCl in 1 l of ddH_2O and autoclave for 20 min at 121 °C in a 2 liter screw cap flask. When cooled add 2 ml of Kan stock (25 mg/ml) to 1 l of LB medium. Using a sterile pipette aseptically remove 10 ml of the medium and transfer to a 50 ml Falcon tube. Inoculate the 50 ml tube with several colonies of the 2 day old 28 °C LB agar plate containing *E. coli* harboring pTV1-OK. Incubate the culture with shaking at 200 rpm at 28 °C. The next day if the 15 ml culture has grown, as indicated by an increase in turbidity, use the entire 15 ml culture to inoculate the remainder of the 1 l of LB containing Kan. Incubate the culture at 28 °C overnight with shaking (200 rpm).

When cells have reached saturation (this may take up to 48 h) collect the cells by dividing into 2 × 500 ml centrifuge bottles and centrifuging for 10 min. at 6000 ×*g*. Plasmid DNA can be recovered by the alkaline lysis method [1] and purified by polyethylene glycol precipitation [1] or CsCl-EtBr density gradient centrifugation [1]. After purification, the plasmids are dissolved in TE buffer (10 mM Tris-HCl, 10 mM EDTA, pH 7.5) to a concentration of 1 μg/μl.

B. Transformation of *S. mutans* with pTV1-OK and generation of mutants

Todd Hewitt broth (THB) is prepared by dissolving 30 g of Todd Hewitt medium in 1 l of ddH_2O which is dispensed into milk dilution bottles in 100 ml aliquots and autoclaved for 20 min at 121 °C. Five ml of the medium is aseptically dispensed by pipette into a sterile 10 ml screw-cap tube. The tube is then inoculated with a single colony of *S. mutans*, and incubated at 37 °C overnight. One ml of heat inactivated horse serum and 5 ml of sterile THB is added to a 10 ml screw capped tube and inoculated with 125 μl of the overnight *S. mutans* culture. The optical density of the culture is monitored spectrophotometrically at 600 nm until the culture has achieved an OD_{600} of between 0.15 and 0.25 (usually 2–4 hours). The cells are then left at room temperature for 10 minutes and 1 µg (1 µl) of pTV1-0K is added to the tube and incubated at 30 °C for 1 h (note temperature change). Cells are then diluted by the addition of 5 ml of THB supplemented with horse serum and incubation is continued for an additional hour. The cells are then harvested by centrifugation in a bench-top centrifuge and suspended in 200 µl of Todd Hewitt-yeast-extract (THYE) prepared by dissolving 30 g of THB medium and 3 g of yeast extract in 1 l of ddH_2O and autoclaving at 121 °C for 20 min. THYE Kan-Erm agar plates are prepared by dissolving 30 g of TH medium plus 3 g of yeast extract in 1 l of ddH_2O and adding 17 g of agar, autoclaving at 121 °C for 20 min and, when cooled sufficiently, supplemented with 2 g of solid kanamycin and 200 µl of Erm stock solution. The agar then is poured in 25 ml aliquots into 100 mm × 15 mm petri plates and allowed to harden. Fifty µl aliquots of the cells are spread onto four of the hardened plates. The plates are inverted and incubated at 30 °C anaerobically. All anaerobic incubations of *S. mutans* are performed in a Gas Pak anaerobic culture jar containing a Gas Pak plus disposable gas generator envelope. Colonies should appear on the plate after three to five days of incubation.

The transformants growing on the THYE plates containing Kan and Erm should harbor a replicating pTV1-OK plasmid. Several single colonies should be used to inoculate separate 5 ml tubes of THYE containing Erm and Kan. The tubes are prepared by adding 20 µl Erm stock and 1.2 ml of Kan stock to 100 ml of THYE and dispensing 5 ml volumes by pipette into 15 ml screw cap tubes. Once inoculated the tubes are incubated at 30 °C for 48 hours. Between 1–5 tubes are chosen as inocula for facilitating mutagenesis. Five, 10, 25 and 50 µl aliquots of each of the selected cultures (that will represent dilutions of 1:1000, 1:500, 1:200, and 1:100 respectively) are used to inoculate new 5 ml tubes of THYE (no antibiotic). The cultures are then incubated overnight in a water bath at 40–45 °C. The temperature range is determined by the heat tolerance of the individual *S. mutans* strain. For example, *S. mutans* strain NG8 is incubated at 41 °C and *S. mutans* strain JH1005 is incubated at 43 °C. These are the highest temperatures at which the organisms are able to grow well.

The culture is incubated at the higher restrictive temperature (40–45 °C) overnight. The cells are then assessed for the aquisition and efficiency of transposition of Tn*917* by virtue of their Erm resistance. To determine the percentage of the culture containing successful transposants a direct plating of 10, 50 and 200 µl of the cultures are spread on THYE plates containing Erm (200 µl Erm stock per liter of THYE agar). The same volumes of cells from each tube are plated on THYE plates containing both antibiotics Erm and Kan, and 10^{-2}–10^{-6} dilutions of the cultures (dilute with THYE broth) are plated on THYE without antibiotics. One ml of sterile glycerol is then added to the balance of the culture tubes and they are stored at –70 °C. These pools of mutants are stored at this point, prior to further analysis, to retain optimal viability. The plates are then incubated anaerobically at 37 °C for 24–48 h and the plates are scored for the number of colonies present. The ratio of Erm-resistant colonies (determined from the number of colonies on the THYE-Erm plates) to the total number of colonies (determined by the number of colonies on the plates devoid of antibiotic), indicates the percentage of the *S. mutans* population containing Tn*917*. The number of Erm- and Kan-resistant colonies represents the ratio of the mutants harboring the entire pTV1-OK plasmid, including Tn*917*, in the chromosome (replicon fusions).

The mutagenesis technique is deemed successful if the total number of Erm-resistant mutants represented in the culture tubes is at least 3000 in total. For using a genetic screen, as exemplified here for selection of acid-sensitive mutants, determine the volume of culture from each tube to be spread on agar plates that will yield 150–200 colonies per plate. If this volume exceeds 200 µl the cells should be concentrated by 5 min of centrifugation at full speed in a bench-top centrifuge and resuspended in a smaller volume of the supernatant so that a suitable density is achieved.

C. Screening the mutants for the desired phenotype

To screen for acid-sensitive mutants, spread out enough cells on THYE-Erm plates so that at least 3000 mutants will be available for screening. The plates are incubated at 37 °C for 2 days and removed from the incubator. Individual colonies are picked using sterile toothpicks and patched in duplicate in the same orientation onto a second set of plates. The first set (master plate) contains THYE Erm (pH 7.0), while the second set (pH-selective plate) contains THYE made 2× by adding 30 g of Todd Hewitt media and 3 g of yeast extract to 500 ml of ddH_2O in a 1 l screw-capped flask. The pH of this solution is adjusted to 5.0 by the addition of 5N HCl prior to autoclaving. A second 1 l flask is prepared containing 500 ml of ddH_2O and 15 g of agar. After autocalving for 20 min at 121 °C the flasks are cooled to aprox. 55 °C and mixed immediately before pouring. Typically, the plates are placed over a grid to aid in orienting the patched colonies between master and pH-selective plates. Routinely, 100 clones are represented on each plate without cross-contamination between colonies. The plates are incubated anaerobically at 37 °C for 2 days and examined. Clones that do not grow, or grow poorly at pH 5.0, are selected from the corresponding master plate and chosen for further study. The selected clones are grown at 37 °C overnight in 15 ml screw-cap tubes containing 5 ml THYE Erm. One ml of sterile glycerol is mixed with the culture and then it is aliquoted and stored at –70 °C. The phenotype and viability of the clone should be re-tested at this time for its ability to grow at pH 7, but not at pH 5, by streaking a loopful of culture onto a THYE Erm plate and a pH 5 THYE plate and confirming the negative or reduced growth at pH 5 after two days of 37 °C anaerobic incubation.

D. Linkage of Tn*917* insertion to the mutant phenotype

Before attempting to recover the inactivated gene the following protocols are recommended: confirmation of a single Tn*917* insertion by Southern blot analysis and transference of the desired phenotypic characteristic by genetic back-cross to a wild-type *S. mutans* strain.

To perform these procedures chromosomal DNA is extracted from *S. mutans* as follows: mutanolysin stocks are prepared by dissolving 7,500 units of solid mutanolysin per ml of TE buffer and frozen at –20 °C in 1 ml volumes. RNase stocks are prepared by dissolving RNase at 20 mg/ml in H_20 and placing in a boiling water bath for 10 min prior to storage at –20 °C. Lysozyme stocks are prepared fresh immediately before use by dissolving 15 mg of solid enzyme in 1 ml of TE buffer. A single colony of the *S. mutans* mutant to be characterized is used to inoculate a 15 ml screw-capped tube containing 5 ml of THB supplemented with 20 mM DL-threonine. After incubation at 37 °C overnight, 10 ml of fresh, prewarmed THB is added and the culture is allowed to grow for an additional 45 min. Glycine (0.75 g) is added and incubation continued for 60 min. The cells are then harvested by centrifugation in a bench-top centrifuge for 5 min at maximum speed and resuspended in 1 ml of ddH_20. Ten µl of mutanolysin, 6 µl of RNase and 70 µl of lysozyme stock solutions are added to the cells and 37 °C incubation is continued for one hour.

The cell suspension is lysed by the addition of 40 µl of 20% SDS followed by gentle inversion of the tube. The lysed cell mixture is then divided between two microfuge tubes and extracted twice with phenol:chloroform:isoamyl alcohol (25:24:1) and twice with chloroform:isoamyl alcohol (24:1). The solution is then transferred to sterile dialysis tubing and the DNA is dialyzed against two 2 l changes of TE buffer at 4 °C overnight. The DNA is quantitated by measuring the absorbance at OD_{260} in a spectrophotometer. The concentration should be between 0.2 and 0.5 µg/ml. If the DNA is dilute it can be concentrated by the addition of a half volume of 7.5 M ammonium acetate, 2.5 volumes of ice-cold absolute ethanol, followed by 10 min of centrifugation at 4 °C, two rinses with 70% ethanol (v:v), 2 h of drying at room temperature and dissolving in a smaller volume of TE buffer.

The BRL photogene kit is used to perform non-radioactive Southern hybridizations. Detailed instructions for its use are provided with the kit. Other non-isotopic or isotopic Southern hybridization techniques can also be implemented for this purpose. Chromosomal DNA (5 µg) is digested to completion using separate reactions containing the restriction endonucleases *Eco*RI and *Xba*I according to the enzyme manufacturer's instructions. These enzymes are chosen since restriction fragments generated by these enzymes harbor chromosomal DNA and the erythromycin resistance gene from Tn*917*.

The digested DNA and biotinylated λ *Hin*dIII standards (BRL) are subjected to electrophoresis on a 0.5% agarose gel and transfer to a Photogene nylon membrane is performed as described by the manufacturer. The membrane is probed with 0.5 µg of biotinylated pTV1-OK which is prepared using the Bionick nick translation kit from BRL. After completion of the procedure the resultant images are examined for single bands in the *Eco*RI lanes. A single band indicates a single chromosomal insertion of Tn*917*. The relative size of the fragments generated by each enzyme is then determined and used to confirm results from the gene recovery experiments.

The next step before proceeding to rescue transposon-flanking chromosomal fragments is a back-cross involving transfer of the mutant phenotype to a wild-type strain of *S. mutans* by transformation using chromosomal DNA from the mutant. For this genetic back-cross experiment, we routinely use *S. mutans* strain NG 8 since it is highly transformable. When the parent strain used in the construction of the original mutant is transformable, it is recommended that it be included to conserve the genetic background of the original mutant. Transformation of *S. mutans* with chromosomal DNA is performed as for transformation with pTV1-OK with the exception that 5 µg of *Eco*R1-digested chromosomal DNA is added to the cultures and incubation is continued at 37 °C rather than 30 °C. After transformation, aliquots of the cells are spread onto THYE Erm plates and incubated anaerobically for 48 h. Erm-resistant colonies are then used to confirm the transference of phenotype back to the parent strain. In our example, the transformants are examined for their diminished growth at pH 5 when compared to the parent strain.

E. Recovery of Tn*917* interrupted genes

When the phenotype of the parental backcross is confirmed to be the same as the mutant of interest the gene can be recovered by one of two proven techniques. The first involves use of pTV21Δ2TetM, a plasmid that, by allelic exchange, replaces the transposon in the *S. mutans* mutant and introduces an *E. coli* origin of replication. This allows excision of the plasmid by restriction endonuclease digestion of chromosomal DNA, followed by religation and transformation of *E. coli* (Figure 3). To utilize pTV21Δ2TetM, the *S. mutans* transposon mutant is transformed with 1 μg of linearized plasmid DNA as described above. The resultant transformants are selected on THYE-Tet plates that are prepared by adding 400 μl of tetracycline stock solution (20 μg/ml final conc.) to THYE agar before pouring. *S. mutans* transformation mixtures are centrifuged, concentrated and spread onto several plates as described previously in this paper. After 2 days of 37 °C anaerobic incubation Tet-resistant colonies are picked and then patched onto master THYE Tet plates and THYE Erm plates. Those clones that are Tet-resistant and Erm-sensitive are then tested for retention of the

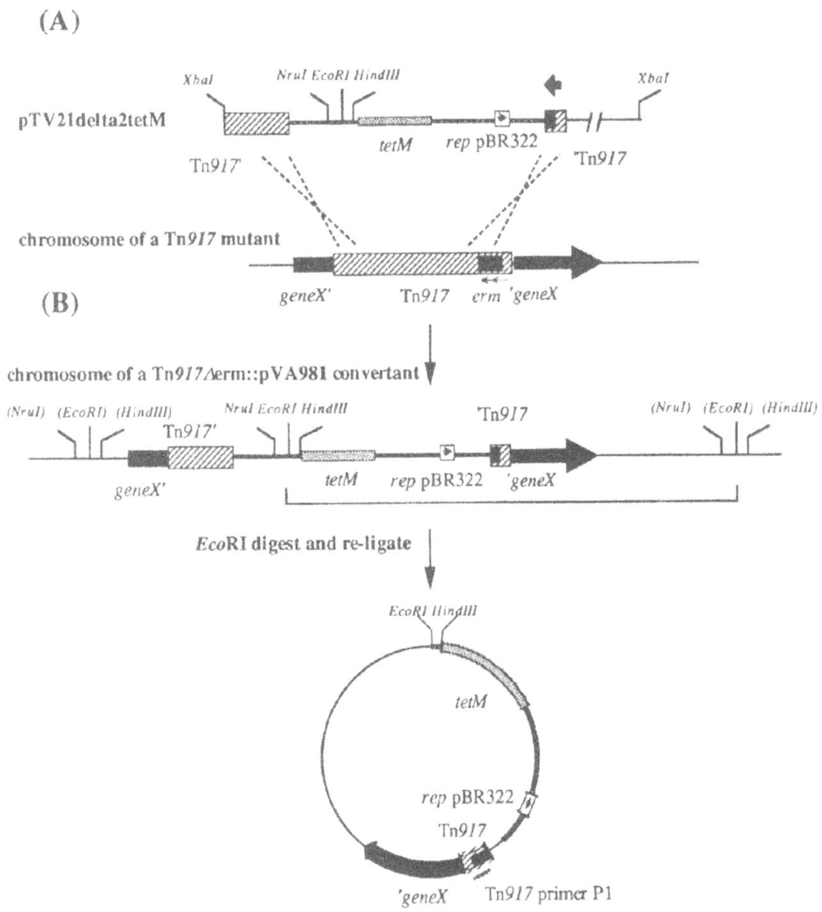

Figure 3. Cloning streptococcal DNA sequences adjacent to a Tn*917* insertion with the recovery vector pTV21Δ2TetM. The two-step process requires, first, allelic exchange of the wild-type Tn*917* with linearized pTV21Δ2TetM (Tn*917*Δ*erm*::pVA981 portion) (A), followed by marker rescue in *E. coli* (B). To clone the DNA bordering the Erm-proximal end of the transposon insertion, chromosomal DNA from the *S. mutans* convertant strain is digested with restriction enzymes that cut outside of the selectable marker *TetM* and the pBR322 *rep*, such us *Bgl*II, *Eco*RI, *Hind*III, *Nru*I or *Xba*I (e.g., *Eco*RI). The digests are then ligated at concentrations that promote self-ligation. Finally, ligation products are precipitated and used to transform *E. coli* selecting for clones containing plasmids expressing tetracycline resistance. Digests with restriction enzymes that do not cut within the Tn*917*Δ*erm*::pVA981 element, such us *Bam*HI, *Xho*I and others as well as partial digests with the above mentioned enzymes can also be used for marker rescue (e.g., partial digest with *Eco*RI). In this case, DNA adjacent to the two ends of the transposon insertion could be cloned. Gene designations are those in the legends to Figures 1 and 2. *geneX*, a Tn*917*-inactivated gene; P1, sequencing primer based on homology to the Erm-proximal sequence (see Procedures). Reprinted with permission of The American Society for Microbiology.

phenotype (in this example reduced growth at pH 5) and are designated as convertants that are subsequently used for gene recovery.

To recover the pTV21Δ2TetM plasmid and flanking DNA, 0.5–1.0 μg of chromosomal DNA from the convertant is prepared and digested in separate reactions with *Eco*RI, *Hin*dIII, *Nru*I, *Xba*I, *Bgl*II, or *Bam*HI in a total volume of 20 μl. Reaction mixtures containing the heat stable enzymes *Hin*dIII, *Nru*I, *Bgl*II must be extracted with an equal volume of phenol: $CHCl_3$: isoamyl alcohol (25:24:1) followed by extraction with an equal volume of $CHCl_3$: isoamyl alcohol (24:1) followed by ethanol precipitation and resuspension in H_2O. For reactions containing the heat-labile enzymes *Eco*RI, *Xba*I, or *Bam*HI the DNA is heated at 65 °C for 20 min. The DNA with inactivated restriction endonuclease is diluted 100-fold (i.e., 10 μl digest added to 90 μl sterile ddH_2O). Diluted DNA is then self-ligated for 16 h at 15 °C in a 100 μl reaction containing 10 μl of digested DNA, 10 μl of 10× ligation buffer, 2 μl of T4 DNA ligase and 78 μl of sterile ddH_2O. The DNA is precipitated with 10 μg of carrier transfer RNA and used to transform $CaCl_2$ competent [1] *E. coli* DH5α. Clones harboring the pTV21Δ2TetM fragment along with flanking *S. mutans* DNA are recovered by selecting Tet-resistant *E. coli* transformants and recovering the plasmid by alkaline lysis followed by PEG purification.

The second method to recover inactivated genes is accomplished by construction of a genomic library with mutant *S. mutans* chromosomal DNA. Two μg of *S. mutans* DNA is digested with *Eco*RI in a total volume of 10 μl. Five μl of this reaction is used in a ligation reaction containing 50 ng of *Eco*RI digested and dephosphorylated pUC18, 1 μl of 10× ligation buffer, 1 μl of T4 DNA ligase in a total volume of 10 μl. The reaction is incubated at 15 °C overnight. The ligation is then used to transform $CaCl_2$ competent *E. coli* MC1061 [1]. The transformed cells are spread onto LB Amp plates (LB agar plates supplemented with 4 ml of ampicillin stock/l) and colonies which arise following overnight incubation are replica-plated onto LB Amp-Erm plates (LB agar plates supplemented with 4 ml of Amp stock and 300 mg solid Erm added per liter before pouring). *E. coli* transformants capable of growth on both antibiotics are selected for plasmid screening and sequencing. Plasmids are prepared for sequencing by alkaline lysis and PEG precipitation [1].

Nucleotide sequencing of streptococcal DNA flanking the transposon insertion site following marker rescue, is carried out utilizing the Tn*917*-based primers, P1, 5′GCAATAACCGTTACCTG3′, an Erm-proximal oligonucleotide and P2, 5′GAAGCATGTATCTCCTAT3′, an Erm-distal primer. DNA recovered with pTV21Δ2TetM are sequenced with P1 only, while DNA recovered in pUC18 are sequenced with both P1 and P2. Our sequencing is performed by the DNA Sequencing Core Laboratory of the University of Florida's Interdisciplinary Center for Biotechnology Research (ICBR). Sequencing is accomplished with Taq DiDeoxy terminators and the DyePrimer Cycling Sequence protocol developed by Applied Biosystems using fluorescently labeled dideoxynucleotides and primers, respectively. Labeled extension products are analyzed on a Biosystems 373A DNA sequencer. Any suitable nucleotide sequencing technique can be utilized.

F. Identification of the recovered gene

The sequence obtained by the described protocol will contain approximately 200 b.p. of Tn*917* sequence and some vector sequence if using the pUC-based rescue technique. The start point of the sequence of interest is determined by finding the termini of the transposon by alignment of the transposon sequence with the recovered sequence. Transposon Tn*917* sequence is available from Genbank (Accession Number M11180) for this purpose. One should be aware that after editing of the Tn*917* sequence there are 5 base pairs of the target chromosomal sequence that are duplicated. These repeated bases must be removed to generate an in-frame splice of the original nucleotide target sequence.

When the vector and transposon sequences have been edited from the recovered sequence the search for previously identified homologs is performed using BLAST (National Center for Biotechnology Information). Access to this analysis program is currently available at www.ncbi.nlm.nih.gov. The BLASTX and BLASTN options are preferred for deduced amino acid and nucleic acid sequences, respectively.

4. Results and discussion

Transformation with pTV1-OK

The use of pTV1-OK for the delivery of transposon Tn*917* to *S. mutans* has been proven to be an extremely useful tool for the isolation of mutants with a variety of phenotypes. The use of the temperature-sensitive plasmid allows the transposon to be delivered to strains that are poorly transformable, as shown with *S. mutans* JH1005. This system allows the transposon to be delivered on a replicating plasmid so that only a single pTV1-OK transformant needs to be isolated to proceed with mutagenesis. In our case only four transformants were obtained with *S. mutans* JH1005 [13].

The initial requirements for using the system are that the strain to be used is able to grow at the permissive temperature of 28–30 °C and that it is sensitive to the antibiotics erythromycin and kanamycin. This facilitates selection of successful pTV1-OK transformants by virtue of growth at 30 °C in the presence of kanamycin and erythromycin. Another requirement of the strain to be mutagenized is that it is able to grow in the restrictive temperature range of 40–45 °C to allow efficient curing of the plasmid. These criteria must be met for the delivery system to be successfully used. In our recent experience two strains of *S. mutans* were used successfully, JH1005 and NG8, however, no attempts have been made to date with other strains.

Transposition frequencies

It was our experience that the critical and most variable part of the experiment occurred during the curing of the plasmid. It is suggested that multiple tubes using a range of dilutions (1:1000, 1:500, 1:200, and 1:100) of the culture be transferred to the higher curing temperature. Usable pools of mutants have been generated successfully at all of the dilutions, yet the highest values obtained do not appear to be consistent with any single dilution factor. One must also be aware that the optimal temperature for efficient plasmid loss may vary between strains. Although the *repATs* mutation results in non-permissive replication at temperatures as low as 35 °C [17] it is suggested that a range of temperatures from 40–45 °C be utilized. Typically, in our laboratories, pools of mutants were successfully generated at a rate of only 50% of all attempts despite rigid control of all the variables involved.

We found that the transposition frequency varied between strains JH1005 and NG8 with the former strain exhibiting a 10-fold higher transposition efficiency over the latter strain. Specifically, with *S. mutans* strain JH1005, in 6 independent, successful experiments, we achieved transposition frequencies of between 2.5×10^{-5} and 0.5×10^{-4}. The mean transposition frequency observed with this strain was 1.0×10^{-4}. This frequency was sufficient to yield between 10^4 and 10^5 mutants per culture tube. The frequency of transposition observed with *S. mutans* strain NG8 in 6 independent, successful experiments varied between 0.5×10^{-6} to 2.2×10^{-5} with a mean frequency of 1.1×10^{-5}. These values typically resulted in over 1000 mutants per tube.

When the erythromycin-resistant mutants are quantitated it is also useful to determine the frequency of kanamycin-resistant clones in the population. kanamycin-resistant clones are the product of the entire delivery vector being integrated into the chromosome (replicon fusion). We found that the percentage of kanamycin-resistant clones varied from 3% to 17% with strain JH1005 and between 3% and 70% with strain NG8. The average values obtained from 6 independent experiments with each strain were 10% and 20% with strains JH1005 and NG8, respectively. Since these mutants can also be recovered, replicon fusions should also be included in mutagenic screens. In fact, AS34 (Table 1) is a replicon fusion mutant.

Table 1. Phenotypes and genes of *S. mutans* mutants isolated using pTV1-OK-mediated Tn*917* mutagenesis

Mutant designation	Relevant phenotype	Affected gene	Function	Reference and accession number
AS5 L22	AS, MUT- (BLIS-) AX adenine	*fhs*	Adenine biosynthesis	[9, 13] U48882 U39612
AS17	AS	*ffh*	Protein secretion	[13] U48883 U88582
AS25	AS	*dfp*	Pantothenate metabolism, DNA biosynthesis[a]	[13] U48885
AS34	AS	*oad*	Oxalacetate decarboxylase[a]	This study
AS36	AS	*gluP*	Glutamate transporter[a]	This study
AX1	AX glutamate	*icd*	Isocitrate dehydrogenase	[10, 13] U48886 U62799
AX3	AX arginine	*argD*	Arginine biosynthesis[a]	[13] U48887
AX5	AX CMM$^-$	*gbpC*	Glucan binding protein	[18] D85031 This study
DM25	MUT$^-$ (BLIS$^-$)	*lanA-lanB*	Lantibiotic synthesis	This study

[a] The function of these gene products has not been confirmed biochemically, but is based on strong homologies to genes isolated from other bacterial species.
Abbreviations and definitions: AS = acid-senstive, reduced or absent growth on pH 5.0 agar plates; AX = auxotrophic for specified amino acid or, as for strain AX5, unable to grow on minimal medium (CMM) [5]; MUT = mutacin or BLIS = Bacteriocin-like inhibitory substance negative phenotype is defective in its ability to inhibit growth of target cells as determined by a deferred antagonism assay [9].

The variability in the transposition and replicon fusion frequencies is likely a reflection of the over- or under-representation of individual clones in the population of transposants. The occurrence of so-called 'jack-pots' must always be considered when using mutagenesis followed by growth in liquid culture. Erythromycin-resistant clones that are the result of an early transposition event or which have higher growth rates than the rest of the population can be over-represented. Conversely, those clones that occur as the result of a later transposition event, or are slow growers will be under-represented among the population. We determined that cultures displaying a transposition frequency and percentage of kanamycin-resistant phenotype closest to the mean values likely contained pools of mutants representing the most heterogeneous mix of mutants and were therefore considered to be the best pools to screen for relevant phenotypes.

Isolation of mutants displaying various phenotypes

Randomness of the transposon insertion in the bacterial chromosome has not been completely assessed, although, based on the experiences in our laboratories resulting in the successful isolation of a variety of mutants with different phenotypes, we believe the degree of randomness is sufficient to isolate several independent mutants, each harboring single copies of the transposon. Three individual phenotypes have been isolated using this technique: the example described in the 'Procedures' section, acid sensitivity (AS), is characterized by reduced growth on solid media at pH 5; auxotrophy (AX) for one or more amino acids as determined, initially, by the inability of the mutant to grow on minimal medium devoid of amino acids (CMM) [5, 10, 13]; and loss of mutacin production as indicated by the mutant's inability to inhibit growth of a target strain of *Streptococcus rattus* BHT-2 as measured in a deferred antagonism assay [9, 12].

The screening for acid-sensitive mutants resulted in the highest frequency of recovered mutants, representing an average of 0.28% of the total pool. This result was not unexpected since a wide variety of mutations are known to cause an acid-sensitive phenotype in other bacteria [20]. The acid protective mechanisms utilized by *S. mutans* appear to be similar to other bacteria and include both constitutive and inducible systems that involve a large subset of genes [20]. The auxotrophs and MUT$^-$ phenotypes were isolated at frequencies of 0.2% and 0.11%, respectively (Figure 4).

Southern analysis and genetic backcross

An important step to be taken before attempts to recover and identify interrupted genes is the demonstration that the observed phenotype is truly the result of a single transposon insertion. Restriction digestion of the chromosomal DNA followed by Southern hybridization with pTV1-OK as a probe will verify the insertion of a single transposon. In all of the mutants examined by our laboratories to date only single transposon insertions have been observed in each of the mutants examined. Although this result may tempt one to bypass this procedure, it is strongly recommended that the Southern hybridization be performed with a variety of restriction digests to give an indication of the expected fragment size to be recovered, especially to determine if the fragment is amenable to recovery by shot-gun cloning into pUC18. For example, if Southern analysis indicates that a restriction fragment harboring the transposon generated by a particular enzyme is found to be very large (> 15 kb), an alternate enzyme or recovery strategy should be considered.

Gene recovery and nucleotide sequence analysis

Two different techniques were used successfully to recover chromosomal DNA flanking Tn*917* insertions in the mutants. The first involved the use of the recovery vector pTV21Δ2TetM. This vector has been used successfully to recover segments of *S. mutans* chromosomal DNA ranging in size from 2.5–9 kb. It is recommended that each of the restriction enzymes suggested in the 'Procedures' section be used to digest the chromosomal DNA before religation. Chances of recovery will be increased due to the fact that smaller products are favored to self-ligate and the greater the number of enzymes used the more likely that a self-ligatable plasmid will be generated. An illustration of the scheme for utilization of this technique is shown in Figure 3. The recovery vector technique has proven effective with each of the strains tested (NG8 and JH1005). Even though the latter strain is poorly transformable, several gene fragments have been recovered using this technique.

A second technique that has proven to be valuable is the construction of genomic libraries in pUC18 and selection of clones with ampicillin and erythromycin resistance in *E. coli* MC1061. The use of this strain is mandatory since it is able to allow expression and selection of the Erm gene from Tn*917* that can be toxic to other *E. coli* strains [13]. While we have successfully recovered clones from Erm-Amp selection plates we have found that plating of the *E. coli* transformation mixture on LB-Amp plates and then replica plating onto LB Erm plates results in a greater recovery of desirable clones.

Upon recovery of clones of interest, the purified plasmid DNA is sequenced using the specified primers. The resultant sequences are often putatively identified by comparison to the Genbank database using BLASTN for nucleotide sequences or BLASTX for deduced amino acid sequences. A list

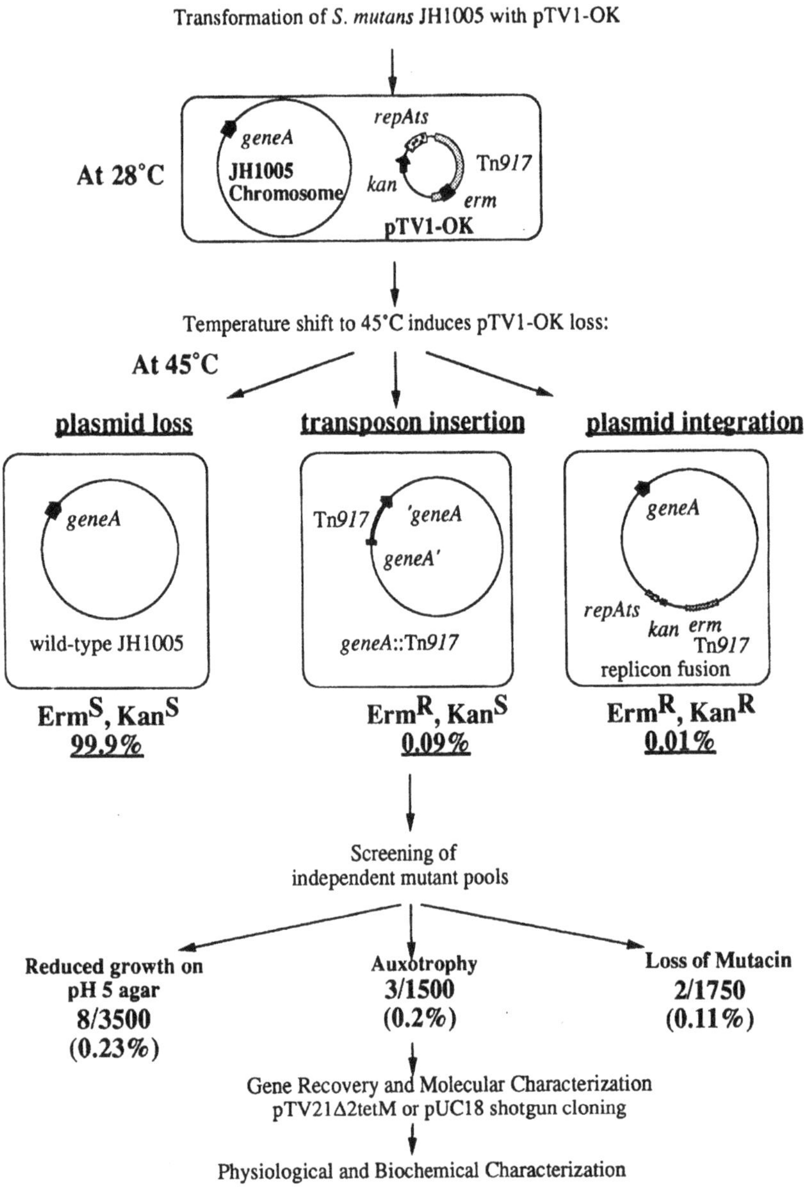

Figure 4. Schematic flow chart of the protocol used to mutagenize *S. mutans* with pTV1-OK. The plasmid is shown in the replicative form at 30 °C and in the subsequent generation of mutants with the respective frequencies of isolation after plasmid curing at 45 °C. The number of mutants isolated following various selection protocols is also listed and described in the text.

of *S. mutans* genes that we have recovered is provided in Table 1. The genes whose function have not been confirmed biochemically and physiologically are indicated. These genes have strong homology to the bacterial genes listed, however further experimentation is necessary to confirm their identity.

If a gene fragment appears to be interesting, the intact wild-type gene can be cloned from a genomic library constructed from *S. mutans* parent strain DNA using the recovered gene fragment as a probe. An alternate strategy that involves complementation of appropriate *E. coli* mutants with chromosomal plasmid libraries constructed from the wild-type *S. mutans* DNA has also proven successful in our hands [9, 10].

The use of pTV1-OK for the generation of *S. mutans* mutants and the subsequent recovery of the genes is now well established. The plasmid has also been used successfully with Group B streptococci [11] and *Streptococcus pyogenes* [15]. The utility of this vector has only recently come to light, but will surely prove useful for a variety of other species.

Acknowledgments

This work was supported by research grants from the National Institutes of Health grants DE04529 (JDH) and DE08007-09 (ASB).

Notes on suppliers

1. Sigma Chemical Company, St. Louis, MI 63178, USA
2. BBL Becton Dickson and Co., Cockeysville, MD 21030, USA
3. Promega Corporation, Madison, WI 53711-5399, USA
4. GIBCO/BRL Life Technologies, Gaithersburg, MD 20884-9980, USA
5. Fisher Scientific Company, Fair Lawn, NJ 07410-0414, USA
6. Becton Dickinson and Company, Lincoln Park, NJ 07035, USA
7. Precision Scientific, Chicago IL 60647, USA
8. Haake Instruments Inc., Paramus, NJ 07652, USA
9. Shimadzu Corporation, Kyoto, Japan
10. Difco Laboratories, Detroit, MI 48232-7058, USA
11. AAPR Alcohol and Chemical Company, Shelbyville, KY 40065, USA
12. Boehringer Mannheim, Indianapolis, IN 46250-0414, USA
13. Pharmacia Biotech, Piscataway, NJ 08855-1327, USA

References

1. Ausebel FM, Brent R, Kingston RE, Moore DD, Seidman JG, Smith JA, Struhl K (eds) (1985–1997). Current protocols in molecular biology. John Wiley and Sons.
2. Babb BL, Collett HJ, Reid SJ, Woods DR (1993). Transposon mutagenesis of *Clostridium acetobutylicum* P262: Isolation and characterization of solvent deficient and metronidazole resistant mutants. FEMS Microbiol Lett 114: 343–348.
3. Burne RA, Quivey RG Jr (1994). Use of transposons to dissect pathogenic strategies of Gram-positive bacteria. Methods Enzymol 235: 405–426.
4. Camilli A, Portnoy DA, Youngman P (1990). Insertional mutagenesis of *Listeria monocytogenes* with a novel Tn*917* derivative that allows direct cloning of DNA flanking transposon insertions. J Bacteriol 172: 3738–3744.
5. Carlsson J (1970). Nutritional requirements of *Streptococcus mutans*. Caries Res 4: 305–320.
6. Cheung AL, Koomey JM, Butler CA, Projan SJ, Fischetti VA (1992). Regulation of exoprotein expression in *Staphylococcus aureus* by a locus (*sar*) distinct from *agr*. Proc Natl Acad Sci (USA) 89: 6462–6466.
7. Clewell DB, Tomich PK, Gawron-Burke MC, Franke AE, Yagi Y, An FY (1982). Mapping of *Streptococcus faecalis* plasmids pAD1 and pAD2 and studies relating to transposition of Tn*917*. J Bacteriol 152: 1220–1230.
8. Cossart P, Vicente MF, Mengaud J, Baquero F, Perez-Diaz JC, and Berche P (1989). Listeriosin O is essential for virulence of *Listeria monocytogenes*: Direct evidence obtained gene complementation. Infect Immun 57: 3629–3236.
9. Crowley PJ, Gutierrez JA, Hillman JD, Bleiweis AS (1997). Genetic and physiologic analysis of a formyltetrahydrofolate synthetase mutant of *Streptococcus mutans*. J Bacteriol 179: 1563–1572.
10. Cvitkovitch DG, Gutierrez JA, Bleiweis AS (1997). Role of the citrate pathway in glutamate biosynthesis by *Streptococcus mutans*. J Bacteriol 179: 650–655.
11. Framson PE, Nizet V, Nittayajarn A, Youngman P, Rubens CE (1995). Mutagenesis of *Streptococcus agalactiae* (Group B Streptococcus) with Tn*917*, abstr. H-152. In: Abstracts of the 95th General Meeting of the American Society for Microbiology, American Society for Microbiology, Washington, DC.
12. Hillman JD, Johnson KP, Yaphe BI (1984). Isolation of a *Streptococcus mutans* strain producing a novel bacteriocin. Infect Immun 44: 141–144.
13. Gutierrez JA, Crowley PJ, Brown DP, Hillman JD, Youngman P, Bleiweis AS (1996). Insertional mutagenesis and recovery of interrupted genes of *Streptococcus mutans* by using transposon Tn*917*: Preliminary characterization of mutants displaying acid sensitivity and nutritional requirements. J Bacteriol 178: 4166–4175.
14. Kok J, Van der Vossen JMBN, Venema G (1984). Construction of plasmid cloning vectors for lactic streptococci which also replicate in *Bacillus subtilis* and *Escherichia coli*. Appl Environ Microbiol 48: 726–731.
15. Li Z, Gutierrez JA, Bleiweis AS, Boyle MDP (1997). Tn*917* mediated transposon mutagenesis of streptococcus pyogenes. Abstract B-48 *in*: Abstracts of the 97th General Meeting of the American Society for Microbiology. Washington: ASM Press.
16. Loesche WJ (1986). Role of *Streptococcus mutans* in human dental decay. Microbiol Rev 50: 353–380.
17. Maguin E, Duwat P, Hege T, Ehrlich D, Gruss A (1992). New thermosensitive plasmid for Gram-positive bacteria. J Bacteriol 174: 5633–5638.
18. Sato Y, Yamamoto Y, Kizaki H (1997). Cloning and sequence analysis of the gene encoding a novel glucan-binding protein of *Streptococcus mutans*. Infect Immun 65: 668–675.
19. Spatafora G, Rohner K, Barnard D, Michalek S (1995). A *Streptococcus mutans* mutant that synthesizes elevated levels of intracellular polysaccharide is hypercariogenic *in vivo*. Infect Immun 63: 2556–2563.
20. Svensater G, Larsson U-B, Greif ECG, Cvitkovitch DG, Hamilton IR (1997). Acid tolerance response and survival by oral bacteria. Oral Microbiol and Immunol (in press).
21. Thomas WD, Archer GL (1989). Identification and cloning of the conjugative transfer region of *Staphylococcus aureus* plasmid pGO1. J Bacteriol 171: 684–691.
22. Tomich PK, An FY, Clewell DB (1978). A transposon (Tn*917*) in *Streptococcus faecalis* which exhibits enhanced transposition during induction of drug resistance. Cold Spring Harbor Symp Quant Biol 43: 1217–1221.
23. Yamada T, Igarashi K, Mitsutomi M (1980). Evaluation of cariogenicity and glycosylsucrose by a

new method of measuring pH under human dental plaque *in situ*. J Dent Res 59: 2157–2162.

24. Yamashita Y, Takehara T, Kuramitsu HK (1993). Molecular characterization of a *Streptococcus mutans* mutant altered in environmental stress responses. J Bacteriol 175: 6220–6228.

Address for correspondence: Arnold S. Bleiweis, University of Florida College of Dentistry, Department of Oral Biology, P.O. Box 100424, Gainesville, FL32610-0424, USA
Phone: (352) 846-0784; Fax: (352) 392-3070
E-mail: bleiweis@dental.ufl.edu

Methods in Cell Science **20**: 13–20 (1998).

Site-specific homologous recombination mutagenesis in group B streptococci

Harry H. Yim & Craig E. Rubens
Department of Pediatrics, University of Washington, Division of Infectious Disease, Children's Hospital and Medical Center, Seattle, Washington, USA

Abstract. We describe the use of suicide vectors to generate site-specific mutations in group B streptococcus. This is accomplished by cloning gene-specific sequences into a temperature sensitive plasmid and selecting for clones which have undergone homologous recombination. The recombinan clones can be easily isolated by selecting for clones which have retained the antibiotic resistance markers that are present on the vector or cloned into the gene-specific sequences. To confirm the fidelity of the recombination events, PCR analysis is performed on chromosomal DNA isolated from the recombinant clones. Using this strategy, we have generated site-specific insertions in several capsule genes and have found that these insertions occurred as expected and that the mutations result in loss of capsule expression. In this report, we specifically detail the construction of *cpsB* mutants by single and double cross-over recombination and demonstrate that the resulting strains are acapsular.

Key words: Capsule synthesis, Mutagenesis, Recombination, Streptococci, Suicide vector

Abbreviations: Amp = Ampicillin; BSA = Bovine Serum Albumin; *cat* = Chloramphenicol Acetyl Transferase; Cm = Chloramphenicol; Cm^R = Chloramphenicol Resistant; *cps* = GBS capsule gene; CSF = Cerebral Spinal Fluid; GBS = Group B Streptococcus; HRP = Horse Radish Peroxidase; LB = Luria-Bertani; NFDM = Non Fat Dry Milk; ori = origin of plasmid replication; PBS = Phosphate Buffered Saline; THA = Todd-Hewitt Agar; THB = Todd-Hewitt Broth; TBS = Tris Buffered Saline; TTBS = Tween-20 Tris Buffered Saline; XGal = 5-Bromo-4-chloro-3-indoyl-β-D-galactopyranoside

1. Introduction

We have begun the use of suicide vectors to generate site-specific mutations and believe that this is a significant advance for the genetic analysis of group B streptococcus. Previously, mutagenesis of the GBS chromosome was performed by the well established method of transposon mutagenesis using Tn*916*ΔE or Tn*917*. While transposon mutagenesis is an effective and efficient means for generating mutant libraries, it lacks specificity and is not suitable for targeting a particular gene on the chromosome. Additionally, transposon mutations are often an initial step to investigating a gene's function and should not be considered definitive proof of a gene/phenotype relationship. Additional proof is essential, and is often obtained by directly mutating a gene via homologous recombination or complementation. Until recently, homologous recombination methods in GBS were not available. However, improved transformation methods and the availability of temperature sensitive plasmids that replicate in GBS have enabled us to devise methods for homologous recombination in GBS.

Although the transformation frequency for GBS by electroporation, has improved dramatically, we have been unable to obtain recombinants by transforming GBS with either linear DNA or non-replicating plasmid DNA. This may be due to poor uptake or stability of linear DNA and/or the requirement that non-replicating DNA recombines immediately after transformation lowering the frequency of obtaining a recombinant.

As an alternative strategy, we first transformed GBS [1] with a conditionally replicating vector pVE6007 [2]. This *cat* gene containing plasmid, can replicate normally at 30 °C, but not at 37 °C [2]. Therefore, as long as a temperature of 30 °C or lower is maintained, we can obtain stable GBS transformants. The transformants are then grown in batch culture to obtain a large population of cells before recombination is induced. When exposed to non-permissive temperatures, plasmid replication is inhibited and will be lost unless it integrates into the chromosome. Such integrants can be easily identified by selecting for chloramphenicol resistance. By cloning a homologous piece of DNA into this vector, we dictate the integration site via homologous recombination. We have used this strategy to generate site-specific insertions in several capsule genes and have found that these integration events result in loss of capsule expression.

In addition to the single cross-over mutations, we have constructed allelic exchange mutants by a modification of the single cross-over recombination technique. Basically, capsule gene sequences are cloned into pVE6007, then a kanamycin omega cassette is cloned into the central portion of the capsule sequence. The plasmid is transformed into GBS and the strain is exposed to conditions that favor two separate recombination events (occurring on each side of the kanamycin cassette). In this way, we can obtain double cross-over mutations and allelic exchange mutants in *cps* genes resulting in unencapsulated GBS. As an illustration of these techniques, we present the details of *cpsB* mutagenesis by single and double cross-over recombination resulting in an acapsular phenotype.

2. Materials

A. Bacterial strains and media
GBS were grown in TH broth (DIFCO Laboratories,[1] Detroit, MI, Cat. #0492-07-8) or TH agar and *E. coli* was grown in LB broth (Sigma Chemical,[2] St. Louis, MO, Cat. #L-3152) or LB agar. COH1 is a type III encapsulated GBS isolated from the CSF of an infected newborn [4]. DH5α [*supE44* Δ*lacU169* (φ80 *lacZ*ΔM15) *hsdR17 recA1 endA1 gyrA96 thi-1 relA1*] (Life Technologies Inc. [GIBCO BRL],[3] Gaithersburg, MD) was used as the *E. coli* host for cloning and propagation of pVE6007. Chloramphenicol resistance (Sigma Chemical,[2] St. Louis, MO, Cat. #C0378) in *E. coli* was selected for at 10 μg/ml and kanamycin resistance (Sigma Chemical,[2] St. Louis, MO, Cat. #K4000) at 40 μg/ml. For GBS, chloramphenicol was used at 10 μg/ml and kanamycin at 400 μg/ml when appropriate. To degrade the cell wall of GBS, mutanolysin (Sigma Chemical,[2] St. Louis, MO, Cat. #M-9901) was added to a final concentration of 0.25–0.5 U/μl.

B. Molecular biology reagents
Purification of DNA was performed using various kits from Qiagen Inc. according to the manufacturers protocols (Qiagen Inc.,[4] Santa Clarita, CA): plasmids from *E. coli* were purified using the QIAprep spin plasmid mini-prep kit (Cat. #27106), PCR products were purified using the QIAquick PCR purification kit (Cat. #28104) and DNA fragments were purified from agarose gels using the QIAquick gel extraction kit (Cat. #28704). For cloning PCR products, we used the T-vector, pT7Blue (Novagen Inc.,[5] Madison WI, Cat. #69835-1). Restriction enzymes (see list below), ligase (Cat. #202L) and Klenow polymerase (Cat. #210S) were purchased from New England Biolabs (New England Biolabs,[6] Beverly, MA). Taq polymerase (Cat. #M1861) was purchased from Promega Co.[7] (Madison, WI). For PCR reactions that produced products over 2.0 kb, the Expand long template PCR system was used according to the manufacturers recommendations (Boeringer Mannheim Corp.,[8] Indianapolis, IN. Cat. #1 681 834). dNTP nucleotides were purchased from Sigma (Sigma Chemical,[2] St. Louis, MO, Cat. #D-7295).
Oligonucleotide primers were produced in-house on an ABI (Perkin-Elmer Cetus Corp.,[9] Foster City CA) DNA synthesizer with trityl-OFF at a 0.2 μMol scale. PCR was performed using a Perkin-Elmer 2400 thermocycler (Perkin-Elmer Cetus Corp.,[9] Foster City CA).

C. Immunoblot materials
0.45 μm, 82 mm nitrocellulose circles (Cat. #20440) were purchased from S&S Inc. (Schleicher and Schuell Inc.,[10] Keene, NH). BSA (Sigma Chemical,[2] St. Louis, MO, Cat. #A7030) was resuspended in TBS (20 mM Tris-HCl, 0.5 M NaCl, pH 7.5) at 3% wt/vol. Non-fat dry milk (Carnation brand purchased at a local grocery store) was resuspended at 5% wt/vol in water. 3MM filter paper was purchased from Whatman Inc.[11] (Maldstone, UK, Cat. #3030917). TTBS is TBS with 0.05% Tween-20 (Sigma Chemical,[2] St. Louis, MO, Cat. #C0378). Primary rabbit antibody specific to type III GBS capsule conjugated to tetanus toxoid was used at a 1:30,000 dilution in 3% BSA in TBS and was a generous gift from Dr. Mike Wessels. Goat anti-rabbit IgG conjugated to horse radish peroxidase (Bio-Rad Laboratories, Hercules,[12] CA, Cat. #170-6463) was used at 1:10,000 dilution in 3% BSA in TBS. SuperSignal chemiluminescent substrate for HRP was used according to manufacturers protocols (Pierce Chemical Co.,[13] Rockford, IL, Cat. #34080). X-ray film was from Kodak (Eastman Kodak, Rochester, NY.[14]).
Restriction Enzymes purchased from NEB:
- *Bam*HI (Cat. #136S)
- *Bgl*II (Cat. #144S)
- *Eco*RI (Cat. #101L)
- *Hin*dIII (Cat. #104S)
- *Kpn*I (Cat. #142S)

3. Procedures

As stated above, we have created gene-specific mutants via a homologous recombination between an intragenic fragment cloned into a temperature-sensitive plasmid and the chromosomal copy of the gene. The basic steps for such a project involve; 1) generating a mutated form of the gene, 2) cloning the altered gene into an appropriate vector, 3) transfer of the recombinant plasmid into the host cell 4) setting up conditions that favor recombination between vector and chromosome 5) isolation of the recombinant strain and 6) confirmation of the genetic structure of integrant. These steps are outlined below and are illustrated by our derivation of the *cpsB* mutations.

A. Construction of the plasmid integration vector

Mutations in any gene can be easily created by using common molecular techniques such as cloning and PCR to delete the 5′ and 3′ ends of a gene. A typical reaction includes: 100 ng of chromosomal DNA from COH1, 25 pmols of each primer, 1× PCR buffer (50 mM KCl, 10 mM Tris-HCl pH 9.0, 0.1% Triton X-100), 2.5 mM $MgCl_2$, 200 μM dNTPs and 2.5 U Taq polymerase in a 100 μl reaction. The PCR cycling protocol was: 1 cycle of 94 °C for 5 minutes, 30 cycles of 94 °C for 30 sec, 53 °C for 30 sec and 72 °C for 1 min and finally 1 cycle of 72 °C for 15 min and a 4 °C hold. The product is purified using the QIAquick PCR purification kit, and ligated to pT7Blue. Ligations were performed according to the manufacturers protocol (50 ng vector, 2 μl of purified PCR product, 1× ligase buffer [50 mM Tris-KCl, 10 mM MgCl, 10 mM Dithiothreitol, 1 mM ATP, 25 μg/ml BSA, pH 7.5] and 200 U of ligase, 3 μl of the ligation mixture was transformed into DH5α, and grown on L-Agar with 100 μg/ml ampicillin, 40 μg/ml XGal. White colonies were grown in LB + 100 ug/ml ampicillin and the plasmids isolated using QIAprep DNA miniprep kit. The plasmids were screened for the presence of the insert by cutting 5 μl of each DNA miniprep with a restriction enzyme specific to the vector and unable to cut the insert fragment. Plasmids containing the PCR product are identified because they are larger than the original pT7Blue plasmid (which is approximately 2.9 kb). The insert is digested from the pT7Blue-based plasmid and ligated into pVE6007, previously digested with the same restriction enzymes. Although there are several methods to directly clone PCR products into a vector, we found that cloning the PCR product first into pT7Blue vector and then subcloning into pVE6007 was more reliable and less time consuming. The new plasmid derivative is then transformed into COH 1, and the presence of the plasmid verified by plasmid mini-prep analysis and agarose gel electrophoresis.

To construct the allelic replacement vector, a gene fragment is first cloned into pVE6007 vector and then digested with an enzyme that cuts once within the cloned sequence. The kanamycin omega cassette is isolated by digesting pC1V2 [5] with *Bam*HI and gel purifying the 2.3 kb band containing the cassette. If the ends of the vector and the ends of the kanamycin cassette are not compatible, they are converted to blunt ends by standard molecular techniques [3]. The kanamycin cassette is ligated into the vector under the following conditions; 50 ng vector, 100 ng insert, 1× ligase buffer and 200 units ligase in a total volume of 15 μl. 5 μl of the ligation mix is transformed into DH5α and transformants grown on selective L-agar plates containing chloramphenicol (5 μg/ml) and kanamycin (40 μg/ml) at 30 °C.

B. Transformation of *E. coli* and GBS by electroporation

E. coli was prepared and transformed according to standard protocols [3]. GBS were transformed with at least 400 ng of plasmid DNA using the electroporation method described by Framson et al. [1]. After electroporation, the cells were incubated at 30 °C for two hours in recovery broth so that plasmid containing cells could express the *cat* gene. Transformants were selected by plating cells on THA + Cm (10 μg/ml) and growing cells at 30 °C.

C. Conditions that favor in vivo recombination between vector and chromosome DNA for single cross-over mutagenesis

We observed that specific growth conditions affected optimal recovery of clones containing single cross-over mutations. After cloning the intragenic fragment of a gene into pVE6007 and transformation into GBS, a chloramphenicol resistant colony is grown overnight in liquid culture at 30 °C with antibiotics. After a 1:50 dilution in fresh broth with antibiotics, the culture is grown for 2–2.5 hrs at the permissive temperature of 30 °C in a shaking incubator at 200 rpm. The culture is then shifted to a shaking incubator at 200 rpm at the non-permissive temperature (37 °C) and allowed to grow for an additional 2–2.5 hours. The cultures are then serially diluted in pre-warmed (37 °C) PBS and spread on pre-warmed TH agar plates containing chloramphenicol (10 μg/ml). We found that without dilution of the original culture, high background growth occurs on the selective agar plates that obscures the identification of true chloramphenicol resistant colonies. A duplicate set of samples were spread on TH agar plates without any antibiotics to determine the number of total viable bacterial cells. After overnight incubation, the relative number of chloramphenicol resistant clones is compared to the total number of viable colonies to estimate the recombination frequency. Typically, the frequency of Cm^r clones (at 37 °C) representing the plasmid integrants is between 0.1% to 0.001% of the total cell population. Chloramphenicol resistant colonies are streak purified and maintained at 37 °C, their chromosomal DNA isolated and analyzed to confirm plasmid integration into the chromosome at the proper gene location (see below).

D. Allelic replacement mutagenesis using a kanamycin omega cassette and the conditions to select for double cross-over recombination

Allelic replacement recombinants are generally

more stable than the single cross-over recombinants. This is because single cross-over insertions have duplicate homologous sequences in close proximity to each other (which can undergo homologous excisional recombination) and secondly, strains with plasmid integrants can be exposed to conditions (such as growth at permissive temperatures) that promotes plasmid replication after excision from the chromosome. If excision occurs by homologous recombination, reversion to wild-type and restoration of the original integrity of the chromosomal gene results. By contrast, allelic replacement mutants don't contain duplicate sequences in close proximity to each other, and don't contain plasmid replicons, resulting in more stable mutations.

As mentioned above, the plasmid construct for allelic exchange mutagenesis was constructed by cloning the kanamycin omega fragment from pCIV2 into the gene of interest on pVE6007, resulting in the kanamycin gene flanked by the cloned gene sequences. These flanking sequences serve as potential homologous recombination sites for cross-over recombinations on either side of the kanamycin cassette during the cross-over events. Double cross-over recombination transfers the cassette to the chromosome within the recipient gene (Figure 1B). The first recombination event is obtained by a single cross-over recombination as described above. There are two possible sites for recombination upstream of the kanamycin gene (site A) and downstream of the kanamycin gene (site B; Figure 1B). The excision of the integrated plasmid is promoted by a second recombination event induced by growth at room temperature without antibiotics (as discussed above, growth of a single cross-over plasmid integrant at the permissive temperature encourages excision). This second recombination event can occur at either site A or site B and results in two possible outcomes. If recombination occurs on the same side that was used during plasmid integration, the strain reverts to wild type and the original plasmid is excised in its entirely (Figure 1B). However, if the second recombination occurs on the side opposite to the original recombination event, an allelic exchange occurs and the omega fragment is left on the chromosome creating an insertion mutation (Figure 1B). Although excision should theoretically occur on either side of the kanamycin gene with equal probability, we have observed that excision occurs more frequently on the side of the original recombination site.

Following excision, the autonomously replicating plasmid (now without the kanamycin cassette) can be eliminated by growing the cells at the non-permissive temperature without antibiotics. Recombinant clones are recovered by plating cultures on THA + kanamycin (400 μg/ml). Surviving clones are then screened for chloramphenicol sensitivity (10 μg/ml). Kan^r, Cm^s strains are strong candidates for clones in which double recombination has occurred. Chromosomal DNA from the resulting clones is isolated and analyzed (see below). Using this method, we have successfully inserted the kanamycin cassette into several capsule genes in the GBS chromosome.

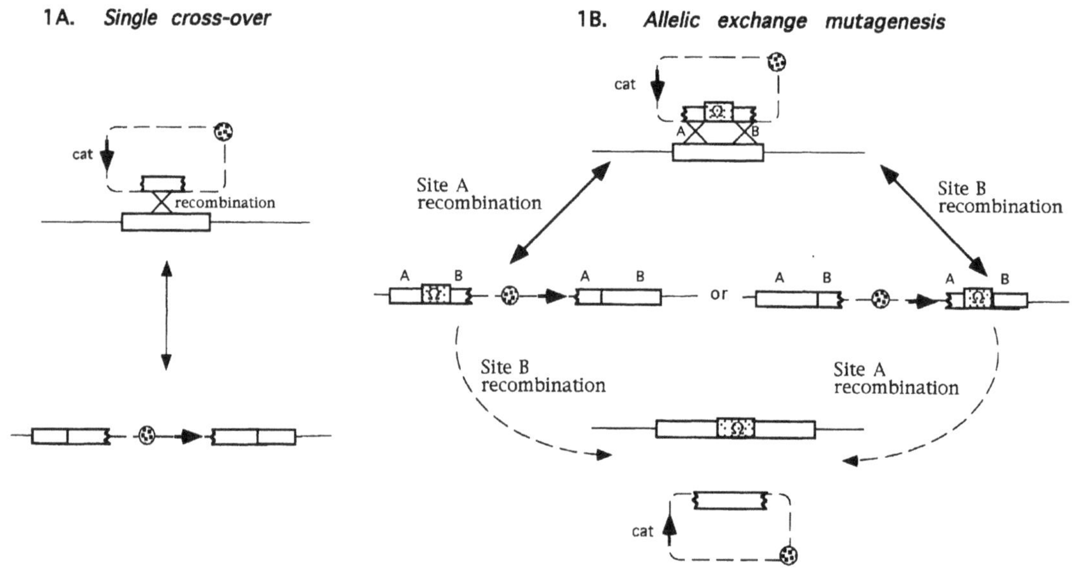

Figure 1. Diagram of single and double recombination events. In panel 1A, insertion of a plasmid into the chromosome is mediated by a single recombination event between the plasmid and the chromosome. In panel 1B, allelic exchange mutagenesis occurs vis a double recombination event. See text for details.

E. Mini-scale plasmid preparation from GBS
GBS were grown overnight in TH broth + 0.2% glycine with the appropriate antibiotic. The culture (approximately 4 ml) was harvested by centrifugation and resuspended in 250 µl protoplasting buffer (20% sucrose, 20 mM Tris-HCl pH 7.0, 10 mM $MgCl_2$, 0.5% Triton X-100). 10 µl of mutanolysin (10 units/µl) was added to the cells and incubated at 37 °C for 15 minutes. The culture was then frozen at –80 °C for 15 minutes and thawed at 37 °C for one minute to aid in bacteria lysis. From the QIAprep spin plasmid mini-prep kit, 250 µl of resuspension solution I was added, 500 µl of lysis solution II and 700 µl of neutralization solution III were added to the lysed GBS. The rest of the procedure involves a typical mini-prep as detailed in the Qiagen protocol.

F. Isolation of chromosomal DNA from GBS
Chromosomal DNA was isolated by the method described by Framson et al. [1].

G. Analysis of chromosomal DNA to determine recombination efficacy
Following isolation of chromosomal DNA from a potential recombinant, the site and orientation of the integrated plasmid on the chromosome must be confirmed. Southern analysis can be used but is often a tedious method especially if many samples need to be analyzed. A quick and easy method to confirm the location of the plasmid site is to use PCR analysis. The analysis requires only four primers; two that are plasmid specific and flank the multiple cloning site in pVE6007 (primers 1 and 2; Figure 2A), and two that are capsule locus specific and not part of the gene cloned into pVE6007 (primers 3 and 4; Figure 2A). By using a plasmid-specific primer together with a chromosome-specific primer in a PCR reaction, a product is generated only if a single cross-over integration event has occurred at the correct location. For example, it is impossible for primers 2 & 3 and/or 1 & 4 to produce a product unless pHY105 has integrated into the chromosome at the correct location and in the proper orientation (Figure 2A). For allelic exchange insertions (Figure 2B), the presence of the kanamycin omega cassette can be determined by performing PCR using primers that flank the cassette. Allelic exchange mutations result in the PCR product being larger by 2.3 kb (the size of the kanamycin cassette) compared to the PCR product produced when using the wild type chromosomal DNA.

H. Detection of type III capsule by colony immunoblot analysis
To detect the presence of capsule on recombinant strains, potential mutants are inoculated onto solid agar plates with the appropriate antibiotic. Following overnight growth, the bacteria are transferred to nitrocellulose filters by placing a filter onto the agar plate. After one minute, the filter is carefully lifted off the plate using gloved fingers or forceps and placed colony side-up onto Whatman 3MM filter paper that has been saturated with TN buffer (0.02 M Tris-HCl, 0.5 M NaCl pH 7.5). When the entire blot is wet, it is transferred (colony side up) to Whatman 3MM filter paper saturated with 70% ethanol. After one minute, the blot is removed and placed on Whatman 3MM paper to dry for at least fifteen

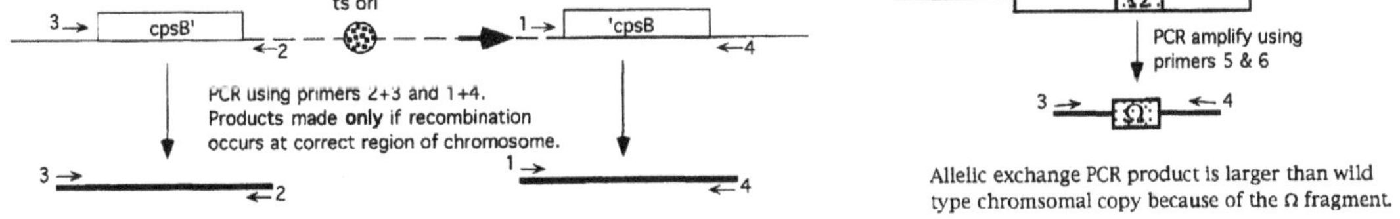

Figure 2. Analysis of recombinant chromosome structure by PCR. Panel 1A, insertion of plasmid into the chromosome can be verified by performing PCR using one primer that anneals to the vector (primers 1 and 2) and a second primer that anneals to the chromosome (primers 3 and 4). Therefore, PCR products from a reaction using primers 1 and 4 or primers 2 and 3, are only possible if recombination has occurred. If a product is produced, this is evidence of recombination at the correct location. To test for an allelic exchange event, chromosome-specific PCR primers are used. These primers are specific to the target gene and flank the kanamycin cassette recombination site. When compared to the wild type chromosome, an allelic exchange mutant will produce a larger product because of the kanamycin cassette. Small arrows represent primers.

minutes. The blot can be used immediately or placed in plastic wrap and stored at 4 °C.

The dried blot is soaked in TBS for 10 minutes and transferred to 5% NFDM for 30 minutes to block nonspecific binding sites on the membrane. The blot is washed in TTBS for 5 minutes twice and then exposed to the rabbit anti-type III GBS capsule antibody (1:30,000 dilution in TBS with 3% BSA). After one or two hours, the blot is washed twice with TTBS for five minutes each. The secondary antibody (goat anti-rabbit IgG conjugated to HRP) is diluted 1:100,000 in TBS with 3% BSA and added to the blot for one hour. The blot is then washed three times with PBS with 5% NFDM (137 mM NaCl, 2.7 KCl, 4.3 mM $Na_2HPO_4 \cdot 7H_2O$, 1.4 mM KH_2PO_4, pH 7.3) and three times with PBS (5 minutes per wash). The blot is removed from the PBS wash, excess liquid is allowed to drain off and the blot is placed into an open seal-a-meal bag. One ml of SuperSignal chemiluminescent substrate for HRP is added so that the entire surface of the blot is covered with substrate. Excess liquid is removed by touching the edge of the blot with a paper towel. Air bubbles are removed, the bag sealed and the blot exposed to X-ray film for autoradiography.

4. Results and discussion

Plasmid integration and allelic replacement vectors recombine into the chromosome by homologous recombination

To produce the plasmid integration vector for mutating the *cpsB* gene, the intragenic portion of the *cpsB* gene was amplified from the COH1 chromosome using the primers #19 (5′-TTGCAACAGCTGGACTTTTCTATAGTC-3′) and #20 (5′-CCTGCACCAACAATAAACCCGATTAG-3′). This PCR reaction produced a 469 bp PCR product that contains only the intragenic portion of *cpsB*. This PCR products lacks 100 nt from the 5′ end of *cpsB* and 120 nt from the 3′ end. The '*cpsB*' product was cloned into pT7Blue to produce the plasmid pHY205 (Figure 3). The intragenic fragment was digested from pHY205 with *Kpn*I and *Hin*dIII and ligated into pVE6007 that had been previously digested with the same restriction enzymes (Figure 3). The new plasmid derivative (pHY105) was transformed into COH-1 and clones were exposed to conditions that promote single cross-over homologous recombination (chloramphenicol resistance at 37 °C). A potential recombinant clone was designated HY102 and its chromosomal DNA isolated for further analysis (see below).

In an analytical PCR reaction, we used the primers 2 and 3 and the HY102 chromosomal DNA as a template as shown in Figure 2A. Since primer 2 anneals to the plasmid, and the primer 3 anneals to the chromosome, a PCR product is only possible if the vector has integrated into the chromosome at the correct region and orientation (Figure 2). PCR products of the correct size were produced from HY102, but, as expected, the control reactions using wild type COH1 chromosomal DNA did not produce a product. The distinct PCR amplification product was observed and was the predicted size, confirming that recombination occurred at the correct location (data not shown).

To generate the plasmid construct for allelic exchange mutagenesis in *cpsB*, we cloned the kanamycin omega fragment from pCIV2 into pHY105 to create pHY113 (Figure 3). After pHY113 was transformed into COH1, a clone was exposed to conditions that promote double homologous recombination. A potential mutant (Kan^r, Cm^s) clone was identified and designated HY105. Chromosomal DNA from this strain was isolated and analyzed (see below).

The potential allelic exchange mutant, HY105, was subsequently isolated as described above. Confirmation of the omega kanamycin cassette in *cpsB* was performed using the primers 3 and 4 in PCR amplification (Figure 2B). These primers hybridized to the chromosomal region upstream and downstream of the 'cpsB' sequences as predicted (and are therefore not present on the plasmids pHY105 or pHY113). Therefore, PCR products can only be obtained when using chromosomal DNA as a template for these primers. As compared to the amplification product that is generated when using the wild type COH1 chromosomal DNA as the template, the PCR product from HY105 is approximately 2.3 kb larger (*data not shown*), corresponding to the size of the kanamycin cassette. These data confirmed that the allelic exchange mutation was successfully introduced into the wild type *cpsB* gene of COH1.

Capsule expression is abolished when cpsB is disrupted in the chromsome

To determine if the mutations in clones HY102 and HY105 affected capsule synthesis on the bacterial surface, immunoblot analysis was performed. Bacteria were streaked onto TH agar plates and incubated overnight. After transfer to nitrocellulose and probing with rabbit, anti-type III GBS capsule antibody, we could not detect capsule from HY102 or HY105 in contrast to the strong signal observed with the parental strain COH1 (Figure 4). These results demonstrate that insertion mutations into *cpsB* inhibits wild type capsule expression.

How these mutations affects capsule gene expression is not clear. Because the kanamycin cassette contains translational and transcriptional stop signals, it probably causes polar mutations in downstream genes. However, it is not clear what effect the single cross-over insertions have on downstream expres-

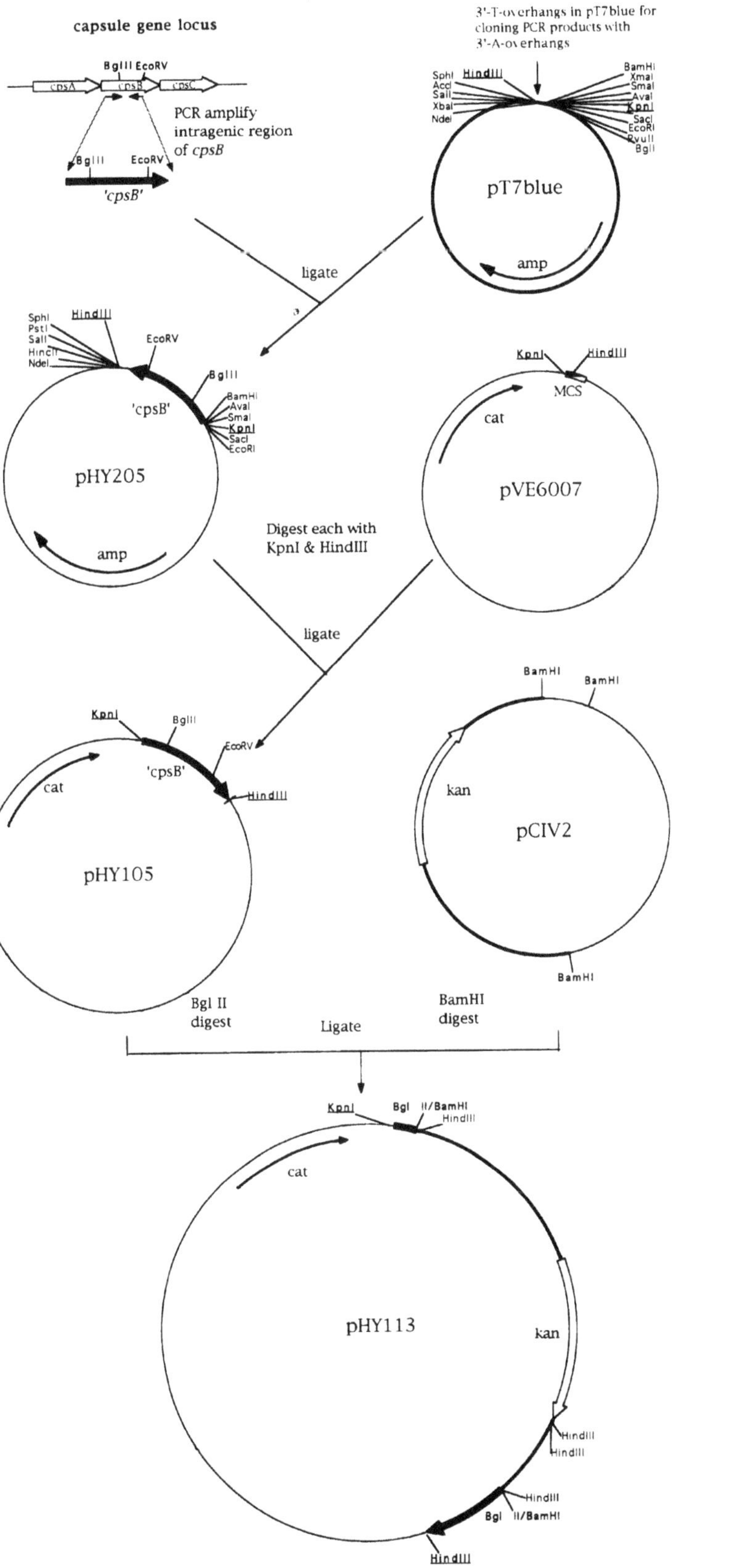

Figure 3. Schematic showing construction of vectors for recombination. See text for details.

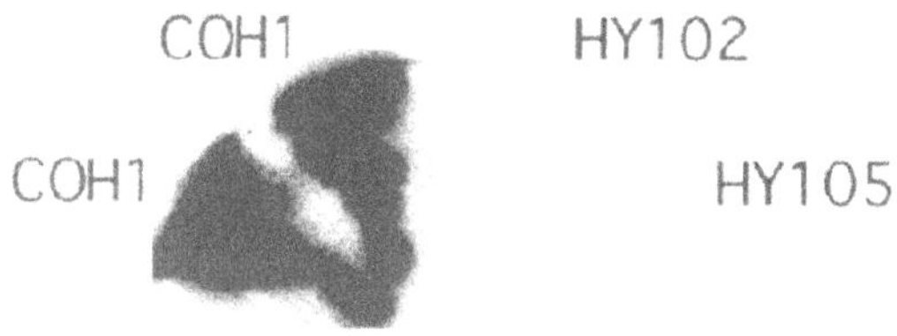

Figure 4. Immunoblot of GBS strains to examine the presence of type III capsule. The wild type and parental strain COH1 are included as a positive control. HY102 is the single cross-over recombinant clone and HY105 is the allelic exchange mutant of *cpsB*. Clones were lifted onto nitrocellulose, prepared as described and exposed to a rabbit, anti-type III capsule antibody. The secondary antibody was goat, anti-rabbit IgG antibody that has been conjugated to horse radish peroxidase. A positive signal was produced by a chemiluminescence and detected by autoradiography.

sion. In our mutant constructs, the *cat* gene on pVE6007 and the *cpsB* gene sequence are oriented in the same direction. Therefore, it is possible that transcription initiated from the *cat* gene is sufficient to transcribe downstream genes. If this occurs, polar effects would be diminished. Further analysis by complementation of the mutations and/or transcriptional analysis can be used to further define the effects of these mutations on distal gene expression.

In conclusion, we were able to obtain capsule gene mutants by single and double cross-over recombination into the *cpsB* gene. We also demonstrated that these clones were unencapsulated as detected by immunoblot. This report demonstrates that this process is a straightforward and feasible method for producing directed mutations in virtually any chromosomal gene in GBS.

Acknowledgments

This work was supported by the National Institutes of Health Initiative for Prevention of Group B Streptococcal Infections in Newborn Infants (grant AI 25152) and the National Institutes of Health (grant AI 22498).

We wish to thank Barbara Droppelman for manuscript preparation.

Notes on suppliers

1. DIFCO laboratories, P.O. Box 331058, Detroit, MI 48232–0758, USA
2. Sigma Chemical, P.O. Box 14508, St. Louis, MO 63103, USA
3. Life Technologies Inc. (GIBCO BRL), 8400 Helgerman Ct., P.O. Box 6009, Gaithersburg, MD 20884, USA
4. Qiagen Inc., 28159 Ave Stanford, Santa Clarita, CA 91355, USA
5. Novagen Inc., 597 Science Drive, Madison, WI 53711, USA
6. New England Biolabs Inc., 32 Tozer Rdoad, Beverly, MA 01915-5599, USA
7. Promega Corp., 2800 Woods Hollow Road, Madison, WI 53711, USA
8. Boehringer Mannheim Corp., 9115 Hague Road, P.O. Box 50414, Indianapolis, IN 46250, USA
9. Perkin-Elmer Cetus Corp., 850 Lincoln Centre Drive, Foster City, CA, USA
10. Schleicher & Schuell Inc., 10 Optical Ave, P.O. Box 2012, Keene, NH 03431, USA
11. Whatman Inc., 9 Bridewell Place, Clifton, NJ 07014, USA
12. Bio-Rad Laboratories, 2000 Alfred Nobel Drive, Hercules, CA 94547, USA
13. Pierce Chemical Co., 3747 N. Meridian Road, P.O. Box 117, Rockford, IL 61105, USA
14. Eastman Kodak, 343 State St., Rochester, NY 14652-3512, USA

References

1. Framson PE, Nittayajarn A, Merry J, Youngman P, Rubens CE (1997). New genetic techiques for group B Streptococcus: High-efficiency transformation, maintenance of temperature-sensitive pWV01 plasmids, and mutagenesis with Tn*917*. App Env Microbiol 63: 3539–3547.
2. Maguin ED, Duwaat P, Hege T, Ehrlich D, Gruss A (1992). New thermosensitive plasmid for gram-positive bacteria. J Bactoriol 174: 5633–5638.
3. Maniatis T, Fritsch EF, Sambrook J (1989). Molecular cloning: a laboratory manual. Cold Spring Harbor, NY: Cold Spring Harbor Laboratory.
4. Martin TR, Rubens CE, Wilson CB (1988). Lung antibacterial defense mechanisms in infant and adult rats: implications for the pathogenesis of group B streptococcal infections in neonatal lung. J Infect Dis 157: 91–100.
5. Okada N, Geist RT, Caparon MG (1993). Positive transcriptional control of *mry* regulates virulence in the group A streptococcus. Mol Microbiol 7: 893–903.

Address for correspondence: Craig E. Rubens, Division of Infectious Disease, Children's Hospital and Medical Center Mailstop CH32, 4800 Sand Point Way NE, Seattle, WA 98105, USA
Phone: (206) 528-2770; Fax: (206) 527-3890
E-mail: cruben@chmc.org

Methods in Cell Science **20**: 21–33 (1998).

Targeted mutagenesis of enterococcal genes

Xiang Qin[1,2,4], Fang Teng[3,4], Yi Xu[3,4], Kavindra V. Singh[1,4], George M. Weinstock[2,4] & Barbara E. Murray[1,2,4]

Division of Infectious Diseases, [1] Department of Medicine, [2] Department of Microbiology and Molecular Genetics, [3] Department of Biochemistry and Molecular Biology, [4] Center for the Study of Emerging and Re-emerging Pathogens, University of Texas Medical School at Houston, Houston, Texas, USA

Abstract. A series of methods was developed for targeted mutagenesis in enterococci. First, a transposon mutagenesis system, miniγδ-200 (mγδ), which was used previously to make insertion mutants in streptococci, was shown to be useful for generation of mutants in enterococci. After mutagenesis of cosmid clones carrying enterococcal DNA inserts in *Escherichia coli* with mγδ, we were able to isolate the mutants by phenotype or to screen for them by immunoblotting or comparison of restriction digestion patterns. Clones with mγδ insertions in targeted enterococcal genes were then introduced into enterococci by electroporation to generate targeted disruption mutations. Allelic replacement en masse, that is, electroporation of enterococci with DNA from a pool of mutagenized cosmid clones, was shown to be an efficient method to obtain mutations in genes with detectable phenotypes in enterococci. We next constructed a vector for mutagenesis using small intragenic fragments of enterococcal genes to disrupt targeted genes. This vector was modified from pBluescript SK ( ) by cloning the kanamycin resistance determinant from mγδ into the *Sca*I site internal to the ampicillin resistance gene. It was then used to generate insertion mutants in *Enterococcus faecalis* with an intragenic fragment as small as about 500 bp. A third method, based on the conjugation system reported by Trieu-Cuot et al. [1], was developed in order to circumvent difficulties in the electroporation of some enterococcal strains and to improve the efficiency with which targeted mutations can be generated in enterococci. This system was capable of mobilizing both small plasmids and large cosmids into enterococci by conjugation, and produced disruption mutations by homologous recombination.

Key words: Conjugation, Electroporation, Enterococcus, mγδ, Targeted mutagenesis, Transposon

Abbreviations: Amp = ampicillin; BHI = brain-heart infusion; Cam = chloramphenicol; Erm = erythromycin; Kan = kanamycin; IPTG = isopropyl thiogalactoside; Nal = nalidixic acid; Spc = spectinomycin; Tet = tetracycline

1. Introduction

Enterococci are among the major causes of hospital-acquired infections, and *Enterococcus faecalis* and *Enterococcus faecium* are the two most common species isolated from enterococcal infections. The treatment of such infections is difficult due to the intrinsic resistance of enterococci to antibiotics, and has become even more difficult in recent years due to their acquisition of resistance to additional antibiotics [2]. Therefore, it is important to understand enterococcal infection mechanisms and to develop new therapeutic approaches for enterococcal infections.

To identify and study the virulence genes of enterococci, it is important to have methods available for the construction of mutants. However, methods for constructing targeted mutations in enterococci are limited. Several approaches to the generation of enterococcal mutants have been described. Among them, electroporation has been commonly used to introduce DNA into enterococci to make insertion mutants [3–8]. Waser et al. successfully made an insertion mutant of a Na-H-antiporter gene (*napA*) in *Enterococcus hirae* via homologous recombination by cloning an Erm resistance gene into the *napA* gene and introducing the construct back into *E. hirae* [8]. One problem with the electroporation method to create mutants of enterococci is that, although some strains are transformable by electroporation, others are not; or the efficiency is so low that recovery of insertion mutants by homologous recombination is difficult. Another disadvantage of this method is that it requires multiple cloning steps and is time-consuming. Enterococcal transposons, Tn*916* and Tn*917*, have also been used to make insertion mutants in enterococci [9–18]. However, Tn*916* and Tn*917* can introduce multiple insertions and are not useful for targeted mutagenesis. In addition, a scorable phenotype is required to identify the expected mutants after mutagenesis. Thus, alternative ways to make targeted enterococcal mutants are of interest. In this paper, we summarize the targeted mutagenesis methods for enterococci that were developed in our laboratory.

2. Materials

A. Chemicals
1. Agarose, No. 15510-019.[9]
2. α^{32}P-dCTP, No. AA 0005.[1]
3. Ampicillin, No. A-9518.[15]
4. Bacto-agar, No. 0140-01.[6]
5. Brain-heart infusion, No. 0037-01-6.[6]
6. Calf thymus DNA, No. D-1501.[15]
7. Cesium chloride, No. BP210-500.[7]
8. Chloramphenicol, No. C-0378.[15]
9. 4-chloro-1-naphthol, No. C-8990.[15]
10. dNTPs, No. U-1240.[14]
11. Erythromycin, No. 104663.[4]
12. Fusidic acid, No. F-0881.[15]
13. Gelatin, No. 0143-17-9.[6]
14. Glycine, No. G46-500.[7]
15. Hybond N+ membrane, No. RPN303B.[1]
16. InstaGene™ kit, No. 732-6030.[3]
17. Isopropyl thiogalactoside, No. BP1620-1.[7]
18. Kanamycin, No. 17924.[17]
19. Klenow fragment of *E. coli* DNA polymerase I, No. 008404.[4]
20. Low-melting-temperature agarose, No. 50123.[8]
21. Nalidixic acid, No. 19436.[17]
22. NitroPlus™ nitrocellulose filters, No. NOTHY 08250.[11]
23. pBluescript SK (–).[16]
24. pCR™ II, No. K2050-01.[10]
25. Protein A-peroxidase, No. P-8651.[15]
26. Spectinomycin, No. S-9007.[15]
27. *Taq* DNA polymerase, No. M-1861.[14]
28. Tetracycline, No. 22105.[17]
29. Todd Hewitt, No. 4311736.[2]
30. Tryptone, No. 0123-17-3.[6]
31. Yeast extract, No. 0127-17-9.[6]

B. Equipment
1. DNA thermal cycler, No. N801-0150.[13]
2. Gene Pulser apparatus, No. 165-2075.[3]
3. Microtiter dish (Cell Wells™), No. 25850.[5]
4. UltraLok tube, No. 3420-2539.[12]

3. Procedures

A. Preparation of stock solutions
1. BYGT medium: 19 g BHI, 5 g yeast extract, 2 g glucose, 100 ml 1 M Tris.HCl (pH 8.0) in 1 liter distilled water.
2. EC lysis solution: 6 mM Tris.HCl (pH 7.6), 1 M NaCl, 100 mM EDTA (pH 7.5), 0.5% Brij 58, 0.2% deoxycholate, 0.5% sodium lauroyl sarcosine, 20 µg of RNase per ml, 1 mg of lysozyme per ml.
3. Electroporation solution: 0.625 M sucrose, 1 mM $MgCl_2$ adjusted to pH 4.0 with 1 N HCl.
4. ESP: 0.5 M EDTA, pH 9 to 9.5, 1% sodium lauroyl sarcosine, 50 µg of proteinase K per ml.
5. GTE: 50 mM glucose, 25 mM Tris and 10 mM EDTA, pH 8.
6. LB medium: dissolve 5 g yeast extract, 10 g tryptone, 5 g NaCl in 1 liter distilled water.
7. PCR buffer: 100 mM Tris.HCl (pH 8.3), 500 mM KCl, 15 mM $MgCl_2$ and 0.01% (w/v) gelatin.
8. PIV buffer: 1 M NaCl, 10 mM Tris.HCl (pH 7.6).
9. SR medium: 10 g tryptone, 5 g yeast extract, 200 g sucrose, 10 g glucose, 25 g gelatin, 15 g agar in 1 liter distilled water; add $MgCl_2$ and $CaCl_2$ to concentration of 2.5 mM, respectively.
10. SSC (20×): 3 M NaCl and 0.3 M sodium citrate, pH 7.0.
11. SSPE (20×): 3 M NaCl, 0.2 M NaH_2PO_4, 25 mM EDTA, adjust to pH 7.4 with 10 N NaOH.
12. TE buffer: 10 mM Tris.HCl (pH 7.4), 1 mM EDTA (pH 8.0).

B. Preparation of cosmid clones for mutagenesis
1. Transposon mutagenesis
To make insertion mutations in *E. coli* for subsequent use in construction of enterococcal mutants by allelic replacement, the previously described transposon miniγδ-200 (mγδ) [19] was chosen to mutagenize pLAFRx cosmid clones in *E. coli* (Figure 1). The Kan resistance determinant (*aph(3')-IIIa*) [19] in this element is expressed in both Gram- positive and Gram-negative hosts. The cosmid vector pLAFRx is a derivative of pLAFR which contains the *oriT* and Tet resistance determinant of RK2 [5] and was used for construction of enterococcal genomic libraries.
a. Preparation of the donor strain: Transform the donor strain, CBK884 (pMGD5, pXRD4043) [19], with the desired pLAFRx cosmid according to the method previously described [20]. Select transformants on LB/Tet (40 µg/ml)/Kan (25 µg/ml) agar plates. The mγδ element which carries the ΩKm-2 interposon (resistance to 25 µg/ml Kan in *E. coli* and 2000 µg/ml Kan in *E. faecalis*) is carried on a conjugative donor plasmid (pMGD5). The plasmid pXRD4043 provides γδ transposase and encodes resistance to 40 µg/ml Cam.
b. Transposon mutagenesis:
1) Grow a single transformant colony (donor strain) in 3 ml of LB/Cam (40 µg/ml) with 0.5 mM IPTG for 3 h (to OD_{660} = 0.3) at 37 °C to induce trans-

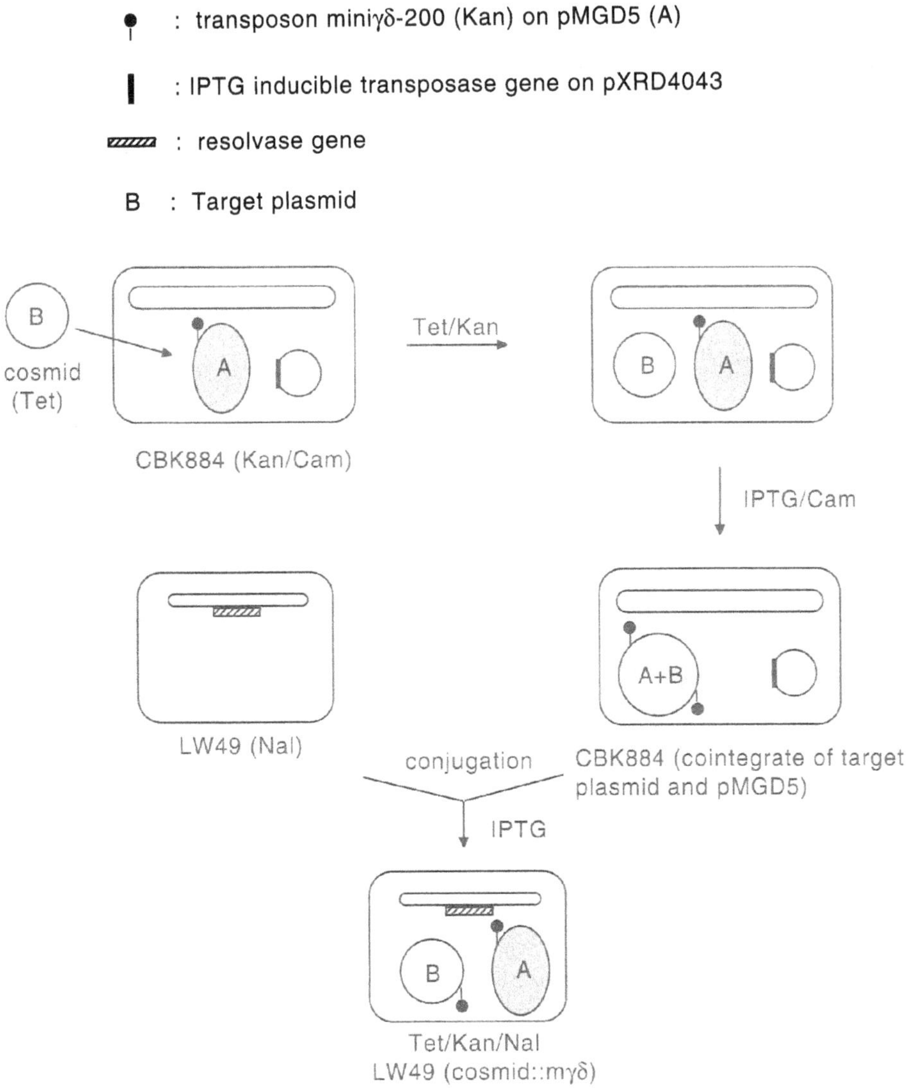

Figure 1. miniγδ-200 transposon mutagenesis. The larger rectangular squares labeled CBK884 represent mutagenesis strain containing a conjugative donor plasmid pMGD5 (the shaded oval) with miniγδ-200, and a plasmid, pXRD4043 (smaller open circle), with transposase gene. The elongated ovals represent the chromosome. LW49 is the recipient with a source of inducible resolvase gene. Target plasmids (cosmids) containing a Tet resistance marker are represented with a medium sized circle. The largest circle with two transposon symbols demonstrates the cointegrate of pMGD5 and target plasmid.

posase production. At the same time, grow the recipient strain, LW49 (Nal^r) [19], in LB/Nal (20 μg/ml) at 37 °C (to OD_{660} = 0.3)

2) Mix 0.5 ml donor cells with 0.2 ml recipient cells in an 18-mm diameter test tube, incubate for 30 min in a 37 °C water bath without shaking. Then, add 5 ml of prewarmed LB broth containing IPTG (0.5 mM), followed by incubation for 3 h without agitation. Select exconjugants on LB/Tet (40 μg/ml)/Kan (25 μg/ml)/Nal (20 μg/ml) plates.

2. Isolation of cosmid clones with transposon insertions in desired genes
   a. Isolation of mutants by phenotype: If a targeted gene encodes a scorable phenotype in *E. coli*, cosmids with insertions in the targeted gene can be isolated by the change in phenotype. For example, pLAFRx cosmid clones, pKV48 and pKV53, from a genomic library of *E. faecalis* OG1RF can complement pyrimidine and purine auxotrophs of *E. coli*, respectively (5). We transformed *E. coli pyr* or *pur* auxotrophs [5] with mutagenized pKV48 or pKV53

cosmids and scored for the loss of complementation to identify the cosmids which had insertions in responsible genes. We used these phenotypes to illustrate the utility of this system, other phenotypes can be used in a similar fashion.

1) Transform *E. coli pyr* or *pur* auxotrophs [5] en masse with DNA from pools of mγδ mutagenized exconjugants of pKV48 or pKV53 as described [20].
2) Plate the transformants on LB/Tet/Kan plates, then transfer single colonies with sterile toothpicks to both M63 minimal agar medium [21] and LB/Tet (40 μg/ml)/Kan (25 μg/ml). Save the colonies that no longer complement the relevant *E. coli* auxotrophs for subsequent analysis and allelic replacement studies.

b. Immunoscreening: If serum or antibody is available for a specific targeted gene product, immunoscreening can be used to detect cosmid clones with insertions in the responsible gene, or in other genes that are required for expression in *E. coli*. For example, pKV4 is a pLAFRx cosmid clone from a genomic library of *E. faecalis* OG1RF [22] which reacts with serum MH-1 (from a human patient with enterococcal endocarditis) [23]. Immunoscreening was applied to identify mutants which had lost their immunoreactivity with MH-1 serum. For immunoscreening:

1) After mutagenizing the desired cosmid clone with mγδ as described above, pick single colonies growing on LB/Tet (40 μg/ml)/Kan (25 μg/ml)/Nal (20 μg/ml) plates into microtiter dish Cell Wells™ and grow them overnight at 37 °C, inoculate the clones onto NitroPlus™ nitrocellulose filters which have been placed on LB agar with appropriate antibiotics (Tet, 40 μg/ml; Kan, 25 μg/ml).
2) After about 6–10 h of incubation at 37 °C, remove the filters and use them for immunoscreening by the method previously described [24]. Use immune serum as the primary antibody, protein A-peroxidase to react with the primary antibody and apply 4-chloro-1-naphthol as a substrate for protein A-peroxidase to produce visible color.

c. Restriction mapping: If the position of a targeted gene is known on a cosmid clone, mutants with mγδ insertions in this region can be identified by restriction mapping and comparing the restriction enzyme digestion patterns of mutagenized cosmids to the patterns of the original cosmid. This approach is especially useful to identify cosmid mutants with mγδ insertions in targeted genes when a scorable phenotype in *E. coli* and an antibody or serum are not available. In a previous study, we mapped the gene encoding the immunoreactivity of pKV4 with MH-1 serum to an 11 kb *Hin*dIII fragment. We used this approach to identify cosmid mutants which had insertions in this 11 kb *Hin*dIII fragment after mγδ mutagenesis. For restriction mapping:

1) After mutagenesis of a cosmid clone with mγδ as described above, pick single colonies growing on LB/Tet (40 μg/ml)/Kan (25 μg/ml)/Nal (20 μg/ml) plates into 5 ml LB medium and incubate them overnight at 37 °C.
2) Isolate cosmid DNA as previously described [25].
3) Digest the cosmid DNA with restriction enzymes according to the instructions of the suppliers and separate DNA fragments on a 0.7% agarose gel by electrophoresis.

C. Preparation of constructs with small intragenic fragments for targeted mutagenesis

1. Preparation of intragenic fragments and mutagenesis constructs

   A mutagenesis vector pTEX4577 [23] was constructed by digesting pBluescript® SK (–) with the restriction endonuclease *Sca*I, followed by blunt-ending with Klenow fragment of *E. coli* DNA polymerase I and ligation to the kanamycin resistance gene (*aph(3')-IIIa*) of Gram-positive bacterial origin; this gene was released from pMGD4 [19] by *Bam*HI digestion and blunt-ended by Klenow fragment. Small fragments internal to desired genes can be cloned into this vector using the multiple cloning sites of pBluescript® SK (–). Small internal fragments and mutagenesis constructs can be generated as following.

   a. Small internal fragments from subclones can be obtained from subclones such as those that may have originally been generated for sequencing or other purposes. To make the mutagenesis constructs:

   1) Digest subclones which contain internal fragments of desired genes with restriction endonucleases whose sites are present in the multiple cloning site of the vector used. Digest pTEX4577 with the same restriction endonuleases or other enzymes that produce compatible ends. *Hin*dIII, *Eco*RV and *Bam*HI cannot be used because the fragment

containing the Kan resistance gene has sites for these enzymes.

In our experiments, pBluescript SK (–) subclones containing a 1.2 kb fragment internal to the autolysin gene of *E. faecalis* OG1RF, and a 480 bp fragment internal to the *E. faecalis* gene encoding a homologue of P54 of *E. faecium*, which were originally generated for sequencing [26], were digested with *Sac*I and *Kpn*I, as was pTEX4577.

2) Separate the fragments by agarose gel electrophoresis and recover the fragments.
3) Mix the the fragments with digested pTEX4577, treat with DNA ligase, and use the ligation mixture to transform *E. coli* DH5α. Select transformants on LB/Kan (25 μg/ml) plates.

b. Small internal fragments of target genes can also be generated by PCR if the sequences are known. An intragenic *efaA* DNA fragment (730 bp, coding for amino acids 24 to 254 of EfaA of *E. faecalis*) was generated by PCR using a primer pair (forward primer, 5′-CGT TAG CTG CTT GCG GGA ATC-3′; reverse primer, 5′-CCA TAC TAC GTT TAT CGA CAC-3′) designed according to the previously published sequence [27]. For PCR and generation of mutagenesis constructs:
1) Mix 1 μl of forward primer (0.25 μg/μl), 1 μl of reverse primer (0.25 μg/μl), 10 μl of chromosomal DNA of OG1RF (2.0 ng/μl), 5 μl of dNTP (2.5 mM each), 5 μl of 10× PCR buffer, 1 μl (1 unit of activity) of *Tap* DNA polymerase and 27 μl of water (final total volume, 50 μl) in a PCR tube, cover the top of the mixture with mineral oil.
2) Heat the mixture at 94 °C for 5 min in a DNA thermal cycler, and then start the PCR cycles (94 °C for 1 min, 56 °C for 2 min, 72 °C for 3 min, total 35 cycles).
3) Purify the PCR product by precipitation with 70% ethanol.
4) Clone the PCR product into direct PCR product cloning vector, pCR™ II by ligating 100 ng of PCR product with 50 ng of vector DNA overnight at 16 °C. Use the ligation mixture to transform *E. coli* DH5α as described [20] and select transformants on LB/ Amp (50 μg/ml) plates.
5) Isolate plasmid DNA as previously described [25]. Release the PCR fragment from pCR™ II by *Eco*RI digestion and clone it into *Eco*RI-digested mutagenesis vector pTEX4577.

D. Electroporation of *E. faecalis* and selection

To generate allelic replacement in *E faecalis*, cosmids with mγδ insertions in target genes or derivatives of pTEX4577 with internal fragments of targeted genes can be used to transform OG1RF using a modification of the electroporation protocol as described [5].

1. Preparation of plasmid DNA

Large scale preparation of cosmid or plasmid DNA from *E. coli* by equilibrium centrifugation in CsCl-ethidium bromide gradients [24] is as follows:

a. Preparation of bacterial cells: Inoculate a 5 ml aliquot of an overnight *E. coli* culture into 500 ml of LB-broth with the appropriate antibiotics. Incubate the culture at 37 °C with shaking at 250 rpm for about 6 h.
b. Isolation of cosmid or plasmid DNA:
1) Harvest *E. coli* cells by centrifugation at 8,500 ×*g* for 15 min at 4 °C. Resuspend the pellet in 20 ml of GTE buffer, then mix gently with 40 ml of NaOH (0.2 N)/SDS (1%) and chill on ice for 5 min. Add 30 ml of KAc solution (5 M acetate, 3 M K^+) and mix gently. Chill the suspension on ice for an additional 5 min.
2) Centrifuge the suspension for 30 min at 4,500 ×*g*, collect the supernatant, add 90 ml cold 95% ethanol.
3) After chilling at –70 °C for 10 min, pellet the mixture at 11,000 ×*g* for 20 min, dry the pellet under vacuum.
4) Resuspend the pellet in 7.5 ml of TE buffer. After the addition of 0.5 ml of ethidium bromide (10 mg/ml), measure the volume and add cesium chloride at 1 g/ml followed by gentle shaking at 37 °C for 10 min to dissolve the cesium chloride.
5) Centrifuge at 25,000 ×*g* (at room temperature) for 15 min to clear the mixture. Transfer the supernatant to UltraLok tubes for quick seal and then centrifuge for 24 h at 230,000 ×*g*.
6) Remove the plasmid band (lower band) with needle and syringe. Extract the ethidium bromide with an equal volume of isoamyl alcohol three times.
7) Precipitate the plasmid DNA by adding two volumes of TE buffer and 2 times of the total volume of 95% ethanol. After placing the mixture at –20 °C for 2 h to overnight, harvest the plasmid DNA by centrifugation for 15 min at 12,000 ×*g*. Wash the DNA pellet with 70% ethanol three times, vacuum-dry

the DNA, and then resuspend in TE buffer or water for future use.

2. Electroporation
   a. Preparation of competent *E. faecalis* cells:
      1) Determination of optimal growth inhibition: Grow *E. faecalis* OG1RF overnight at 37 °C in BYGT medium with various glycine concentrations (1% to 10%). Use a glycine concentration that gives 70–90% reduction in the OD_{660} of an overnight culture, compared to that without glycine. The glycine concentration needed for this optimal growth inhibition may vary from strain to strain [28]. It is necessary to determine the glycine concentration for optimal growth inhibition each time when making competent cells, even if the same strain has been used before. Usually, about 6% glycine produces optimal growth inhibition for *E. faecalis* OG1RF.
      2) Preparation of competent cells: Dilute the overnight culture which shows optimal growth inhibition by 10-fold into fresh medium with the same concentration of glycine and incubate for one hour at 37 °C. Chill the culture on ice, harvest the cells by centrifugation at 5000 ×*g* for 12 min, and wash twice with 1/3 the original volume of chilled electroporation solution. Resuspend the cells in 1/30 of the original volume of electroporation solution and then chill on ice for 30–60 min (or save at –70 °C for later use).
   b. Electroporation:
      1) Mix 1 to 5 μg of cosmid or plasmid DNA (in < 20 μl of distilled H_2O or low salt buffer) prepared by equilibrium centrifugation in CsCl-ethidium bromide gradients with competent cells (100–200 μl, > 5×10^6 cfu/ml) on ice. Add the mixture into a chilled 0.2-cm electroporation cuvette and electroporate immediately with a Bio-Rad Gene Pulser apparatus at a capacitance of 25 μF, resistance of 200 Ω and peak voltage of 2.5 kV (field strength 8,750 to 10,000 V/cm for *E. faecalis*).
      2) Incubate the electroporated cells in 1 ml of BYGT medium supplemented with 0.25 M sucrose for 90–120 min at 37 °C. Optimal results will be obtained only if greater than 30 percent of OG1RF competent cells survive (usually, pulsing twice gave better results without an effect on the survival of OG1RF competent cells).
   c. Selection:
      1) After electroporation, spread the cell aliquots on selective SR agar plates (with Kan 2000 μg/ml or other appropriate antibiotics) and incubate at 37 °C for 48 to 72 h. Wrap slightly damp paper towels around the plates to avoid excessive drying.
      2) Restreak colonies which appear after 24 h to 72 h onto BHI/Kan (2000 μg/ml) plates. Colonies which grow well on restreaking have generally acquired the Kan resistance gene. Save those colonies for further study.
   d. Determination of transformation efficiency: Transformation efficiency can be measured by introducing 1 μg of pWM401 (29) DNA prepared from OG1 (pWM401) into OG1RF competent cells.
      1) Electroporate OG1RF competent cells with 1 μg of pWM401 as mentioned above.
      2) Make a series of 10 fold dilutions in BHI from the pulsed cells. Plate 25 μl of each dilutions on BHI/Cam (8 μg/ml) plates and then incubate the plates at 37 °C for 24 h. The colony counts from different plates will be used for the calculation of transformation efficiency.

3. Allelic replacement en masse
   For genes that have scorable phenotypes in enterococci, it is possible to generate targeted mutations by direct electroporation with a pool of DNA from mγδ-mutagenized cosmids containing the targeted gene. A gelatinase mutant and a *pyr* auxotroph of OG1RF were generated by this approach.
   a. Mutagenize the cosmid clone containing the targeted gene with mγδ according to the method described above.
   b. After selection of transconjugants on LB/Tet/Kan/Nal plates, pool about 500 to 2000 colonies and prepare cosmid DNA from the pool as described above.
   c. Transform competent OG1RF cells by electroporation with the pool of cosmid DNA and select for transformants on SR/Kan (2000 μg/ml) [22].
   d. Score transformants that grow on SR/Kan plates for a change of phenotype on appropriate media, e.g., score *pyr* auxotrophs in a defined broth (e.g., DMMS) or agar [22] medium, and gelatinase mutants on Todd Hewitt (TH) agar containing 3% gelatin [30].

E. Mutagenesis by conjugation and selection

In previous studies, shuttle vectors carrying an origin of conjugative transfer (*oriT*) were mobilized from *E. coli* to Gram-positive hosts, including the *E. faecalis* strain BM4110 [1]. The mobilization of these vectors is due to transfer functions provided by an integrated IncP plasmid on the chromosome of donor *E. coli* S17-1 cells [1]. We modified this conjugation system by constructing vectors that carry the *oriT* from pAT18 [1] and an antibiotic selection marker, but which cannot replicate in enterococci. After transfer into enterococci, the conjugative plasmid carrying an intragenic fragment of an enterococcal gene, or enterococcal DNA with a transposon or other insertion mutation, can integrate into the enterococcal chromosome by homologous recombination and the resulting mutants can be selected for antibiotic resistance.

Two new vectors with *oriT* are shown in Figure 2. pTEX5235 (pFT1) is a derivative of pBluescript SK (–) [31]. The spectinomycin resistance gene can be expressed in Gram-negative and Gram-positive bacteria [32]. The *oriT* fragment (≈ 760 bp) was amplified by PCR from pAT18 and cloned into the *Xmn*I site of pBluescript SK (–). The pTEX5235 vector cannot replicate in Gram positive bacteria but can be transferred to them by conjugation to make mutants when an intragenic fragment of a Gram-positive bacterial gene has been inserted into it. The *oriT* sequence was also cloned into the cosmid vector pBeloBAC11 [26, 33] to generate pTEX5236 (pFT2) [31]. The pTEX5236 vector can be used to clone large inserts and, after transposon mutagenesis, the cosmid can be transferred from donor *E. coli* S17-1 to Gram-positive bacteria to make targeted insertion mutations by double cross-over events.

1. Construction of *E. faecalis* mutants by conjugation using plasmid vector pTEX5235 with small intragenic fragments
   a. Generating mutagenesis construct:
      1) Small internal fragments can be obtained from subclones or PCR as described above. Small inserts in the PCR product cloning vector pCR™II, or other vectors, can be released using restriction endonucleases whose restriction sites are available in the multiple cloning sites of pBluescript SK (–), except *Kpn*I and *Xba*I. In an experiment designed to test the utility of this system, we used *Bam*HI/*Xho*I to digest pTEX5235 and a pBluescript SK (–) subclone containing an internal fragment of an autolysin gene of *E. faecalis* OG1RF [26] for cloning.
      2) Clone the fragment into pTEX5235 and use the recombinant plasmid to transform *E. coli* S17-1 as described [20].
   b. Conjugation:
      1) Grow the donor, *E. coli* S17-1 containing the mutagenesis construct, and an enterococcus recipient (e.g., OGIRF) in LB and BHI to about 10^8 cells per ml, respectively. Mix 1 ml of the donor cell culture with 1 ml of the recipient cell culture. Pellet the cells by centrifugation, wash them twice with BHI, and then resuspend in 50 µl BHI.
      2) Spot the mixture on a nitrocellulose filter, and incubate (cells facing up) on BHI agar overnight at 37 °C as described [1].
      3) Use BHI agar containing 25 µg/ml Kan

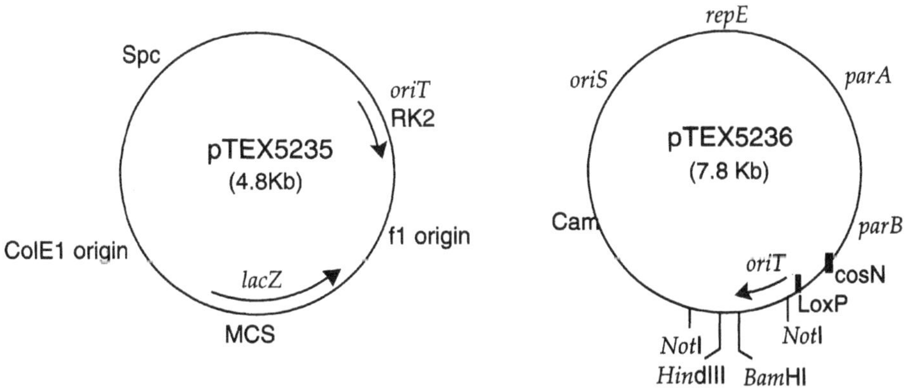

Figure 2. Two vectors with *oriT* of RK2. Plasmid pTEX5235 is a derivative of pBluescript SK (–). As a result of inserting the *spc* gene in the *Sca*I site in the *bla* gene, the plasmid no longer confers ampicillin resistance and the *Kpn*I and *Xba*I sites are no longer unique. *oriT* was cloned into pBluescript SK (–) using the *Xmn*I site. MCS, multiple cloning sites of pBluescript SK (–). Plasmid pTEX5236 is a derivative of the pBeloBAC11 vector. The plasmid has two unique cloning sites, *Bam*HI and *Hin*dIII. The two *Not*I sites outside of *Bam*HI and *Hin*dIII sites can be used to excise the cloned insert. Part of the *lac*Z coding region was delected when inserting the *oriT* sequence between the *Bam*HI site and *Not*I site on the left.

to select against *E. coli* S17-1, and 2000 μg/ml spectinomycin to select for the insertion of the plasmid into the chromosome of OG1RF. The integration of the plasmid into the chromosome by a single crossover is expected to disrupt the targeted autolysin gene. Save Spc resistant *E. faecalis* colonies for further characterization.

b. Generation of *E. faecalis* mutants by conjugation using large mobilizable cosmids

1) Mutagenize pLAFRx cosmid clones containing the desired genes with transposon mγδ, isolate the cosmids with insertions in the desired genes as described above.

Cosmid pBEM219 (about 45 kb in size) is a derivative of pLAFRx [5], which has the *oriT* from RK2. It also carries the *pyr* gene cluster from *E. faecalis* OG1RF with an insertion of the Kan^r transposon mγδ in the *pyrC* gene. We used this cosmid to determine whether insertion mutants in enterococci could be generated by conjugation using a large mobilizable cosmid via a double cross-over event. pTEX5236 containing the desired genes with transposon insertions, or other insertions, may be used in a similar fashion, although use of mγδ with pTEX5236 was not attempted because both pXRD4043 (in the donor strain, CBK884) and pBeloBAC11 were derived from the bacterial F factor.

2) Transform *E. coli* S17-1 with the mutagenized cosmids (e.g., pBEM219) as described above.

3) Transfer the mutagenized cosmids containing mγδ insertions in desired genes from *E. coli* S17-1 to OG1RF by filter matings as described above. Select insertion mutants of OG1RF on BHI/Kan (2000 μg/ml). If a double crossover occurs between the cosmid and the chromosome of enterococci, the desired gene is expected to be disrupted by the transposon mγδ.

F. Characterization of mutant strains

1. PCR characterization of enterococcal mutants

PCR can be applied to verify the correct insertion in *E. faecalis* using the genomic DNA from the putative enterococcal mutant clones as templates. To determine the position of the mγδ insertion, since the orientation of the mγδ insertion is not known, a primer (mγδ-R, 5′-GAT TTA GGA TAC ACG GAA TTT CG-3′) which is complementary to one end of mγδ in combination, in separate reactions, with each of two other primers which are complementary to the 5′ and 3′ ends of the targeted gene may be used for PCR. A positive product with one of the targeted gene primers can confirm the insertion and the orientation of mγδ. Sequencing of the PCR products will determine the exact position and orientation of the mγδ insertion.

a. Mini-preparation of bacterial DNA from *E. faecalis* for PCR:

1) Grow putative targeted mutants on BHI agar with appropriate antibiotics overnight at 37 °C.

2) Pick an isolated bacterial colony, resuspend it in 1 ml of sterile water in a microcentrifuge tube and then pellet the cells at 8,000 to 11,000 ×*g* for 1 to 2 min. Mix the pellet with 200 μl of InstaGene matrix and incubate the mixture at 56 °C for 15–30 min. Vortex the mixture at high speed for 10 sec and then place the tube in a 100 °C heat block or a boiling waterbath for 8 min. After vortexing another 10 sec at high speed, centrifuge the mixture at 8,000 to 11,000 ×*g* for 2–3 min. The supernatant is ready for PCR (using 20 μl the resulting supernatant per 50 μl PCR reaction).

b. PCR: Perform PCR reaction as above except that the annealing temperature may vary according to the primers used.

2. Southern blots and hybridization

Southern blots and hybridization can be used to map the positions of insertions in the OG1RF chromosome using fragments from cosmid clones containing the targeted genes as the probes, especially when the sequences of the genes encoding the targeted products are not known and PCR is not applicable.

a. Preparation of chromosomal DNA from *E. faecalis* for Southern blot: The method for isolation of chromosomal DNA from putative enterococcal mutants is a modification of the method described by Smith and Cantor [34].

1) Grow putative mutant clones overnight in 5 ml of BHI at 37 °C.

2) Harvest the cells by centrifugation and resuspend the cells in 1 ml of PIV buffer. Mix a portion (0.6 ml) of this cell suspension with 0.6 ml of 1.6% low-melting-temperature agarose in water at 40 to 50 °C and then pipette the mixture into a plug mold and allow the agarose mixture to solidify at 4 °C for 10 min.

3) For lysis, place the plugs of each mutant strain in 10 ml of fresh EC lysis solution.

After incubation overnight at 37 °C with gentle shaking, replace the solution with 10 ml of ESP and then incubate the plugs overnight at 50 °C with gentle shaking. Wash the plugs three times for 30 min each with 15 ml of TE buffer and then store them at 4 °C.

4) For digestion, place a small slice (5 × 5 mm) of an agarose plug in a microcentrifuge tube with 200 µl of distilled water followed by 25 µl of reaction buffer and 2 µl of the relevant restriction enzyme; incubate the reaction mixture for 12 h at 37 °C. Wash the slices with 1 ml TE buffer for 1 h at 37 °C and then melt them at 55 to 65 °C. Load the melted slices into the wells of regular agarose gels for electrophoresis.

b. Southern blot and hybridization: Southern transfer of DNA from agarose gels to membranes by capillary action is performed as described [24].

1) Expose the DNA in the gel to UV light for 80 sec and then denature the DNA in the gel in 0.4 N NaOH for 20 min. Cut a Whatman 3 mm paper wick slightly longer than the solid gel support (gel tray), wet it with transfer buffer (0.4 N NaOH) and then place it on top of the gel tray in a glass baking dish filled with transfer buffer. Place the gel on top of the 3 mm paper, then place a piece of Hybond N+ membrane cut to the size of the gel and pre-wet in 0.4 N NaOH on top of the gel and remove all air bubbles. Place another two pieces of pre-wet Whatman 3 mm paper on top of the membrane, then a stack of paper towels (4–7 cm high) cut to the size of the gel over the 3 mm paper, a glass plate over the paper towels and with a 500 g weight on top. Allow transfer to proceed overnight.

2) The next day, carefully remove the Hybond N+ membrane from the blotting apparatus, wash the membrane in 0.5 M Tris HCl (pH 7.0) for 5 min and then neutralize in 2× SSC (24) for 2 min. Dry the membrane at 80 °C for 2 h to fix the DNA to the membrane.

3) Incubate the filter with DNA to be probed in a sealable plastic pouch for 12 h at 42 °C in prehybridization solution (50% formamide, 5× Denhardt's solution, 5× SSPE (24) , 0.1% SDS, and 100 µg/ml heat-denatured calf thymus DNA).

4) After prehybridization, heat-denature the probe labeled with α^{32}P-dCTP (about 10^7 cpm per filter) with 0.2 ml calf thymus DNA (100 µg/ml) for 5 min at 100 °C, chill the probe on ice for 5 min, then add the probe to the prehybridization pouch and continue incubation at 42 °C for 12 h.

5) After hybridization, wash the membrane with 2× SSPE and 0.1% SDS with slow shaking twice for 10 min at room temperature, followed by two 15 min washes with 0.1× SSPE and 0.1% SDS at 50 °C. Air-dry and expose the membrane to X-ray film.

4. Results and discussion

We have shown that, following transposon mutagenesis in *E. coli*, it is possible to generate insertion mutants of *E. faecalis* by electroporation and homologous recombination. Several approaches were used to isolate the transposon insertion mutants in *E. coli*. For the enterococcal genes that have detectable phenotypes in *E. coli*, mutants could be easily isolated by their phenotypes. Disruption mutations in enterococcal pyrimidine and purine genes were identified in this manner, which were then used to generate *E. faecalis* OG1RF auxotrophs [5]. For enterococcal genes which do not encode detectable phenotypes in *E. coli*, restriction mapping of the positions of the mγδ insertions were used to identify mutants with insertions in known genes, or genes for which restriction maps are available [23]. For enterococcal genes for which antibodies directed against their products are available, immunoscreening can be a useful approach to isolate mutants with insertions in antigen-encoding genes or in loci that affect expression of the antigen-encoding genes [23]. One potential problem with immunoscreening is that some insertions in the antigen-encoding genes may only be immuno-diminished and may be difficult to detect. We encountered this problem when screening for mutants of one antigen gene in *E. coli*. We isolated only one immunonegative mutation in the pKV4 cosmid after immunoscreening 69 mγδ insertion derivatives, while 9 insertions were found in the 11 kb *Hin*dIII fragment thought to encode this antigen when restriction mapping was used to screen 100 derivatives. Seven of these 9 mγδ insertion derivatives were immuno-diminished. For the genes which have scorable phenotypes in enterococci, allelic replacement en masse also proved useful for the generation of enterococcal mutants after mγδ mutagenesis in *E. coli*. It simplifies the procedures for allelic replacement, as demonstrated by the generation of a gelatinase mutant and a pyrimidine mutant in *E. faecalis* OG1RF [5, 23]. The efficiency of mutant generation by electroporating cosmid DNA into *E. faecalis* OG1RF is listed in Table 2.

Table 1. Strains and plasmids

Strain or plasmid	Relevant characteristics	Source or reference
Escherichia coli strains		
LW49	mγδ mutagenesis recipient strain. Nal^r	[19]
CBK884	mγδ mutagenesis donor strain, containing pMGD5 and pXRD4043. Kan^r, Cam^r	[19]
DH5α	Recipient strain	Stratagene
LE392	pLAFRx competent strain	[5]
S17-1	RP4 derivative integrated in chromosome	[1]
Enterococcus strains		
OG1RF	*E. faecalis* strain[a]	[22]
SE34	*E. faecium* strain[a]	Laboratory collection
Plasmids		
pBluescript SK (–)	2.9 kb cloning vector. Amp^r	Stratagene
pLAFRx	A 21.6 kb cosmid vector with *mob* site and *oriT* of RK2. Tet^r	[5])
pAT18	Erm^r, *oriT* of RK2	[1]
pSN13	1.1 kb *spc* gene cloned into *NdeI/AccI* sites of pUC19. Spc^r	[32]
pBeloBAC11	A modified cosmid vector of pBAC108L. Cam^r	[26, 33]
pKV4	pLAFRx cosmid clone of OR1RF library, reacting with human patient serum	[23]
pKV48	pLAFRx containing a 22 kb fragment from OG1RF with *pyr* gene cluster	[5]
pBEM220	pKV48 with mγδ insertion in *pyrD*	[5]
pBEM219	pKV48 *pyrC*::mγδ 219. Tet^r, Kan^r	[5]
pKV53	pLAFRx containing a 24 kb fragment from OG1RF with *pyr* gene cluster	[5]
pTEX5062	pKV53 with mγδ insertion in *purL*	[5]
pTEX4577	A pBluescript SK (–) derivative with *aph(3')-IIIa* from mγδ inserted into the ampicillin resistance gene, Kan^r	[23]
pTEX4581	pTEX4577 with a 1.2 kb intragenic fragment of an *E. faecalis* autolysin gene	[35]
pTEX4578	pTEX4577 with an intragenic 480 bp of the *E. faecalis* P54 gene	[23]
pTEX5124	pTEX4577 with an intragenic 730 bp of the *efaA* gene of *E. faecalis*	[23]
pTEX5235	pBluescript SK (–) with *spc* gene from pSN13 and *oriT* of RK2. Spc^r	[31]
pTEX5236	pBeloBAC11 with *oriT* of RK2. Cam^r	[31]
pTEX5237	Internal part of an autolysin gene cloned into pTEX5235	[31]

[a] OG1RF and SE34 both have low level kanamycin resistance. 25 μg/ml kanamycin was used to select against *E. coli* S17-1 in this study.

Among the enterococcal mutants created by introducing cosmid DNA with mγδ insertions, all were generated by double crossovers between cosmid and chromosome. By Southern blot analysis of the chromosomal DNA using the cosmid as probe, we did not observe any mutants that arose from a single crossover between the cosmid and the host chromosome. The chance of mγδ transposition in the chromosome is very low, since neither transposase nor resolvase are supplied in mγδ.

Another approach to generate insertion mutants in *E. faecalis* by electroporation is to use insertion-duplication mutagenesis if the sequences of the genes are known. Autolysin [35], P54 [23] and *efaA* [23] mutants were generated by this method. The efficiency of this method is shown in Table 2. The smallest intragenic fragment that we used to generate a mutant was about 500 bp. Intragenic fragments of the targeted genes can be obtained by subcloning (e.g., the autolysin gene and P54 gene) or by PCR (e.g., *efaA*). The vector that we used in this work was pTEX4577 (pBluescript/Kan^r). However, if vectors with other antibiotic resistant genes are needed for selection, they can also be generated by similar approaches. One potential disadvantage of insertion-duplication mutagenesis is the possible excision of the insertion due to the duplication. However, we have not encountered this as a significant limitation.

Mutagenesis following conjugation into *E. faecalis* OG1RF was an efficient alternative method for targeted mutagenesis in enterococci. With this system, we were able to generate an autolysin mutant by transferring pTEX5235 carrying an intragenic fragment of the autolysin gene and a *pyrC* mutant

Table 2. Efficiency of targeted mutations of enterococci

Method	Plasmid	Gene (size of insert)	Recipient	Efficiency[a]	Correct allelic replacement[b]
Cosmid with mγδ	Electroporation of a mutagenized cosmid				
	pBEM220	*pyrD* (22 kb)	OG1RF	20/5 μg DNA	2/2
	pTEX5062	*purL* (24 kb)	OG1RF	30/5 μg DNA	2/2
	pKV4::mγδ	pKV4 antigen (23 kb)	OG1RF	50/5 μg DNA	3/3
	Electroporation of a pool of mutagenized DNA				
	gelE::mγδ	*gelE* (22 kb)	OG1RF	51/5 μg DNA	1/51
	pKV48::mγδ	*pyr* gene cluster (22 kb)	OG1RF	23/5 μg DNA	1/23
Small insert in pTEX4577	pTEX4578	P54 (480 bp)	OG1RF	3/5 μg DNA	3/3
	pTEX4581	autolysin gene (1.2 kb)	OG1RF	28/5 μg DNA	3/3
	pTEX5124	*efaA* (730 bp)	OG1RF	30/5 μg DNA	2/3
Conjugation	pAT18	(shuttle vector)	OG1RF	5×10^{-7}	
			SE34	1×10^{-7}	
	pTEX5237	autolysin gene (1.0 kb)	OG1RF	1×10^{-8}	3/5
	pBEM219	*pyrC* (22 kb)	OG1RF	5×10^{-9}	10/10

[a] Efficiency was calculated by the colonies which appeared on antibiotic selection plates.
[b] Correct allelic replacement was verified by either phenotypes or Southern blot analysis.

by transferring a cosmid clone with a mγδ insertion in the *pyrC* gene.

The efficiency of the conjugation method is also illustrated in Table 2. Typically, many more exconjugants could be recovered, compared to electroporation in which, at best, tens of mutants were isolated. Compared to electroporation, conjugation is also easier and faster since DNA isolation is not required. Also, conjugation may overcome the restriction systems in the recipient cells which can potentially interfere with the transformation and electroporation [36]. We also found that pAT18 could be transferred from *E. coli* S17-1 to *E. faecium* SE34 (Table 2), which has a very low transformation efficiency by electroporation (less that 10 transformants per μg DNA).

In an effort to increase the efficiency of mutant generation, we used glycine to enhance the conjugation efficiency. Glycine (1–2.5%) in the medium used to prepare conjugation plates was found to increase the conjugation efficiency by about 5-fold (data not shown).

In summary, while methods for direct transposon mutagenesis in *E. faecalis*, such as Tn*917* and Tn*916*, have been used previously for mutagenesis studies, the procedures described here should have particular applicability for the generation of mutants once DNA has been cloned into *E. coli*.

Acknowledgments

This work was supported in part by USPHS grant Nos. AI42399 and AI33516 from the NIH as well as the Texas ARP (Advanced Research Program).

Note of suppliers

1. Amersham Life Science Inc., Arlington Heights, IL, USA
2. Becton Dickinson and Company, Cockeysville, MD, USA
3. Bio-Rad Laboratories, Hercules, CA, USA
4. Boehringer Mannheim Corp., Indianapolis, IN, USA
5. Corning Glass Works, Corning, NY, USA
6. Difco Laboratories Inc., Detroit, MI, USA
7. Fisher Scientific Corp., Fair Lawn, NJ, USA
8. FMC BioProducts, Rockland, ME, USA
9. GIBCO BRL, Life Technologies, Grand Island, NY, USA
10. Invitrogen, San Diego, CA, USA
11. Micron Separations Inc., Westborough, MA, USA
12. Nalgene, Rochester, NY, USA
13. Perkin-Elmer Corp., Norwalk, CT, USA
14. Promega Corp., Madison, WI, USA
15. Sigma Chemical Co., St Louis, MO, USA
16. Stratagene, La Jolla, CA, USA
17. United States Biochemical Corp., Cleveland, OH, USA

References

1. Trieu-Cuot P, Carlier C, Poyart-Salmeron C, Courvalin P (1991). Shuttle vectors containing a multiple cloning site and a *lacZα* gene for conjugal transfer of DNA from *Escherichia coli* to Gram-positive bacteria. Gene 102: 99–104.
2. Murray BE (1990). The life and times of the enterococci. Clin Microbiol Rev 3: 46–65.
3. Cruz-Rodz AL, Gilmore MS (1990). High efficiency introduction of plasmid DNA into glycine treated *Enterococcus faecalis* by electroporation. Mol Gen Genetics 224: 152–154.
4. Friesenegger A, Fiedler S, Devriese LA, Wirth R (1991). Genetic transformation of various species of *Enterococcus* by electroporation. FEMS Microbiol Lett 63: 323–327.
5. Li X, Weinstock GM, Murray BE (1995). Generation of auxotrophic mutants of *Enterococcus faecalis*. J Bacteriol 177: 6866–6873.
6. Shepard BD, Gilmore MS (1995). Electroporation and efficient transformation of *Enterococcus faecalis* grown in high concentrations of glycine. Methods Mol Biol 47: 217–226.
7. Solioz M, Waser M (1990). Efficient electrotransformation of *Enterococcus hirae* with a new *Enterococcus-Escherichia coli* shuttle vector. Biochimie 72: 279–283.
8. Waser M, Hess-Bienz D, Davies KMS (1992). Cloning and disruption of a putative NaH-antiporter gene of *Enterococcus hirae*. J Biolog Chem 267: 5396–5400.
9. Bensing BA, Dunny GM (1993). Cloning and molecular analysis of genes affecting expression of binding substance, the recipient-encoded receptor(s) mediating mating aggregate formation in *Enterococcus faecalis*. J Bacteriol 175: 7421–7429.
10. Casey J, Daley C, Fitzgerald G (1991). Chromosomal integration of plasmid DNA by homologous recombination in *Enterococcus faecalis* and *Lactococcus lactis subsp. lactis* hosts harboring Tn*919*. Appl Environ Microbiol 57: 2677–2682.
11. Christie PJ, Kao SM, Adsit JC, Dunny GM (1988). Cloning and expression of genes encoding pheromone-inducible antigens of *Enterococcus (Streptococcus) faecalis*. J Bacteriol 170: 5161–5168.
12. Ehrenfeld EE, Clewell DB (1987). Transfer functions of the *Streptococcus faecalis* plasmid pAD1: Organization of plasmid DNA encoding response to sex pheromone. J Bacteriol 169: 3473–3481.
13. Handwerger S (1994). Alterations in peptidoglycan precursors and vancomycin susceptibility in Tn*917* insertion mutants of *Enterococcus faecalis* 221. Antimicrob. Agents Chemother 38: 473–475.
14. Ike Y, Clewell DB, Segarra RA, Gilmore MS (1990). Genetic analysis of the pAD1 hemolysin/bacteriocin determinant in *Enterococcus faecalis*: Tn*917* insertional mutagenesis and cloning. J Bacteriol 172: 155–163.
15. Pontius LT, Clewell DB (1992). Conjugative transfer of *Enterococcus faecalis* plasmid pAD1: Nucleotide sequence and transcriptional fusion analysis of a region involved in positive regulation. J Bacteriol 174: 3152–3160.
16. Pontius LT, Clewell DB (1992). Regulation of the pAD1-encoded sex pheromone response in *Enterococcus faecalis*: Nucleotide sequence analysis of *traA*. J Bacteriol 174: 1821–1817.
17. Tomich PK, An F, Clewell DB (1980). Properties of erythromycin-inducible Tn*917* in *Streptococcus faecalis*. J Bacteriol 141: 1366–1374.
18. Trotter K, Dunny G (1990). Mutants of *Enterococcus faecalis* deficient as recipients in mating with donors carrying pheromone-inducible plasmids. Plasmid 24: 57–67.
19. Fogg GC, Gibson CM, Caparon MG (1994). The identification of *rofA*, a positive-acting regulatory component of *prtF* expression: Use of a mγδ-based shuttle mutagenesis strategy in *Streptococcus pyogenes*. Mol Microbiol 11: 671–684.
20. Chung CT, Niemela SL, Miller GH (1989). One-step preparation of competent *Escherichia coli*: Transformation and storage of bacterial cells in the same solution. Proc Natl Acad Sci USA 86: 2172–2175.
21. Miller JH (1972). Experiments in molecular genetics. Cold Spring Harbor, NY: Cold Spring Harbor Laboratory.
22. Murray BE, Singh KV, Ross RP, Heath JD, Dunny GM, Weinstock GM (1993). Generation of restriction map of *Enterococcus faecalis* strain OG1 and investigation of growth requirements and regions encoding biosynthetic function. J Bacteriol 175: 5216–5223.
23. Singh KV, Qin X, Weinstock GM, Murray BE (1997). Generation and testing of mutants of *Enterococcus faecalis* in a mouse peritonitis model. J Infect Dis 178: 1416–1420.
24. Sambrook J, Fritsch EF, Maniatis T (1989). Molecular cloning. A laboratory manual, 2nd Edition. Cold Spring Harbor, NY: Cold Spring Harbor Laboratory Press.
25. Birnboim HC, Doly J (1979). A rapid alkaline extraction procedure for screening recombinant plasmid DNA. Nucleic Acids Res 7: 1513–1523.
26. Xu Y, Jiang LX, Murray BE, Weinstock GM (1997). *Enterococcus faecalis* antigens in human infections. Infect Immun 65: 4207–4215.
27. Lowe AM, Lambert PA, Smith AW (1995). Cloning of an *Enterococcus feacalis* endocarditis antigen: Homology with adhesins from some oral streptococci. Infect Immun 63: 703–706.
28. Dunny GM, Lee LN, LeBlanc DJ (1991). Improved electroporation and cloning vector system for gram-positive bacteria. Appl Environ Microbiol 57: 1194–1201.
29. Wirth R, An FY, Clewell DB (1986). Highly efficient protoplast transformation system for *Streptococcus faecalis* and a new *Escherichia coli-S. faecalis* shuttle vector. J Bacteriol 165: 831–836.
30. Su YA, Sulavik MC, He P, et al. (1991). Nucleotide sequence of the gelatinase gene (*gelE*) from *Enterococcus faecalis* subsp. *liquefaciens*. Infect Immun 59: 415–420.
31. Teng F, Murray BE, Weinstock GM (1998). Conjugal transfer of plasmid DNA from *Escherichia coli* to enterococci: A method to make insertion mutations. Plasmid 39: 182–186.
32. LeBlanc DJ, Lee LN, Inamine JM (1991). Cloning and

nucleotide base sequence analysis of a spectinomycin adenyltransferase AAD(9) determinant from *Enterococcus faecalis*. Antimicrob Agents Chemother 35: 1804–1810.
33. Shizuya H, Birren B, Kim U-J, et al. (1992). Cloning and stable maintenance of 300-kilobase-pair fragments of human DNA in *Escherichia coli* using an F-factor-based vector. Proc Natl Acad Sci USA 89: 8794–8797.
34. Smith CL, Canter CR (1987). Purification, specific fragmentation, and separation of large DNA molecules. Methods Enzymol 155: 449–467.
35. Qin X, Singh KV, Weinstock GM, Murray BE (1998). Effect of interruption of the gene encoding autolysin of *Enterococcus faecalis* strain OG1RF. Antimicrob Agents Chemother 42: 2883–2888.
36. Matsushima P, Broughton MC, Turner JR, Baltz RH (1994). Conjugal transfer of cosmid DNA from *Escherichia coli* to *Saccharopolyspora spinosa*: Effects of chromosomal insertions on macrolide A83543 production. Gene 146: 39–45.

Address for correspondence: Barbara E. Murray, Center for the Study of Emerging and Re-emerging Pathogens, Division of Infectious Diseases, Department of Medicine, University of Texas Medical School, 6431 Fannin Street, Houston, Texas 77030, USA
Phone: (713)-500-6767; Fax: (713)-500-5495
E-mail: iminfdis@heart.med.uth.tmc.edu

Methods in Cell Science **20**: 35–50 (1998).

A lactococcal pWV01-based integration toolbox for bacteria

Kees Leenhouts, Gerard Venema & Jan Kok
Department of Genetics, Biomolecular Sciences and Biotechnology Institute, University of Groningen, Haren, The Netherlands

Abstract. A conditionally replicating lactococcal vector system is described, based on pWV01, that is used for chromosomal integration in *Lactococcus lactis*. The system consists of plasmids that are all based on the broad-host-range lactococcal replicon pWV01 which has been deprived of its gene encoding the replication initiation protein RepA. These so-called pORI plasmids can only replicate if RepA is provided *in trans*. Special *Escherichia coli*, *Bacillus subtilis* and *L. lactis* helper strains, producing RepA *in trans* (RepA$^+$), are used as intermediate hosts for the construction of pORI integration plasmids. The presence of a lactococcal chromosomal DNA fragment in a pORI plasmid enables its chromosomal integration by homologous recombination in an *L. lactis* strain that does not produce RepA (RepA$^-$). A set of special purpose pORI vectors are available: (i) pORI280, designed to mutate or delete genes or to insert new genes in the chromosome in such a way that no heterologous DNA or antibiotic resistance markers are left in the recombinant strain, here designated as silent gene replacement; (ii) pORI19, suitable for random mutagenesis of the chromosome, and (iii) pORI13, constructed to make random single copy transcriptional fusions in the chromosome to search for (environmentally regulated) promoters. Some of these vectors have also been succesfully applied for integration in *E. coli* and *B. subtilis*. We strongly believe that the systems described here can be used in various other bacterial species.

Key words: Conditional replication, Gene replacement, Integration, Mutagenesis, Transcriptional fusions

1. Introduction

The development of tools for the analysis and modification of bacterial chromosomal genes and their expression signals has resulted in a large number of integration vectors and strategies. The essential parts of the most versatile and frequently employed vectors are usually derived from transposable elements, plasmids or combinations thereof. The integration event either relies on *rec*-independent transposition or on *rec*-dependent homologous recombination. Transposition has been a valuable tool for bacteria like *Escherichia coli* and *Bacillus subtilis* [1, 31]. However, transposable elements often have a limited host range or may have one or more transposition hot spots. In *Lactococcus lactis*, a Gram-positive mesophilic lactic acid bacterium and subject of our studies, this problem has also been encountered. Heterologous conjugative transposons like Tn*916* and Tn*919* have been used to some extent in *L. lactis*, but their use has been restricted either because transposition was site-specific or high-efficiency conjugal transfer systems were required [5, 22]. Derivatives of the *Enterococcus faecalis* transposon Tn*917* were successfully used in *L. lactis* to obtain random transposition using pLTV1 or pTV32, both of which contain a temperature-sensitive replicon of pE194 [31]. In e.g., *B. subtilis*, the temperature-sensitive property of this type of plasmid was used to establish a culture of cells containing the replicative form of the plasmid at 37 °C and to cure the cells of the plasmid at 42 °C, thereby selecting for transposition events. Since the maximum growth temperature for *L. lactis* is about 37.5 °C, a similar temperature upshift can not be used in this organism. Nevertheless, efficient transposition was obtained and cells were cured for plasmids at 30 °C by shifting a culture from medium with one antibiotic, chloramphenicol, to medium containing another antibiotic, erythromycin. The molecular basis for this phenomenon has yet to be resolved. Nevertheless, the process was efficient enough to establish a collection of environmentally regulated chromosomal promoters [7, 8].

Lactococcal transposable elements that have been used succesfully in the analysis of the *L. lactis* chromosome belong to the ISS*1* family of insertion sequences [19, 21, 23, 25]. Replicative transposition of an ISS*1* element present on a plasmid results in the integration of the entire plasmid in the chromosome, flanked by two copies of the IS element. This system has been used in combination with an *E. coli* replicon for insertional mutagenesis purposes [3, 24] and for the construction of physical and genetic maps of lactococcal chromosomes [11, 12]. A more versatile ISS*1* vector, pGh:ISS*1*, has recently been constructed by making use of a thermosensitive derivative of the lactococcal vector pWV01. This vector replicates in the target organism at 28 °C but is lost at 37.5 °C. The use of a thermosensitive replicon uncouples transformation and transposition events which results in an important improvement of transposition frequencies. High-frequency random transposition (> 0.5%), allowing for efficient random gene inactivation, has been demonstrated with pGh:ISS*1* in *L. lactis*, *Streptococcus thermophilus* and *E. faecalis* [19]. The wide host range of pWV01-based replicons, which replicate both in Gram-negative and Gram-positive bacteria [9], offers the possibility to use this tool in a great variety of bacteria. Its successful application will depend on the thermosensitivity of the pWV01 replicon and the transposition activity of the ISS*1* element in the target organism.

The *rec*-dependent plasmid integration strategies rely on the presence in the vector of one or more DNA fragments with homology to the chromosome of the target organism. In principle any heterologous plasmid which is unable to replicate in the bacterium of interest can be used for this purpose. This approach has also been successfully taken for *L. lactis* to obtain single and double cross-over integrations [17]. However, the nature of several (potential) applications of chromosomally modified lactococcal strains demanded the development of more sophisticated integration vectors, allowing to do the necessary steps of vector construction in a homologous strain. The broad host-range lactococcal plasmid pWV01 proved to be an excellent substrate for the construction of such versatile integration vectors. Two conditionally replicating vector systems have been developed: (i) the above mentioned temperature sensitive (Ts) pWV01 derivative pG⁺host, which was developed in the laboratory of D. Ehrlich,

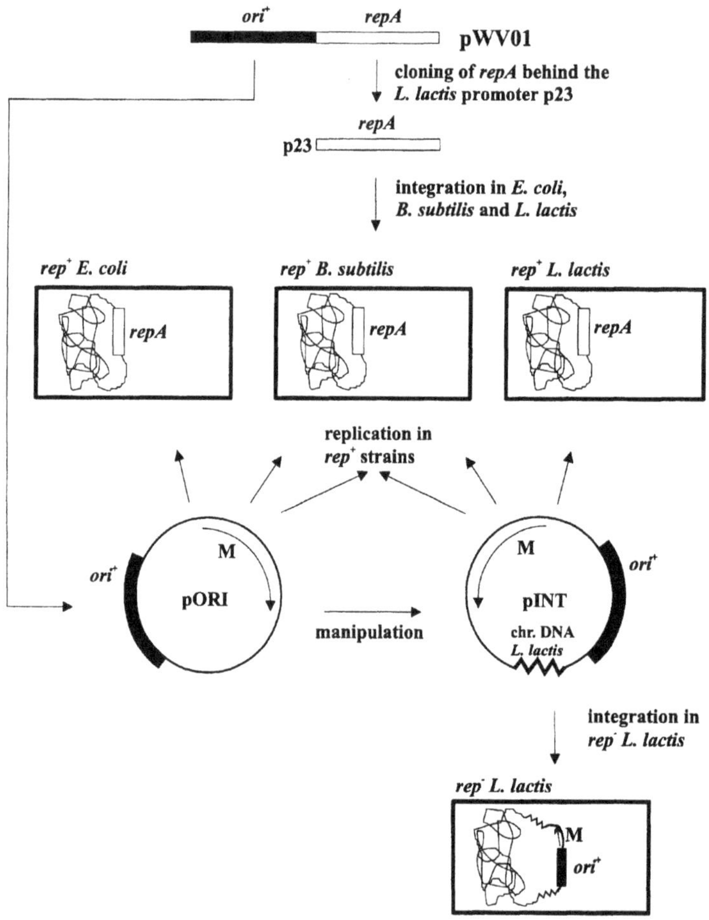

Figure 1. Schematic representation of the Ori⁺-system. Ori⁺ (black box): fragment containing the plus origin of replication of pWV01; RepA (white box): fragment encoding the replication initiation protein of pWV01; rep⁺: RepA-producing strain; p23: constitutive lactococcal promoter, also active in *E. coli* and *B. subtilis*; pORI: Ori⁺-vector that does not contain lactococcal chromosomal DNA; pINT: Ori⁺-vector containing lactococcal chromosomal DNA; rep⁻: strain that does not produce RepA; chr. DNA (wavy lines): chromosomal DNA; M: selectable marker.

Jouy en Josas, France [18]. The Ts-system has the important advantage that it uncouples the transformation and integration events, which bypasses the need for high transformation frequencies in the target organism. Nevertheless, in some applications it may have the disadvantage that the mutant strains have to be kept at temperatures above 37.5 °C. For more details on the Ts-system the reader is referred to the original literature [2, 18, 19]; (ii) a system in which pWV01-derivatives deprived of *repA*, the gene essential for plasmid replication, are multiplied in strains which provide RepA *in trans* and are integrated in strains which lack the *repA* gene of pWV01[Ori⁺-system; 15, 17]. Figure 1 summarises the characteristics of the Ori⁺-system. The pWV01 *repA* gene under the control of the lactococcal promoter P_{23}, which is expressed in Gram-positive as well as in Gram-negative bacteria [30], was integrated into the chromosomes of strains of *E. coli*, *B. subtilis* and *L. lactis* [10, 13–15]. These $RepA^+$ strains produce the RepA protein *in trans*, thus sustaining the replication of pWV01-derivatives lacking *repA*, but still carrying the recognition site for the initiation of replication (pORI plasmids). The Rep^+ helper strains are used for the construction of pORI-derivatives carrying chromosomal DNA of the target strain (pINT plasmids). Such pINT vectors fail to replicate in strains without a functional *repA* (Rep^- strains) and can thus be used to direct integration of cloned DNA. The Ori^+-system has the advantage that it operates independently from temperature shifts. However, in some applications this system may suffer from the disadvantage that relatively high transformation frequencies are required.

The pORI-vectors and strategies tailored for generating silent gene replacements, random chromosomal mutants and random chromosomal transcriptional fusions will be discussed here. A combined strategy will also be addressed, in which the Ts-plasmid is used as a helper plasmid in the Ori^+-system, to benefit from the advantages of both systems. We strongly believe that the nature of the conditionally replicating vectors described here implies that they can be successfully used in various other species of bacteria.

2. Materials

A. Equipment
1. Balances no. 1409 and BA160P.[16]
2. Gene pulser no. 165-2077 and Pulse controller no. 165-2098.[4]
3. Incubator (Heraeus) no. B5060E.[13]
4. Microcentrifuge no. 5417C.[7]
5. pH meter (Hanna Instruments) no. 8520.[13]
6. Pressure cooker canner (Presto) no. VL511080.[13]
7. Research pipettes no. 3110.[7]
8. Table top centrifuge no. 5810R.[7]
9. Vacuum dryer SpeedVac SC110-240 and Refrigerated Vapor Trap RVT100.[17]
10. Vortex IKA-Vibrofix VF1 Electronic no. 7102775.[14]

B. Supplies
1. Centrifuge tubes 50 ml no. 227.261.[9]
2. Electroporation cuvettes no. 165-2086.[4]
3. Microcentrifuge tubes, safe-lock (1.5 ml and 2 ml) no. 0030.120.086 and 0030.120.094.[2]
4. Pasteur capillary pipettes (W.U. Mainz) no. 222001.[14]
5. Petri dishes no. 631102.[9]
6. pORI13 + EC 1000.[11]
7. pORI19 + EC 101.[11]
8. pORI280.[11]
9. pVE6007.[12]
10. Research pipette tips 100 µl no. 0030.036.000.[7]
11. Research pipette tips 1 ml no. 808518.[14]

C. Media and chemicals
1. Acetic acid, glacial no. 63.[13]
2. Agar no. 4311849.[2]
3. $CaCl_2$ no. 2382.[13]
4. Chloramphenicol no. C-0378.[18]
5. Chloroform no. A3505E.[10]
6. Dimethylformamide no. 822275[13]
7. Double distilled H_2O (ddH_2O).
8. EDTA no. 2802145.[3]
9. Erythromycin E-6376.[18]
10. Ethanol no. 983.[13]
11. Glucose no. 8342.[13]
12. Glycerol no. 4094.[13]
13. Glycine no. 190.[13]
14. HCl no. 317.[13]
15. Isoamylalcohol no. 999.[13]
16. Lysozyme no. 5281.[13]
17. M17 no. 1856-17.[6]
18. $MgCl^2$ no. 5832.[13]
19. NaAc no. 6268.[13]
20. NaCl no. 6404.[13]
21. NaOH no. 6498.[13]
22. Phenol no. 206.[13]
23. Proteinase K no. 24568.[13]
24. RNase no. 109.193.[5]
25. SDS no. 1667.289.[5]
26. Sucrose no. 17714-0010.[1]
27. Tris no. 708976.[5]
28. Xgal no. OP-0020-10.[8]

3. Procedures

Some of the procedures for the pORI-system involve cloning steps in *E. coli* and general DNA techniques which are described in standard handbooks [26]. Here we describe only the *L. lactis*-specific procedures.

Silent gene replacements. One of the procedures which avoids the use of an antibiotic resistance marker to mutate a gene of interest is a two-step procedure described by Hamilton et al. [4]. In general, in the first step of homologous recombination between the delivery vector and the chromosome a cointegrate is formed which is selected for by positive selection. The second step is usually based on negative selection and, depending on the nature of the delivery vector, consists of a temperature shift or a period of non-selective growth. Resolution of the cointegrate results either in reversion to the parental chromosomal structure or in gene replacement. Bioassays or analyses of the chromosome are required to distinguish between the two possibilities. We developed a vector, pORI280 (Figure 2), which was designed for use in such a strategy (Figure 3).

A. Preparation of media and solutions

1. GM17 growth medium: dissolve 3.7 g M17 in 100 ml ddH_2O, sterilize by heat treatment using a pressure cooker or autoclave (15 min 120 °C) and store at room temperature in the dark. Add 2.5 ml 20% glucose solution just prior to use.
2. $GM17E^5$ growth medium: as in 1, add 50 µl of an erythromycin solution (10 mg/ml) before use.
3. GM17 and $GM17E^5$ agar medium: as in 1 and 2 but add 1.5 g agar before sterilization.
4. GM17 Xgal and $GM17E^5$Xgal agar medium: as in 3 but add 200 µl 4% Xgal before use.
5. GSM17gly growth medium: dissolve 3.7 g M17, 17.1 g sucrose and 1.9 g glycine in 80 ml ddH_2O, adjust the volume to 100 ml, sterilize as before and store at room temperature in the dark. Add 2.5 ml 20% glucose just prior to use. Note: the optimal amount of glycine in the medium differs among lactococcal strains e.g. for MG1363 it is 1.9 g per 100 ml while for IL1403 its 1 g per 100 ml.
6. GSM17MCery recovery medium: dissolve 3.7 g M17 and 17.1 g sucrose in 80 ml ddH_2O, adjust the volume to 100 ml, sterilize and store as before. Add 2.5 ml 20% glucose, 2 ml 1 M $MgCl_2$, 200 µl 1 M $CaCl_2$ and 500 µl erythromycin (10 µg/ml) just prior to use. Note: the 50 ng/ml erythromycin in this medium is required for induction of the expression of the erythromycin resistance gene. If selection for another antibiotic is used, erythromycin should be omitted.
7. $GSM17E^5$ agar medium: as in 3 but add 10 g

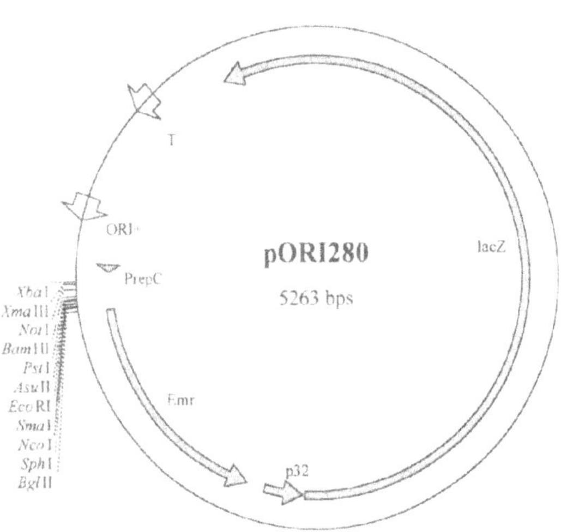

Figure 2. Plasmid map of pORI280 (5.3 kb). The restriction enzyme sites indicated are unique. Em^r, erythromycin resistance gene; *lacZ*, β-galactosidase gene of *E. coli* expressed under control of lactococcal promoter P_{32} (p32); T (open arrow), terminator of the lactococcal proteinase gene *prtP*; open arrow (ORI+), origin of replication of lactococcal plasmid pWV01; PrepC, promoter of the *repC* gene of plasmid pWV01 [16].

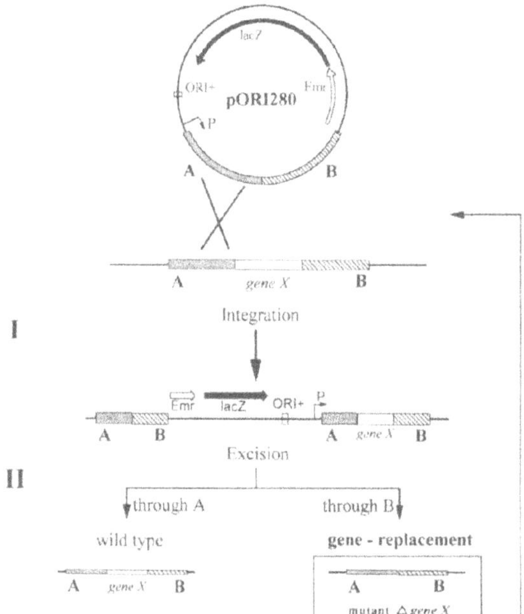

Figure 3. Scheme of a two-step procedure to obtain gene-replacement recombination. A (grey box) and B (hatched box): two fragments flanking the *geneX* to be deleted from the chromosome and through which recombination can take place. In step I depicted here, only recombination via A is visualized. The end result in II would be the same if an integration would take place through B. P: promoter P_{repC} of pWV01; ORI+ (open box): origin of replication of pWV01; lacZ (black arrow): β-galactosidase gene of *E. coli* expressed under control of lactococcal promoter P_{32}; Emr (open arrow): erythromycin resistance gene; *gene X*: target gene; Δ: deletion.

sucrose before sterilization to 80 ml ddH_2O and adjust after dissolving to 100 ml.
8. 20% glucose: dissolve 20 g glucose in 80 ml ddH_2O, adjust the volume to 100 ml and sterilize as before. Store at room temperature.
9. Erythromycin (10 mg/ml): dissolve 1 g erythromycin in 100 ml ethanol. Store at –20 °C.
10. Erythromycin (10 μg/ml): dissolve 1 mg erythromycin in 100 ml ethanol. Store at –20 °C.
11. X-gal (4%): dissolve 400 mg in 10 ml dimethylformamide and store at –20 °C.
12. 1 M $MgCl_2$: dissolve 20.3 g $MgCl_2$ in 100 ml ddH_2O and sterilize as before. Store at room temperature.
13. 1 M $CaCl_2$: dissolve 16.8 g $CaCl_2$ in 100 ml ddH_2O and autoclave as before. Store at room temperature.
14. 0.5 M sucrose/10% glycerol: dissolve 85.5 g sucrose and 57.5 ml glycerol (87%) in 400 ml ddH_2O, adjust volume to 500 ml and autoclave as before. Store at 4 °C.
15. 10 N NaOH: dissolve 40 g NaOH in 90 ml ddH_2O. Adjust the volume to 100 ml and store at room temperature.
16. 0.5 M EDTA: resuspend 14.6 g EDTA in 80 ml ddH_2O. Slowly add, while stirring, 10 N NaOH until solution becomes clear and reaches 8. Adjust volume to 100 ml and store at room temperature.
17. 1 M Tris pH 8: dissolve 12.1 g Tris in 80 ml ddH_2O. Slowly add, while stirring, concentrated hydrochloric acid untill the pH reaches 8. Adjust volume to 100 ml and store at room temperature.
18. Lysis solution: dissolve 20 g sucrose and 300 mg NaCl in 80 ml ddH_2O, add 1 ml 1 M Tris (pH8) and 2 ml 0.5 M EDTA. Adjust volume to 100 ml and autoclave as before. Store at 4 °C.
19. 10% SDS: dissolve 10 g SDS in 90 ml ddH_2O. Adjust volume to 100 ml and store at room temperature.
20. Proteinase K (20 mg/ml): dissolve 20 mg proteinase K in 1 ml ddH_2O. Store at –20 °C.
21. Chl/IAA: mix 96 ml chloroform with 4 ml isoamylalcohol. Store at room temperature.
22. 3 M NaAc pH 5.2: dissolve 24.6 g NaAc in 80 ml ddH_2O. Add, while stirring, acetic acid until the pH reaches 5.2.
23. TE: add 1 ml of Tris (pH 8) and 200 μl of 0.5 M EDTA to 99 ml ddH_2O.
24. RNase (10 mg/ml): dissolve 10 mg RNAse in 1 ml ddH_2O and incubate the sample at 100 °C for 10 min to inactivate DNase activity.

B. Preparation of cells for electrotransformation of strain MG1363 according to the method of Holo and Nes [6]
1. Inoculate strain in 5 ml GSM17gly and grow overnight at 30 °C (standing culture).
2. Add the 5 ml of overnight culture to 200 ml of fresh GSM17gly medium and grow at 30 °C to an OD_{600} of 0.4.
3. Pellet the cells by centrifugation at 4 °C in a cooled table top centrifuge using sterile 50 ml tubes (5,000 rpm, 5 min). Keep the cells on ice in between all further steps.
4. Wash the cells three times with 50 ml ice-cold 0.5 M sucrose/10% glycerol. Pellet the cells at 4 °C at 6,000 rpm for 10 min. Note: during the washing steps the cells pellet less well and higher centrifugation speeds and/or longer centrifugation times may be required to pellet all cells.
5. Resuspend the cells in 1 ml 0.5 M sucrose/10% glycerol and transfer the mixture to a 2 ml microcentrifuge tube.
6. Pellet the cells at 4 °C in a cooled microcentrifuge at 12,000 rpm for 3 min.
7. Resuspend the cells to a total volume of 1 ml using 0.5 M sucrose/10% glycerol.
8. Store aliquots of 50 μl at -80 °C.

C. Electroporation
1. Chill a 2 mm electroporation cuvette and the plasmid DNA preparation on ice.
2. Thaw electro-competent cells on ice; use an aliquot of 50 μl per transformation.
3. Prepare one sterile 2 ml microcentrifuge tube with 1 ml GSM17MCery per transformation.
4. Mix the cells with the plasmid DNA and transfer the mixture to the chilled cuvette.
5. Expose the cells immediately to a single electrical pulse (2.5 kV, 25 μF, 200 Ω).
6. Mix the cells immediately after the discharge with the GSM17MCery as follows: transfer the medium into the cuvette using a Pasteur pipette, mix and transfer the mixture back into the microcentrifuge tube.
7. Incubate the cells at 30 °C for 2 h.
8. Plate 100 μl aliquots per $GSM17E^5$ agar plate. If necessary, plate dilutions (made in GSM17MCery).
9. Incubate plates at 30 °C.

D. Gene replacement in MG1363 using pORI280 [14]
1. Insert two flanking fragments of the region to be deleted in the multiple cloning site (mcs) of pORI280 and transform to *E. coli* EC1000 ($RepA^+$). The sizes of the flanking fragments may range from 500 to 2,500 bp. It is recommended to use similar sized fragments.
2. Isolate a correct plasmid from EC1000 and transform *L. lactis* MG1363 ($RepA^-$) by electrotransformation with 0.1 to 1 μg of the pORI280 construct in a volume of 1 to 3 μl TE (step I in Figure 3).
3. Select the transformants on $GSM17E^5$ agar

plates. Colonies should be visible after 48 h of incubation at 30 °C.

4. Transfer a number of the colonies to $GM17E^5$Xgal plates. The colonies should stain blue after overnight incubation at 30 °C.
5. Inoculate a number of blue colonies in 2 ml $GM17E^5$ each for the isolation of chromosomal DNA. Use standard Southern hybridization or PCR techniques to check whether the plasmid has integrated at the expected chromosomal location. Note: steps 3 to 5 are optional. However, we recommend to go through these steps in order to get a better understanding of the process.
6. Grow one (or more) of the integrants overnight in 2 ml $GM17E^5$. Note: at this stage it may be advantageous to use those integrants which were created by integration of pORI280 through the chromosomal fragment that demonstrated the lowest recombination frequency of the two fragments present in the vector (A and B in Figure 3).
7. Dilute the overnight culture 10^6 times. A lactococcal overnight culture in GM17 medium contains about 2×10^9 cells. By inoculating 100 µl of the 10^6-fold diluted culture in 100 ml GM17 (without erythromycin), a cell density of 1 to 10 cells per ml is obtained.
8. Grow the diluted culture overnight at 30 °C. In this way nonselective growth of approximately 35 generations has been achieved. Note: overnight growth at 37 °C at this stage may enhance the excision-frequency of the integrated plasmid up to 10-fold (step II in Figure 3).
9. Plate dilutions of the overnight culture on GM17Xgal plates and incubate at 30 °C. Ideally not more than approximately 500 colonies should appear per plate (∅ 9 cm). Alternatively ∅ 15 cm plates can be used which should contain not more than approximately 2000 colonies. Ten ∅ 9 cm plates are usually enough to identify a sufficient number of white colonies.
10. Growth for 24 to 48 h on these plates should be sufficient to distinguish white from blue colonies.
11. Select white colonies and grow each of them overnight in 2 ml GM17 at 30 °C for chromosomal DNA isolation. Southern hybridization or PCR strategies should be used to distinguish between wild-type colonies and strains in which gene replacement has occurred. Alternatively, white colonies can be grown for bioassays if these allow distinction between the two types of strains.

E. Chromosomal DNA isolation

1. Inoculate strain in 2 ml GM17 and grow overnight at 30 °C (standing culture).
2. Transfer 1 ml overnight culture to a 2 ml microcentifuge tube and pellet the cells using a microcentrifuge (14,000 rpm, 3 min).
3. Wash the cells with 1 ml ddH_2O and pellet as before.
4. Resuspend the cells in 500 µl lysis solution to which lysozyme has been added (end concentration 5 mg/ml).
5. Incubate 15 min at 50 °C.
6. Add 25 µl proteinase K and mix by inversion.
7. Add 25 µl 10% SDS and mix by inversion.
8. Incubate for 1 h at 60 °C. The solution should now be completely clear.
9. Add 250 µl phenol and mix well by inversion.
10. Add 250 µl Chl/IAA and mix well by inversion.
11. Separate the two phases by centrifugation (14,000 rpm, 4 min).
12. Carefully transfer 450 µl of the aqueous (DNA-containing) upper phase to a clean 2 ml microcentrifuge tube. Do not take any of the inter- and/or phenol phase.
13. Add 100 µl ddH_2O to the DNA solution and mix by inversion.
14. Repeat steps 9 to 12.
15. Add 45 µl 3 M NaAc pH 5.2 to the DNA solution and mix by inversion.
16. Add 1 ml ice cold 96% ethanol and mix by inversion.
17. Pellet the DNA by centrifigation (14000 rpm, 5 min).
18. Rinse the pellet with 1 ml ice cold 70% ethanol.
19. Remove all ethanol carefully with a Pasteur pipette and resuspend the DNA pellet in 100 µl TE.
20. Add 5 µl RNase.
21. Store the DNA at 4 °C.

Random mutagenesis. A vector containing a chromosomal DNA fragment can inactivate a gene after integration in the chromosome by homologous recombination if the fragment is an internal part of that gene. In the pORI vector, pORI19, random chromosomal DNA fragments can be cloned in the mcs in α*lacZ* (Figure 4). The vector allows easy assessment of the cloning efficiency by screening for α-complementation in the $RepA^+$ *E. coli* EC101. The pORI19 library is subsequently integrated in *L. lactis* MG1363 with the help of the temperature sensative plasmid pVE6007, which is used to uncouple the transformation and integration events (Figure 5). The same temperature sensative plasmid, pVE6007, is used to excise and rescue the integration plasmid from an identified mutant (Figure 6).

A. Preparation of media and solutions

1. $GM17E^5C^5$ growth medium: dissolve 3.7 g M17 in 100 ml ddH_2O, sterilize (15 min 120 °C) and store at room temperature in the

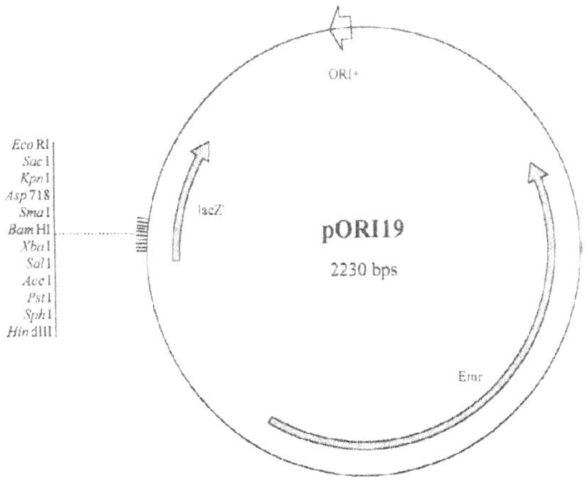

Figure 4. Plasmid map of pORI19 (2.3 kb). The restriction enzyme recognition sites indicated are unique.ORI+ (open arrow): origin of replication of pWV01; lacZ': gene encoding the α-fragment of LacZ; Emr: erythromycin resistance gene.

dark. Add 2.5 ml of a sterile 20% glucose solution, 50 μl of an erythromycin solution (10 mg/ml) and 50 μl of a chloramphenicol solution (10 mg/ml) just prior to use.

2. GM17E^5C^5 agar medium: as in 1 but add 1.5 g agar before sterilization.
3. GSM17E^5C^5 agar medium: as in 2 but add 10 g sucrose before sterilization.
4. Chloramphenicol (10 mg/ml): dissolve 1 g chloramphenicol in 100 ml ethanol. Store at –20 °C.

B. Random mutagenesis in MG1363 using pORI19 [10]

1. Insert a pool of chromosomal fragments ranging in size from 500 to 1,500 bp in a suitable restriction enzyme recognition site of the pORI19 mcs (Figure 4) and transform to *E. coli* EC101 ($RepA^+$) (Figure 5). Check the efficiency of the ligation by: (i) screening for α-complementation; more than 90% of the colonies should be white or light blue; (ii) confirming the presence and determining the length of the inserts in 100 colonies by PCR, using standard pUC sequencing primers; the average insert length should be within 500 to 1500 bp. Calculate the number of colonies (N) that is required to have 99% probability (P = 0.99) of having a given DNA sequence in the library:

   $N = \ln(1 - P)/\ln(1 - f)$, in which f is the average size of the insert divided by the size of the genome (2.5×10^6 bp for *L. lactis*).

   Note: the chromosomal DNA fragments can be generated by using a restriction enzyme that generates a majority of fragments smaller than 500 bp. By varying the enzyme reaction conditions such as time, temperature or salt concentration, the desired pool of partially digested fragments can be obtained [26]. Alternatively, the chromosomal DNA can be broken by using mechanical shearing techniques. In *L. lactis*, fragments smaller than 500 bp result in very low recombination frequencies [2]. On the other hand the fragments should not be too big, since only a DNA fragment internal to a gene will inactivate that gene upon integration of the vector.
2. Flood the plates with the *E. coli* colonies containing the pORI19 library with 2 ml *E. coli* growth medium each. Scrape the cells from the plates using a glass rod. Pool the cells from all plates in one tube.
3. Isolate the plasmid content without further propagation of the library.
4. Use 0.1 to 1 μg of the pORI19 library (maximum volume: 3 μl in TE) to electrotransform *L. lactis* MG1363 (pVE6007).
5. Incubate the transformation mixture for 90 min at 30 °C in the presence of 50 ng erythromycin to induce expression of the Em^r gene of pORI19 (step 6 in the protocol for the electroporation as described under the procedures for silent gene replacements, paragraph C).
6. Increase the erythromycin concentration in the transformation mixture to 5 μg/ml and incubate for a further 90 min at 30 °C to ensure proper replication of the plasmid bank.
7. Incubate the transformation mixture at 37 °C for at least 3 h to stop replication of pVE6007 and the pORI19 library.
8. Select transformants on GSM17E^5 agar plates and incubate overnight at 37 °C.
9. Transfer the plates to 30 °C and incubate for another 24 h. Colonies should now be visible.
10. Isolate chromosomal DNA from cultures of a number of colonies to check for random insertion by Southern hybridization.
11. Flood the plates with the *L. lactis* library of integrants with 2 ml GM17 medium each and collect all cells in one tube.
12. Centrifuge at 6,000 rpm for 5 min.
13. Discard the supernatant and replace it with the same volume of GM17.
14. Add glycerol to an end concentration of 10%.
15. Store aliquots of 1 ml at –80 °C.

C. Screening of mutants and rescue of the integration plasmid (Figure 6)

1. Thaw an aliquot of the *L. lactis* library of integrants and plate dilutions that enable to discriminate between the wild-type and the mutant phenotype.
2. After identification of a mutant, prepare

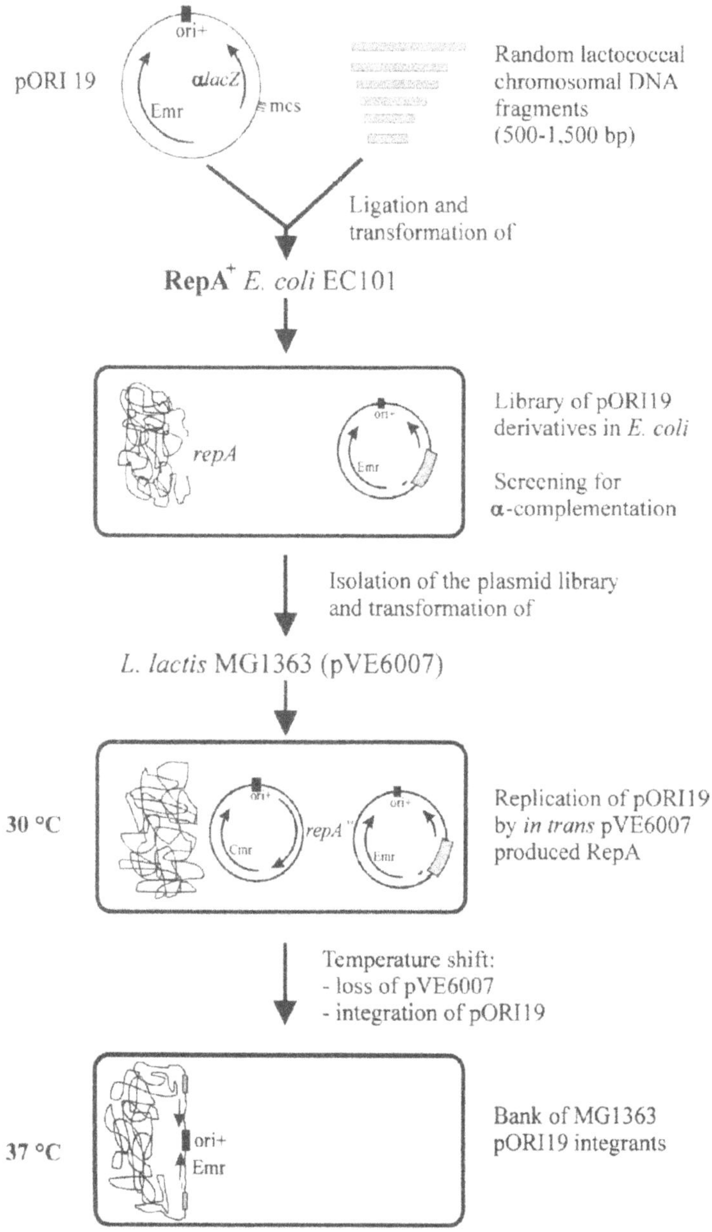

Figure 5. Scheme of using pORI19 for generating random chromosomal insertions in *L. lactis*. In the first step (I) a library of pORI19 containing random lactococcal chromosomal DNA fragments is established in *E. coli* EC101 (RepA$^+$). In the second step (II) *L. lactis* MG1363 (pVE6007) is transformed with the pORI19 library. The integration of the pORI19 derivatives is effectuated by making use of the temperature-sensitive character of pVE6007 (RepA^{ts}). See text for details.

ORI+ (black box): origin of replication of pWV01; Em^r: erythromycin resistance gene; α*lacZ*: gene encoding the α-fragment of LacZ; grey boxes: random lactococcal chromosomal fragments; RepA$^+$: strain producing RepA of pWV01; *repA* (open box): gene encoding RepA of pWV01; *repA^{ts}*: gene encoding a temperature-sensitive RepA of pWV01; cm^r: chloramphenicol resistance gene.

mutant competent cells according to the protocol described under silent gene replacements, paragraph B.

3. Introduce pVE6007 in the mutant strain as described under silent gene replacements, paragraph C. Select transformants by growth at 30 °C for at least 48 h on the same type of agar plates that were used to identify the

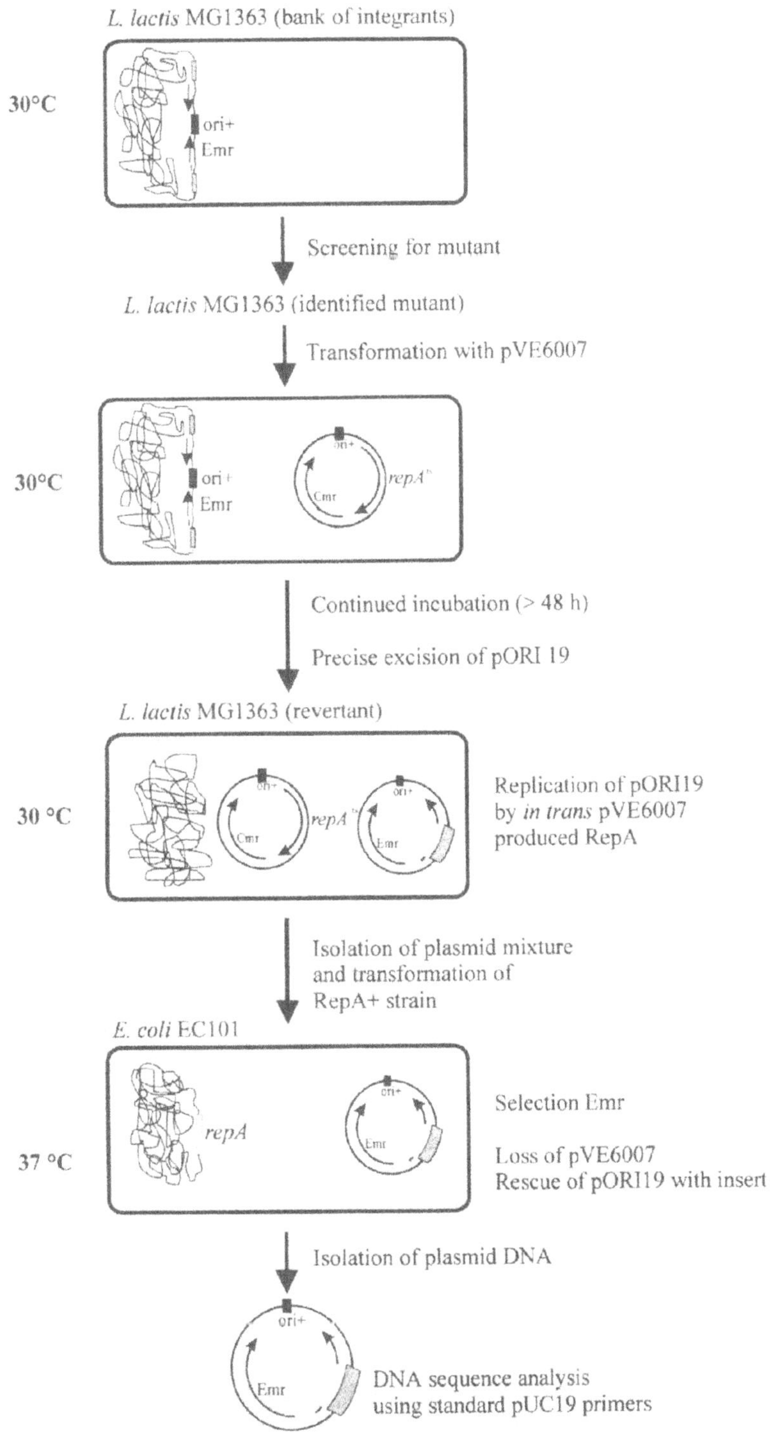

Figure 6. Rescue of an integrated plasmid from an identified mutant using pVE6007 ($RepA^{ts}$). See text for details. ORI+ (black box): origin of replication of pWV01; Em^r: erythromycin resistance gene; α*lacZ*: gene encoding the α-fragment of LacZ; grey boxe: lactococcal chromosomal DNA fragment; $RepA^+$: strain producing RepA of pWV01; *repA* (open box): gene encoding RepA of pWV01; $repA^{ts}$: gene encoding a temperature-sensative RepA of pWV01; cm^r: chloramphenicol resistance gene.

mutant but include chloramphenicol (end concentration 5 μg/ml).
4. Select a transformant that has reverted to the wild-type phenotype and grow the strain overnight in 3 ml GM17E^5C^5 at 30 °C.
5. Isolate plasmid DNA according to the 'miniprep' procedure.
6. Use mixture of the two plasmids to transform a $RepA^+$ strain, e.g., *E. coli* EC101.
7. Select only for the presence of the Em^r gene in pORI19 and incubate the plates at 37 °C. In this way pVE6007 is lost.
8. Isolate the pORI19 derivative that contains part of the gene of interest. This plasmid can be used for sequencing with the standard pUC sequencing primers.

D. 'Miniprep' plasmid DNA isolation procedure by means of the alkaline lysis method
1. Inoculate the strain in 3 ml GM17E^5C^5 and grow overnight at 30 °C (standing culture).
2. Transfer 2 ml overnight culture to a 2 ml microcentrifuge tube and pellet the cells using a microcentrifuge (14,000 rpm, 3 min).
3. Remove the supernatant. Note: it is important to remove all traces of supernatant.
4. Resuspend the cells in 200 μl lysis solution to which lysozyme has been added (end concentration 5 mg/ml).
5. Incubate 15 min at 50 °C.
6. Add 400 μl SDS/NaOH to the solution and mix well by inversion. The mixture should become clear and should not be stored longer than 5 min.
7. Add 300 μl 3 M NaAc pH5.2 and mix well by inversion.
8. Pellet the precipitate by centrifugation (14,000 rpm, 10 min).
9. Pour the supernant into a clean 2 ml microcentrifuge tube.
10. Add 400 μl phenol to the solution and vortex.
11. Add 400 μl Chl/IAA to the mixture and vortex.
12. Separate the two phases by centrifugation (14,000 rpm, 4 min).
13. Carefully transfer 650 μl of the aqueous (DNA-containing) upper phase to a clean 2 ml microcentrifuge tube. Do not take any of the interphase and/or phenol phase.
14. Add 1.3 ml ice cold 96% ethanol to the DNA solution and vortex.
15. Pellet the DNA by centrifugation (14,000 rpm, 10 min).
16. Remove the ethanol using a Pasteur pipette.
17. Rinse the pellet with 1 ml ice cold 70% ethanol.
18. Centrifuge 1 min at 14,000 rpm.
19. Remove the ethanol using a Pasteur pipette.
20. Dry the pellet using a vacuum dryer.
21. Dissolve the DNA in 20 μl TE.
22. Add 1 μl RNase and incubate 10 min at 37 °C.
23. Store at −20 °C.

Random transcriptional fusions. In most cases promoter screening systems are plasmid-based. Multiple copies of a regulated promoter on a plasmid may interfere with the regulation mechanism by titration of regulatory proteins. In addition, plasmid copy numbers may vary. Therefore, we developed the integrative pORI13 transcriptional fusion vector which allows to assess the expression of (regulated) genes in a single copy fusion with a reporter gene in the chromosome. The strategies to obtain a library of pORI13 integrants and the rescue of the integration plasmid from an integrant of interest are the same as described for random mutagenesis (Paragraphs B and C; Figures 5 and 6).

A. Preparation of media and solutions
As described under silent gene replacements and random mutagenesis.

B. Random transcriptional fusions in MG1363 using pORI13 [29]
1. Insert a pool of chromosomal DNA fragments in a suitable restriction enzyme recognition site of the pORI13 mcs (Figure 7) and transform *E. coli* EC1000 ($RepA^+$). This strain lacks *lacZ*. Generate the fragments in a similar way as described for the random mutagenesis procedure paragraph B. It is not necessary to use small chromosomal fragments to obtain transcriptional fusions after integration of the plasmid in the chromosome, as large fragments also may result in transcriptional fusions so long as no transcriptional terminator is present between the promoter and the reporter gene. The chromosomal fragment does not have to carry a promoter sequence. If the fragment carries only part of an open reading frame, the reporter gene in pORI13, after the integration event, will be placed under the control of the promoter present in the upstream chromosomal DNA.
2. Isolate the pORI13 library and follow the procedures of the random mutagenesis (paragraph B) to establish and stock a library of pORI13 integrants.

C. Screening for environmentally regulated genes and rescue of the integration plasmid
1. Thaw an aliquot of the *L. lactis* library of integrants and plate appropriate dilutions onto X-gal containing plates. Apply an environmental condition that will activate a certain set of genes.
2. Select blue colonies and transfer them to GM17E^5Xgal plates and incubate at 30 °C.
3. Identify colonies that are white or light blue under 'normal' conditions (GM17E^5Xgal at

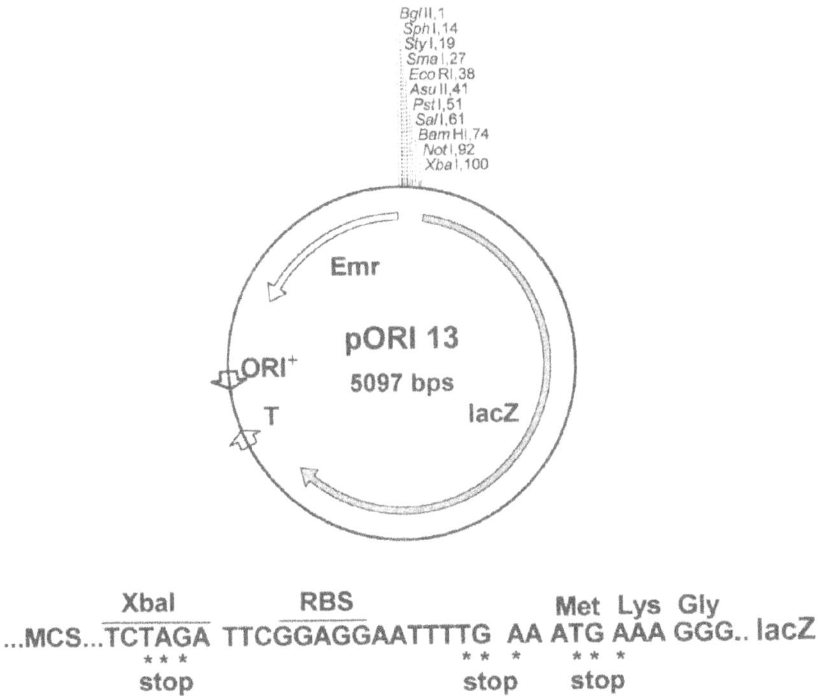

Figure 7. Plasmid map of pORI13 (5.1 kb). ORI+ (open arrow): plus origin of replication of pWV01; Em^r: erythromycin resistance gene; T (open arrow): terminator of *prtP*; *lacZ*, promoterless *E. coli* β-galactosidase gene fused to the ribosome binding site (RBS) and translational start codon of lactococcal *orf-32* (shown in detail). Stop codons are indicated by asterisks. Unique restriction enzyme recognition sites present in the multiple cloning site (MCS) are shown.

30 °C) and that are blue under 'stress' conditions.

4. Excise and rescue the integration plasmid from the integrant of interest as described before (random mutagenesis, paragraph C).

4. Results and discussion

Silent gene replacements. The vector pORI280 (Figure 2) has been constructed to replace genes in a two-step strategy and allows the isolation of mutants which do not carry a selectable marker (Figure 3). The mcs of pORI280 contains several unique restriction enzyme recognition sites that allow insertion of chromosomal DNA fragments that flank the gene to be deleted or mutated. For *L. lactis* a minimum size of 500-bp for each flanking region is recommended. It has been observed that recombination frequencies in *L. lactis* increase when the sizes of the chromosomal fragments increase, although no further stimulation of recombination was observed for fragments larger than 2.5-kb [2]. To detect resolution of plasmids integrated in genes for which no simply detectable phenotypic difference exists between the wild-type and mutant copy, the *E. coli lacZ* reporter gene is present in pORI280. The *lacZ* gene is under control of the lactococcal promoter P_{32} which is recognized in Gram-positive and Gram-negative bacteria. Strains containing an integrated copy of pORI280 are erythromycin resistent (Em^r) and stain blue on agar plates containing X-gal. After excision of the plasmid from the chromosome by homologous recombination through one of the flanking regions, the *lacZ* reporter and Em resistance genes (and Ori$^+$-fragment) are lost. Consequently, colonies which arise from cells in which this process has occurred stain white on non-selective agar plates containing X-gal, a readily scorable phenotype among the majority of blue colonies which retained the integrated plasmid. Resolution of the integrated plasmid either restores the wild-type gene or results in a gene replacement. Standard DNA analysis techniques or, if available, an activity assay can be used to discriminate between the two possibilities.

To illustrate the feasibility of the strategy, integration studies using the *pepX* X-prolyl-dipeptidyl amino peptidase) gene region of *L. lactis* MG1363 will be described. We selected this region because in previous gene-replacement studies we observed that recombination frequencies in *pepX* of this strain are relatively low with values as low as 10^{-6} per generation. Two 1.5-kb fragments of *pepX* were cloned in pORI280. Fragment A carried the promoter and the 5′-end of the gene and fragment B the 3′-end. In the final construct, pORI280-*pepX*, *pepX* lacks an internal 716-bp fragment. pORI280-*pepX* was used to transform *L. lactis* MG1363 by the two-step strategy outlined in Figure 2. Colonies were screened for LacZ activity but not for PepX activity during this procedure to mimic the case in which a bioassay is lacking. We first determined through which fragment pORI280-*pepX* had integrated. Southern blots were used to analyse 28 transformants: nineteen had integrated through fragment A (MG1363::*pepX-A*) and nine through fragment B (MG1363::*pepX-B*). One transformant of each type was taken and grown for 35 generations under nonselective conditions to allow resolution of the cointegrate structure. Approximately 20,000 colony forming units (4,000 cfu per ∅ 15 cm plate) of each culture were plated onto agar medium containing X-gal. Five $LacZ^-$ colonies were detected among the MG1363::*pepX-A* colonies and nine among those derived from MG1363::*pepX-B*. All $LacZ^-$ colonies were Em^s and by Southern blot analysis (data not shown) it was shown that the numbers of gene-replacements were one and three for the MG1363::*pepX-A* and MG1363::*pepX-B* cultures, respectively. The mutant nature of the colonies was confirmed by the PepX plate assay. Taken together, the frequencies by which gene replacements were generated in the MG1363::*pepX-A* and MG1363::*pepX-B* cultures ranged between 1.4×10^{-6} and 4.2×10^{-6} per generation. Recombination through the promoter-containing fragment A was consistently more efficient than through the promoterless fragment B. We also observed this phenomenon in other replacement studies and it is in agreement with results described earlier by Biswas et al. [2].

The pORI280-system has been used in our group to replace over ten different chromosomal genes. The recombination frequencies observed for these genes range from approximately 10^{-4} to 10^{-5} per generation. We noted that recombination in the second step of the procedure could be stimulated about five to ten times by increasing the growth temperature to 37 °C, a temperature that induces a heat-shock response in *L. lactis*. The system has also been successfully applied to combine several different gene replacements in a single strain. For instance, a strain was constructed from which seven different peptidase genes were removed [20, M. A. Hellendoorn, unpublished data]. During this work it was observed that the promoter P_{repC} of pORI280 is active and can drive the transcription of genes located downstream of the plasmid insertion site (Figure 3). This is an important feature if the gene to be removed from the chromosome is located upstream of (an) essential gene(s) in an operon structure. Thus, polar effects due to the insertion of the pORI vector can be avoided by employing the transcriptional activity of P_{repC} [14].

The pORI280-system has also been successfully used in *B. subtilis* [14]. It is very likely that the vectors can be used in many other bacterial hosts. An exception may be some (lactococcal) strains harbouring pWV01-like plasmids, as these would provide RepA *in trans*. In such a case, a plasmid curing step prior to the use of pORI280 will be necessary. Adjustment of the vector for use in some bacterial species may be required, such as insertion of another selectable marker and replacement (of the promoter) of the reporter gene. A pORI-vector with a tetracycline resistance marker, pORI240, is available. Other adjustments can easily be achieved by virtue of the modular design of the vectors. The pORI plasmids have the advantage over other nonreplicative vectors that they allow cloning of the target gene (fragment) in Gram-positive or Gram-negative backgrounds. This may minimise cloning problems. If such problems should persist, construction of a homologous $RepA^+$ background may be considered. The availability of a *repA* expression cassette is convenient for this purpose. A drawback of the method is that transformation frequencies are required which are sufficiently high to obtain at least a few integrants. Therefore, it may be necessary to optimise transformation protocols for poorly transformable strains.

The pORI280 system uses *lacZ* expression to visualize recombination events in the second step of the procedure by a simple blue/white screening of colonies. As this system is based on negative selection, it may result in extensive screening of colonies if recombination frequencies in the target gene are very low ($< 10^{-6}$ per generation). The reporter gene may be replaced by a gene which allows positive selection, if available. Nevertheless, the pORI280-system was found to be highly efficient in all gene replacements so far. The fact that no antibiotic resistance marker is left on the chromosome of a mutated strain makes the strain not only more desirable for applied purposes but also leaves open the possibility to introduce more desired mutations (or genes) in the strain, which is of special interest for metabolic pathway engineering purposes.

Random mutagenesis. A pORI-system suitable for generating random chromosomal insertions makes use of pORI19 (Figure 4). This vector contains an Em^r gene as selectable marker and the pUC19 *lacZα* gene and mcs in which random chromosomal DNA fragments can be inserted. To inactivate as many genes as possible upon integration of the plasmid bank, a few requirements with respect to the plasmid bank have to be met: (i) the inserts in the plasmid should be relatively small. A gene will be inactivated only when the fragment used for the homologous recombination is internal to the gene. Since the average size of genes is about 1 to 1.5 kb, the cloned chromosomal DNA fragments should preferentially be smaller. However, recombination frequencies of fragments smaller than 500 bp are very low in *L. lactis*. Therefore, DNA fragmentation conditions should generate fragments in the range of 500 to

1,500 bp. This can either be done by using a proper DNA restriction enzyme or by DNA shearing techniques; (ii) ligation of the random chromosomal fragments in the integration vector should be efficient. For this purpose *lacZα* is present in the vector which allows rapid assessment of the cloning efficiency by testing part of the ligation mixture for α complementation in the RepA$^+$ derivative of *E. coli* JM101, strain EC101 (Figure 5).

For fragmentation of chromosomal DNA of *L. lactis* MG1363 the restriction enzyme *Alu*I was selected. Several partial *Alu*I digests were made in which most fragments ranged in size from 100–1,500 bp. These digest were pooled and the mixture was ligated into the dephosphorylated *Sma*I site of pORI19 and used to transform *E. coli* EC101. More than 90% of the transformants were white or pale blue on agar plates containing Xgal. All of the white colonies analysed contained inserts, as did several of the blue and pale blue colonies, indicating that in-frame insertions had occured in those cases. By PCR, the estimated average insert size in pORI19 of 100 randomly picked colonies was 650 bp. The number of colonies required for 99% certainty that all 0.65-kb fragments of the *L. lactis* MG1363 chromosome had been cloned is 18,000. The plasmid bank was isolated from approximately 30,000 *E. coli* colonies without further propagation of the cells.

Alternatively, the ligation mixture can be introduced in a RepA$^+$ helper strain of one of the other bacterial species to isolate the pORI19 plasmid bank. *E. coli* EC101 can then be merely used to assess the cloning efficiency. One reason for following the latter approach is that in a homologous system (in this case a RepA$^+$ *L. lactis*) a pORI19 library would probably be more complete since it is likely that many lactococcal DNA fragments can not be cloned in *E. coli*. However, we reasoned that a high percentage of those fragments which are unclonable in *E. coli* probably contain (strong) promoters or complete (lethal) genes. Such fragments will not result in mutants since the cloned fragments need to be internal to a transcriptional unit to result in a mutant phenotype. Therefore, constructing the library in the RepA$^+$ *E. coli* strain may increase the percentage of plasmids containing an internal gene fragment, which would enhance the efficiency of the library to generate mutations.

In the next step, the pORI19 plasmid bank was used to transform the RepA$^-$ *L. lactis* strain MG1363. Under optimal conditions this resulted in approximately 10^3 Em^r colonies per μg of pORI 19 plasmid bank DNA. Southern hybridization analysis of chromosomal DNAs of several of the transformants revealed that all carried an integrated copy of pORI19 at different locations. A library of pORI19 integrants in which a maximum number of different genes are inactivated should contain at least 18,000 colonies, if not more. To increase the number of integrants, the transformation event can be separated from the integration event by using a Ts version of a pWV01 derivative which enables the replication of both the Ts and the pORI plasmid at the permissive temperature (see above). Therefore, MG1363 containing the pWV01 Ts-variant pVE6007 (Cm^r) was transformed with the pORI19 plasmid bank (Figure 5). A subsequent temperature shift from 30 °C to 37 °C causes loss of pVE6007 and integration of the pORI19 derivatives at the sites on the chromosome from which the inserts originated. An 1-μg sample of the pORI19 chromosomal DNA bank was used to transform *L. lactis* MG1363 (pVE6007). The transformation mixture was incubated at 30 °C for 90 min in the presence of 50 ng of Em per ml to induce expression of the Em^r gene. Subsequently, the Em concentration was increased to 5 μg/ml and incubation was continued at 30 °C for a further 90 min to ensure proper replication of the plasmid bank in *L. lactis* prior to the temperature shift. Plating the bank at this point and incubating overnight at the non-permissive temperature for pVE6007 (37 °C) did not cure the plasmid, as 40% of the colonies at this stage were Em^{r}Cm^r and thus harbored both plasmids. To ensure total curing of pVE6007, it was necessary to incubate the transformation mixture at 37 °C for at least 3 h following the 3 h period at 30 °C before plating on GM17 Em agar plates and incubation overnight at 37 °C. Overnight incubation at 30 °C following this 6-h treatment reduced the percentage of colonies harboring both replicating plasmids to 10%, while at 37 °C all Em^r colonies, recovered at a frequency of 10^4/μg of DNA, were cm^s. The chromosomal DNAs of several integrants were analysed in Southern hybridizations and these indicated that pORI19 had integrated at different sites [10].

To assess the feasibility of the system to generate stable random chromosomal mutations in *L. lactis*, the bank of integrants was screened for a mutation in the cell wall hydrolysing system. The target gene, *acmA*, was selected because: (i) it is nonessential; (ii) a simple bioassay is available, and (iii) the gene is monocistronic and of average size (1.3 kb). The size of the target fragment is even smaller if it is taken into account that the three repeats located in the C terminus of the hydrolase can be removed without loss of enzyme activity. Therefore, integration within the first 700 bp of the gene is required in order to inactivate it. About 5000 *L. lactis* transformants were spread onto glucose-M17 plates in which autoclaved *Micrococcus lysodeikticus* cells had been included, allowing approximately 40 colonies per plate. One transformant without a halo was detected and analysis of its chromosomal DNA learned that, indeed, pORI19 had integrated in an internal 5′-fragment of *acmA*.

The pORI19 system was further tested by selection of mutants in the maltose metabolic pathway, as isolation of stable Mal$^-$ mutants had previously been shown to be unsuccessful using other insertional methods. The bank of lactococcal integrants was plated on maltose indicator plates with a density of approximately 1000 colonies per plate. Following 24 hours of incubation at 30 °C, 1 in 10,000 colonies was found to be unable to ferment maltose (approx. 30,000 colonies were screened). One of the *L. lactis* Mal^- colonies was streaked for single colonies on maltose selective agar and after 3 days of incubation at 30 °C, all of the colonies were still white, indicating that the mutation was stable. A culture of this strain maintained its maltose-negative phenotype even after incubation overnight at 30 °C in GM17 without Em and subsequent plating on maltose selective agar.

The maltose-negative strain was used to develop a strategy for the easy isolation of the integrated plasmid from a defined mutant. Such a strategy would allow to rapidly identify the disrupted gene (Figure 6). *L. lactis* Mal^- was endowed with pVE6007 ($RepA^{ts}$) and plated on maltose indicator agar containing Cm and Em. After 24 h of incubation at 30 °C all transformants were still unable to ferment maltose, whereas after 48 h approximately 8% were faintly yellow, indicating that cells in these colonies had reverted to wild type. After colony purification it was found that 20% had regained maltose-fermenting ability, most probably by precise excision of the integrated plasmid. A Mal^+ *L. lactis* single colony isolate contained two plasmids, pVE6007 and pORI19 containing an insert (pMAL). To rescue pMAL, the plasmid mixture was used to transform *E. coli* EC101 ($RepA^+$) at 37 °C with selection for Em^r only, resulting in the separation of pVE6007 and pMAL. Upon introduction of pMAL into *L. lactis* MG1363 ($RepA^-$), all integrants were maltose negative, as expected. Southern hybridisation analysis of five transformants revealed that pMAL had integrated at the same site on the lactococcal chromosome as in *L. lactis* Mal^-.

The nucleotide sequence of the *Alu*I insert in pMAL was determined by making use of the standard pUC19 sequencing primers. A continuous open reading frame was present on this 564-bp fragment, and its deduced amino acid sequence of 189 amino acids showed high homology to ATP-binding proteins of several sugar uptake systems, among which the inner membrane MalK proteins of the maltose uptake systems of *Enterobacter aerogenes* and *E. coli*. These results suggested that, indeed, a gene involved in the maltose metabolic pathway of *L. lactis* had been targeted and that part of it had been cloned [10]. The 5′- and 3′-end of the gene may now be cloned by restriction of the chromosomal DNA of the Mal^- integrant with suitable enzymes and recovering the integration plasmid with additional flanking DNA in one of the $RepA^+$ helper strains.

Although the method described here is efficient in generating chromosomal insertions, the isolation of genes is, like most insertional methods, restricted to non-essential ones and to genes that allow detection of phenotypic negatives. Another disadvantage of the method is that rather high transformation frequencies are required if the Ts-helper plasmid can not be used, which may be the case in a number of bacterial species. Critically important is also the construction of a proper pORI19 library in one of the $RepA^+$ helper strains. However, cloning efficiencies can be easily assessed by the availabilty of the *E. coli* helper strain EC101. Other important advantages of the method are that: (i) stability of the mutants is high because of the absence of residual activity of transposases or (Ts) replication proteins; (ii) screening of mutants can be performed at optimal growth temperatures; (iii) the integration plasmid can be readily recovered by a simple and rapid procedure, and (iv) the availability of $RepA^+$ *L. lactis*, *B. subtilis*, and *E. coli* helper strains minimizes difficulties (lethality and deletions) in cloning of the targeted gene in incompatible host backgrounds.

Random transcriptional fusions. An Ori^+-integration vector, pORI13, was developed to screen for (environmentally regulated) gene expression signals in *L. lactis* and to assay for transcriptional activity of genes in single copy. The plasmid carries a promoterless *E. coli lacZ* gene preceded by a start codon, a lactococcal ribosome binding site, and a mcs. It contains translational stop codons in all three reading frames upstream of or overlapping the start codon of *lacZ* to prevent translational fusions (Figure 7). Plasmid pORI13 did not produce detectable β-galactosidase activity when replicating in *L. lactis* $RepA^+$ helper strains.

To produce chromosomal transcriptional fusions using pORI13 there are no restrictions with respect to the size of the fragment and, also, promoter elements do not have to be present on the cloned chromosomal fragment. Any fragment of a transcriptional unit cloned upstream of *lacZ* will produce a transcriptional fusion after integration of pORI13, provided that the transcriptional terminator of that unit is lacking. Several partial *Sau*3A digests of *L. lactis* MG1363 chromosomal DNA, with the majority of the fragments ranging in size from 1 to 10 kb, were pooled and the mixture was ligated into the alkaline phosphatase-treated *Bam*HI restriction enzyme site of pORI13. The ligation mixture was used to transform *E. coli* EC1000 ($RepA^+$), a derivative of strain MC1000 (Δ*lacZ*). More than 10,000 transformants were collected from agar plates and their plasmid DNA content was isolated. The resulting pORI13 plasmid DNA bank was used for integration in the chromosome of *L. lactis* MG1363 by employing the same strategy as described for pORI19, with pVE6007 as a means to uncouple transformation and integration events. The system was tested by plating the transformants, immediately after the temperature shifts, onto selective agar plates containing 0.3 M

NaCl as the stress inducing agent. Southern hybridisation analysis of the chromosomal DNAs of more than 10 integrants showed that pORI13 had inserted in all cases at different positions in the chromosome of these clones. One hundred and ninety five colonies showed various levels of blue staining in the presence of NaCl and Xgal after prolonged incubation at 30 °C, which is also indicative of *lacZ* insertion in different transcribed regions.

These colonies were transferred to GM17Xgal agar plates with or without 0.5 M NaCl. Various phenotypes could be distinguished: colonies staining light blue in the absence of NaCl and staining dark blue in the presence of NaCl; colonies which were blue in the absence of NaCl but light blue in the presence of NaCl. One selected clone, designated *L. lactis* NS3, produced blue colonies on plates with NaCl but was white on plates without added NaCl. Interestingly, β-galactosidase activity increased proportionally from 0 to 50 Miller units with NaCl concentrations in the medium increasing from 0 to 500 mM NaCl. The pORI13 derivative integrated in NS3 (pNS3) was rescued as described above using pVE6007 to excise the plasmid from the chromosome. Plasmid pNS3 still showed the NaCl-dependent phenotype when replicating in *L. lactis* ($RepA^+$), indicating that the regulatory elements were present in the 10-kb insert upstream of *lacZ* [29]. Detailed sequence and biochemical analyses learned that *lacZ* had been transcriptionally fused to an operon that is regulated by chloride ions, glutamate and low pH. Furthermore, the salt inducible promoter is preceded by a constitutively expressed gene encoding a positive regulator which acts on the salt inducible promoter in a still unknown way [27].

The results described above demonstrate that pORI13 can be used to study environmentally regulated gene expression signals in a single copy situation. The relatively large size of the chromosomal DNA fragments used for integration increases the probability that, upon integration of pORI13, in addition to the disrupted copy, a wild-type gene continues to be present in the integrant. pORI13 can also adventageously be used to trace transcription regulatory elements: after integration via the 3′-end of an operon, *lacZ* will be under control of a promoter upstream of the fragment used for recombination. The pORI13 system is suitable both for single copy screening of regulated promoters and the construction of targeted transcriptional fusions without gene disruption [28].

In conclusion, we have shown that the combined use of two conditionally replicating vectors derived from the same broad-host-range plasmid pWV01 is an important and extremely valuable research strategy in the analysis of the lactococcal chromosome. Also, we strongly believe that the vectors described here may represent useful tools in other species of bacteria.

Acknowledgments

Part of this work was supported by Unilever Research Laboratorium, Vlaardingen, The Netherlands. J.K. is the recipient of a fellowship from the Royal Netherlands Academy of Arts and Sciences (KNAW). The authors like to thank Jan Willem Sanders, Jean Law and Igor Mierau for their contributions to the development and use of the pORI system.

Notes on suppliers

1. Acros Organics, B-2440 Geel, Belgium
2. Becton Dickinson and Co., Cockeysville, MD 21030, USA
3. BDH Laboratories, Poole BH15 1TD, England
4. BIORAD Laboratories, Hercules California, CA 94547, USA
5. Boehringer Mannheim GmbH, Mannheim, Germany
6. Difco Laboratories, Detroit, MI 48232-7058, USA
7. Eppendorf- Netheler-Hinz-GmbH, D-22331 Hamburg, Germany
8. EuroGenTec, B-4102 Seraing, Belgium
9. Greiner GmbH, D-7443 Frickenhausen, Germany
10. Lab-Scan Ltd., Dublin, Ireland
11. Leenhouts K., Biological Centre, Department of Genetics, Kerklaan 30, NL-9751 NN Haren, The Netherlands
12. Maguin E., Génétique Microbienne, INRA, F-78352 Jouy en Josas, France
13. Merck, D-64271 Darmstadt, Germany
14. Omnilabo international, NL-4800 DX Breda, The Netherlands
15. Phoenix Biomedical, Ontario L5S 1R7, Canada
16. Sartorius Filtratie BV, NL-3439 MN Nieuwegein, The Netherlands
17. Savant Instruments Inc., Farmingdale, NY, USA
18. Sigma Chemical Co., St. Louis, MO 63178-9916, USA

References

1. Berg CM, Berg DE, Groisman EA (1989). Transposable elements and the genetic engineering of bacteria. In: Berg DE, Howe M (eds), Mobile DNA, pp 879–925. Washington DC: American Society for Microbiology.
2. Biswas I, Gruss A, Ehrlich SD, Maguin E (1993). High-efficiency gene inactivation and replacement system for gram-positive bacteria. J Bacteriol 175: 3628–3635.
3. Dinsmore PK, Romero DA, Klaenhammer TR (1993). Insertional mutagenesis in *Lactococcus lactis* subsp. *lactis* mediated by IS*946*. FEMS Microbiol Lett 107: 43–48.
4. Hamilton CM, Alsea M, Washburn BK, Babitzke P, Kushner SR (1989). New method for generating deletions and gene replacements in *Escherichia coli*. J Bacteriol 171: 4617–4622.
5. Hill C, Daly C, Fitzgerald GF (1991). Isolation of chromosomal mutations of *Lactococcus lactis* biovar.

diacetylactis 18-16 after introduction of Tn*919*. FEMS Microbiol Lett 81: 135–140.

6. Holo H, Nes IF (1989). High-frequency transformation by electroporation of *Lactococcus lactis* subsp. *cremoris* grown with glycine in osmotically stabilized media. Appl Environ Microbiol 55: 3119–3123.
7. Israelsen H, Hansen EB (1993). Insertion of transposon Tn*917* derivatives into the *Lactococcus lactis* subsp. *Lactis* chromosome. Appl Environ Microbiol 59: 21–26.
8. Israelsen H, Madsen SM, Vrang A, Hansen EB, Johansen E (1995). Cloning and partial characterisation of regulated promoters from *Lactococcus lactis* Tn*917-lacZ* integrants with the new promoter probe vector, pAK80. Appl Environ Microbio 61: 2540–2547.
9. Kok J, van der Vossen JMBM, Venema G (1984). Construction of plasmid cloning vectors for lactic streptococci which also replicate in *Bacillus subtilis* and *Escherichia coli*. Appl Environ Microbiol 48: 726–731.
10. Law J, Buist G, Haandrikman A, Kok J, Venema G, Leenhouts K (1995). A system to generate chromosomal mutations in *Lactococcus lactis* which allows fast analysis of targeted genes. J Bacteriol 177: 7011–7018.
11. Le Bourgeois P, Lautier, Mata M, Ritzenthaler P (1992). New tools for the physical and genetic mapping of *Lactococcus* strains. Gene 111: 109–114.
12. Le Bourgeois P, Lautier M, van den Berghe L, Gasson MJ, Ritzenthaler P (1995). Physical and genetic map of the *Lactococcus lactis* subsp. *cremoris* MG1363 chromosome: Comparison with that of *Lactococcus lactis* subsp. IL1403 reveals a large genome inversion. J Bacteriol 177: 2840–2850.
13. Leenhouts K, Bolhuis A, Venema G, Kok J (1998). Construction of a food-grade multiple-copy integration system for *Lactococcus lactis*. Appl Microbiol Biotechnol 49: 417–423.
14. Leenhouts K, Buist G, Bolhuis A, Berge A ten, Kiel J, Mierau I, Dabrowska M, Ven8ema G, Kok J (1996). A general system for generating unlabelled gene replacements in bacterial chromosomes. Mol Gen Genet 253: 217–224.
15. Leenhouts KJ, Kok J, Venema G (1991). Lactococcal plasmid pWV01 as an integration vector for lactococci. Appl Environ Microbiol 57: 2562–2567.
16. Leenhouts KJ, Tolner B, Bron S, Kok J, Venema G, Seegers JFML (1991). Nucleotide sequence and characterisation of the broad-host-range lactococcal plasmid pWV01. Plasmid 26: 55–66.
17. Leenhouts KJ, Venema G (1993). Lactococcal plasmid vectors. In: Hardy KG (ed), Plasmids. A practical approach, 2nd edn, pp 65–94. New York: Oxford University Press.
18. Maguin E, Duwat P, Hege T, Ehrlich SD, Gruss A (1992). New thermosensitive plasmid for gram-positive bacteria. J Bacteriol 174: 5633–5638.
19. Maguin E, Prévot H, Ehrlich SD, Gruss A (1996). Efficient insertional mutagenesis in lactococci and other gram-positive bacteria. J Bacteriol 178: 931–935.
20. Mierau I, Kunji ERS, Leenhouts KJ, Hellendoorn MA, Haandrikman AJ, Poolman B, Konings WN, Venema G, Kok J (1996). Multiple-peptidase mutants of *Lactococcus lactis* are severely impaired in their ability to grow in milk. J Bacteriol 178: 2794–2803.
21. Polzin KM, Shimizu-Kadota M (1987). Identification of a new insertion element, similar to gram-negative IS*26*, on the lactose plasmid of *Streptococcus lactis* ML3. J Bacteriol 169: 5481–5488.
22. Renault P, Heslot H (1987). Selection of *Streptococcus lactis* mutants defective in malolactic fermentation. Appl Environ Microbiol 53: 320–324.
23. Romero DA, Klaenhammer TR (1990). Characterisation of insertion sequence IS*946*, an iso-IS*S1* element, isolated from the conjugative lactococcal plasmid pTR2030. J Bacteriol 172: 4151–4160.
24. Romero DA, Klaenhammer TR (1992). IS*946*-mediated integration of heterologous DNA into the genome of *Lactococcus lactis* subsp. *lactis*. Appl Environ Microbiol 58: 699–702.
25. Romero DA, Klaenhammer TR (1993). Transposable elements in lactococci: A review. J Dairy Sci 76: 1–19.
26. Sambrook J, Fritsch EF, Maniatis T (eds) (1989). Molecular cloning. A laboratory manual, 2nd edn. New York: Cold Spring Harbor Laboratory Press.
27. Sanders JW, Leenhouts K, Burghoorn J, Brands JR, Venema G, Kok J (1998). A chloride-inducible acid resistance mechanism in *Lactococcus lactis* and its regulon. Mol Microbiol 27: 299–310.
28. Sanders JW, Leenhouts KJ, Haandrikman AJ, Venema G, Kok J (1995). Stress response in *Lactococcus lactis*: Cloning, expression analysis, and mutation of the lactococcal superoxide dismutase gene. J Bacteriol 177: 5254–5260.
29. Sanders JW, Venema G, Kok J, Leenhouts K (1998). Identification of a sodium chloride-regulated promoter in *Lactococcus lactis* by a single copy chromosomal fusion with a reporter gene. Mol Gen Genet 257: 681–685.
30. Van der Vossen JMBM, van der Lelie D, Venema G (1987). Isolation and characterisation of *Lactococcus lactis* subsp. *cremoris* Wg2-specific promoters. Appl Environ Microbiol 53: 2452–2457.
31. Youngman P (1993). Transposons and their applications. In: Sonenshein AL, Hoch JA, Losick R (eds) *Bacillus subtilis* and other gram-positive bacteria, pp 585–596. Washington, DC: American Society for Microbiology.

Address for correspondence: Dr J. Kok, Department of Genetics, Biomolecular Sciences and Biotechnology Institute, University of Groningen, Kerklaan 30, NL-9751 NN Haren, The Netherlands
Phone: +31 50 3632111; Fax: +31 50 3632348
E-mail: kokj@biol.rug.nl

Methods in Cell Science **20**: 51–57 (1998)

Vectors containing streptococcal bacteriophage integrases for site-specific gene insertion

W. Michael McShan[1], Robert E. McLaughlin[1], Annika Nordstrand[2] & Joseph J. Ferretti[1]
[1] *Department of Microbiology and Immunology, The University of Oklahoma Health Sciences Center, Oklahoma City, USA*; [2] *Department of Clinical Microbiology, Umeå University, Umeå, Sweden*

Abstract. Two new shuttle-suicide plasmid vectors, pWM245 and p71NT, capable of site-specific integration have been developed for gene insertion or allelic replacement of inactivated genes in streptococci. These vectors contain the integrase gene (*int*) and phage attachment site (*attP*) from the temperate bacteriophage T12 of *Streptococcus pyogenes*. Additionally, these plasmids contain the *ermB* gene specifying resistance to erythromycin in both *Escherichia coli* and streptococci. The plasmid origin of replication, however, is only functional in *E. coli*, forcing integration of the plasmid after introduction into *S. pyogenes* at the phage T12 bacterial attachment site (*attB*), a serine tRNA gene. This tRNA gene is conserved in a number of streptococcal species, thus these integration vectors may serve as broad host-range insertion vectors. The phage excisionase gene (*xis*) is not present in these vectors, thus, integration is highly stable and permanent. These vectors will provide important new tools for genetic analysis of group A and possibly related streptococci.

Key words: Gene insertion, Plasmid vectors, Steptococci

1. Introduction

The use of integration shuttle plasmids for insertional mutagenesis and linkage analysis has proven to be a very useful tool in group A streptococci (GAS or *S. pyogenes*) as well as many other streptococcal species [3, 24, 31, 32]. Such vectors contain an origin of replication (*ori*) that allows replication in *E. coli* for maintenance of the plasmid but that is not functional in streptococci, and thus the plasmid cannot replicate in these species. However, if a region of streptococcal DNA is cloned into the vector, integration into the GAS chromosome can occur by single cross-over, homologous recombination with the genomic copy. The presence of an antibiotic resistance gene in these plasmids functional in both microorganisms provides a convenient means of positive selection. As useful as these vectors are, they are sometimes limited in their ability to introduce new DNA into the streptococcal chromosome for allelic replacement or gene insertion studies. Additionally, it is not possible to control precisely the site of integration of these vectors following homologous recombination, increasing the number of clones that may need to be screened. Therefore, we have created a new set of streptococcal vectors to meet these needs by employing the naturally occurring site-specific integration systems of temperate bacteriophages.

Temperate bacteriophages recombine with their host bacterial chromosome at a specific sequence as an integral part of their biological cycle; the molecular mechanisms that control this event in the best-studied of the viruses, bacteriophage Lambda of *Escherichia coli*, have been covered in an excellent review [26]. These phages, upon infection of the bacterial cell, produce an integrase protein (Int) that mediates recombination between the bacterial chromosome at a specific site (bacterial attachment site or *attB*) and a corresponding region on the phage chromosome (phage attachment site or *attP*). There is 'core' sequence of 15–100 bp that is identical in *attB* and *attP* and serves as the site of recombination between the two genomes [4]. Integration of the phage requires only *attP*, *attB*, Int, and a host-encoded protein, integration host factor (IHF); however, excision of the integrated phage from the bacterial chromosome requires two additional proteins, the phage encoded excisionase (Xis) and the host protein Fis [2, 33]. The integration reaction is highly specific and a single nucleotide change in either *attP* or *attB* in the region of sequence identity will diminish the frequency of integration by several orders of magnitude [33].

Recently we described the genetic elements from bacteriophage T12 of *S. pyogenes* that were responsible for integration of the phage chromosome into a host gene encoding a serine tRNA. The phage attachment site contains a 96 bp duplication of the bacterial chromosome, providing the downstream portion of the tRNA gene after integration [23]. As a demonstration of the capacity of the cloned phage T12 *int* to specify a functional integrase protein, we constructed suicide plasmid pWM139 containing *int* and *attP* as well as a selectable marker for erythromycin resistance. This plasmid was found to integrate into *attB* reproducibly and at a high frequency, and its potential as a cloning vector for site-specific integration of new genes into *S. pyogenes* was immediately apparent. Integration vectors that use bacteriophage-derived *int* genes for such integration have been constructed for use in a number of bacteria including, for example, *E. coli* [8], *S. aureus* [6, 15], lactococcus [7, 13, 18] and lactobacillus [1, 10, 11] species, mycobacterial species [16], and various streptomyces species [12, 19, 30]. In this report, we describe two integration vectors for *S. pyogenes* and possibly other streptococcal species.

2. Materials

1. Todd-Hewitt dehydrated broth, Cat. No. 0492-17-6.[1]
2. Bacto-Yeast Extract, Cat. No. 0127-01-7.[1]
3. Bacto-Agar, Cat. No. 0140-01.[1]
4. Bacto-Tryptone, Cat. No. 0123-01.[1]
5. Normal, pooled, sterile-filtered horse serum, Cat. No. S2-0109.[2]
6. Erythromycin, Cat. No. E-6376.[3]
7. Lysozyme, Cat. No. L-2879.[3]
8. N-Lauroyl sarcosine (sarkosyl), Cat. No. L-5125.[3]
9. Sodium chloride, Cat. No. BP358-10.[3]
10. Sucrose, Cat. No. S-9378.[3]
11. Tris base.[3]
12. Electroporation cuvettes plus apparatus, Cat. No. 610.[4]
13. Genius DNA labeling kit, Cat. No. 1175033.[5]
14. Ethylenediaminetetraccetate, disodium salt (EDTA), Cat. No. BP120-500.[6]
15. Isopropanol, Cat. No. A464-4.[6]
16. Isoamyl alcohol, Cat. No. A393-500.[6]
17. Bio-Rad *E. coli* Pulser™, Cat. No. 165-2102.[7]

3. Procedures

Media and reagents. Strains of *S. pyogenes* are grown in Todd-Hewitt broth supplemented with yeast extract (THY). Per liter, dissolve 30 g Todd-Hewitt dehydrated broth (Difco Laboratories) and 2 g Bacto yeast extract in distilled or deionized water (THY broth). THY-HS broth is THY broth supplemented with 50 ml/l of heat-inactivated horse serum (65 °C for 30 min). Strains of *E. coli* are propagated in B broth (per liter: combine 10 g Bacto-tryptone, 5 g Bacto-yeast extract, and 5 g NaCl, and adjust the pH to 7 by the addition of approximately 2.2 ml of 1 M NaOH [17]). For solid media, 15 g Bacto-agar are added per liter to both media for streptococci and *E. coli*. All media are sterilized by autoclaving using standard conditions. Erythromycin is prepared as a 20 mg/ml stock solution in 95% ethanol and added to the media after autoclaving at a concentration of 3 μg/ml for *S. pyogenes* and 300 μg/ml for *E. coli*.

The 0.5 M sucrose solution is prepared by dissolving 171.2 g sucrose in distilled water to a final volume of 1 liter. Sterilize by autoclaving and store at 4 °C. TE is 10 mM Tris-HC1, 1 mM EDTA, pH 8.0. TEN is 10 mM Tris-HC1, 1 mM EDTA, 50 mM NaCl, pH 8.0. GES for lysis of *S. pyogenes* is prepared by combining 60 g guanidium thiocyanate, 20 ml 0.5 M EDTA, pH 8, and 20 ml distilled water. Heat the mixture at 65 °C until the guanidium thiocyanate is completely dissolved. Allow the solution to cool to room temperature and add 5 ml of 10% (v/v) sarkosyl. Adjust the final volume to 100 ml. Filter the solution through a 0.45 μm filter (Nalgene) and store at room temperature [25].

Plasmid construction and DNA purification. Each of the integration vectors described in this paper possesses unique restriction endonuclease sites for cloning for insert DNA. A standard reference (or the manufacturer's instructions) should be consulted for methods using these sites for the cloning of the DNA of interest [27]. In any electroporation protocol, the purity of the transforming DNA will effect the number of transformants obtained. We routinely perform electrotransformations with plasmid DNA prepared with a commercially available kit and following the manufacturer's protocol (QIAprep-spin, Qiagen, Chatsworth, CA). Standard plasmid preparation protocols that employ a cesium chloride density gradient for purification may also be used to maximize electrotransformation efficiencies [27]. No matter what method of plasmid isolation is used, the transforming DNA must be in the supercoiled, circular form. The energy used by phage integrases for integration is derived from the supercoiling of the plasmid (or phage genome in natural infections) [33]. Linear DNA will not be a substrate for site-specific integration using this system, and methods of plasmid purification that give preparations with significant amounts of linear or nicked DNA will result in decreased electrotransformation efficiencies.

Electrotransformation. The described protocol for electrotransformation has been optimized for *S. pyogenes* NZ131, a highly electrocompetent type M49 laboratory strain [29]. For other strains of *S. pyogenes*, modifications may be necessary, especially in the incubation time to achieve the early logarithmic phase of growth. We employ a 0.5 M sucrose solution for electrotransformation; other investigators have used alternative solutions resulting in essentially the same transformation efficiencies [20]. Whether a given strain will transform well with a given buffer may have to be determined experimentally. The competence for electrotransformation of some strains is also increased by the addition of cell wall weakening agents such as glycine or threonine [5, 9]; additionally, the amount of hyaluronic acid in the capsule of group A streptococci can influence electrocompetence [29]. A complete review of these issues has been presented previously by our laboratory [20].

Inoculate 2–5 ml of THY broth with an isolated colony of NZ131 and incubate overnight at 37 °C. Add 0.4 ml of this overnight culture to 20 ml of prewarmed THY-HS broth and grow the cells at 37 °C for 3 hours to an A_{600} between 0.2 and 0.25 (early logarithmic phase). Harvest the cells by centrifugation at 7600 ×*g* for 5 min at 4 °C (8000 rpm in a Sorval SS34 roto). Resuspend the cell pellet in 1 ml of ice-cold, sterile 0.5 M sucrose prepared in distilled water by vortex mixing, and transfer to a 1.5 ml snap-capped microcentrifuge tube. Harvest the cells by centrifugation for 15 seconds at 14,000 ×*g* (full speed in most standard desktop microcentrifuges). Remove the supernatant by vacuum aspiration and wash the cells three more times using the same conditions. Minimize the time the cells are not kept on ice. Finally, resuspend the cells in 100 µl of cold 0.5 M sucrose. Add the integration plasmid (1 µg) to the cells in no more than a 10 µl volume. Transfer the mixture to a BTX Electroporation Cuvette Plus™ or equivalent cuvette with a 1 mm gap. Using a Bio-Rad *E. coli* Pulser apparatus, expose the cells to a single electrical pulse of 1.8 kV. In this unit, the capacitance is pre-set to 25 µF, and the pulse controller is pre-set to 200 Ω resistance. A safety interlocking system is in place on this Bio-Rad unit to prevent accidental operator exposure to high voltage; no attempt should be made to defeat this system. Alternatively, a cuvette with 2 mm electrode gap may be used, and the voltage should be set to 2.5 kV. After the electric pulse, immediately place the cells on ice and cool for 5 min. Add 0.9 ml of THY-HS to the cells and incubate for 2 hours at 37 °C. Plate 0.1 ml aliquots on THY agar plates supplemented with the appropriate antibiotic. Colonies are usually visible after 24 hours of incubation, and typically, electrotransformation with 1 µg of integration vector plasmid DNA will result in $>10^3$ transformants.

Isolation of chromosomal DNA and screening for plasmid integration. Streptococcal chromosomal DNA is prepared by the method of Pitcher et al. [25]. Inoculate 20 ml of THY broth with an isolated colony of the *S. pyogenes* strain of interest and incubate overnight at 37 °C. Pellet the cells by centrifugation (1000 ×*g* for 15 min) and remove the supernatant by vacuum aspiration. Resuspend the cell pellet in 1 ml TEN, transfer the suspension to a 1.5 ml microcentrifuge tube and again pellet the cells by centrifugation (14,000 ×*g* or full speed in the microcentrifuge for 1 minute). Suspend the pellet in 100 µl of freshly prepared lyzozyme (50 mg/ml in TE) plus 50 units of mutanolysin, and place the mixture at 37 °C for 30 min. Lyse the cells by adding 0.5 ml GES. Mix by repeated inversion of the tube and cool the mixture on ice for 5 min. Lysis should be complete at this point. Add 0.25 ml cold 7.5 M ammonium acetate, mix by repeated inversion, and incubate on ice for 10 min. Add 0.5 ml chloroform:isoamyl alcohol (24:1) and mix thoroughly until a complete emulsion is achieved. Separate the phases by centrifugation (14,000 ×*g* for 10 min). A thin, white interface should separate the phases. Collect the upper phase and estimate its volume. Add 0.6 volumes of isopropanol and mix by inversion for at least one minute. Typically, a fibrous mass of DNA is visible at this point. Collect the DNA by centrifugation (7,000 ×*g* for 1 minute). If the DNA is not a visible mass after the addition of the isopropanol, then centrifugation should be performed at 14,000 ×*g* for 5 min. Wash the pellet five times with 0.5 ml 70% ethanol using 7,000 ×*g* for 1 minute. Completely remove any traces of the supernatant by vacuum aspiration and dry the DNA pellet *in vacuo* for 5 min. Dissolve the DNA in 100 µl sterile, distilled water or TE.

The DNA probe for *attB*, cloned on pWM130 [23], was prepared using the Boehringer Mannheim Genius™ digoxigenin-dUPT (DIG-dUTP) labeling kit following the manufacturer's protocols. Agarose gel electrophoresis, Southern DNA transfer to nylon membrane, and hybridization to DIG-dUTP DNA probes are conducted according to standard protocols [27].

4. Results and discussion

Site-specific integration vectors. The construction of pWM139, containing the integrase and phage attachment site of bacteriophage T12, has been described previously (Figure 1A). The essential features of this vector and its derivatives are a selectable antibiotic resistance marker that is expressed both in *E. coli* and streptococci, an origin of replication that is functional in *E. coli* but not in streptococci, and the presence of the T12 *int* and *attP*. Importantly, the phage *xis* is not present on this plasmid so that integration is not reversible. The region from upstream of the *Cla*I restriction endonuclease site contains the *lacZ* promoter; passage of the plasmid in *E. coli* results in the isolation of clones containing the plasmid with spontaneous deletions of this region. The promoter and the coding region for *int* as well as *attP* are unaffected by this deletion. We hypothesize that the *lacZ* promoter causes overexpression of the phage gene product in the *E. coli*, and that this overproduction is toxic to the bacterium. The spontaneous derivative of pWM139, pWM245 (Figure 1B), is stable in *E. coli* and able to integrate efficiently into the *S. pyogenes* genome at T12 *attB*. This plasmid is

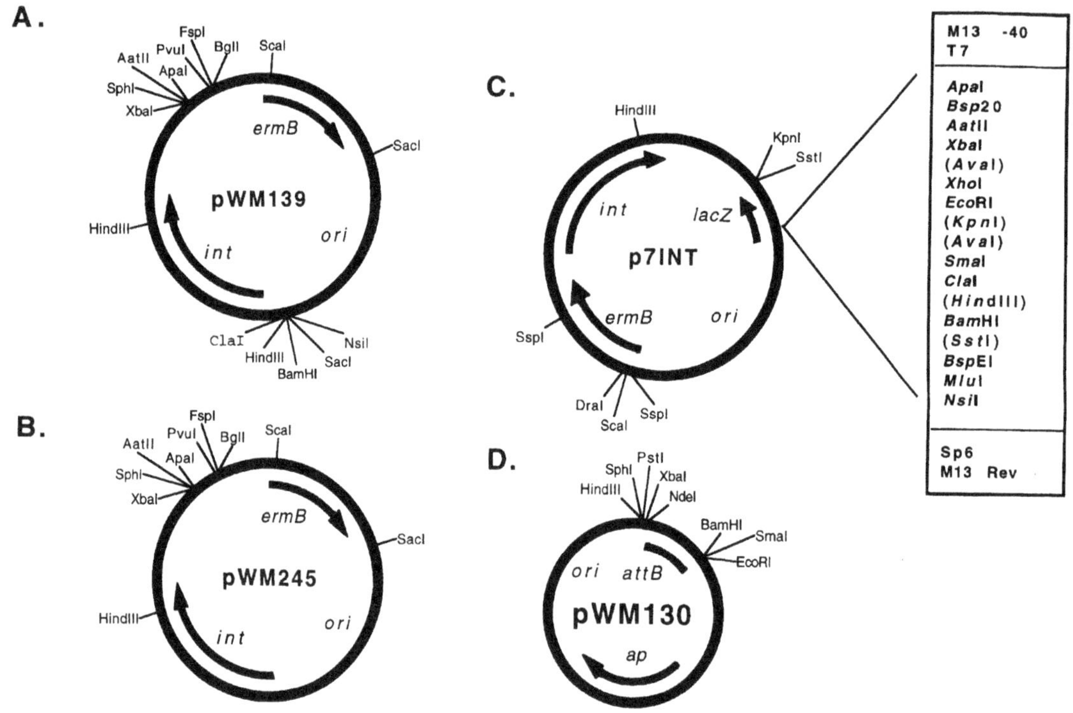

Figure 1. Integration vectors derived from bacteriophage T12. Shown are four plasmids containing the genetic elements derived from bacteriophage T12 that are responsible for mediating site-specific recombination with the *S. pyogenes* chromosome. These plasmids contain the origin of replication from pUC18 and thus can be maintained in *E. coli*. However, this origin does not function in streptococci, forcing the plasmids to integrate into the bacterial attachment site (*attB*), a serine tRNA that exists as a single copy in the bacterial chromosome. The antibiotic resistance gene is functional in both *E. coli* and streptococci. The construction of plasmid pWM139 (4378 bp, panel A) and its capacity to integrate into the bacterial chromosome have been previously reported [23]. Upon passage in *E. coli* DH5αF′, pWM139 spontaneously deletes about 100 bases of the plasmid backbone region upstream of the integrase gene to give pWM245 (4270 bp, panel B). This deletion results in the loss of the *E. coli lacZ* promoter as well as several unique restriction endonuclease cleavage sites. In order to make a more versatile integrative cloning vector, p7INT was constructed (4374 bp, panel C). This plasmid has a multiple cloning site (MCS) in a functional copy of the *E. coli lacZ* gene, allowing blue-white screening of recombinants as well as the phage T12 integrase and attachment site and the *ermB* gene to select for erythromycin resistance. The MCS also contains the sequences complementary to the standard commercially available M13–40, T7, Sp6, and M13 reverse sequencing primers. pWM130 (3257 bp, panel D) is derived from the PCR product cloning vector, pT7Blue, and contains the serine tRNA gene from the *S. pyogenes* chromosome that serves as *attB*. This plasmid is used to make probles for DNA hybridization to *attB* and is not for introduction into streptococci since it contains a functional β-lactamase gene, encoding ampicillin resistance.

somewhat limited in its usefulness because it contains relatively few unique cloning sites and lacks blue-white screening of cloned inserts. An improved vector with these characteristics is p7INT (Figure 1C). This plasmid has a multiple cloning site present in the promoter of *lacZ* containing a number of unique restriction endonuclease cleavage sites that interrupt *lacZ* expression when used for cloning DNA inserts. Additionally, there is no instability of this plasmid in our hands as was the case with pWM139. Both of these vectors use erythromycin resistance as the selectable marker. These vectors and their derivatives will be useful tools for allelic replacement. One such construct has been used to introduce a functional and expressed streptokinase gene from *Streptococcus equisimilis* (*skc*) into a strain of *S. pyogenes* that lacks a functional copy of the group A streptokinase gene (*ska*) [A. Nordstrand, manuscript in preparation]. Finally, plasmid pWM130, containing the serine tRNA that serves as *attB* and its surrounding region from the *S. pyogenes* chromosome, is useful as a probe for determining whether integration has occurred. Insertion of either a phage genome or an integration vector into the phage T12 *attB* of *S. pyogenes* is detected by hybridization of *Hind*III digested chromosomal DNA. *attB* is contained on one *Hind*III fragment when it is unoccupied by a phage or integrative plasmid (Figure 2, lane 2). However, integration of a plasmid into *attB* alters this pattern and creates two hybridizing *Hind*III fragments generated by the plasmid-associated restriction endonuclease sites (lane 3).

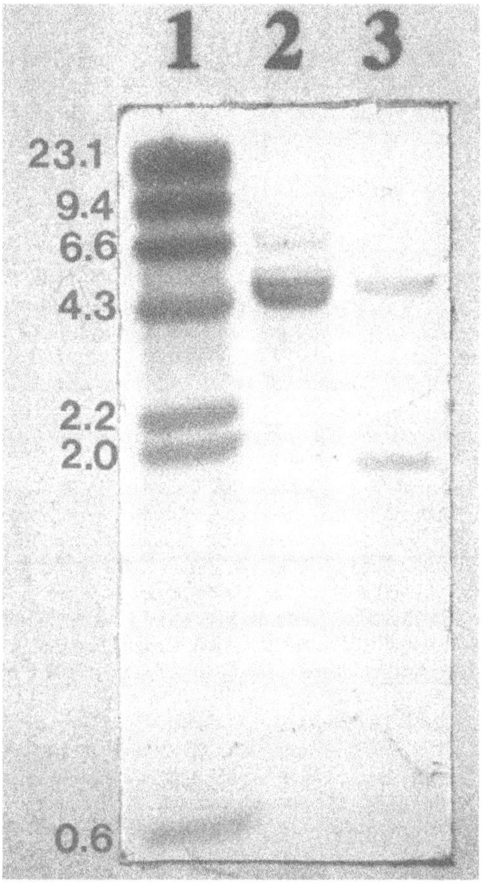

Figure 2. Integration of pWM245 into the phage T12 *attB* site. Insertion of pWM245 into *attB* of *S. pyogenes* NZ131 was detected by hybridization of *Hin*dIII digested chromosomal to a probe specific for the serine tRNA that serves as the phage T12 attachment site. The phage T12 *attB* is contained on one *Hin*dIII fragment of approximately 5,000 bp when it is unoccupied by either a phage or integrative plasmid (lane 2). However, integration of the plasmid into *attB* alters this pattern and creates two hybridizing *Hin*dIII fragments of approximately 5,100 bp and 1,300 bp generated by the plasmid-associated restriction endonuclease sites (lane 3). *Hin*dIII digested bacteriophage Lambda DNA was used as a molecular weight standard (lane 1). The size of the Lambda fragments are labeled in kilobases.

Because of the instability in pWM139, this plasmid will not be available for release. The two other vectors, pWM245 and p7INT, as well as pWM130 will be made available to the scientific community upon request.

Applications and possible limitations. The serine tRNA that serves as the bacterial attachment site appears to be conserved among a number of streptococcal and related species, and thus it may be that these phage-derived vectors may be useful as wide host-range vectors as has been done with mycobacterial phage integration vectors [16]. We have detected the presence of this gene by both polymerase chain reaction and DNA hybridization in *Streptococcus agalactiae, S. mutans, S. salivarius, S. downei, S. gordonii, S. macacae, S. sanguis, S. oralis, S. mitis, S. equisimilis, S. pneumoniae, Lactococcus lactis*, and *Enterococcus faecalis* [22]. The integration vectors should theoretically be functional in any of these species that contain the appropriate target sequence. Several factors will determine whether they may be used successfully in practice, including competence for transformation or electrotransformation, presence of required accessory factors, and sequence fidelity of the *attB* core region. Published protocols for introducing plasmid DNA into a number of these bacteria have been reported in the literature [20], but the actual frequency of a particular isolate of interest may deviate considerably from the published values, and it may well be that some isolates, especially natural isolates, will prove to be difficult or impossible to transform. Additionally, factors influencing integration may determine the success or failure in non-group A streptococcal species. As discussed above, the host-encoded integration host factor (IHF), a DNA binding protein, is required for integration for phage-derived systems. While a requirement for IHF has not been rigorously demonstrated for bacteriophage T12 integration, consensus sequences for IHG binding homologous to the sequences from *E. coli* are present in the appropriate positions in the T12 *attP* region [23]. Variations in IHF and its associated target sequence in non-group A species could influence the efficiency of integration of the phage T12-derived vectors. Finally, single base changes in the core region of the *attB* homologues in the non-group A streptococci could decrease the efficiency of integration significantly [33]. We have sequenced the *attB* region from *S. mutans, S. agalactiae, S. downei*, and *E. faecalis*, and have found it be identical to the sequence from *S. pyogenes* in all cases [W. M. McShan, unpublished results]. Therefore, one would anticipate that these organisms would be a suitable hosts for these vectors. No sequence information is yet available for any of the other species. In one preliminary experiment, transformants were obtained with pWM245 in a strain of *S. macacae* and in the naturally competent *S. gordonii* Challis strain. Both of these species contain a region in their genomes that hybridizes with the probe for the phage T12 *attB*, and one anticipates that these regions will also have DNA sequences that are identical to the one from S. pyogenes.

Only one copy of the attachment site for T12 exists in the *S. pyogenes* genome [23]. Alteration of the 'core' sequence of a phage attachment site by only one nucleotide results in a 100- to 1000-fold reduction in integration frequency [14, 28, 33]. Given that electroporation is a relatively inefficient process and that the plasmid cannot replicate independently in streptococci suggests that the chances for even having two copies of the plasmid present in a cell before integration is highly unlikely. Thus, (1) integration at the target site is highly favored, and (2) tandem duplication by the plasmid at the *attB* site is unlikely. However, it is crucial to determine whether a resident phage is occupying the T12 *attB* in a given strain of *S. pyogenes* before proceeding with any experiments. It should not be assumed that because a strain does not contain the gene for the erythrogenic toxin (*speA*) that no phage occupies *attB*. We have recently reported that it is possible to isolate phages that share *int* and *attP* with phage T12, but do not have *speA* [21]. Tandem insertions of the plasmid along an integrated phage are possible because the sequence of *attB* is not altered by integration, and such events may give unexpected patterns upon DNA hybridization. Further, integration of plasmid DNA in tandem with a resident phage may decrease the stability of integration because of the presence of the phage excisionase gene. Homologous recombination between the flanking regions of an inactivated chromosomal gene and the plasmid borne replacement allele could theoretically occur before integration; however, such an event would only occur at very low frequency. Analysis of putative transformants for integration at the bacterial attachment site easily confirms the desired event. The chromosomal DNA sequence containing *attB* is contained on one *Hin*dIII fragment when it is unoccupied by either a phage or integrative plasmid (Figure 2, lane 2). However, integration of pWM245 or the other integration vectors into *attB* alters this pattern and creates two *Hin*dIII fragments that hybridized with the *attB* probe because of the duplication in *attP* (Figure 2, lane 3). This alteration in restriction pattern is diagnostic for integration and should be used to confirm may putative transformants. Further, if eliminating homologous recombination is a particular concern for a given experiment, a *recA* strain of *S. pyogenes* could be employed [31].

Acknowledgments

This work was supported by Public Health Service grants AI19304 and AI38406 from the National Institutes of Health.

Notes on suppliers

1. Difco Laboratories, Detroit, MI 48232-7058, USA
2. Lampire Biological Laboratories, P.O. Box 270, Pipersville, PA 18947, USA
3. Sigma Chemical, P.O. Box 14508, St Louis, MO 63178, USA
4. BTX, 11199-A Sorrento Valley Rd., San Diego, CA 92121-1334, USA
5. Boehringer Mannheim Biochemicals, P.O. Box 50414, Indianapolis, IN 46250, USA
6. Fisher Scientific, 711 Forbes Ave., Pittsburgh, PA 15219-4785, USA
7. Bio-Rad Laboratories, 2000 Alfred Nobel Drive, Hercules, CA 94547, USA

References

1. Auvray F, Coddeville M, Ritzenthaler P, Dupont L (1997). Plasmid integration in a wide range of bacteria mediated by the integrase of *Lactobacillus delbrueckii* bacteriophage mv4. J Bacteriol 179: 1837–1845.
2. Ball CA, Johnson RC (1991). Efficient excision of phage λ from the *Escherichia coli* chromosome requires the Fis protein. J Bacteriol 173: 4027–4031.
3. Barletta RG, Michalek SM, Curtiss III R (1988). Analysis of the virulence of *Streptococcus mutans* serotype c *gtfA* mutants in the rat model system. Infect Immun 56: 322–330.
4. Campbell AM (1992). Chromosomal insertion sites for phages and plasmids. J Bacteriol 174: 7495–7499.
5. Caparon MG, Scott JR (1991). Genetic manipulation of pathogenic streptococci. Meth Enzymol 204: 556–586.
6. Carroll D, Kehoe MA, Cavanagh D, Coleman DC (1995). Novel organization of the site-specific integration and excision recombination functions of the *Straphylococcus aureus* serotype F virulence-converting phages ϕ13 and ϕ42. Mol Microbiol 16: 877–893.
7. Christiansen B, Johnsen MG, Stenby E, Vogensen FK, Hammer K (1994). Characterization of the lactococcal temperate phage TP901-1 and its site-specific integration. J Bacteriol 176: 1069–1076.
8. Diederich L, Rasmussen LJ, Messer W (1992). New cloning vectors for integration in the Lamba attachment site attB of the *Escherichia coli* chromosome. Plasmid 28: 14–24.
9. Dunny GM, Lee LN, LeBlanc DJ (1991). Improved electroporation and cloning vector system for Gram-positive bacteria. Appl Environ Micobiol 57: 1194–1201.
10. Dupont L, Boizet-Bonhoure B, Coddeville M, Auvray F, Ritzenthaler P (1995). Characterization of genetic elements required for site-specific integration of *Lactobacillus delbrueckii* subsp. *bulgaricus* bacteriophage mv4 and construction of an integration-proficient vector form *Lactobacillus plantarum*. J Bacteriol 177: 586–595.
11. Fremaux C, Antoni GLD, Raya RR, Klaenhammer TR (1993). Genetic organization and sequence of the region encoding integrative functions from *Lactobacillus gasseri* temperate bacteriophage ϕadh. Gene 126: 61–66.
12. Gabriel K, Schmid H, Schmidt U, Rausch H (1995). The actinophage RP3 DNa integrates site-specifically into the putative $tRNA^{Arg}$ (AGC) gene of *Streptomyces rimosus*. Nucleic Acids Res 23: 58–63.
13. van der Guchte M, Daly C, Fitzgerald GF, Arendt EK (1994). Identification of *int* and *att*P on the genome of lactococcal bacteriophage Tuc2009 and

their use for site-specific plasmid integration in the chromosome of Tuc2009-resistant *Lactococcus lactis* MG1363. Appl Environ Microbiol 60: 2324–2329.
14. Kim S, Landy A (1992). Lambda Int protein bridges between higher order complexes at two distant chromosomal loci *attL* and *attR*. Science 256: 198–203.
15. Lee CY, Buranen SL, Ye ZH (1991). Construction of single-copy integration vectors for *Straphylococcus aureus*. Gene 103: 101–105.
16. Lee MH, Pascopella L, Jacobs Jr WR, Hatfull GF (1991). Site-specific integration of mycobacteriophage L5: integration-proficient vectors for *Mycobacterium smegmatis, Mycobacterium tuberculosis*, and bacille Calmette-Guerin. Proc Natl Acad Sci USA 88: 3111–3115.
17. Lennox ES (1955). Transdunction of linked genetic characters of the host by bacteriophage P1. Virology 1: 190–206.
18. Lillehaug D, Nes IF, Birkeland NK (1997). A highly efficient and stable system for site-specific integration of genes and plasmids into the phage phiLC3 attachment site (*attB*) of the *Lactococcus lactis* chromosome. Gene 188: 129–136.
19. Matsuura M, Noguchi T, Yamaguchi D, Aida T, Asayama M, Takahashi H, Shirai M (1996). The *sre* gene (ORF469) encodes a site-specific recombinase responsible for integration of the R4 phage genome. J Bacteriol 178: 3374–3376.
20. McLaughlin RE, Ferretti JJ (1995). Electrotransformation of streptococci. In: Nickoloff JA (ed), Electroporation protocols for microorganisms, vol. 47, pp 185–193. Totowa, NJ: Humana Press.
21. McShan WM, Ferretti JJ (1997). Genetic diversity in temperate bacteriophages of *Streptococcus pyogenes*: Identification of a second attachment site for phages carrying the erythrogenic toxin A gene. J Bacteriol 179: 6509–6511.
22. McShan WM, Ferretti JJ (1997). Genetic studies of erythrogenic toxin carrying temperate bacteriophages of *Streptococcus pyogenes*. Adv Exp Med 418: 971–974.
23. McShan WM, Tang YF, Ferretti JJ (1997). Bacteriophage T12 of *Streptococcus pyogenes* integrates into the gene encoding a serine tRNA. Mol Micobiol 23: 719–728.
24. Perry D, Nilsen LJ, Kuramitsu HK (1985). Mapping of a cloned glucosyltransferase gene in *Streptococcus mutans*. Infect Immun 50: 130–135.
25. Pitcher DG, Saunders NA, Owen RJ (1989). Rapid extraction of bacterial genomic DNA with guanidium thiocyanate. Lett Appl Micro 8: 151–156.
26. Ptashne M (1992). A genetic switch: phage Lambda and higher organisms, 2nd edn. Cambridge, MA: Blackwell.
27. Sambrook J, Fritsch EF, Maniatis T (1989). Molecular cloning: a laboratory manual, 2nd edn. Cold Spring Harbor, NY: Cold Spring Harbor Laboratory Press.
28. Shimada K, Weisberg RA, Gottesman ME (1972). Prophage lambda at unusual chromosomal locations. I. Location of the secondary attachment sites and the properties of the lysogens. J Mol Biol 63: 483–503.
29. SimonD, Ferretti JJ (1991). Electrotransformation of *Streptococcus pyogenes* with plasmid and linear DNA. FEMS Microbiol Lett 82: 219 224.
30. Sladkova IA, Orekhov AV (1994). Actinomycete plasmid and integrative vectors based on DNA of the temperate Phi C31 actinophage, determining limitation of lytic development of phage Phi C31, not dependent on repressor. Antibiot Khimioter 39: 3–11.
31. Tao L, Hollingshead SK, Suvorov AN, Ferretti JJ, McShan WM (1995). Construction of a *Streptococcus pyogenes recA* mutant via insertional inactivation, and cloning and sequencing of the complete *recA* gene. Gene 162: 59–62.
32. Tao L, LeBlanc DJ, Ferretti JJ (1992). Novel streptococcal-integration shuttle vectors for gene cloning and inactivation. Gene 120: 105–110.
33. Weisberg RA, Landy A (1983). Site-specific recombination in phage lambda. In: Roger JWR, Hendrix W, Stahl W, Weisberg RA (eds), Lambda II, pp 211–250. Cold Spring Harbor Press, NY: Cold Spring Harbor.

Address for correspondence: Dr Joseph J. Ferretti, Department of Microbiology and Immunology, The University of Oklahoma Health & Science Center, Oklahoma City, OK 73190, USA
Phone: (405) 271 2133; Fax: (405) 271 0000
E-mail: joe-ferretti@uokhsc.edu

Methods in Cell Science **20**: 59–64 (1998).

Streptococcal integration vectors for gene inactivation and cloning

Lin Tao*
Department of Oral Biology, School of Dentistry, University of Missouri-Kansas City, Kansas City, Missouri, USA
*(*Present address: Department of Oral Biology, College of Dentistry, University of Illinois, Chicago, Illinois, USA)*

Abstract. Five streptococcal integration vectors containing different antibiotic resistance-encoding genes capable of expressing in both *Streptococcus* sp. and *Escherichia coli* are introduced. These plasmids can replicate in *E. coli*, but not in streptococci, because the plasmids lack a streptococcal origin of replication. If these plasmids carry a fragment of streptococcal DNA they can specifically integrate into the streptococcal chromosome via Campbell-like, single crossover homologous recombination. Methods are described to use these vectors for gene inactivation and cloning in streptococci.

Key words: Cloning, Inactivation, Integration, Plasmid, *Streptococcus*

Abbreviations: Ap = ampicillin; bp = base pair(s); Cm = chloramphenicol; Em = erythromycin; kb = kilobase(s) or 1000 bp; Km = kanamycin; *ori* = origin of DNA replication; ^{R} = resistant/resistance; Sp = spectinomycin; Tc = tetracycline; TSS = transformation and storage solution

1. Introduction

Streptococcal integration plasmids, also called suicidal plasmids, have been used successfully for linkage analysis [13–15], gene cloning [10, 18–21], characterization of chromosomal conjugative elements [21], and insertion-duplication mutagenesis among various streptococcal species [1, 2, 5, 11, 19, 20]. An important characteristic of these plasmids is that they can replicate in *Escherichia coli*, but not in streptococci because they lack a streptococcal origin (*ori*). Additionally, they have an antibiotic resistance gene from streptococci or other Gram-positive organisms that can be expressed both in *E. coli* and streptococci. These plasmids transform streptococci by integrating into the chromosome via Campbell-like, single crossover homologous recombination. This type of transformation is 10^4 times more efficient than transformation with linear DNA [16], which requires a double crossover recombination. Because the integration plasmids have an *E. coli ori*, the genes flanking an inserted plasmid in the streptococcal chromosome can be retrieved and cloned in *E. coli*. This offers a method for cloning genes by chromosomal walking. This article introduces five streptococcal integration vectors that contain different antibiotic resistance-encoding genes and the procedures for using these vectors in gene inactivation and cloning.

2. Materials

A. Equipment
1. Candle Jar. A glass jar with a metal screw cap and a short candle.
2. Microcentrifuge, model MC-140, Tomy.[1]
3. Superspeed centrifuge, Sorvall model RC2-B, DuPont.[2]
4. Gene Pulser, No. 165-2105, Bio-Rad.[3]
5. Gene Pulser Cuvettes, No. 165-2088.[3]
6. Incubator shaker, No. M1024-1000, New Brunswick.[4]

B. Growth media and reagents.
1. THB medium
   - Todd-Hewitt Broth, No. 0492-17-6, Difco.[5]
   - Agar, No. 0138-17-6.[5]
   - Horse serum, No. H1270, Sigma.[6]
2. LB medium
   - Yeast extract, No. 0886-17-0.[5]
   - Tryptone, No. 0123-17-3.[5]
   - Sodium chloride, No. S671-500, Fisher.[7]
   - Sodium hydroxide, No. S318-500.[7]
3. TSS (Transformation and storage solution)
   - Polyethylene glycol (PEG) 8000, No. BP233-1.[7]
   - Dimethyl sulfoxide (DMSO), No. BP231-1.[7]
   - Magnesium chloride hexahydrate, No. BP214-500.[7]
4. Chemicals and enzymes
   - Glycine, No. G46-500.[7]
   - Sucrose, No. S5-500.[7]
   - Sodium acetate, No. S210-500.[7]
   - Sodium dodecyl sulfate (SDS), No. S529-500.[7]

- Tris(hydroxymethyl)aminomethane, No. T395-500.[7]
- Ethylenediamine tetraacetic acid (EDTA), No. E478-500.[7]
- Chloroform, No. C606-1.[7]
- Phenol, No. A92-500.[7]
- Lysozyme, No. L6876.[6]
- Ribonuclease A (RNase), No. R4875.[6]

5. Antibiotics
   - Chloramphenicol, No. C0378.[6]
   - Erythromycin, No. E6376.[6]
   - Kanamycin, No. K4000.[6]
   - Penicillin G, No. PEN-K.[6]
6. Restriction enzymes and ligase
   - *Bam*HI, No. R6021, Promega.[8]
   - *Eco*RI, No. R6011.[8]
   - *Pst*I, No. R6111.[8]
   - *Sau*3A1, No. R6191.[8]
   - T4 ligase, No. M1801.[8]

C. Glassware and plastics
1. Glass tubes, No. 14-959-25A.[7]
2. Nalgene Oak Ridge centrifuge tubes, No. 05-529-1D.[7]
3. Microcentrifuge tubes, No. 05-406-16.[7]
4. Petri dishes, No. 08-757-14G.[7]

D. DNA purification kits
1. GeneClean Kit, No. 1001-000, Bio-101 Inc.[9]
2. QIAprep Spin Miniprep Kit, No. 27106. Qiagen.[10]

E. DNA labelling kit
1. BioNick labelling system, No. 18247-015, Life Technologies.[11]

F. Plasmid vectors (Figure 1)
1. pVA891 [8], Macrina.[12]
2. pEVP3 [4], Morrison.[13]
3. pSF151 [17], Tao.[14]
4. pSF152 [17].[14]
5. pSF143 [17].[14]

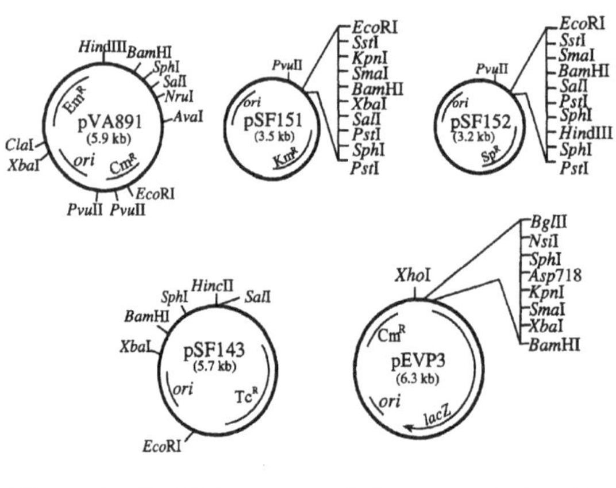

Figure 1. Restriction maps of streptococcal plasmids pVA891, pSF143, pSF151, pSF152 and pEVP3. The construction of these plasmids is described in references [4, 8, 17].

3. Procedures

A. Preparation of culture media and solutions:

1. LB broth
   Dissolve 5 grams yeast extract, 5 grams Tryptone, and 5 grams sodium chloride into 999 ml distilled or deionized water. Add 1 ml of 1 N sodium hydroxide into the solution (pH 7.5) and sterilize by autoclaving.
2. LB selective agar
   Add 15 grams agar into 1 liter of LB broth and sterilize by autoclaving. Before the agar is solidified, add antibiotics into the medium to appropriate final concentration [chloramphenicol (Cm), 15 μg/ml; ampicillin (Ap) 60 μg/ml; or kanamycin (Km), 50 μg/ml], and pour the agar plates.
3. TSS medium
   To 20 ml LB broth, add 2 grams PEG 8000, 1 ml DMSO, and 0.2 gram magnesium chloride. Sterilize by filtration.
4. THB broth
   Dissolve 30 grams Todd Hewitt Broth solid in 1 liter distilled or deionized water. Distribute 4 ml portions into glass test tubes and sterilize by autoclaving.
5. THBS (THB with 5% horse serum) medium
   Incubate horse serum in a water bath at 65 °C for 1 h and add the horse serum into the THB broth to a final concentration of 5%.
6. THB selective agar.
   Add 15 grams agar into 1 liter THB broth and sterilize by autoclaving. Before the agar is solidified, add antibiotics into the medium to appropriate final concentration [erythromycin (Em), 20 μg/ml; or kanamycin (Km), 350 μg/ml].
7. Sucrose solution (0.5 M)
   Dissolve 171.15 grams sucrose into 1 liter distilled water, and sterilize by autoclaving.
8. Lysis buffer.
   Dissolve 25 grams sucrose in 80 ml TE buffer (10 mM Tris, 1 mM EDTA, pH 8.0), and bring it up to 100 ml with additional TE buffer.
9. RNase preparation.
   Dissolve 10 mg RNase in 1 ml deionized water in a microcentrifuge tube. Heat the tube in boiling water for 15 min to eliminate DNase activity.

B. Genetic transformation

1. Transformation of *E. coli* cells [3]
   Inoculate a single colony of *E. coli* JM109 to 2 ml LB broth in a glass test tube. Incubate the culture at 37 °C with vigorous shaking for 16 h. Transfer 1 ml of the *E. coli* culture into 20 ml LB broth in a 200 ml flask. Incubate the culture at 37 °C with vigorous shaking for 3 h. Harvest the cells by centrifugation (7,000 ×*g*, 5 min, 4 °C) and resuspend the pellet in

2 ml ice-cold TSS medium. Distribute the cells into multiple microcentrifuge tubes. Add 1–5 μl plasmid DNA to 100 μl competent *E. coli* cells, and incubate at 4 °C for 30 min. Add 0.9 TSS or LB broth, and transfer to a glass test tube. Incubate the culture at 37 °C for 1 h with vigorous shaking. Plate the cells on LB selective agar plates.

2. Transformation of streptococcal cells
   a. Natural transformation
   Inoculate a single colony of a streptococcal strain, such as *Streptococcus gordonii* Challis, into 2 ml THBS medium in a glass test tube. Incubate the culture at 37 °C for 16 h. Transfer 0.1 ml of the culture into 4 ml prewarmed THBS medium. Incubate the diluted culture at 37 °C for 2 h to achieve competence. In the case of *Streptococcus mutans* or *Streptococcus milleri*, the incubation time for achieving competence is 3 to 3.5 h [13–15, 18]. After adding 1–5 μl plasmid DNA to the competent cells, incubate the culture at 37 °C for 1 h. Spread the cells on the THB selective agar plates.
   b. Electrotransformation.
   Inoculate a single colony of a streptococcal strain that cannot be transformed naturally, such as *Streptococcus pyogenes* NZ131 [16], into 2 ml THBS medium in a glass test tube. Add glycine to the broth at different concentrations (0.3 to 1.5%) if other streptococcal species are used [6]. Incubate the culture at 37 °C for 16 h. Transfer 1 ml culture into 40 ml of prewarmed THBS medium in a glass bottle and continue to incubate for 3 h. Harvest the cells by centrifugation (7,000 ×*g*, 5 min, 4 °C) and resuspend the cells in 1 ml ice-cold sterile 0.5 M sucrose solution. Transfer to a 1.5-ml microcentrifuge tube, washing twice by resuspension in 1 ml 0.5 M sucrose, and pelleting for 30 seconds in a microcentrifuge. Resuspend the cells in 200 μl 0.5 M sucrose, and distribute into 5 microcentrifuge tubes with 40 μl each. After mixing with 1 or 2 μl DNA solution, transfer the cell suspension to a chilled Gene Pulser cuvette (4 millimeter electrode gap) (Bio-Rad), and expose to one electric pulse with the Bio-Rad Gene Pulser (peak voltage, 2.5 kV; capacitance, 25 F; pulse controller, 200 Ω). Immediately dilute the culture into 1 ml THBS and incubate for 2 h at 37 °C. Spread the cells on the THB selective agar plates.

C. Isolating genomic DNA from streptococci
Inoculate a single colony of a streptococcal strain, such as *S. gordonii* Challis, into 2 ml THB broth in a glass test tube. Incubate the culture at 37 °C for 16 h. Transfer the culture into 100 ml prewarmed THB broth. Incubate the culture at 37 °C for 4 to 5 h to reach mid-exponential phase. Add 1,000 U penicillin G to the culture, and continue to incubate for 2 h. Harvest the cells by centrifugation (7,000 ×*g*, 10 min), and wash the cells with deionized water. Suspend the cells in 5 ml lysis buffer supplemented with 5 mg lysozyme and 50 μg RNase, and incubate at 37 °C for 1 h. Add 0.5 ml 10% SDS and incubate at 55 °C for 20 min. Add 0.6 ml 3 M sodium acetate, 3 ml phenol and 2 ml chloroform to the viscous lysate. Gently mix the lysate, and centrifuge at 15,000 ×*g* for 10 min. Transfer the upper phase to a new tube and repeat the procedure with 5 ml chloroform. Transfer the upper phase again to a new tube and add 10 ml 95% ice-cold ethanol. Centrifuge at 15,000 ×*g* for 10 min. Wash the pellet with 10 ml 70% ethanol. Dry the pellet for 30 min at room temperature, and dissolve it in 200 μl TE buffer.

D. Inactivate streptococcal genes with an integration plasmid
   1. Inactivation by insertion duplication (Figure 2)
      a. Digest a cloned streptococcal gene with the restriction enzyme *Sau*3AI, and purify the internal gene fragment in the agarose gel with the GeneClean kit after performing electrophoresis.
      b. Digest the streptococcal integration vector pVA891 with *Bam*HI that produces the same cohesive ends as *Sau*3AI. Remove the enzyme and buffering salt by the GeneClean kit.
      c. Ligate the internal gene fragment and the plasmid with T4 ligase.
      d. Transform the ligation mixture into *E. coli* JM109, and select for Cm resistant transformants.

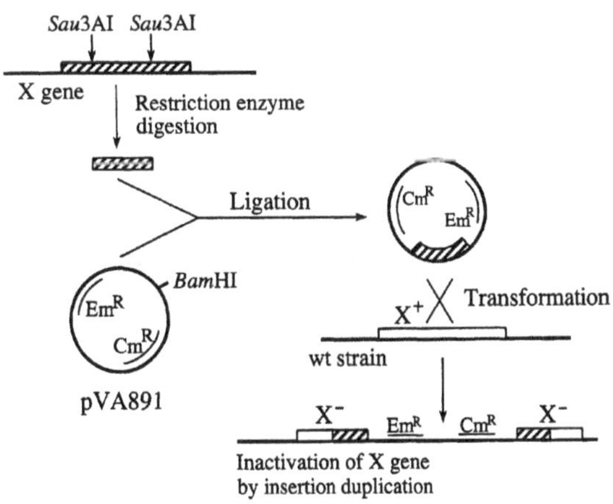

Figure 2. Gene inactivation by insertion duplication.

e. Screen for a recombinant plasmid that carries the correct insert among the transformants by performing miniprep with the Qiagen kit, restriction enzyme digestion, and agarose gel electrophoresis.
f. Transform the recombinant plasmid into the streptococcal strain either by natural transformation or electroporation.
g. Incubate the transformed culture at 37 °C in a candle jar for 24 to 48 h on the THB agar plates containing Em (20 µg/ml) to select for transformants.

2. Inactivation by allelic exchange (Figure 3)
   a. Digest the cloned streptococcal gene with one or two restriction enzymes that cut inside the gene but not the cloning vector pUC18.
   b. Digest the streptococcal integration vector pSF151 with the same restriction enzyme(s) that cuts a unique site in the vector.
   c. Purify the two digested DNA fragments from agarose gel with the GeneClean kit, and ligate them with T4 ligase.
   d. Transform the ligation mixture into *E. coli* JM109.
   e. Plate the transformed cells onto LB agar plates containing both Ap (60 µg/ml) and Km (50 µg/ml) to select for transformants that carry the recombinant plasmid.
   f. Screen for the recombinant plasmid in which the integration vector has inserted into the cloned streptococcal gene by miniprep with the Qiagen Kit and by restriction enzyme analysis.
   g. Linearize the recombinant plasmid with *Sca*I that cuts a unique site in the pUC18 Ap^R gene but does not cut pSF151.
   h. Transform the streptococcal strain with the linear DNA in which the integration vector is inside the cloned streptococcal gene.
   i. Incubate the transformed culture at 37 °C in a candle jar for 24 to 48 h on the THB agar plates containing Km (350 µg/ml) to select for transformants.

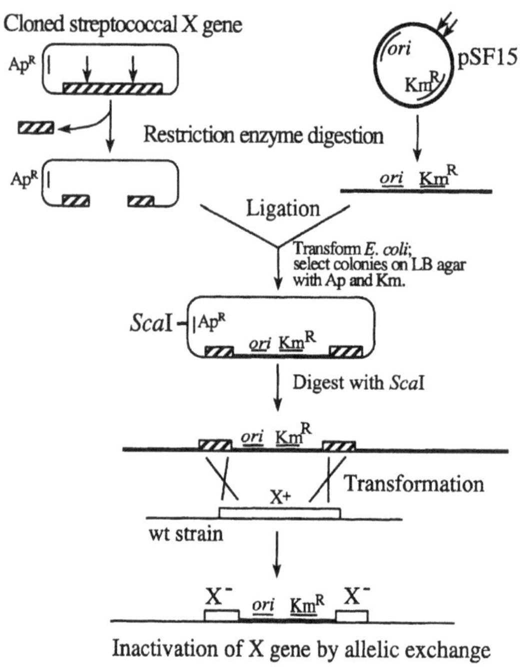

Figure 3. Gene inactivation by allelic exchange.

3. Confirm the plasmid insertion
   a. Southern hybridization
   Integration of a plasmid into a bacterial chromosome can shift the electrophoretic position of the target gene. This can be detected by Southern hybridization with the labeled integration vector and the cloned streptococcal gene [9, 20].
   b. Phenotypic analysis
   Inactivation of a functional gene may cause a physiological, biochemical and/or morphological change in the mutant. This may be detectable by phenotypic analysis. For example, the *S. mutans dexA* mutant generated by insertion of pVA891 showed an altered colonial morphology on sucrose agar and diminished dextranase activity [5].

E. Clone the inactivated gene (Figure 4)

1. Identify a pVA891 insertion mutant (Figure 2) that displays an altered expression of a specific gene (e.g., X gene).
2. Isolate the chromosomal DNA from this mutant.
3. Digest the DNA with *Hin*dIII that cuts a unique site on the left of the inserted plasmid.
4. Perform self-ligation of the DNA fragments with T4 ligase.
5. Transform the ligation mixture into *E. coli* JM109.
6. Select for the Cm resistant transformants, and isolate the plasmid with the Qiagen miniprep kit.
7. Transform the plasmid back to the wild-type streptococcal strain; select for a transformant that displays the wild-type phenotype in which the plasmid is inserted near the complete gene X.
8. Isolate chromosomal DNA from the transformant.
9. Digest the DNA with *Sph*I that cuts at a unique site on the right of the inserted plasmid.
10. Perform self-ligation with T4 ligase.
11. Transform the ligation mixture into *E. coli* JM109.
12. Select for Cm resistant transformants in which the X gene is cloned.

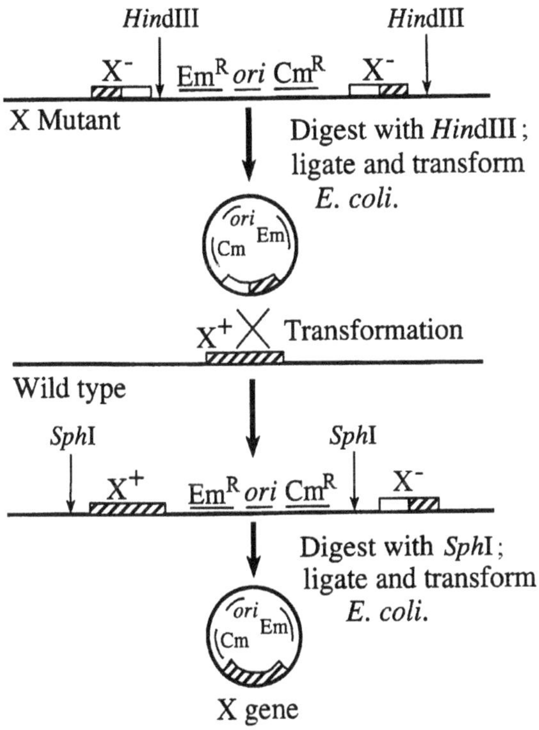

Figure 4. Cloning of the inactivated gene from streptococci.

4. Results and discussion

The integration plasmids pVA891, pSF143, pSF151, and pSF152 have been used successfully for gene inactivation and cloning in many streptococcal species, including *S. mutans*, *S. gordonii*, *S. pneumoniea* and *S. pyogenes* [1, 2, 5, 11, 18–21]. However, the newly constructed plasmid pEVP3 is specifically designed for use in *S. pneumoniae* [4], but this plasmid may also work in other streptococcal species because its chloramphenicol resistance gene (Cm^R) can express in the Gram-positive *S. pneumoniae*. Moreover, pEVP3 has a promoterless *lacZ* gene for selecting promoters or studying gene regulation in streptococci.

Although streptococcal integration vectors are commonly used for inactivating genes by a Campbell-like, single crossover recombination, a mutant produced in this manner may revert to the wild type due to segregation of the inserted plasmid. This has been found to occur at a low rate of 10^{-6} to 10^{-7} [7]. Therefore, this type of mutant needs to be maintained with antibiotics in the growth medium. However, a mutant produced by allelic exchange will not revert because a portion of the target gene has been deleted.

The streptococcal integration vectors have been used successfully to clone genes that do not express in *E. coli* or whose expressions cannot be readily detected in *E. coli*. These include *S. mutans rodD* [19] and *dexA* [5], and *S. pyogenes recA* [20]. Using the cloning strategy described above (Figure 4), a small gene may be readily cloned in one step. However, if a gene is large, one cloning step may not obtain the complete gene. A second round of chromosomal walking may be required with a different restriction enzyme, or the cloned gene fragment can be used as a probe to obtain the complete gene via chromosomal walking [20].

Confirmation of the integration of these vectors into the chromosome of a streptococcal strain can be accomplished by Southern-blot hybridization [9, 20]. Although these plasmids can be propagated successfully in *E. coli*, deletions or rearrangements may occur in plasmids that carry DNA fragments from streptococci, especially *S. pneumoniae* or *S. gordonii* [data not shown]. One way to overcome this problem is to directly transform the ligation mixture into competent streptococcal strains, skipping the propagation step in *E. coli*. In this case, more plasmid and chromosomal DNA will be required to prepare the ligation mixture.

In addition to gene inactivation and cloning of the inactivated gene, these vectors can also be used to clone genes flanking the integration site [17], to inactivate multiple genes in the same strain [17], to analyze linkages among several genetic loci [13–15], and to clone a large gene fragment into a heterologous streptococcal host [18]. Therefore, these integration vectors are useful tools for genetic studies in streptococci.

Notes on suppliers

1. Tomy Tech USA, Inc., Palo Alto, CA, USA
2. DuPont Company, Wilmington, DE, USA
3. Bio-Rad Laboratories, Hercules, CA, USA
4. New Brunswick Co., Inc., Edison, NJ, USA
5. Difco Laboratories, Detroit, MI, USA
6. Sigma Chemical Company, St. Louis, MO, USA
7. Fisher Scientific, Pittsburgh, PA, USA
8. Promega Corporation, Madison, WI, USA
9. Bio-101, Inc., Vista, CA, USA
10. Qiagen, Inc., Chatsworth, CA, USA
11. Life Technologies, Geithersburg, MD, USA
12. Francis L. Macrina, Department of Microbiology and Immunology, Box 980678, MCV, Virginia Commonwealth University, Richmond, VA 23298, USA
13. Donald A. Morrison, Laboratory for Molecular Biology, Department of Biological Sciences, University of Illinois, 900 South Ashland, Chicago, IL 60612, USA
14. Lin Tao, Department of Oral Biology, University of Illinois, College of Dentistry, 801 South Paulina Street, Chicago, IL 60612, USA

References

1. Barletta RG, Michalek SM, Curtiss III R (1988). Analysis of the virulence of *Streptococcus mutans* serotype c *gtfA* mutants in the rat model system. Infect Immun 56: 322–330.
2. Berry AM, Lock RA, Hansman D, Paton JC (1989). Contribution of autolysin to virulence of *Streptococcus pneumoniae*. Infect Immun 57: 2324–2330.
3. Chung CT, Niemela SL, Miller RH (1989). One-step preparation of competent *Escherichia coli*: Transformation and storage of bacterial cells in the same solution. Proc Natl Acad Sci USA 86: 2172–2175.
4. Clavery JP, Dintilhac A, Pestova EV, Martin B, Morrison DA (1995). Construction and evaluation of new drug-resistance cassettes for gene disruption and mutagenesis in *Streptococcus pneumoniae*, using an *ami* test platform. Gene 164: 123–128.
5. Colby SM, Whiting GC, Tao L, Russell RRB (1995). Insertional inactivation of the *Streptococcus mutans dexA* (dextranase) gene results in altered adherence and dextran catabolism. Microbiology 141: 2929–2936.
6. Dunny GM, Lee LN, LeBlanc DJ (1989). Improved electroporation and cloning vector system for Gram-positive bacteria. Appl Environ Microbiol 57: 1194–1201.
7. Leenhouts KJ, Kok J, Venema G (1991). Replacement recombination in *Lactococcus lactis*. J Bacteriol 173: 4794–4798.
8. Macrina FL, Evans RP, Tobian JA, et al. (1983). Novel shuttle plasmid vehicles for *Escherichia-Streptococcus* transgeneric cloning. Gene 25: 145–150.
9. Maniatis T, Fritsch EF, Sambrook J (1982). Molecular cloning: A laboratory manual. Cold Spring Harbor, NY: Cold Spring Harbor Laboratory.
10. Mejean V, Claverys JP, Vasseghi H, Sicard AM (1981). Rapid cloning of specific DNA fragments of *Streptococcus pneumoniae* by vector integration into the chromosome followed by endonucleolytic excision. Gene 15: 289–293.
11. Morrison DA, Trombe MC, Hayden MK, Waszak GA, Chen JD (1984). Isolation of transformation-deficient *Streptococcus pneumoniae* mutants defective in control of competence, using insertion-duplication mutagenesis with the erythromycin resistance determinant of pAMβ1. J Bacteriol 159: 870–876.
12. Munro C, Michalek SM, Macrina FL (1991). Cariogenicity of Streptococcus mutans V403 glucosyltransferase and fructosyltransferase mutants constructed by allelic exchange. Infect Immun 59: 2316–2323.
13. Perry D, Kuramitsu HK (1989). Genetic linkage among cloned genes of *Streptococcus mutans*. Infect Immun 57: 805–809.
14. Perry D, Kuramitsu HK (1990). Linkage of sucrose-metabolizing genes in *Streptococcus mutans*. Infect Immun 58: 3462–3464.
15. Perry D, Nilsen LJ, Kuramitsu HK (1985). Mapping of a cloned glucosyltransferase gene in *Streptococcus mutans*. Infect Immun 50: 130–135.
16. Simon D, Ferretti JJ (1991). Electrotransformation of *Streptococcus progenes* with plasmid and linear DNA. FEMS Microbiol Lett 82: 219–224.
17. Tao L, LeBlanc DJ, Ferretti JJ (1992). Novel streptococcal-integration shuttle vectors for gene cloning and inactivation. Gene 120: 105–110.
18. Tao L, Sutcliffe IC, Russell RRB, Ferretti JJ (1993). Cloning and expression of the multiple sugar metabolism (*msm*) operon of *Streptococcus mutans* in heterologous streptococcal hosts. Infect Immun 61: 1121–1125.
19. Tao L, Tanzer JM, Kuramitsu HK, Das A (1993). Identification of several *rod* loci and cloning of the *rodD* locus of *Streptococcus mutans*. Gene 126: 123–128.
20. Tao L, Hollingshead SK, Suvorov AN, Ferretti JJ, McShan WM (1995). Construction of a *Streptococcus pyogenes recA* mutant via insertional inactivation, and cloning and sequencing of the complete *recA* gene. Gene 162: 59–62.
21. Vijayakumar MN, Priebe SD, Pozzi G, Hageman JM, Guild WR (1986). Cloning and physical characterization of chromosomal conjugative elements in streptococci. J Bacteriol 166: 972–977.

Address for correspondence: Dr L. Tao, Department of Oral Biology, College of Dentistry, University of Illinois, 801 South Paulina Street, Chicago, IL 60612, USA
Phone: 312-996-7732; Fax: 312-996-6044

Methods in Cell Science **20**: 65–70 (1998)

Induction of transformation in streptococci by synthetic competence stimulating peptides

Peter Gaustad[1] & Donald A. Morrison[2]
[1] *Microbiological Institute, National Hospital, Oslo, Norway;* [2] *Laboratory for Molecular Biology, Department of Biological Sciences, University of Illinois at Chicago, Chicago, Illinois, USA*

Abstract. Competence in genetic transformation is well known in oral streptococci as well as in the pathogenic pneumococcus. But the application of this genetic phenomenon as a tool to study the biology and the virulence of these organisms has been restricted to a limited number of strains, both by the stringency of the conditions required for competence development, and by the fact that wild types or encapsulated pneumococci transform poorly, or not at all. Many of these limitations have been alleviated by the use of synthetic competence stimulating peptides. These peptides are small, stable and inexpensive molecules inducing competence under a wide variety of conditions in strains belonging to the same pherotype. Their use circumvents many of the limitations to the expression of transformability in oral streptococci and the pneumococcus, and expands the opportunities for application of tools of molecular genetics to many strains of the species without prior genetic manipulation.

Key words: Competence, Competence stimulating peptide, Oral streptococci, Pheromone, Pneumococcus, Transformation

1. Introduction

Natural genetic transformation of bacteria depends on a physiological state, competence, permitting the binding and active uptake into the bacteria of extracellular, free DNA and its incorporation in the genome. This process can result in new heritable traits. Natural competence in streptococci arises at a specific stage of growth, typically at early or mid-log growth phases and persists for a limited period of time [22, 30]. Sterile culture filtrates from competent streptococci contain an extracellular competence factor [9, 22, 31]. The competence factors exhibit strain specificity [9, 24]. The isolation and the characterization of the competence factor were not successfully accomplished for many years because of its lability. A competence factor was isolated from *Streptococcus pneumoniae* for the first time in 1995, and characterized as a 17-amino-acid, unmodified peptide [16], and was designated competence stimulating peptide (CSP). The *S. pneumoniae* CSP is synthesised in vivo as a precursor peptide of 41 amino acids, including the 24 N-terminal amino acids as a double-glycine type leader peptide. CSP induces competence when its concentration in the cultivation medium reaches a critical level [22, 30, 31], at which the signal is perceived by a membrane bound CSP receptor [18].

In this article, we provide some useful information on the experimental use of synthetic CSPs.

2. Materials

Experimental conditions of transformation

Competence appears at a specific cell density in cultures of streptococci growing in appropriate media. This cell density is reproducible from experiment to experiment but varies with the medium composition as well as the bacterial strain used. Culture pH is also an important variable in pneumococci [6]. Addition of serum to the medium enhances competence and transformation in pneumococci and *Streptococcus sanguis (gordonii)* [14, 22].

The use of synthetic CSP circumvents most of the difficulties of finding a suitable transformation medium. When using synthetic CSP, the cells can be induced to competence independently of cell density, although the quality of competence declines as cells approach stationary phase.

In most oral streptococci Todd Hewitt broth (Oxoid[5]) with 2.5–5% horse serum (THS) adjusted to pH 7.4 with NaOH, can be used in transformation experiments [9].

In pneumococcal transformation more specific media have been developed, such as the complete transformation medium (CTM) described below [20]. The medium is composed of N-Z-Amine A (ICN[2]) (10 g/l), NaCl (5 g/l), Bacto-Tryptone (Difco[1]) (5 g/l), Bacto-Yeast Extract (Difco[1]) (1 g/l), glucose (2 g/l), K_2HPO_4 (1/60 M/l), bovine serum albumin (2 g/l), and $CaCl_2$ (1 mM/l) [26]. Frozen cell stocks are prepared from cultures grown in this medium to

OD_{550} = 0.2, and stored at –80 °C after adjusting to 10% glycerol.

Competent cells
Many strains, wild as well as collection strains, of the species *Streptococcus intermedius, Streptococcus constellatus, Streptococcus anginosus, Streptococcus mitis, Streptococcus oralis, Streptococcus crista, S. sanguis* and *S. gordonii* become competent spontaneously and produce CSP in quantities sufficient to transform without adding CSP from other sources. Table 1 shows collection strains in which natural competence has been demonstrated and the sequence of a CSP has been found. Other strains belonging to these species have to be induced to competence by adding CSP, either from culture filtrate or as chemically synthesized material [9, 19].

Among the pneumococci rough strains have traditionally been use in transformation experiments, but with the availability of synthetic CSP, transformation also has been achieved in many capsulated strains [28]. Various laboratories have reported different critical cell densities (2×10^6 to 10^8 cells/ml) for the appearance of competence in pneumococci [see reference 4] and other oral streptococci. Competence occurs usually during exponential growth and the optimal cell density has to be determined for each strain and medium.

For oral streptococci competent cells can be prepared by a 10-fold subculture in prewarmed THS of an 18 hour THS culture (OD_{625} = 0.8). Incubation is continued for 150 to 210 min until OD_{625} reaches 0.8–0.9. The competent cells can be prepared in this way for each experiment or by using frozen cells (frozen during their competence peak and stored at –80 °C in 10% glycerol until use).

Table 1. Collections strains of the mitis and anginosus group naturally competent and reference for sequence of the competence stimulating peptide

Species	Collection strain	Reference for CSP
S. pneumoniae	Rx	16
S. mitis	NCTC[a] 12661	19
S. oralis	NCTC 11427	19
S. gordonii	NCTC 7865	18
	NCTC 3165	19
	NCTC 7868	18
S. sanguis	NCTC 7863	13
S. crista	NCTC 12479	18
S. anginosus	NCTC 10713	19
S. intermedius	NCDO[b] 2227	19
S. constellatus	NCTC 11325	19
S. milleri	NCTC 10708	19

[a] NCTC, National Collection of Type Culture, 61 Colindal Avenue, London NW9 5HT, England.
[b] NCDO, National Collection of Food Bacteria, Agricultural and Food Research Counsil, Institute of Food Research, Norwich, United Kingdom.

DNA
Both chromosomal and plasmid DNA can be used in transformation experiments. Most commonly, antibiotic resistance is chosen as a marker that can be selected readily in rich media (examples include streptomycin [9] or novobiocin [20] resistance). Streptococci are transformable with many different plasmids [1, 23], but chromosomal DNA, especially chromosomal DNA from the same species, gives a higher transformation yield than plasmids.

For use of chromosomal DNA, streptomycin resistant mutants from oral streptococci can be obtained by plating a sensitive strain on blood agar plates (5% blood) containing 500 mg/l streptomycin. When direct plating does not yield mutants, it may be helpful to seed onto blood agar plates and allow 3–6 hours growth before transferring the agar to agar bottoms (4 ml Brain Heart Infusion agar with, Difco,[1] 600 µg/ml streptomycin) giving a final concentration of 100 mg/l streptomycin. DNA has to be present when the cells are induced to competence, but as streptococci usually do not have any DNase, DNA can be added before the cells reach the critical cell density. A concentration of DNA during the experiments of 1 µg/ml is usually above saturation.

3. Procedures

Naturally transformable streptococci

Occurrence of natural competence in vitro was first described in *S. pneumoniae* by Dawson [8]. Later the phenomenon was described in different species of oral streptococci, such as *S. sanguis, S. gordonii, Streptococcus mutans*, and *Streptococcus milleri* group [6, 9, 21, 25]. Natural competence in streptococci arises at a specific stage of growth in culture, but does not always develop spontaneously in all strains of a competent species. By the use of culture filtrates from competent cells, competence can be induced even outside the period of spontaneous competence. Strains induced to competence by the same competence factor belong to the same transformation group or same pherotype [11, 12, 19]. In naturally competent streptococci the competence regulation operon consists of three genes, *com*C encoding the CSP precursor, *com*D a histidine kinase and *com*E which encodes its cognate response regulator [5, 16, 17, 27, 28]. CSP is secreted and processed by a secretion apparatus consisting of ComA and ComB [15, 17, 33]. By screening strains with PCR, using primers complementary to highly conserved genes flanking the *com*C gene [18, 27], it has been possible to characterize the CSP from oral streptococci belonging to the anginosus- and the mitis-groups. The *com*C gene has been found in all species of the two groups with the exception of one (*Streptococcus parasanguis*), and the strains have been naturally transformable [19].

Table 2 shows members of streptococcal groups for which evidence for natural transformation has been demonstrated and the presence of the *com*C gene has been established. When *com*C gene is present CSP can be synthesized for use in transformation experiments.

Induction of competence

Competence can be induced either by the use of sterile culture filtrates or by the use of synthetic CSP. The use of sterile culture filtrates requires knowledge of the peak of competence in the donor as well as the recipient strain and the use of an appropriate transformation medium. In *S. sanguis* and *S. gordonii*, both the donor strain of CSP as well as the induced strain may produce a competence factor inactivator (CFI) during exponential growth. The CFI's destroy the activity of CSP, but may be eliminated by heating of the culture filtrate [10].

In pneumococcus, competence terminates after a short time, independent of such external factors [7].

Competence induction by synthetic CSP in oral streptococci

The majority of laboratory strains as well as wild strains of the mitis and milleri groups are transformable. They release CSP during exponential growth. In addition, competence can be induced by synthetic CSP of the same transformation group.

Table 2. Naturally competent streptococci and the presence of the *com*C gene[a]

Phylogenetic group	Species	Natural competence	*com*C gene
Mitis group:	*S. pneumoniae*	+	+
	S. mitis	+	+
	S. oralis	+	+
	S. gordonii	+	+
	S. sanguis	+	+
	S. crista	+	+
Mutans group:	*S. mutans*[b]	+	–
Anginosus group:	*S. anginosus*	+	+
	S. intermedius	+	+
	S. constellatus	+	+

[a] In the salivarius group, pyogenic group and bovis group neither natural transformation nor the *com*C gene has been found [19].

[b] In *S. mutans* natural transformation occurs in the serogroups c,e and f [32].

Competence can be achieved by growing the strains in THS until OD_{625} (about 3×10^6 cells/ml) (see above, competent cells). The peak of competence for many strains is reached after 90 to 120 min of incubation, but can occur later.

Before the competence peak, after 60 min growth CSP (200 ng/ml) (in excess to avoid elimination by CFI) and *str-R* DNA (20 μg/ml) are added. Treatment with DNase (10 μg/ml) when the DNA uptake has taken place is optional.

Further incubation allows for integration and expression of new alleles, and their selection, for example by the use of blood agar plates with 100 mg/l streptomycin.

When selection for transformants by this procedure does not succeed, the transferring of seeded blood agar plates (2–4 h growth) onto agarbottoms containing streptomycin (final concentration 100 mg/l) can be used to ensure the phenotypic expression of streptomycin resistance [2].

Competence induction by synthetic CSP in S. pneumoniae

Strains descended from the pneumococal strain Rx [29] both respond readily to CSP stimulation and release large amounts of the peptide into culture fluids during competence. A simple procedure for transformation of these strains uses frozen cell stocks prepared from cultures grown to OD_{550} = 0.2, adjusted to 10% glycerol, and stored at –80 °C. For competence induction with CSP, stock cells are diluted 100-fold in the transformation medium (CTM) adjusted to a pH below 7 by addition of HCL to 5 mM. Growth proceeds with a doubling time of approximately 40 min at 37 °C. When the culture reaches approximately OD_{550} = 0.06, peptide is added to induce competence. The culture is chilled to 0 °C and 100 ng/ml CSP is added. Portions of the cold culture are distributed into tubes containing donor DNA, and warmed to 37 °C for induction (20 min), transformation (5 min), and expression of new alleles (30–90 min, depending on the marker). Plating on suitable selective media completes the process.

While cells also respond well to the stimulating peptide at lower densities, this density was chosen for routine work as the highest density at which maximal competence is induced in 100% of the cells. The chilling step is not required for competence induction, but it is included for convenience when multiple transformation assay reactions are desired. Less CSP is usually sufficient, but an excess does not reduce competence.

. **How to obtain CSP**

CSP from culture filtrate

Isolation of useful amounts of CSP from cultures of competent cells has not been possible, but sterile culture filtrates containing CSP or synthetic CSP can be used. In oral streptococci culture filtrates can be prepared by filter sterilization of the THS during the growth of competent cells at the time of peak competence. The filtrate obtained can be kept frozen at –20 °C until use, after heating to 95 °C for 20 min to destroy the CFI [10]. In transformation experiments, 1 to 1 dilution of filtrate and THS with cells grown to competence gives a sufficient concentration of CSP to induce competence. Pneumococcal CSP can be prepared in the same way, but with less or no heating; typical preparations are active in dilutions as high as 1:100.

Known synthetic CPSs

In the species of the mitis and milleri groups of streptococci several different CSP's and their amino-acid sequences have been described, and these can also be synthesized for use in transformation experiments. So far, the amino acid sequences of CSP from the following species have been published: two from *S. pneumoniae* [16, 28], three from *S. gordonii* [18, 19], one from *S. sanguis* [13], five from *S. mitis*, one from *S. oralis*, one from *S. crista* [19], and two from the milleri group [18]. The milleri group has one CSP common to the three known species, *S. intermedius, S. constellatus* and *S. anginosus*. In addition, another CSP was described in a strain called *S. milleri* (NCTC[4] 10708) [18]. In the species of the mitis group, the CSP's are highly specific as they only induce competence in strains of their own pherotype [11, 12, 19]. All bacteria induced to competence by a particular CSP belong to the same transformation group or pherotype. Each CSP, with its unique primary structure is recognised by the signalling domain of the downstream histidine kinase, ComD.

Looking for new CSP's

It has been shown that the *com*CDE operon encodes competence regulatory elements in streptococci of the mitis and anginosus groups [5, 16, 18, 27, 28]. The *com*CDE sequences are so divergent in the different pherotypes that it is not possible to construct common PCR primers within these genes. However, the use of primers complementary to the more highly conserved Arg- and Glu-tRNA genes flanking the *com*CDE operon [18, 27] can amplify PCR fragments of approx. 2.6 kb containing the whole operon. By sequencing about 350 bp from the 5′-ends of the 2.6kb PCR fragments obtained, the complete sequence of the *com*C can be determined and the CSP can be synthesised. Synthetic CSP can be stored at –20 °C when reconstituted in aqueous solution. The operon has been demonstrated in about half of the strains tested to date [19]. Since the amplification of the *com*CDE operon depends on the flanking tRNA genes, its detection is not possible among strains in which the competence regulation locus resides elswhere on the chromosome.

4. Discussion

Enhancement of transformation in oral streptococci and pneumococcus

Among the oral streptococci, the practical significance of synthetic CSPs is by the fact that they increase the number of strains for which transformation can be readily accomplished, and they decrease the effort required for optimization of culture conditions. Among oral streptococci, even species such as *S. sanguis* and *S. gordonii* where endogenous induction of competence is widespread, the use of synthetic CSPs will facilitate studies on the molecular biology of streptococci. The practical significance is also quite clear in the case of the important pathogen, *S. pneumoniae*. Despite the long record of transformation in this species, the number of strains reported to exhibit a high level of endogenously induced competence in vitro is quite small, and even among those strains, the presence of a normal amount of capsular polysaccharide appears to interfere severely with such competence. Recent experience has been that exogenously supplied synthetic CSP increases greatly the proportion of strains that can be transformed, decreases the need to tailor transformation media to individual strains, and reduces the need to work with rough (uncapsulated) derivatives for genetic manipulations [28].

Studies of the molecular biology of the streptococci

The application of genetic transformation as a tool to study of the biology and virulence of competent streptococci has been restricted to a few strains by the fact that wild type strains in some species transform poorly or not at all, and by the stringency of the conditions required for development of competence for transformation. Many of these limitations have been alleviated by the ability to synthesise the CSPs which have proven to be small, stable and inexpensive peptides that induce competence under a wide variety of conditions. The use of synthetic CSPs circumvents most of the limitations to the expression of transformability in oral streptococci, including pneumococci, and therefore expands the opportunities for application of tools of molecular genetics.

Speciation and typing of streptococci

In diagnostic microbiology the differentiation of species of oral streptococci is difficult. The use of synthetic CSP specific for species or pherotypes may contribute to a more exact speciation than obtained today with biochemical and physiological tests. Thus, synthetic CSP may also be a valuable tool in epidemiological studies.

Notes on suppliers

1. Difco, PO Box 331058, Detroit, MI 48232-7058, USA
2. ICN, ICN Biomedicals, 3300 Hyland Ave., Costa Mesa. CA 92626, USA
3. NCDO, National Collection of Food Bacteria, Agricultural and Food Research Council Institute of Food Research, Norwich, England
4. National Collection of Type Cultures, 61 Colindal Avenue, London NW9 5HT, England
5. Oxoid, Wade Road, Basingstoke, Hampshire, RG24 OPW, England
6. Sigma Chemical Co., P.O. box 14508, St. Louis, Missouri 63178-9916, USA

References

1. Alloing G, Granadel C, Morrison DA, Claverys J-P (1996). Competence pheromone, oligopeptide permease, and induction of competence in *Streptococcus pneumoniae*. Mol Microbiol 21: 471–478.
2. Bøvre K, Frøholm LO (1972). Competence in genetic transformation related to colony type and fimbriation in three species of *Moraxella*. Acta Pathol Microbiol Immunol Scand Sect B 80: 649–659.
3. Caufield PW, Shah G (1995). Transformation of naturally competent *Streptococcus mutans* with replicative and non-replicative Tn*916*-containing plasmids: implications for a mechanism of transposition. In: Ferretti JJ, Gilmore MS, Klaenhammer TR, Brown F (eds), Genetics of streptococci, enterococci and lactococci, vol 85: Dev Biol Stand, pp 19–25. Basel: Karger .
4. Chen J-D, Morrison DA (1987). Modulation of competence for genetic transformation in *Streptococcus pneumoniae*. J Gen Microbiol 133: 1959–1967.
5. Cheng Q, Campell EA, Naughton AM, Johnson S, Masure HR (1997). The com locus controls genetic transformation in *Streptococcus pneumoniae*. Mol Microbiol 23: 683–692.
6. Colman G (1969). Transformation of viridans-like streptococci. J Gen Microbiol 28: 275–286.
7. Coomaraswamy, G (1996). Induction of genetic transformation in *Streptococcus pneumoniae* by a pheromone peptide and its synthetic analogs. PhD Thesis, University of Illinois, Chicago, IL, USA.
8. Dawson, MH, Sia RHP (1931). In vitro transformation of pneumococcal types. I. A technique for inducing transformation of pneumococcal types in vitro. J Exp Med 54: 681–699.
9. Gaustad P (1979). Genetic transformation in *Streptococcus sanguis*. Distribution of competence and competence factor in a collection of strains. Acta Pathol Microbiol Immunol Scand Sect B 87: 123–128.
10. Gaustad P (1981). Genetic transformation in *Streptococcus sanguis*. Competence factor and competence factor inactivator. Acta Pathol Microbiol Immunol Scand Sect B 89: 67–73.
11. Gaustad P (1985). Genetic transformation in *Streptococcus sanguis*. Effects on genetic transformation by culture filtrates of *S. sanguis* (serogroups H and W) and *Streptococcus mitis (mitior)* with reference to identification. Acta Pathol Microbiol Immunol Scand Sect B 93: 283–287.
12. Gaustad P (1993). Genetic transformation and competence factors in the identification of *Streptococcus sanguis*. In: Bella E, Berencsi G, Szentirmai A (eds), DNA transfer and gene expression in microorganisms – proceedings of the 11th European Meeting on Genetic Transformation, pp 51–54. UK: Intercept, Hants.
13. Gaustad P, Håvarstein LS (1997). Competence-pheromone in *Streptococcus sanguis*. Identification of the competence gene *com*C and the competence pheromone. In: Horaud T, Bouvet A, Leclercq R, de Montclos H, Sicard M (eds), Advances in experimental medicine and biology, vol. 418: Streptococci and the Host, pp 1019–1021. New York and London: Plenum Press.
14. Hotchiss RD, Ephrussi-Taylor H (1951). Fed Pro 10: 200.
15. Hui FM, Morrison DA (1991). Genetic transformation in *Streptococcus pneumoniae*: nucleotide sequence analysis shows *com*A, a gene required for competence induction, to be a member of the bacterial ATP-dependant transport protein family. J Bacteriol 173: 372–381.
16. Håvarstein LS, Coomaraswamy G, Morrison DA (1995a). An unmodified heptadecapeptide pheromone induces competence for genetic transformation in *Streptococcus pneumoniae*. Proc Natl Acad Sci USA 92: 11140–11144.
17. Håvarstein LS, Diep DB, Nes IF (1995b). A family of bacteriocin ABC transporters carry out proteolytic processing of their substrates concomitant with export. Mol Microbiol 16: 229–240.
18. Håvarstein LS, Gaustad P, Nes IF, Morrison DA (1996). Identification of the streptococcal competence-pheromone receptor. Mol Microbiol 21: 863–869.
19. Håvarstein LS, Hakenbeck R, Gaustad P (1997). Natural competence in the genus *Streptococcus*: Evidence that streptococci can change pherotype by interspecies recombinational exchanges. J Bacteriol 179: 6589–6594.
20. Morrison DA, Trombe M, Hayden M, Waszak G, Chen J-D (1984). Isolation of transformation-deficient *Streptococcus pneumoniae* mutants defective in control of competence, using insertion-duplication mutagenesis with the erythromycin resistance determinant of pAMβ1. J Bacteriol 159: 870–876.
21. Pakula R, Hulanicka E, Walcsac (1959). Transformation reactions between different species of streptococci, pneumococci and staphylococci. Schweiz Z allgem Pathol u Bakteriol 22: 202–214.
22. Pakula R, Walczak W (1963). On the nature of

transformable streptococci. J Gen Microbiol 31: 125–133.

23. Pearce BJ, Naughton AM, Masure HR (1994). Peptide permeases modulate transformation in *Streptococcus pneumoniae*. Mol Microbiol 12: 881–892.
24. Perry D, Slade HD (1966). Effects of filtrates from transformable and nontransformable streptococci on the transformation of streptococci. J Bacteriol 91: 2216–2222.
25. Perry D, Kuramitsu HK (1981). Genetic transformation of *Streptococcus mutans*. Infect Immun 31: 1295–1297.
26. Peterson JM, Guild WR (1968). Fractionated strands of bacterial DNA. III. Transformation efficiencies and rates of phenotypic expression. J Bacteriol 96: 1991–1999.
27. Pestova EV, Håvarstein LS, Morrison DA (1996). Regulation of transformability in *Streptococcus pneumoniae* by an auto-induced peptide pheromone and two-component regulatory system. Mol Microbiol 21: 853–862.
28. Pozzi G, Masala L, Iannelli F, Manganelli R, Håvarstein LS, Piccoli L, Simon D, Morrison DA (1996). Competence for genetic transformation in encapulated strains of *Streptococcus pneumoniae*. Two allelic variants of the peptide pheromone. J Bacteriol 178: 6087–6090.
29. Ravin AW (1959). Reciprocal capsular transformations of pneumococci. J Bacteriol 77: 296–309.
30. Tomasz A, Hotchkiss PD (1964). Regulation of the transformability of pneumococcal cultures by macromolecular products. Proc Natl Acad Sci USA 51: 480–487.
31. Tomasz A (1965). Control of the competent state in pneumococcus by a hormone-like product: an example of a new type of regulatory mechanism in bacteria. Nature 208: 155–159.
32. Westergren G, Emilson CG (1983). Prevalence of transformable *Streptococcus mutans* in human dental plaque. Infect Immun 41: 1386–1388.
33. Zhou L, Hui FM, Morrison DA (1995). Competence for genetic transformation in *Streptococcus pneumoniae*: organization of a regulatory locus with homology to two lactococcin A secretion genes. Gene 153: 25–31.

Address for correspondence: Peter Gaustad, Microbiological Institute, National Hospital, 0027 Oslo, Norway
Phone: 47.22869510; Fax: 47.22869490
E-mail: peter.gaustad@labmed.uio.no

Methods in Cell Science **20**: 71–78 (1998)

Characterization of the lactococcal conjugative element pRS01 using IS*946*-mediated mutagenesis

David A. Mills,[1] Trevor G. Phister,[1] Kathleen A. Baldwin,[2] Gary M. Dunny[1] & Larry L. McKay[2]

[1] *Department of Microbiology, University of Minnesota, Minneapolis;* [2] *Department of Food Science and Nutrition, University of Minnesota, St Paul, Minnesota, USA*

Abstract. Conjugation is a common mode of genetic transfer among the lactic acid bacteria. In an effort to exploit conjugation as a means of lactococcal strain development, we have characterized the transfer regions of pRS01 by insertional mutagenesis via IS*946*-mediated cointegration with the lactococcal-*Escherichia coli* shuttle plasmid pTRK28. In this work we will discuss our use of pTRK28-mediated cointegration to identify, clone, and characterize the pRS01 transfer regions including the conjugative origin of transfer.

Key words: Conjugation, *oriT*, Mobilization, Lactic acid bacteria, Lactococci

Abbreviations: Cm^r = resistant to chloramphenicol; Erm^r = resistant to erythromycin; IS = insertion sequence; LAB = lactic acid bacteria; *oriT* = origin of transfer; PBS = phosphate-buffered saline; Rec^+ = recombination proficient; Rif^r = resistant to rifampicin; Spc^r = resistant to spectinomycin; Str^r = resistant to streptomycin

1. Introduction

The use of lactic acid bacteria (LAB) in the production of foods predates recorded history. The LAB which are used as starter cultures in the majority of meat, vegetable, cereal and milk fermentations belong to five genera: *Lactococcus, Leuconostoc, Lactobacillus, Pediococcus* and *Streptococcus*. Due to the economic significance of these fermented foods and the desire for their consistent quality, extensive research has focused on bacterial traits that are essential for production of fermented foods. Recent advances in the understanding of starter culture genetics and physiology have led to the desire to engineer specific starter cultures with a multitude of metabolic capabilities to ensure a consistent fermentation. This desire presents a problem, however, because there is some ambiguity in the federal guidelines regarding the use of genetically engineered starter cultures for food production. In addition, negative public perception of food products generated by use of 'genetically engineered' bacteria creates another challenge for the food manufacturer. These factors have resulted in a focus on starter culture development through 'natural' genetic processes, thereby reducing the use of recombinant DNA methodology and eliminating non-starter culture DNA as a means of strain development. Since bacterial conjugation is a natural method of gene transfer, we have sought to develop conjugation as a food grade means of delivering specific genetic traits among lactic acid bacterial starter cultures. Within the genus *Lactococcus*, conjugation has been exploited for strain development [16, 17], but in those cases the specific genetic trait involved was fortuitously positioned in association with a conjugative element, thereby allowing mobilization into other lactococci. Currently a means to deliver specifically engineered genes remains to be developed. Several groups have developed mobilization systems in lactococci [6, 14], however, these methods are not commercially viable due to the presence of non-lactococcal DNA as a key component of the delivery system. Given the economic importance of fermented foods, the creation of a food grade mechanism to conjugally deliver specific genes to various lactic acid bacterial starter cultures is a necessity for future strain improvement regimes.

In an effort to exploit conjugation as a means of lactococcal strain development, we have focused on the conjugative element pRS01 from *Lactococcus lactis* subsp. *lactis* ML3. pRS01 has been shown to form large cointegrate plasmids in transconjugants, exhibit high frequency transfer, and confer a cell aggregation (Clu) phenotype [20]. Anderson and McKay [1] carried out preliminary mapping of the regions of pRS01 responsible for the Clu and Tra phenotypes. In this communication, we will discuss the identification and characterization of the pRS01 transfer regions by insertional mutagenesis using IS*946*-mediated cointegration with the 11-kb plasmid pTRK28 [12]. In addition, we will discuss the broader implications for pTRK28 cointegration-based mutagenesis as a tool for analysis of conjugative elements in LAB.

2. Materials

A. Bacterial strains and media. *Escherichia coli* strains were grown in Luria broth (LB) medium (No. 0414-07-3[1]) [15] with agitation. Selective media for *E. coli* strains contained tetracycline (No. T-3383[2]) at 15 µg/ml or chloramphenicol (No. C-0378[2]) at 30 µg/ml. *L. lactis* strains were grown in GM17 (M17 medium [No. 1856-17-4[1]] [19] containing 0.5% glucose [No. G-6152[2]], at 30 °C without agitation. *L. lactis* NCK168 (pTRK28, pRS01) is resistant to erythromycin due to a resistance gene within pTRK28. *L. lactis* MMS370 is resistant to streptomycin, and *L. lactis* LM2345 is resistant to spectinomycin and rifampicin. Selective media for *L. lactis* strains contained antibiotics at the following concentrations: erythromycin (No. E-6376[2]), 10 µg/ml; chloramphenicol, 5 µg/ml; rifampicin (No. R-3501[2]), 50 µg/ml; streptomycin (No. S-6501[2]), 600 µg/ml; spectinomycin (No. S-9007[2]), 300 µg/ml. All plating media contained 1.5% Bacto-agar (No. 0140-02-9[1]). Milk-agar plates contained 5% skim milk powder (No. 902887[3]), 1% glucose (No. G-6152[2]), and 1.5% Bacto-agar.

B. Reagents for DNA manipulation. Restriction endonucleases, T4 DNA ligase (No. 15224-017[4]), the Klenow fragment of DNA polymerase 1 (No. 18012-021[4]), and calf intestinal phosphatase (No. 18009-0127[4]) were purchased from Gibco-BRL Life Technologies Inc., Gaithersburg, MD, and used as described by the manufacturers.

3. Procedures

A. Cloning and DNA manipulation. Plasmid isolation from *L. lactis* was performed as described by O'Sullivan et al. [9]. Rapid plasmid isolation from *E. coli* was accomplished using the alkaline lysis method [2]. Large scale plasmid isolation from *E. coli* was performed according to the PEG precipitation procedure [2]. DNA fragments were isolated from agarose gels using the GeneClean II kit (No. 1001-400[5]) as indicated by the manufacturer.

B. Transformation. Electroporation of *E. coli* or *L. lactis* was performed using a Bio-Rad Gene Pulser apparatus (No. 165-2105[6]). *L. lactis* cells grown in GM17 to an optical density of 0.5 (λ = 600 nm) are harvested at 4000 × *g* and washed twice in an equal volume of sterile cold distilled water. Cells are then suspended in $^{1}/_{50}$th the original volume, purified plasmid is added (a total of 100 ng to 500 ng DNA), and the cell/DNA solution mixed gently and transferred to a cold 0.1-cm electroporation cuvette (No. 165-2099[6]). The electroporation is conducted at a voltage of 1.7 kV/cm and a capacitance setting of 100 ohms, and the cells are immediately transferred to a tube containing 0.5 ml of M17 broth containing 0.5 M sucrose and incubated for 2 to 3 hours at 30 °C. The cells are then plated on GM17 plates containing the appropriate antibiotic and incubated overnight at 30 °C. Typically, transformants arise within 24 to 48 hours. *E. coli* transformation was performed as described by Dower et al. [4] in a 0.1-cm cuvette at a voltage of 1.7 kV/cm and a capacitance setting of 100 ohms.

C. Swab matings. To rapidly screen for transfer phenotype, an overnight culture of *L. lactis* MMS370 strains containing a pTRK28::pRS01 cointegrate plasmid (Erm^r) is streaked onto GM17 medium using sterile cotton swabs and cross-streaked with an overnight culture of the recipient strain *L. lactis* LM2345 (Spc^r, Rif^r). Following overnight growth at 30 °C, the streak intersection growth is collected using a sterile cotton swab and swabbed onto the surface of a selective GM17 agar plate containing erythromycin, spectinomycin and rifampicin. As controls, isolated (initial) portions of the donor and recipient streaks are also swabbed onto selective GM17 agar. Strains possessing high frequency transfer (Tra^{high}) phenotypes typically produced a confluent lawn on selection plates, while stains defective in transfer ability produced few or no colonies. With the pRS01 element, we have found that swab matings are a rough indicator of transfer frequency, thereby allowing us to gauge potential frequencies and reduce the number of plates needed to determine transfer frequencies in subsequent plate matings. This was helpful, as the potential transfer frequency for pTRK28::pRS01 cointegrates can range from 1 to $<1 \times 10^{-9}$ transconjugants per input donor cell. In general, donor strains which produced a confluent lawn of growth on selection plates commonly possessed transfer frequencies of 1 to 1×10^{-4} transconjugants per input donor cell. In contrast, donor strains which produced isolated colonies or no growth at all on the swab transconjugant plates typically possessed transfer frequencies from 1×10^{-5} to 1×10^{-8} or $<1 \times 10^{-9}$ transconjugants per input donor cell, respectively. For a visual depiction of a lactococcal swab mating we refer the reader to the review by Steele and McKay [18].

D. General plate mating procedure

1. Inoculate 10 ml of GM17 containing the appropriate antibiotic (erythromycin for *L. lactis* MMS30 [pTRK28::pRS01]) with a 2%

inoculation from an overnight culture of donor cells. Inoculate 10 ml of GM17 containing the appropriate antibiotic (spectinomycin and rifampicin for *L. lactis* LM2345) with a 2% inoculation from an overnight culture of recipient cells.

2. Incubate cultures for 6 hours at 30 °C without shaking.
3. Pellet cells by centrifugation, remove supernatant, and resuspend donor and recipient cells in residual GM17 broth (typically around 150 to 300 µl total volume).
4. For determination of input donor and recipient cell numbers, dilute a small aliquot of concentrated donor and recipient cultures in phosphate-buffered saline (PBS) and plate dilutions of 10^{-7} to 10^{-9} (in triplicate) on GM17 agar containing the appropriate antibiotic. Incubate the plates at 30 °C overnight.
5. For each mating pair, mix 75 µl of recipient and 75 µl of donor cells (from step 3), vortex, and spread 100 µl on a single milk-agar plate. Different ratios of donor and recipient cells may also be employed (typically one may use 1:1, 1:5 and 1:10 donor to recipient ratios). Spread 50 µl of recipient and 50 µl of donor(s) separately on milk-agar plates as controls. Incubate mating plates overnight at 30 °C.
6. Add 1 ml sterile PBS to milk agar plates, carefully scrape cells off the plate surface, collect the cells with a pipette (500 to 700 µl), and place them in a sterile microfuge tube.
7. Dilute the mating mixture, and then plate it onto selective medium (in matings with *L. lactis* MMS370 [pTRK28::pRS01] and *L. lactis* LM2345 this is medium containing erythromycin, spectinomycin and rifampicin). Dilute the mating mixture depending on the expected transfer frequency as determined from prior swab matings. If frequencies between < 10^{-9} to 10^{-7} are expected, plate all cells from the mating mixture onto six or more selection plates. If frequencies between 10^{-7} to 10^{-5} are expected, plate triplicates at dilutions of 10^{-2} to 10^{-3}, if frequencies between 10^{-4} to 1 are expected, plate triplicates at dilutions of 10^{-3} to 10^{-5}. Plate 100 µl of undiluted controls directly onto selection plates.
8. Incubate 3 to 4 days at 30 °C to ensure growth and expression of transconjugant phenotypes.
9. Calculate transfer frequencies as number of transconjugants per input donor or input recipient.

D. General mapping of pTRK28 insertions. Plasmid pTRK28 possesses two internal *Pvu*II restriction sites each approximately 3.6-kb from the IS*946* sequence (Figure 1). Localization of pTRK28 inserts is obtained by subtracting the junction pTRK28 DNA (total of 4.4 kb including the IS*946* sequence) from the novel *Pvu*II-*Pvu*II junction fragments. Further localization of the pTRK28 insertion site can be achieved by using additional restriction sites internal to pTRK28 (*Sph*I, *Bam*HI, *Hin*dIII, *Xba*I, *Eco*RI, *Sca*I; Figure 1).

Mapping of pTRK28 insertion sites within pTRK28::pRS01 cointegrates. Plasmid pRS01 is cleaved into five distinct fragments by a *Pvu*II, *Sph*I restriction digest (fragments A, B, C, D, E; Figure 2). Disruption of one of the five pRS01 fragments by insertion of pTRK28 results in the generation of two novel *Pvu*II-*Pvu*II (or *Pvu*II-*Sph*I) junction fragments and two *Pvu*II-*Sph*I fragments internal to the pTRK28 plasmid (see Figure 1). Localization of pTRK28 inserts in pRS01 is obtained by subtracting the junction of pTRK28 DNA (4.4-kb, see above) from the novel *Pvu*II-*Pvu*II or *Pvu*II-*Sph*I junction fragments. Verification of the pTRK28 insertions is achieved by Southern hybridization analysis using pTRK28 plasmid as a DNA probe (data not shown). Confirmation of the location of pTRK28 inserts within pRS01 fragments C, D, and E (Figure 2) is accomplished by additional *Bam*HI or *Xba*I restriction digest.

E. Subcloning of Tra region DNA into *E. coli*. The plasmid pTRK28 contains the *E. coli* plasmid pACYC184 origin of replication in addition to tetracycline and chloramphenicol resistance genes. The presence of pACYC184 within pTRK28 enables the subcloning of DNA flanking any pTRK28 cointegrate insertion, in either direction, by restriction, self-ligation and transformation into *E. coli*. As indicated in Figure 1, pTRK28 contains several unique sites on either side of the pACYC184 origin of replication for use in this subcloning strategy.

To isolate Tra 1 and Tra 2 region DNA, cointegrate plasmid pM2035 is cleaved with *Sph*I, self-ligated and transformed into *E. coli* DH5α. The resulting plasmid, pE351 contains approximately 17.4 kb of pRS01 DNA spanning the Tra 1 and Tra 2 regions (Figure 2). To isolate the Tra 3 and Tra 4 region DNA, plasmid pM1001 is digested with *Sca*I or *Xba*I, and the reaction products self-ligated and transformed into *E. coli* DH5α. The resulting plasmids, pE231 and pE560, each contain approximately 16 kb of pRS01 DNA comprising the Tra3 and Tra4 region DNA, respectively (Figure 2).

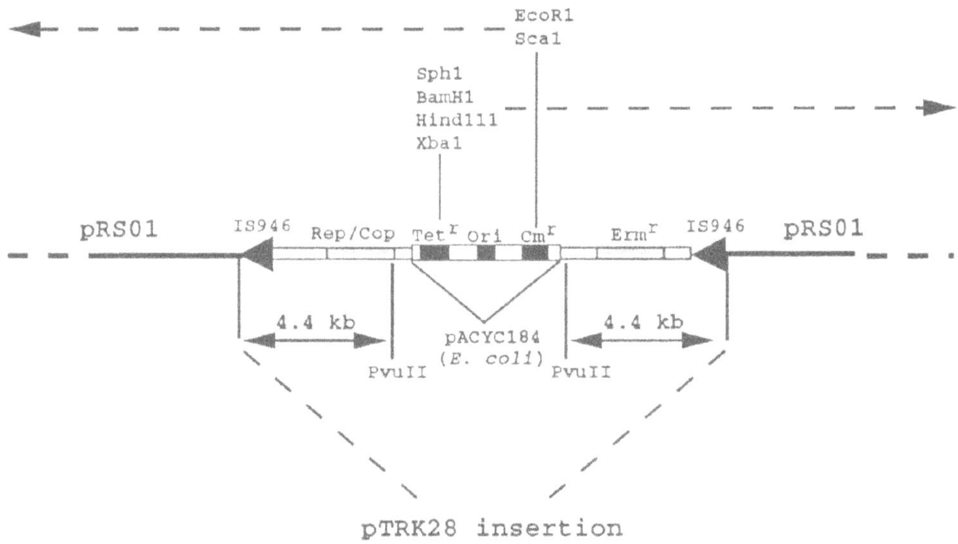

Figure 1. Mapping of pTRK28 insertions within pRS01. Internal *Pvu*II sites which reside 4.4-kb from the ends of IS*946* facilitated the mapping of pTRK28 insertion sites within pRS01 in pTRK28::pRS01 cointegrates. Specific restriction of pTRK28::pRS01 cointegrate plasmids followed by self-ligation and transformation into *E. coli* DH5α allows for the subcloning of pRS01 DNA flanking pTRK28 insertions. Rep/Cop, replication region for maintenance in lactococci; Tet^r, tetracycline resistance determinant from pACYC184; Ori, pACYC184 origin of replication; Cm^r, chloramphenicol resistance determinant from pACYC184; Erm^r, erythromycin resistance determinant (for lactococci).

4. Results and discussion

Generation of transfer-deficient pTRK28::pRS01 cointegrates. Romero & Klaenhammer [12] previously demonstrated the ability of pTRK28, a small *E. coli-L. lactis* shuttle plasmid containing an iso-IS*S1* insertion sequence (IS), IS*946*, to integrate randomly within pRS01. The cointegration of pTRK28 with pRS01 was shown to be a Rec-independent recombination event with a strict requirement for the presence of IS*946*. The pTRK28::pRS01 cointegrates are stable and are maintained at a high copy number due to the plasmid pIP501-derived origin of replication [3] contained within the pTRK28 plasmid. For these reasons, we used pTRK28 to identify the transfer regions of pRS01 by insertional inactivation.

A transconjugant pool containing pTRK28::pRS01 cointegrate plasmids was generated from six independent matings of *L. lactis* NCK168 (pTRK28, pRS01) and the plasmid-free strain *L. lactis* MMS370. pTRK28::pRS01-containing transconjugants were then assayed for secondary transfer in cross-streak matings with *L. lactis* LM2345 as the recipient. Of 558 MMS370 (pTRK28::pRS01) transconjugants assayed, 120 strains (21%) exhibited a reduced transfer phenotype in comparison to the remaining Tra$^+$ transconjugant pool. Thirty-four of these strains were subjected to plate matings using LM2345 as a recipient, in order to determine transfer frequencies. For comparison, the transfer frequencies of 13 randomly selected Tra$^+$ transconjugants were also obtained. An examination of transfer frequencies led to an arbitrary grouping of transconjugant strains into four categories: Tra^{high}, $1 - 1 \times 10^{-3}$ transconjugants per donor cell; Tra^{med}, $1 \times 10^{-4} - 1 \times 10^{-6}$ transconjugants per donor cell; Tra^{low}, $1 \times 10^{-7} - 1 \times 10^{-9}$ transconjugants per donor cell; and Tra$^-$, $<1 \times 10^{-9}$ transconjugants per donor cell (no detectable transfer).

Mapping of the 34 pTRK28::pRS01 cointegrate plasmids altered in transfer ability revealed that four general regions (Tra1, Tra2, Tra3, Tra4) of pRS01 were involved in conjugation (Figure 2). One region, Tra3, was localized within the 15-kb *Pvu*II-*Sal*I fragment previously identified as a transfer region by Anderson and McKay [1]. The clusters of insertions within Tra3 led to a further division of this region into subregions A, B and C. Tra4 is defined by two insertions which mapped within a 4.3-kb *Pvu*II-*Kpn*I fragment (Figure 2) which was previously shown to be associated with a cell aggregation phenotype [1]. Tra1 and Tra2 are two previously unidentified transfer regions located approximately 9 kb away from Tra3.

Complementation of Tra1 and Tra3 region insertions. In an effort to develop mobilization systems for lactococci, we sought to identify the pRS01 *oriT*. One characteristic of conjugative *oriTs* is their ability to confer mobilization, *in cis*, on previously non-conjugative plasmids, provided the remaining transfer

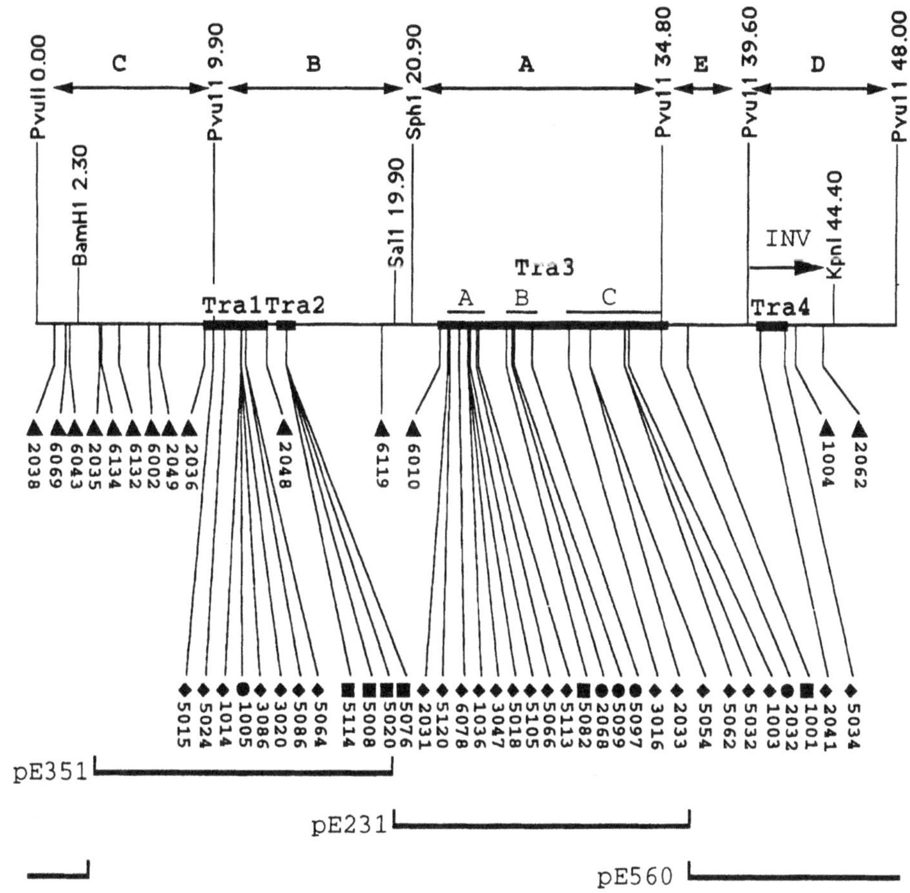

Figure 2. Plasmid pTRK28 insertions within the 49 pTRK28::pRS01 cointegrate plasmids. Individual cointegrate plasmids are denoted at the site of pTRK28 insertion into pRS01 by the final four numbers of the pM-series (e.g. pM1001 to pM6078) cointegrate plasmids. Assignment of Tra^{high}, ▲, 1 – 1 × 10^{-3}/donor cell; Tra^{med}, ■, 1 × 10^{-4} – 1 × 10^{-6}/donor cell; Tra^{low}, ●, 1 × 10^{-7} – 1 × 10^{-9}/donor cell; and Tra^{-}, ◆, < 1 × 10^{-9}/donor cell phenotypes are based on calculated transfer frequencies. Subclones of pRS01 generated in *E. coli* DH5α are indicated below the figure. Plasmids pE560 and pE231, containing the Tra3 and Tra4 regions, respectively, were derived from cointegrate plasmid pM1001. Plasmid pE351, containing regions Tra1 and Tra2, was derived from cointegrate plasmid pM2035.

functions are present, *in trans*. We hypothesized that complementation of the pRS01 transfer region insertions with the cognate Tra region DNA would reveal the pRS01 *oriT* by subsequent mobilization of the complementing plasmid.

To complement insertions within Tra3, three different subclones (pLE14, pLE16, pLE33; Figure 3B) were generated in the *E. coli-L. lactis* shuttle plasmid pLE1 [8]. Plasmid pLE33, which possessed a 6.8-kb *Sca*I-*Stu*I fragment encompassing both Tra3A and Tra3B, complemented all Tra3A and Tra3B insertions for transfer. Mobilization of the complementing plasmid, pLE33, was not observed in any mating, suggesting that the 6.8-kb *Sca*I-*Stu*I fragment does not encode the pRS01 *oriT*. Plasmid pLE14, which was generated by subcloning a 9.5-kb *Hga*I fragment from pE231 and contained all of Tra3C, functionally complemented Tra3C insertional mutants (Figure 3B). In contrast, pLE16, which contained a 5.2-kb *Sca*I-*Stu*I fragment did not complement Tra3C mutants for transfer. Neither plasmid pLE14 nor pLE16 were mobilized by pTRK28::pRS01 cointegrates, indicating the pRS01 *oriT* does not reside within Tra3C.

To complement Tra1 region insertions, a 7.5-kb *Pst*I fragment encompassing the Tra1 region was cloned from plasmid pE351 (Figure 2) into pLE1, resulting in the plasmid pLE12 (Figure 3A). All Tra1 insertion mutations were complemented for transfer with pLE12 *in trans*. In addition, mobilization of pLE12 vector was observed with all matings (Figure 3A). The mobilization of plasmid pLE12 indicated the 7.5-kb *Pst*I fragment contained a *cis*-acting *oriT*. The complete sequence of the Tra1 and Tra2 region was generated, and subsequent sequence analysis revealed six substantial open reading frames, *ltrC, ltrD, ltrE, ltrBE1, ltrA*, and *ltrBE2*, in the same orientation (Figure 3A). In order to localize the *oriT*,

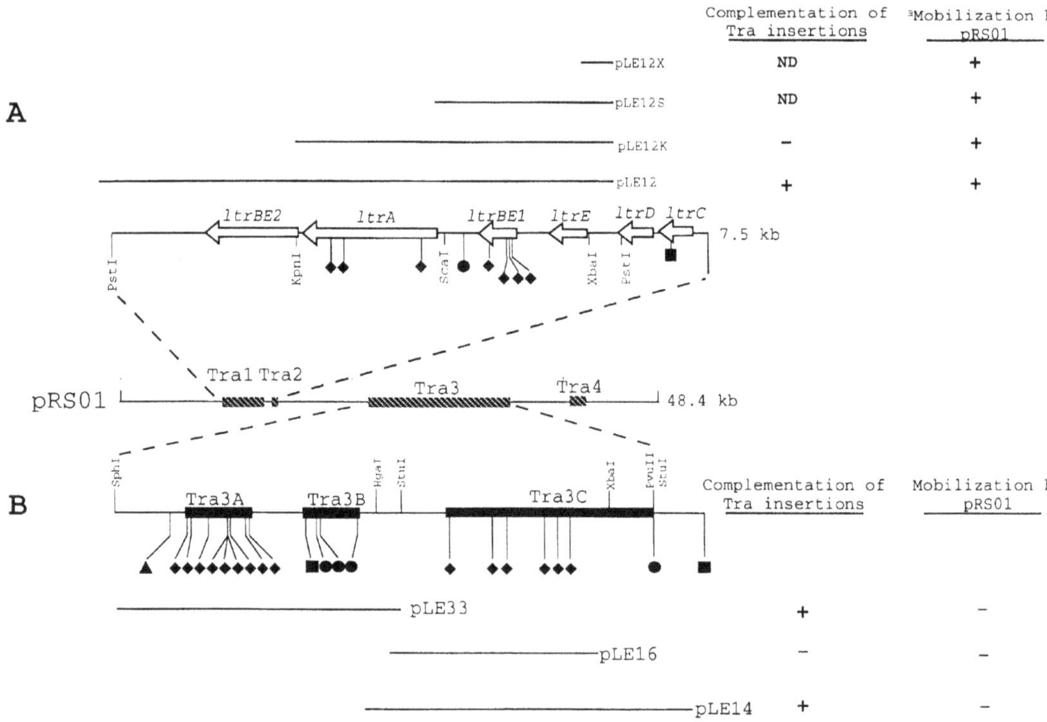

Figure 3. Complementation and mobilization analysis of the Tra1 and Tra3 regions. Transfer phenotype of pTRK28 insertions is depicted by the symbol: ▲, Tra^{high}; ■, Tra^{med}, ●, Tra^{low}; ◆, Tra^{-}. (**A**): Genetic map of the Tra1 and Tra2 regions of pRS01 with accompanying results of pLE12-derivative complementation and mobilization analysis. Plasmid pLE12 contains the complete Tra1 region. Plasmids pLE12K, pLE12S and pLE12X are deletion derivatives of pLE12 generated through *Kpn*I, *Sca*I, and *Xba*I cleavages, respectively, and self-ligation. [a]Mobilization analysis of pLE12 and pLE12-based derivatives was accomplished using a Tra^{high} strain *L. lactis* DM2036 [8] (*L. lactis* MMS370 possessing a Tra^{+} pTRK28::pRS01 cointegrate). (**B**): Genetic map of the Tra3 region with accompanying results of complementation and mobilization analysis. Plasmids pLE33, pLE14, and pLE16 were generated by subcloning from plasmid pE231 as described in the text. ND, not determined.

we analyzed a series of pLE12 deletion derivatives for mobilization in the presence of a Tra^{+} cointegrate plasmid. As shown in Figure 3A, all derivatives possessing the small 0.4-kb *Pst*I-*Xba*I fragment, which contains the intergenic region between *ltrD* and *ltrE*, were mobilized by a Tra^{+} pTRK28::pRS01 cointegrate. Sequence analysis of the intergenic region between *ltrD* and *ltrE* revealed clusters of inverted and direct repeat structures similar to other bacterial *oriT* regions [7, 21].

In this work we demonstrate the use of IS*946*-mediated cointegration between the shuttle plasmid pTRK28 and the conjugal element pRS01 as a method for insertional mutagenesis of pRS01. While we have not used pTRK28 cointegration to map other conjugal elements, this method was quite reliable for analysis of pRS01. The resolution of mapping the pTRK28 insertions using *Pvu*II-*Sph*I restriction analysis (Figure 2) has proven to be quite accurate. To date, we have sequenced the exact site of pTRK28 insertion within 23 cointegrate plasmids, and found that all insertions were within 0.5-kb of the location predicted by restriction mapping (of the 60-kb cointegrate plasmid). This accuracy was facilitated by the fact that pRS01 is cleaved by *Pvu*II into DNA fragments which are easily separated by gel electrophoresis. Since other conjugal elements may not be cleaved by *Pvu*II, other sites internal to the two *Pvu*II sites in pTRK28 could be used in conjunction with *Pvu*II to map cointegrate junctions.

The success of this method relies on the formation of stable conintegrates between the target conjugal element and the IS-containing vector within the recipient. Typically, resolution of cointegrate plasmids occurs through homologous recombination between duplicated IS-elements. We have screened numerous pTRK28::pRS01 cointegrates within the Rec^{-} lactococcal strain MMS370 and have not observed resolution of the constituent plasmids. In contrast, Anderson and McKay identified several types of resolution products derived from cointegrates formed between the lactococcal plasmid pSK08 and pRS01 in the Rec^{+} lactococcal strain LM2301 [1]. Thus, one possible way to circumvent cointegrate plasmid resolution may be to generate cointegrates within a Rec^{-} recipient strain.

The mating procedures described here are standard for lactococcal matings. We have found that the growth phase of the respective donor and recipient cultures at the time of mating can dramatically affect the transfer frequency. Donor and recipient cultures growing exponentially, prior to mixing on the mating plate, appear to produce the highest transfer frequencies. This effect becomes increasingly important in inter-species or inter-generic matings where donor and recipients possess different growth rates [10].

IS-directed cointegration between non-conjugative vectors and conjugal elements is a relatively common event in bacteria [11]. Given that the LAB have been shown to harbor a wide variety of IS elements [13] we believe that IS-directed cointegration may provide an additional tool for mutagenesis of conjugal elements within the LAB, or perhaps, be employed as a screen for conjugative elements which are not associated with an observable phenotype. The use of the pTRK28-mediated mutagenesis of pRS01 described here will give insight into the mechanisms of conjugative transfer among LAB and enable conjugation to be more universally employed as a method of LAB strain development. In addition, the identification of the pRS01 *oriT* will enable the generation of mobilizable vectors for the lactococci. In the past decade, electroporation has become the method of choice for transforming various lactic acid bacteria [5]. Unfortunately, many lactic acid bacteria, including commercial strains of lactococci, are difficult to transform by electroporation, resulting in a tedious and sometimes unsuccessful optimization of electroporation conditions. In contrast to electroporation, conjugative mobilization of pRS01 *oriT*-containing derivatives may offer an alternative for genetic delivery into lactococci and, potentially, other LAB.

Acknowledgments

This work was supported by grants from the Minnesota-South Dakota Dairy Foods Research Center (G.M.D.) and the Kraft General Foods Chair (L.L.M.). D.A.M. was supported by a NIGMS Predoctoral Biotechnology Training Grant.

Notes on suppliers

1. Difco Laboratories, Detroit, MI, USA
2. Sigma Chemical Co., St Louis, MO, USA
3. ICN Biochemicals, Cleveland, OH, USA
4. GibcoBRL Life Technologies, Gaithersburg, MD, USA
5. BIO 101, Inc., Vista, CA, USA
6. Bio-Rad Laboratories, Richmond, CA, USA

References

1. Anderson DG, McKay LL (1984). Genetic and physical characterization of recombinant plasmids associated with cell aggregation and high-frequency conjugal transfer in *Streptococcus lactis* ML3. J Bacteriol 158: 954–962.
2. Ausubel FM, Brent R, Kingston RE, et al. (1995). Current protocols in molecular biology. New York: Wiley.
3. Behnke D, Gilmore MS, Ferretti J (1981). Plasmid pGB301, a new multiple resistance streptococcal cloning vehicle and its use in cloning of the gentamycin/kanamycin resistance determinant. Mol Gen Genet 182: 414–421.
4. Dower WJ, Miller JF, Ragsdale CW (1988). High efficiency transformation of *E. coli* by high voltage electroporation. Nucleic Acid Res 16: 6127–6145.
5. Holo H, Nes IF (1995). Transformation of *Lactococcus* by electroporation. Methods Mol Biol 47: 195–199.
6. Langella P, Le LY, Ehrlich SD, Gruss A (1993). Efficient plasmid mobilization by pIP501 in *Lactococcus lactis* subsp. *lactis*. J Bacteriol 175: 5806–5813.
7. Lanka E, Wilkins BM (1995). DNA processing reactions in bacterial conjugation. Ann Rev Biochem 64: 141–169.
8. Mills DA, Choi CK, Dunny GM, McKay LL (1994). Genetic analysis of regions of the *Lactococcus lactis* subsp. *lactis* plasmid pRS01 involved in conjugative transfer. Appl Environ Microbiol 60: 4413–4420.
9. O'Sullivan DJ, Klaenhammer TR (1993). Rapid mini-prep isolation of high-quality plasmid DNA from *Lactococcus* and *Lactobacillus* spp. Appl Environ Microbiol 59: 2730–2733.
10. Phister TG, Mills DA, Cocconcelli PS, McKay LL (1996). Host range analysis of lactococcal conjugal element pRS01. In: Abstracts of the 96th General Meeting of the American Society for Microbiology, abstr. H-90, p 498. Washington, DC: American Society for Microbiology.
11. Reimmann C, Haas D (1993). Mobilization of chromosomes and nonconjugative plasmids by cointegrative mechanisms. In: DB Clewell (ed), Bacterial conjugation, pp 137–188. New York: Plenum Press.
12. Romero DA, Klaenhammer TR (1990). Characterization of insertion sequence IS*946*, an iso-ISS*1* element, isolated from the conjugative lactococcal plasmid pTR2030. J Bacteriol 172: 4151–4160.
13. Romero DA, Klaenhammer TR (1993). Transposable elements in Lactococci: a review. J Dairy Sci 76: 1–19.
14. Romero DA, Slos P, Robert C, Castellino I, Mercenier A (1987). Conjugative mobilization as an alternative vector delivery system for lactic streptococci. Appl Environ Microbiol 53: 2405–2413.
15. Sambrook J, Fritsch EF, Maniatis T (1989). Molecular cloning: a laboratory manual, 2nd ed. Cold Spring Harbor, NY: Cold Spring Harbor Laboratory Press.
16. Sanders ME, Leonhard PJ, Sing WD, Klaenhammer TR (1986). Conjugal strategy for construction of fast acid-producing bacteriophage-resistant lactic strepto-

cocci for use in dairy fermentations. Appl Environ Microbiol 52: 1001–1007.

17. Sing WD, Klaenhammer TR (1993). A strategy for rotation of different bacteriophage defenses in a lactococcal single-strain starter culture system. Appl Environ Microbiol 59: 365–372.
18. Steele JL, McKay LL (1989). Conjugal transfer of genetic material in Lactococci: a review. J Dairy Sci 72: 3388–3397.
19. Terzaghi BE, Sandine WE (1975). Improved medium for lactic streptococci and their bacteriophages. Appl Microbiol 29: 807–813.
20. Walsh PM, McKay LL (1981). Recombinant plasmid associated with cell aggregation and high-frequency conjugation of *Streptococcus lactis* ML3. J Bacteriol 146: 937–944.
21. Wilkins B, Lanka E (1993). DNA processing and replication during plasmid transfer between Gram-negative bacteria. In: D Clewell (ed), Bacterial conjugation, pp 105–129. New York: Plenum Press.

Address for correspondence: Larry L. McKay, Department of Food Science and Nutrition, University of Minnesota, 1334 Eckles Avenue, St Paul, MN 55108, USA
Phone: (612) 624-3090; Fax: (612) 625-5272
E-mail: lmckay@che2.che.umn.edu

Methods in Cell Science **20**: 79–84 (1998).

Use of electroporation in genetic analysis of enterococcal virulence

Helmut Hirt, Yi Chen, Patrick M. Schlievert & Gary M. Dunny
Department of Microbiology, University of Minnesota, Medical School, Minneapolis, Minnesota, USA

Abstract. We present two methods for electroporation for the gram positive bacterium *Enterococcus faecalis* that can also be used as guidelines for work with other gram positive species. We demonstrate the use and the advantages of this technique for investigating genes, both chromosomal and plasmid-linked, encoding surface structures. Electroporation was used to deliver constructs created on shuttle vectors for insertional inactivation of a chromosomal gene involved in binding substance formation as well as for the expression of Aggregation Substance in strains with different chromosomal backgrounds. The influence of defects in lipoteichoic acid synthesis and the expression of Aggregation Substance on virulence was shown in a rabbit endocarditis model.

Key words: Aggregation substance, Electroporation, Enterococcal binding substance, *Enterococcus faecalis*, Virulence

Abbreviations: AS = aggregation substance; EBS = enterococcal binding substance

1. Introduction

Analysis of the sex pheromone conjugation system of *Enterococcus faecalis* provides an excellent example of the use of genetic manipulation, by methods such as electroporation, to increase the level of understanding of a complex regulatory system. We describe results obtained on surface structures involved in bacterial cell-cell contact as well as contributions of these surface components to the virulence of *E. faecalis*. This organism has become increasingly important as an opportunistic pathogen [13, 17].

The sex pheromone system is a high efficiency plasmid transfer system, to our knowledge restricted to Enterococci [8]. Transfer frequencies up to 10^{-1} in liquid culture can be observed, as compared with other conjugative plasmids which transfer with a frequency of around 10^{-6} [5]. The high efficiency of plasmid transfer is facilitated by induction of a plasmid encoded surface protein, Aggregation Substance (AS), that ensures a tight contact between plasmid donor cells and recipient cells, resulting in characteristic macroscopic visible clumps.

Several sex pheromone plasmids and corresponding pheromones have been described. The best characterized plasmids are pAD1 and pCF10; pCF10 is the focus of our group. The course of induction and plasmid transfer is depicted schematically in Figure 1 for pCF10. Plasmid free recipient cells secret several small, 7–8 amino acid hydrophobic peptides into the medium. Corresponding peptides are detected by a OppA homologue (PrgZ, in the case of pCF10) of the donor cell encoded by the cognate plasmid (A). Aggregation substance is induced and expressed on the cell surface, enabling donor and recipient cells to form a stable mating pair (B). Plasmid transfer occurs (C), resulting in a recipient cell harboring the transferred plasmid and therefore converted to a donor cell (D).

The gene for AS is highly conserved between different pheromone plasmids [11, 12] and encodes a 137 kDa protein with a typical gram-positive signal peptide and cell wall anchor sequence LPXTG. The protein does not, in contrast to many other gram positive cell surface proteins, contain repetitive sequences and is thought to have a more globular shape. Besides the function as an adhesion molecule for bacterial cells, the protein also contains two RGD sequence motifs [10, 14] that could promote adhesion to eucaryotic integrins, suggesting that AS may be involved in virulence.

Analysis of the receptor for AS on recipient cells, termed Enterococcal Binding Substance (EBS), suggests that lipoteichoic acid is the main component. Transposon mutagenesis with Tn*916* (described elsewhere in this volume) together with the use of an enterococcal phage allowed the isolation of the mutant strain INY3000 that acted poorly as a recipient for plasmid transfer [18]. Analysis of the lipoteichoic acids in this strain revealed a shortened chain length and an altered saturation level of the lipoteichoic acids of this strain.

For further characterization at the molecular level of the INY3000 mutant, and for other genetic analysis in the organism, the development of an efficient transformation system was extremely useful. The first described transformation method was based

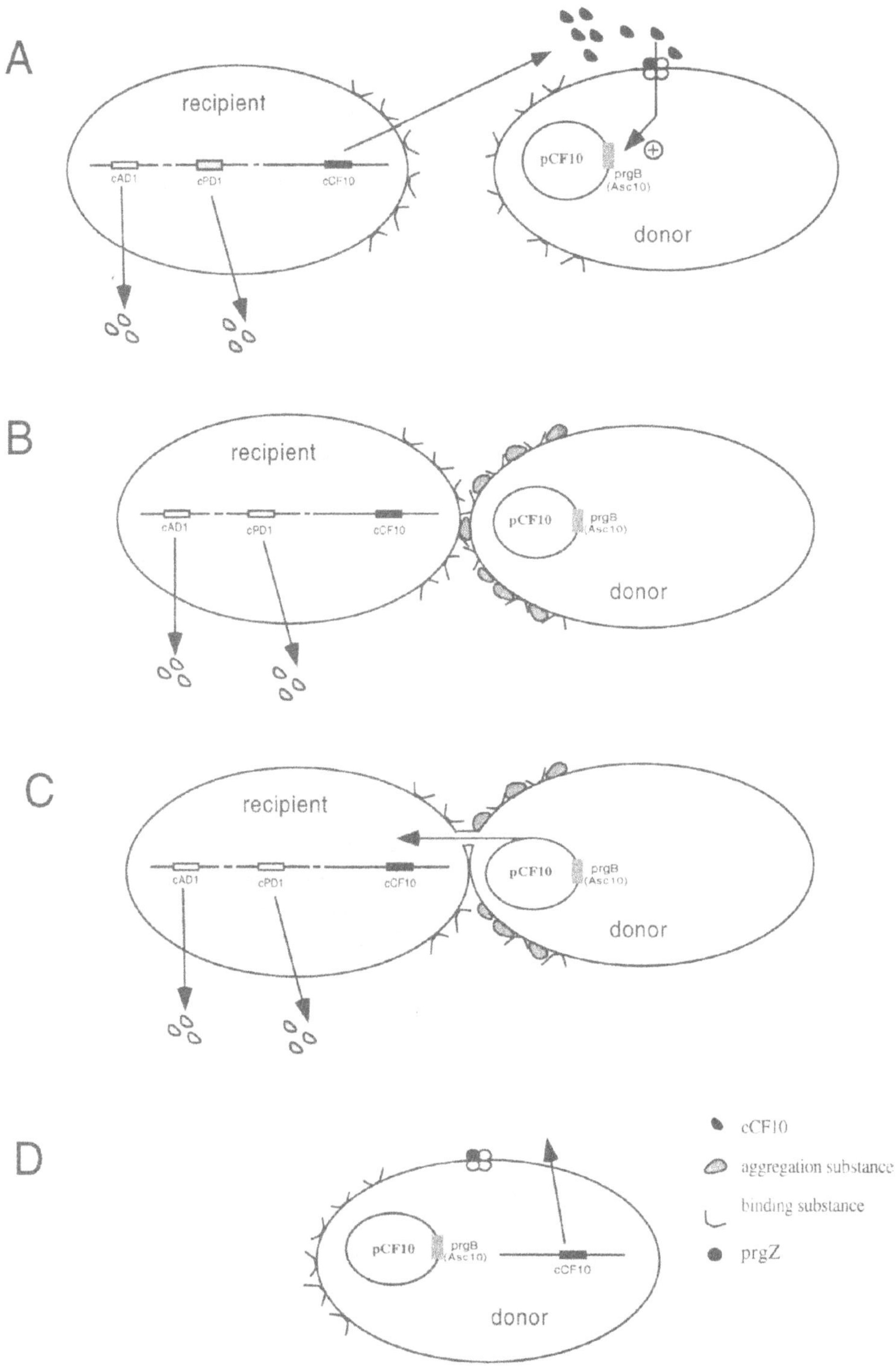

Figure 1. Model of the sex pheromone plasmid transfer system in *Enterococcus faecalis*.

on protoplast transformation [19] but was soon replaced by the simpler method of electroporation. We present two methods used for electroporation in this gram positive species, and results obtained by using this method to study the surface structures EBS and AS in *E. faecalis*. No single method for electroporation exists which is uniformly effective, for streptococci and lactococci as well as enterococci.

We present, therefore, two methods that are used in our laboratory. Results may vary and depend on the strains used, uncharacterized clinical isolates may be difficult to transform.

2. Material

A. Electroporation

Method I: [6, 7]

- Medium compositions
- M9-YE:
  - Yeast extract 3 g/l (#0127-01-7, Difco[1]).
  - Casamino acids (#0230-17-3, Difco[1]) 10 g/l.
  - 10× M9 salts* all chemicals from Fisher Scientific[2] (gram/liter: Na_2HPO_4, 60, S373-500.
  - KH_2PO_4, 30, P382-500; NaCl, 5, S640-3.
  - NH_4Cl, 10, A661-500) 1/10 volume.
  - 20% glucose (D16-500)* 1/100 volume.
  - 1 M $MgSO_4$ (M65-500)* 1/500 volume.
  - 1 M $CaCl_2$ (C77-500)* 1/10000 volume.
  - 20% glycine (G46-500).

    * Add after autoclaving.

Method II: Modified from [4] and [9]

- 1 M sucrose.
- 20% glycine.
- 4× Todd Hewitt Broth (#0492-17-6, Difco[1]).
- Selective plates THB containing 0.5 M sucrose.
- GenePulser™ and Pulse Controller (165-2076 and 165-2098, Biorad[3]), 0.2 mm cuvettes (#4307-000-593, Eppendorf[4]).

B. Endocarditis

- American Dutch Belted Rabbits (Birchwood Farm, Wisconsin).
- New Zealand White rabbits (Birchwood Farm, Wisconsin).
- Syringes, 18 gage needle, catheter outside diameter 1.27 mm (No. 427420, Becton Dickenson[5]).
- Ketamine (NDC 57319-291-02, Phoenix Pharmaceutical[6]), xylazine (NDC 57319-210-26, Phoenix Pharmaceutical[6]).

3. Procedures

A. Electroporation

Method I:

1. Grow cells for 12 to 15 h in BYGT or M9-YE medium plus glycine. We recommend a range between 0.5 and 6.0%. Measure the A_{660} in comparison to a culture grown without glycine. Use the culture with a growth reduction between 70 and 90%.
2. Dilute the culture in fresh medium (containing the same concentration of glycine) to a A_{660} between 0.05 and 0.08. Incubate the cells for 60 min at 37 °C.
3. Chill the culture on ice; harvest the cells and wash the cells in 1/3 volume of chilled EP buffer.
4. Harvest the cells and resuspend in 1/100 of the original volume. Aliquot in 40 μl portions and store at –70 to –80 °C.
5. For electroporation, thaw the cells on ice for a few minutes. Add 10–20 μl of DNA in H_2O (0.3–1 μg) and incubate for 5 min. Add mixture to chilled electroporation cuvette and electroporate immediately, using the settings 25 μF, 200 Ω and 2500 V. After the pulse resuspend the cells and place them on ice for 5 min. Incubate at 37 °C for 120 min. Spread the cells on selective plates.

Method II:

1. Grow cells for 25 h in THB, 0.5 M sucrose containing glycine. The amount of glycine depends on the strain used. Examples: OG1SSp: 5%, OG1RF: 3%, JH2-2: 6%. For example for 5% (10 ml): 2.5 ml 4× THB, 2.5 ml 20% glycine, 5 ml 1 M sucrose
2. Wash the cells in chilled EP buffer in 100, 50 and 10% of the original culture volume.
3. Resuspend in 1/100 of the original culture volume, aliquot and store at –70 to –80 °C.
4. Follow step 5 in method I.

B. Endocarditis

Intraventricular Injection model (American Dutch Belted):

1. Grow test organism to a density of approximately 10^8 CFU/ml.
2. Use 25 mg/kg ketamine to calm the rabbits, do not anesthetize.
3. Turn rabbit on its back, front legs over the head.
4. Go up two or three ribs on the left side of the rabbit.
5. Insert syringe filled with 1 ml of the test organism and a 18 gauge needle.
6. Wait until blood comes pumping into the syringe and inject organisms, retreat needle.
7. Monitor the rabbits for two days, euthanize and evaluate results.

Transaortic Catherization (New Zealand White):

1. Anesthetize rabbits with ketamine (25 mg/kg) and xylazine (20 mg/kg).
2. Shave the throat of the rabbit. Expose the left carotid artery, and stop the blood flow with two threads knotted around the artery approximately 2 cm apart. The next part should be carried out by two persons. Lift the artery up on the threads, slit open the artery with a scalpel or razor blade and insert the catheter into the artery. Push the catheter into the heart. Close the catheter by introducing a bend and use a thread to keep it in place.
3. Keep the rabbits anesthetized for two hours.
4. Remove the catheter, seal off the artery with

the thread in place and close the opening with sutures.
5. Inject 2 ml of the test organism in a marginal ear vein. Bacterial cell density of the injected solution should be around 10^9.
6. Monitor the rabbits over the course of three days. Sacrifice the rabbits on day three and assess the results.

4. Results and discussion

Transfer frequencies observed for electroporation are shown for Method I in Table 1. It is generally advisable to propagate plasmids in *E. coli* before transforming them into *Enterococcus*. When the electrocompetent cells are prepared, we recommend to keep the cells chilled during the whole washing and centrifugation procedure. Prechill the tubes used for storage of the aliquots.

Although the cells can be used immediately after the washing procedure, freezing them for 1 h before use can increase the efficiency. The cells should be put into cuvettes only shortly before the electroporation and resuspended immediately after the pulse, since it was found, that long contact with the metal electrodes decreases transformation efficiency. Optimal time constants for the described methods are between 4.0 and 4.6 ms. The DNA used should be resuspended preferentially in H_2O or a low salt buffer. Too high amounts of DNA, poor quality or high salt concentrations in the buffer can lead to a short-circuit with time constants below 1 ms, drastically reducing the probability of recovering transformants. Dilution of the cells 1:1 with 10% cold glycerol can help to avoid short-circuits.

The locus for AS on pCF10 was first characterized by transposon mutagenesis. Two adjacent EcoRI restriction fragments containing the positive control region and the structural genes for surface exclusion and AS are cloned in a shuttle vector [3]. A map of this clone is shown in Figure 2A. Since the negative control elements of pCF10 are lacking in this construct, expression of the aggregation substance is constitutive. Insertion of Tn*5* (INY4515 and INY4561) confirmed the functions of the surface proteins Sec10 (*prgA*) and Asc10 (*prgB*) [14, 15].

Table 1. Transformation frequencies of different bacterial strains using the described method I

Organism	Plasmid	Transformants/μg DNA
E. faecalis OG1RF	pDL414	5×10^4
E. faecalis OG1RF	pWM401	1.2×10^5
E. faecalis OG1SSp	pDL414	1.6×10^4
E. faecalis JH2–2	pDL414	2.5×10^4
E. faecalis UV202	pDL414	4×10^3
S. pyogenes DW1009*	pDL414	1.6×10^3
S. agalactiae H36B (type Ib)	pDL414	4.8×10^3
S. sanguis FW213	pDL414	1.4×10^4

* No glycine for preparation of competent cells. Modified from [7].

The availability of a transformation system allowed further characterization of the Tn*916* insertions in the INY3000 chromosome. The four insertions were separated genetically by protoplast transformation.

None of the individual Tn*916* insertions showed a phenotype; no diminished recipient ability was detected [18]. This suggested that multiple loci were involved in synthesis of EBS. We have begun to analyze each of the loci separately. Work on one of the loci in strain INY3039 greatly relied on an efficient transformation system. The INY3039 locus is shown in Figure 2B. The transposon was used as selective marker for marker rescue of EcoRI-cut INY3039 chromosomal DNA in *E. coli*. Sequence analysis showed that the transposon inserted in the intergenic region between two divergently transcribed genes, *ebsA* and *ebsB* [1]. Further transposon mutagenesis of the cloned wild-type region with Tn*5* revealed that an insertion in the reading frame for *ebsC* abolished the ability of the fragment to complement the INY3000 phenotype. Sequence similarities for *ebsC* showed homology to a negative regulatory protein. This led to formulation of a model suggesting that in wild-type *E. faecalis ebsC* negatively regulated expression of *ebsA* or *ebsB*. Overexpression of one of the latter genes from a promoter within Tn*916* resulted in a disruption or degradation of wild-type EBS. To test the model, *ebsC* was inactivated in the vector pGHost in *E. coli* by deleting an internal fragment of the gene and replacing it with a chloramphenicol resistance marker flanked on both sides by about 1 kb of the wild-type *E. faecalis* chromosomal sequences. This construct was then transformed using electroporation in the wild-type strain OG1SSp and the cells were screened for the double crossover event. A mutant strain carrying the resistance gene in the disrupted *ebsC* gene was isolated. Plasmid transfer experiments revealed a diminished ability of the mutant strain to act as recipient in matings [2]. These results are remarkable, since the original Tn*916* insertion in that locus did not show a phenotype. It is also important to note that the allelic replacement experiment that confirmed the model involved a gene (*ebsC*) not targeted by the original Tn*916* insertion. This is an important consideration in using Tn*916* for random mutagenesis.

Investigating the contribution of the two surface structures lipoteichoic acid and AS to virulence also became possible. Rabbit endocarditis was chosen as a model system. The combination of strain comprised in these experiments were the binding substance

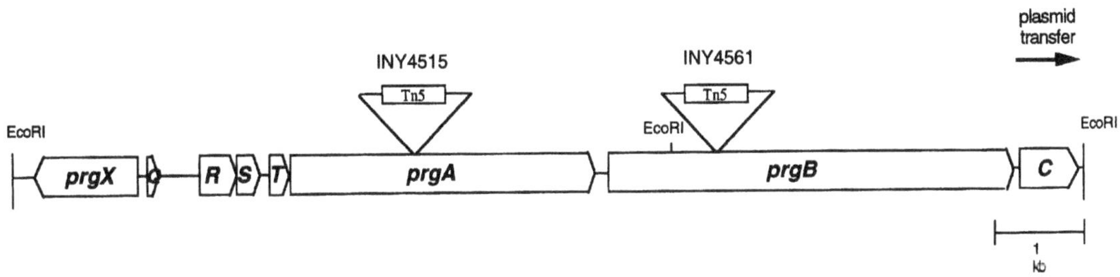

Figure 2A. The positive control region in pCF10. Cloning of the adjacent EcoRI fragments in a shuttle vector lead to the constitutive expression of the *prgB* gene resulting in a clumpy phenotype. Insertion of Tn*5* in the genes *prgA* (INY4515) and *prgB* (INY4561) confirmed their function, surface exclusion and aggregation substance, respectively.

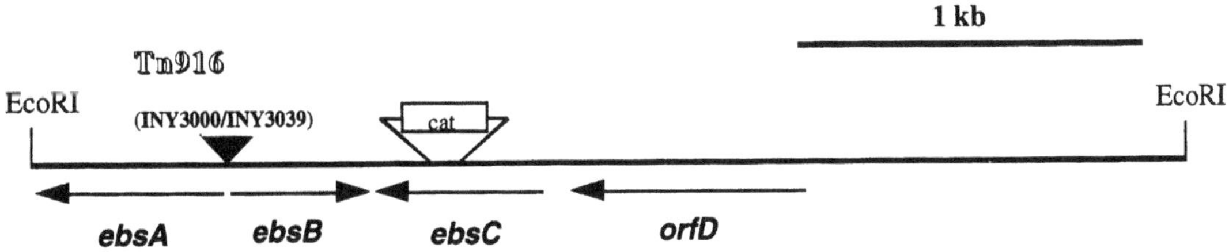

Figure 2B. The *ebs* locus in *E. faecalis*. Location of the initial Tn*916* insertion separated from INY3000 into INY3039. Insertional inactivation of the *ebsC* gene in wild-type OG1SSp.

mutant INY3000 (EBS$^-$) with the cloning vector pWM401 (AS$^-$) as negative control, the wild-type strain OG1SSp (EBS$^+$) with pWM401 and INY3000 or OG1SSp containing the plasmid pINY1801(AS$^+$), constitutively expressing the otherwise inducible Aggregation Substance in the cloning vector pWM401. All strains were transformed by electroporation. All combinations of the two surface structures (EBS$^-$ AS$^-$; EBS$^-$ AS$^+$; EBS$^+$ AS$^-$ and EBS$^+$ AS$^+$) were therefore available for use in this model system. The experiments involved intraventricular injection as well as catheterization for the assessment of the virulence of the chosen strains (Table 2).

In both models the binding substance mutant INY3000 showed no signs of morbidity or inflammation, whereas the wild-type strain OG1SSp showed moderate virulence [16]. This shows clearly that the existence of an intact lipoteichoic acid structure is necessary for virulence of *E. faecalis*. The lack of virulence of the strain INY3000 could be overcome by the presence of the constitutive AS-expressing plasmid pINY1801. The symptoms induced by this strain resemble the plasmid-free wild-type. The highest grade of virulence however, was reached when both lipoteichoic acid and AS were expressed in strain OG1SSp:pINY1801 (EBS$^+$, AS$^+$). All animals challenged with this strain died and showed destruction of tissue in heart and lung. These results demonstrate that not only the Aggregation Substance on the pheromone plasmids has a major influence on the virulence of *E. faecalis*, but the overall architecture of the cell wall is a deciding factor for establishment of an infection. While the presence of the Aggregation Substance could at least partially overcome the inability of INY3000 to induce virulence, the altered lipoteichoic acid expression in this strain clearly makes that strain avirulent either by abolishing the ability to survive and to establish in the host, or by changing the cell wall structure such, that other chromosomally-encoded adhesins are not properly displayed on the cell surface.

Our examples show that the development of electroporation allows better in-depth analysis of the biology of bacteria species not naturally competent for transformation. As the analysis of the transposon

Table 2. Influence of enterococcal binding substance (EBS) and Aggregation Substance (AS) on the virulence of *E. faecalis* in rabbit endocarditis, intraventricular injection model EBS$^-$: INY3000; EBS$^+$: OG1SSp; AS$^-$: pWM401; AS$^+$: pINY1801. Modified from [16]

	EBS$^-$, AS$^-$	EBS$^-$, AS$^+$	EBS$^+$, AS$^-$	EBS$^+$, AS$^+$
Morbidity	No	Yes	Yes	Yes
Mortality	0/6	1/6	0/3	9/9

insertion in INY3039 demonstrates, transposon mutagenesis is useful for initial analysis, however the phenotypic effects shown might not reflect the situation correctly. Insertional inactivation provides a more accurate tool and with the use of electroporation this becomes available for many gram positive species.

Acknowledgments

The work described in this paper was supported by NIH grants HL51987 and GM49530.

Notes on suppliers

1. Difco Laboratories Inc., P.O. Box 331058, Detroit, MI 48232, USA
2. Fisher Scientific Corp., 711 Forbes Ave., Pittsburgh, PA 15219, USA
3. Bio-Rad Laboratories, Life Science Group, 2000 Alfred Nobel Dr., Hercules, CA 94547, USA
4. Eppendorf Scientific Inc., 6524 Seybold Rd., Madison, WI 53719, USA
5. Becton Dickinson, 7 Loveton Circle, Sparks, MD21152, USA
6. Phoenix Pharmaceutical Inc., 4621 Easton Rd., St. Joseph, MO 64503, USA

References

1. Bensing BA, Dunny GM (1993). Cloning and molecular analysis of genes affecting expression of binding substance, the recipient-encoded receptor(s) mediating mating aggregate formation in *Enterococcus faecalis*. J Bacteriol 175: 7421–7429.
2. Chen Y (1997). Genetic analysis of genes involved in *Enterococcus faecalis* binding sustance formation. Masters Thesis, University of Minnesota.
3. Christie PJ, Kao S-M, Adsit JC, Dunny GM (1988). Cloning and expression of genes encoding pheromone-inducible antigens of *Enterococcus (Streptococcus) faecalis*. J Bacteriol 170: 5161–5168.
4. Cruz-Rodz AL, Gilmore MS (1991). Electroporation of glycine-treated *Enterococcus faecalis*. In: Dunny GM, Cleary PP, McKay LL (eds), Genetics and molecular biology of streptococci, lactococci and enterococci, p 300. Washington DC: ASM Press.
5. Dunny GM, Brown BL, Clewell DB (1978). Induced cell aggregation and mating in *Streptococcus faecalis*: Evidence for a bacterial sex pheromone. Proc Natl Acad Sci USA 75: 3479–3483.
6. Dunny GM (1991). Electroporation of Enterococci, Streptococci and Bacilli. In: Dunny GM, Cleary PP, McKay LL (eds), Genetics and molecular biology of streptococci, lactococci and enterococci, p 302. Washington DC: ASM Press.
7. Dunny GM, Lee LN, LeBlanc DJ (1991). Improved electroporation and cloning vector system for gram-positive bacteria. Appl Environ Microbiol 57: 1194–1201.
8. Dunny GM, Leonard BA, Hedberg PJ (1995). Pheromone-inducible conjugation in *Enterococcus faecalis*: Interbacterial and host-parasite chemical communication. J. Bacteriol 177: 871–876.
9. Fiedler S, Wirth R (1991). Transformation of *Enterococcus faecalis* and *Enterococcus faecium* by elecrtroporation. In: Dunny GM, Cleary PP, McKay LL (eds), Genetics and molecular biology of streptococci, lactococci and enterococci, p 301. Washington DC: ASM Press.
10. Galli D, Lottspeich F, Wirth R (1990). Sequence analysis of *Enterococcus faecalis* aggregation substance encoded by the sex pheromone plasmid pAD1. Mol Microbiol 4: 895–904.
11. Galli D, Wirth R (1991). Comparative analysis of *Enterococcus faecalis* sex pheromone plasmids identifies a single homologous DNA region which codes for aggregation substance. J Bacteriol 173: 3029–3033.
12. Hirt H, Wanner G, Galli D, Wirth R (1993). Biochemical, immunological and ultrastructural characterization of aggregation substances encoded by *Enterococcus faecalis* sex-pheromone plasmids. Eur J Biochem 211: 711–716.
13. Jett BD, Huycke MM, Gilmore MS (1994). Virulence of enterococci. Clin Microbiol Rev 7: 462–478.
14. Kao SM, Olmsted SB, Viksnins AS, Gallo JC, Dunny GM (1991). Molecular and genetic analysis of a region of plasmid pCF10 containing positive control genes and structural genes encoding surface proteins involved in pheromone-inducible conjugation in *Enterococcus faecalis*. J Bacteriol 173: 7650–7664.
15. Olmsted SB, Kao SM, van Putte LJ, Gallo JC, Dunny GM (1991). Role of the pheromone-inducible surface protein Asc 10 in mating aggregate formation and conjugal transfer of the *Enterococcus faecalis* plasmid pCF10. J Bacteriol 173: 7665–7672.
16. Schlievert PM, Gahr PJ, Assimacopoulos AP, et al. (1998). Aggregation and binding substances enhance pathogenicity in rabbit models of *Enterococcus faecalis* endocarditis. Infect Immun 66: 218–223.
17. Tailor SA, Bailey EM, Rybak MJ (1993). Enterococcus, an emerging pathogen. Ann Pharmacother 27: 1231–1242.
18. Trotter KM, Dunny GM (1990). Mutants of *Enterococcus faecalis* deficient as recipients in mating with donors carrying pheromone-inducible plasmids. Plasmid 24: 57–67.
19. Wirth R, An FA, Clewell DB (1986). Highly efficient protoplast transformation system for *Streptococcus faecalis* and a new *Escherichia coli-Streptococcus* shuttle vector. J Bacteriol 165: 831–836.

Address for correspondence: Gary Dunny, Department of Microbiology, Medical School, Box 196 UMHC, 1460 Mayo Memorial Building, 420 Delaware Street SE, Minneapolis, MN 55455, USA
Phone: 612-625-9930; Fax: 612-626-0623
E-mail: gary-d@biosci.cbs.umn.edu

Methods in Cell Science **20**: 85–93 (1998).

Genetic transfer methods for *Streptococcus sobrinus* and other oral streptococci

Donald J. LeBlanc[1], Yi-Ywan Chen[2], Nicole D. Buckley[3] & Linda N. Lee[4]

[1] *Department of Oral Biology, Indiana University School of Dentistry, Indianapolis, Indiana, USA;* [2] *Department of Dental Research, University of Rochester, Rochester, New York, USA;* [3] *Department de Biochimie (Sciences), Universite Laval, Quebec, Canada;* [4] *Department of Medicine (Infectious Diseases), University of Texas Health Science Center at San Antonio, San Antonio, Texas, USA*

Abstract. *Streptococcus mutans* and *Streptococcus sobrinus*, members of the mutans group of streptococci, are the major etiologic agents of human dental caries. Several properties of these two bacterial species have been proposed as virulence traits, and many of the genes encoding such traits have been cloned from both species, and their sequences determined. However, assessments of the contributions of these genes and their products to the cariogenicity of the human mutans streptococci requires the ability to replace or complement the original wild-type genes with genetically altered genes Whereas numerous strains of *S. mutans* can achieve a state of natural competence for transformation, or can be transformed by electroporation, which renders them amenable to genetic manipulation, there are no published procedures for the transformation of any strain of *S. sobrinus*. This report describes two methods for the transfer of plasmid vectors, and recombinant DNA molecules, to *S. sobrinus*. The first method utilizes an intermediate streptococcal host, a conjugative plasmid, and a vector molecule derived from a plasmid of mutans streptococcal origin, to mobilize an *E. coli/Streptococcus* shuttle vector into *S. sobrinus*. The second method involves the production of *S. sobrinus* cells competent for transformation by electroporation. Both protocols have been used for the construction of isogenic mutants of *S. sobrinus*, and should be applicable to other species and strains of streptococci not otherwise susceptible to genetic manipulation.

Key words: Conjugation, Mobilization, Shuttle vector, *Streptococcus sobrinus*, Transformation

1. Introduction

The ability of the mutans streptococci to produce dental caries is dependent, to a large extent, on the expression of several sucrose metabolic enzyme systems [12]. These systems may be involved in the production from sucrose of (1) extracellular polymers that contribute to the attachment of the bacteria to enamel surfaces; (2) lactic acid, which contributes to the demineralization of tooth enamel; (3) energy, which fuels all cellular functions; and (4) intracellular polymers that serve as carbon and energy stores during periods of starvation. At any given point in time, and under a particular set of environmental conditions, the activity of one enzyme system may account for a portion of one or more of the above products of metabolism, and two or more such systems may contribute to the total amount of any one product present. Clearly, the application of such molecular-genetic tools as allelic replacement and promoter-reporter gene fusions will be required in order to assess those environmental conditions under which each of the numerous sucrose catabolic enzyme systems may function. Allelic replacement techniques permit the construction of isogenic mutants deficient in only one, or a few specific gene functions at a time, and promoter-reporter gene fusions involve the placement under the control of regulatory elements of interest of genes whose products are easily assayed and quantified, while maintaining the original gene under control of the same regulatory elements. Application of such techniques requires the availability of genetic transfer systems by which manipulated genes can be delivered to target bacterial species. The expression of natural competence for transformation by many strains of *S. mutans* [22–24] has made possible the genetic and molecular manipulation of this mutans streptococcal species [10, 12]. Results of the application of these techniques have led to a better understanding of the roles of such sucrose catabolic enzymes as glucosyl- and fructosyl- transferases in the adhesion to teeth and cariogenicity of *S. mutans*. On the other hand, no such gene transfer systems have been available for strains of *S. sobrinus*. Natural competence has never been described in this mutans streptococcal species, and an earlier report of the successful electrotransformation of the 6715 strain of *S. sobrinus* [11], has yet to be published.

Two methods for the transfer of plasmid vectors,

as well as recombinant DNA derivatives of them, to *S. sobrinus* are described here. The first method is based on a combination of observations related to the genetics and plasmid biology of the oral streptococci. The transfer on solid surfaces, by a mechanism resembling conjugation, of the plasmid, pAMβ1, from a streptococcal donor to different species of oral streptococci, including *S. sobrinus*, and between strains of the latter, was first reported in 1978 [15]. Later, a small cryptic plasmid, pVA380-1, originally isolated from a strain of the mutans streptococcal species, *Streptococcus ferus* [21], was used as the streptococcal replicon in the construction of a number of different *E. coli/Streptococcus* shuttle vectors [8, 17, 20]. The nucleotide base sequence of the entire pVA380-1 molecule was determined and revealed the presence of a mobilization gene, *mob* [13, 17], the function of which was confirmed when pAMβ1 [7] was shown to mobilize it between streptococcal species, including *S. sobrinus* [13]. Subsequently, a new *E. coli/Streptococcus* shuttle vector, pDL289, was constructed that incorporated the pVA380-1 *mob* gene, and which also was shown to be mobilized by pAMβ1 [1]. Finally, a derivative of the above shuttle vector, pDL289Δ202 (see Figure 1), was constructed that facillitated the integration of fragments of *S. sobrinus* chromosomal DNA into their homologous regions on the *S. sobrinus* genome [2]. This vector, and plasmid mobilization, were used in the construction of isogenic mutants of *S. sobrinus* by allelic replacement [2]. The second method for the transfer of plasmid DNA to *S. sobrinus* is electrotransformation, and was adapted from a report of such transfer by Lassiter and Doyle at the Fourth International Symposium on Streptococcal Genetics in Santa Fe, New Mexico, in 1994 [11].

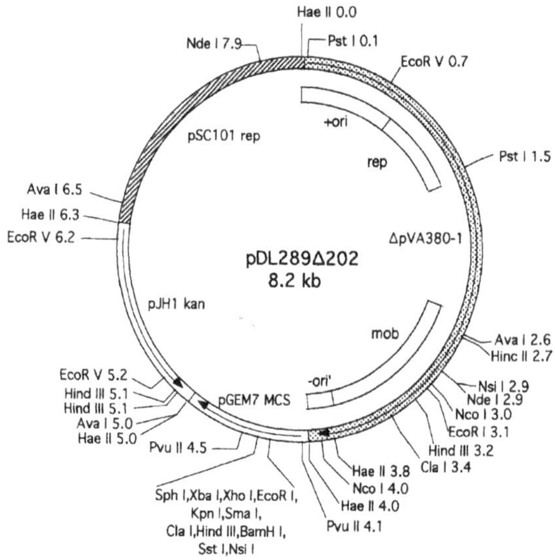

Figure 1. Detailed restriction endonuclease map of *E. coli/Streptococcus* shuttle vector, pDL289Δ202. DNA clockwise from coordinates *Hae*II 0.0, to *Hae*II 4.0 includes the plus origin of replication (*+ori*), replication protein (*rep*), mobilization gene (*mob*), and partially deleted minus origin (*–ori'*) of the streptococcal plasmid, pVA380-1 [J, K, H). Also clockwise, from coordinates *Hae*II 4.0 to *Hae*II 5.0, is the *lac*Zmcs region of the *E. coli* vector, pGEM7zf(–) (Promega). DNA counter clockwise from coordinates *Hae*III 6.3 to *Hae*II 5.0 contains the *kan* gene from pDL413 [V]. DNA clockwise from coordinates *Hae*II 5.0 to *Hae*II 0.0 contains the replication origin of pSC101, subcloned from pGB2 [U].

2. Materials

A. Chemicals

- Kanamycin monosulfate (Km), K-4000, Sigma.[1]
- Erythromycin (Em), E6376, Sigma.[1]
- Spectinomycin dihydrochloride (Sp), S9007, Sigma.[1]
- Rifamycin SV, sodium salt (Rf), R8626, Sigma.[1]
- Fusidic acid, sodium salt (Fa), F0881, Sigma.[1]
- Chloramphenicol (Cm), C0378, Sigma.[1]
- Glucose, G5250, Sigma.[1]
- Glycerol, G9012, Sigma.[1]
- L-Threonine, T8534, Sigma.[1]
- $MgCl_2$, M8266, Sigma.[1]
- NaOH, S8045, Sigma.[1]

B. Media and Reagents

- BHI, Brain Heart Infusion broth, 0037-17-8, Difco[2] (make up according to manufacturer's instructions).
- BHIG, BHI + 20 mM glucose (1 M glucose stock filter sterilized and added after autoclaving.
- BHIG + 10% heat inactivated (56 °C for 30 min) horse serum, H1270, Sigma.[1]
- BHI + 10% glycerol (add 10 ml 100% autoclaved glycerol solution per 90 ml BHI (sterile).
- THB, Todd Hewitt Broth, 0492-17-6, Difco[2] (make up according to manufacturer's instructions).
- THBYE, THB + 0.2% yeast extract, 0127-01-7, Difco[2] (add yeast extract to THB prior to autoclaving).
- BHIG agar & THBYE agar, 15 g Bacto agar, 0140-01, Difco[2] per liter of medium (1.5%).
- Bile-esculin agar, 0879-15-1, Difco[2] (make up according to manufacturer's instructions).
- BBL GasPak Generators, 43-71040 (H_2CO_2), 43-70308 (CO_2), Becton Dickinson.[3]
- EPM, Electroporation Mix [3], (272 mM glucose, 1 mM $MgCl_2$, pH 6.5), dissolve 4.9 g of glucose in ~90 ml of distilled water; add 0.1 ml of 1 M $MgCl_2$; add 1 mM NaOH until pH = 6.5 (~3 ml); adjust volume to 100 ml (do not back titrate).

- Antibiotic Stock Solutions, filter sterilize and
  - add to media after autoclaving;
  - Km or Sp, 50 mg/ml in distilled water;
  - Rf or Fa, 2.5 mg/ml in distilled water;
  - Cm, 1 mg/ml in distilled water;
  - Em, 1 mg/ml (dissolve in 1/10th to 1/5th final volume of absolute ethanol and bring to final volume with distilled water.

C. Equipment
- 37 °C water bath.
- Spectronic 20D$^+$, Fisher.[4]
- Refrigerated table top centrifuge, Sorvall RT6000B, Dupont.[5]
- –80 °C and –20 °C freezers.
- 1–20, 20–100, and 100–1,000 μl capacity pipettes with tips, Oxford BenchMate continuously adjustable pipettes, Oxford Labware.[6]
- GasPak jar, 11-814-21 (12 plate capacity) or 11-816 (36 plate capacity), Becton Dickinson.[3]
- 37 °C incubator, Lab Line, Imperial III, VWR Scientific Products.[7]
- Vortex Genie, Fisher.[4]
- Microcentrifuge, Sorvall MC12V, Dupont.[5]
- CO_2 incubator, Heraeus BB16, Heraeus.[8]
- Gene Pulser, 165-2105, Biorad.[9]

D. Glass and plasticware
- Screw cap test tubes, 16 mm × 125 mm (T1354-13), 13 mm × 100 mm (T1354-11), Scientific Products.[9]
- 250 ml capacity Pyrex bottles, 1395-250, Corning.[10]
- 12 mm × 75 mm Falcon snap cap polystyrene tube, 14-959A, Fisher.[4]
- 30 mm × 115 mm Falcon 50 ml capacity conical centrifuge tubes, 14-432-24, Fisher.[4]
- Microcentrifuge tubes, 0.5 ml capacity (05-408-16), 1.5 ml capacity (05-408-10), Fisher.[4]
- 0.1 cm electrode gap electroporation cuvettes, 1652089, Biorad.[10]
- Multi-flex microcapillary pipette, 107003, PGC Scientific.[11]
- Plastic disposable inoculating loop, 1906-95, Difco.[2]

E. Bacterial strains
- *Streptococcus gordonii* (Challis) strain DL1 [14].
- *Streptococcus sobrinus* 6715 and 6715RF [13].
- *Enterococcus faecalis* JH201 (pAMβ1) [9].
- *Escherichia coli* DH5α, Life Technologies.[12]

F. Bacterial plasmids and vectors
- pDL278 [17].
- pDL287 [13].
- pDL289 [1].
- pDL289Δ202 [2].
- pAMβ1 [7].

3. Procedures

A. Construction of donor strain for mobilization

1. Preparation of pre-competent stocks of *Streptococcus gordonii* strain DL1
   Transfer 10 to 40 μl of a frozen glycerol stock of the Challis strain of *S. gordonii* [strain DL1; 14] to 10 ml of BHI without any added carbohydrate in a 16 × 125 mm screw cap test tube and incubate overnight at 37 °C. Transfer the entire overnight culture to a 250 ml-capacity bottle containing 200 ml of BHIG. Mix cell suspension and transfer 5 ml to a 13 mm × 100 mm screw cap tube. Incubate bottle and tube in a 37 °C water bath. Follow growth by measuring the OD_{600} of the culture in the screw cap tube to between 0.5 and 0.6. Transfer the 5 ml culture to the bottle and chill the entire culture in an ice-water bath. Transfer the culture to 4 pre-chilled, sterile 50 ml capacity conical centrifuge tubes. Centrifuge for 15 min, at 4 °C, 2,500 rpm in a Sorvall refrigerated table top centrifuge. Discard the supernatant fluids and resuspend the cell pellets in a total volume of 10 ml chilled BHI + 30% glycerol. Aliquot the cell suspension into freezer tubes (150 μl each) and freeze quickly in a dry ice-ethanol bath. Store at –70 °C to –80 °C.
2. Transformation of the Challis strain of *S. gordonii* with a mobilizable plasmid
   Thaw a freezer tube containing pre-competent strain DL1, and transfer 0.1 ml to 10 ml BHI (= 10^{-2} dilution). Mix the cell suspension and transfer 0.1 ml to 10 ml BHIG + 10% heat inactivated horse serum. Final dilution of pre-competent culture = 10^{-4}. **Important:** BHI for this step should be freshly prepared (within the week of use). Incubate the 10^{-4} dilution of the precompetent culture 30–60 min at 37 °C. Actual time of incubation to maximum competence is dependent on the lot of horse serum. Generally, 100 ml bottles of each of 2 to 3 different lots are obtained from the supplier with a request to hold 9 others of each until the lots can be tested. Once the best lot (highest transformation frequencies with same plasmid) is identified, shipment of the remaining 9 bottles from that lot is requested. The contents of each bottle are heat inactivated and stored at –20 °C (in 1–5 ml aliquots). Transforming DNA (up to 10 μl; ~100 ng) is transferred to a 5 ml capacity Falcon tube, to which 100 μl of competent strain DL1 are added. Transforming DNA, in this case, is mobilizable plasmid pDL289, pDL289Δ202, or a recombinant derivative of one of these (see Results and discussion). Cap the tube loosely, and incubate at 37 °C, anaerobically

(Gas Pac H_2CO_2 generator) for 4 h. Spread 0.1 ml of the undiluted transformation mixture, as well as of 10-fold dilutions up to 10^{-4}, on BHI agar plates containing 500 μg/ml of Km to select for transformants. If the vector has an insert with a selectable marker, e.g., the Sp^R (^{R} = resistance, ^{S} = susceptible) gene from *E. faecalis*, spc [18], then the presence of this marker can either be selected for directly, or subsequent to direct selection for Km^R. Choose a few colonies to be tested for the presence of intact plasmid DNA by standard plasmid isolation and restriction endonuclease digestion procedures.

3. Transfer of pAMβ1 to the Challis strain of *S. gordonii* containing a mobilizable plasmid
Since pAMβ1may suffer deletion of a portion of its transfer region upon prolonged storage and/or growth of strain DL1 harboring it [19], it is best to transfer this plasmid by conjugation (deletion also may occur as a consequence of transformation) to strains already harboring the mobilizable plasmid. Obtain overnight cultures of strains DL1 containing the mobilizable plasmid and *E. faecalis* strain JH201 (pAMβ1) by transfer of 40 μl of frozen glycerol stock to 4.0 ml of BHI without added carbohydrate, but with appropriate antibiotic for plasmid selection (Km for the mobilizable plasmid and 10 μg/ml Em for pAMβ1) and incubation at 37 °C. Transfer 0.25 to 0.50 ml of each overnight culture to 5 ml BHIG (no antibiotic) in 13 mm × 100 mm screw cap tubes. Incubate in a 37 °C water bath and follow growth to OD_{600} of 0.5. Chill the first culture to reach an OD_{600} of 0.5 on ice until the second culture also reaches the same OD. Transfer 0.1 ml of donor culture and 0.1 ml of recipient culture to a 1.5 ml capacity microfuge tube containing 0.8 ml of BHI. Controls contain 0.1 ml of donor culture (or recipient culture) plus 0.9 ml BHI. Vortex mating and control cultures and centrifuge in a microfuge for 5 min to pellet cells. Remove the supernatant fluids with sterile tips for a 1 ml capacity pipettor. Resuspend the cell pellet in residual liquid (~10 μl) and spot the entire suspension on a BHI agar plate. Incubate 4 h at 37 °C anaerobically (GasPak H_2CO_2 generator). Scrape cells off the plate with a sterile plastic disposable inoculating loop, resuspend in 0.5 ml BHI, and vortex. Spread 0.1 ml aliquots onto BHI agar plates containing Em + Km (for presence of pAMβ1 in strain DL1 containing mobilizable plasmid), and incubate anaerobically at 37 °C for 24 to 48 h. Pick a few representative colonies to test for inability to hydrolyze bile esculin (*E. faecalis* donor strain will produce black colonies on bile esculin agar plates, whereas *S. gordonii* recipient strain, and transconjugants will not) and to check for the presence of both plasmids by standard techniques.

B. Mobilization of plasmid DNA from *S. gordonii* to *S. sobrinus* strain 6715. The matings are conducted in the same manner as in A. 3) above, except that *S. gordonii* containing pAMβ1 plus the mobilizable plasmid is used as the donor, and a strain of *S. sobrinus* resistant to Rf and Fa [*S. sobrinus* 6715RF; 2] serves as the recipient. BHIG agar containing 25 μg/ml each of Rf and Fa are used to select for the recipient strain, and for selection of specific transconjugants Km, Sp, or both, are also included, as described below. Km is used for selection of the mobilizable plasmid (or, e.g., Sp, if the mobilizable plasmid contains a *S. sobrinus* gene that has been inactivated by insertion of the *E. faecalis* spc gene). Only those transconjugants of *S. sobrinus* that have received the mobilizable plasmid will be able to grow in the presence of Km and/or Sp, or in which the insertionally inactivated gene has integrated into the recipient chromosome via a single (resistant to both Km and Sp) or double (resistant to Sp only) crossover event.

C. Transformation of *S. sobrinus* 6715 by electroporation. Transfer 25 μl of a frozen glycerol stock of *S. sobrinus* to 2.5 ml (1:100 dilution) of THBYE, and incubate at 37 °C ~6–8 h. Transfer 0.25 ml of the growing culture (even if only barely turbid) to 5 ml (1:20 dilution) THBYE + 300 mM L-threonine and incubate overnight at 37 °C in the presence of 10% CO_2. Transfer the overnight culture to 47.5 ml of pre-warmed (37 °C) THBYE + 300 mM L-threonine in a 50 ml capacity screw cap disposable conical centrifuge tube. Mix and transfer 5 ml to a 13 mm × 100 mm screw cap tube. Incubate both tubes in a 37 °C water bath. Follow growth of the culture by measuring the OD_{600} of the 5 ml culture at ~30–60-min intervals. At an OD_{600} of 0.2 to 0.3, transfer the 5 ml culture to the 45 ml culture and chill in an ice-water bath for 10 min. Pellet the cells in a table-top centrifuge at 4,000 rpm, for 15 min, at 4 °C. Pour off the supernatant fluid and immediately transfer the tube to an ice-water bath (if the pellet is not tight, centrifuge again, at a higher speed if necessary). Resuspend the pellet in 10 ml of chilled EPM, insuring that all clumps are broken up. Add 15 ml of chilled EPM and mix well. Centrifuge as above. Repeat resuspension of the cells in 10 + 15 ml of chilled EPM and centrifugation one time, and then resuspend the pellet in 10 ml of chilled EPM (for a final wash) and centrifuge as above. After discarding the final supernatant fluid, add 0.5 ml of chilled EPM to residual EPM in the tube and resuspend the cells well. Chill in an ice-water bath for 45 min. Place a pre-chilled electroporation cuvette (0.1 cm electrode gap) in ice and spot 1 to 2 μl of transforming DNA at the top of the cuvette. Wash the DNA down the side of the cuvette with 20 μl of chilled cell suspension. Apply a pulse at 1.6 kV, 400 Ω, 25 μFad. Record time constant (should be ~10 msec). Immediately after pulsing, return the cuvette to ice. Transfer the cells from the cuvette, with a sterile multi-flex microcapillary pipette tip, to a sterile 1.5 ml capacity microfuge tube containing 0.9 ml THBYE. Incubate in a 37 °C water bath for 90 min. Centrifuge approximately 1 min to pellet the cells, pour off the medium, and allow the tube to drain on its side. Resuspend the cells in 100 μl of THBYE and spread 50 μl on THBYE agar containing the appropriate selective antibiotic (Sp for pDL278, Km for pDL289 or pDL289Δ202, and/or antibiotic necessary to select for insertionally inactivated gene, see above). Incubate at 37 °C in the presence of 5–10% CO_2. Transformant colonies should be visible after 36 h, but may require up to 72 h of incubation. Always include a no DNA control treated in exactly the same manner as the transformation culture.

4. Results and discussion

The ability to mobilize plasmid DNA into *S. sobrinus* was demonstrated with a derivative of pVA380-1 in which the *kan* gene of pDL413 [16] had been cloned into a restriction endonuclease site (*Sca*I) located in a non-essential region of the plasmid. This derivative, pDL287, was transferred from *S. gordonii* Challis (pAMβ1 + pAM287) to *S. sobrinus* 6715-RF at frequencies in the range of 10^{-5} per recipient colony forming units (CFU) [13]. Less than 2% of transconjugants selected for the presence of pDL287 (resistant to Km) did not contain pAMβ1 (susceptible to Em). Thus, transconjugants containing only the mobilizable plasmid could be obtained quite readily, without the need to cure such transconjugants of the mobilizing plasmid. The next step toward making *S. sobrinus* amenable to recombinant DNA technology was the construction of a suitable mobilizable *E. coli/Streptococcus* shuttle vector.

The construction of pDL289, the *E. coli/ Streptococcus* shuttle vector mobilizable by pAMβ1, has been described [1], as has the deletion derivative of it, pDL289Δ202 [2]. However, complete physical maps of these plasmids, that would permit immediate assessment of useful restriction endonuclease sites, were never published. Such a map of pDL289Δ202 is presented in Figure 1. The original shuttle vector, pDL289, was constructed by the sequential addition of *Hae*II fragments containing the pSC101 origin of replication (1.9 kb) obtained from pGB2 [6], a *kan* gene (1.4 kb) obtained from pDL413 [16], the *lacZ*-mcs (0.9 kb) from pGEM7Zf(–) from Promega, and intact pVA380-1 (4.2 kb) [13, 17, 21]. The latter fragment was obtained by partial digestion of pVA380-1 with *Hae*II, due to the presence of two sites for this enzyme in the pVA380-1 molecule (see Figure 1). The 8.2 kb derivative, pDL289Δ202, was derived from pDL289 (8.4 kb) by Exonuclease III digestion of the latter, in both directions, after linearization by digestion with *Stu*I, for which there was a single site located near the center of the *–ori* of pVA380-1 [17]. Three of the sites in the mcs of pDL289 and pDL289Δ202, *Eco*RI, *Cla*I, and *Hin*dIII, also occur in the *mob* gene of the respective plasmids, and cannot be used for cloning into either plasmid without destroying the mobilizability of the vector.

When strains of *S. gordonii* Challis containing pAMβ1 plus either pDL289 or pDL289Δ202 were used as donors in conjugation experiments with *S. sobrinus* 6715-RF, Km resistant transconjugants were obtained at frequencies of 3×10^{-6} and 1×10^{-7} per recipient CFU, respectively. Less than 3% and 12% of the transconjugants containing pDL289 and pDL289Δ202, respectively, contained pAMβ1 [2]. Results of subsequent experiments in which either vector containing cloned DNA fragments was mobilized into *S. sobrinus*, revealed that the mobilizing plasmid, pAMβ1, was present in anywhere from 1 to 10% of transconjugants selected only for the mobilized plasmid. The reasoning behind the construction of pDL289Δ202 was that interuptions of the minus origin of the pVA380-1 portion of pDL289 would yield a much less stable plasmid in streptococcal hosts, due to the accumulation of single-stranded replicative intermediates [17]. It was reasoned that this would render the plasmid more readily curable, and that the presence of an abundance of single stranded replicative intermediates would facillitate recombination between the host chromosome and homologous DNA that had been cloned into the vector. A higher rate of curing of pDL289Δ202 over pDL289 has been demonstrated (3 to 10-fold), but the effects of the presence of single-stranded DNA containing sequences with homology to the host chromosome, on the frequency of homologous recombination, have yet to be assessed.

One example of the utility of pDL289Δ202 for the construction of isogenic mutants of *S. sobrinus* is described briefly here, and in greater detail in Buckley et al. [2]. A *spc* gene [18] was inserted into an *Eco*RV digested 2.7 kb *Hin*dIII fragment of *S. sobrinus* that contained the *scrA* gene encoding the sucrose permease, EII^{suc} [4]. This resulted in the insertional inactivation of the *scrA* gene by replacement of the 226 bp *Eco*RV fragment internal to *scrA* with the *spc* gene, as illustrated in Figure 2. The inactivated gene was subcloned into pDL289Δ202, and the resultant recombinant plasmid was used to transform *S. gordonii* Challis. After introduction of

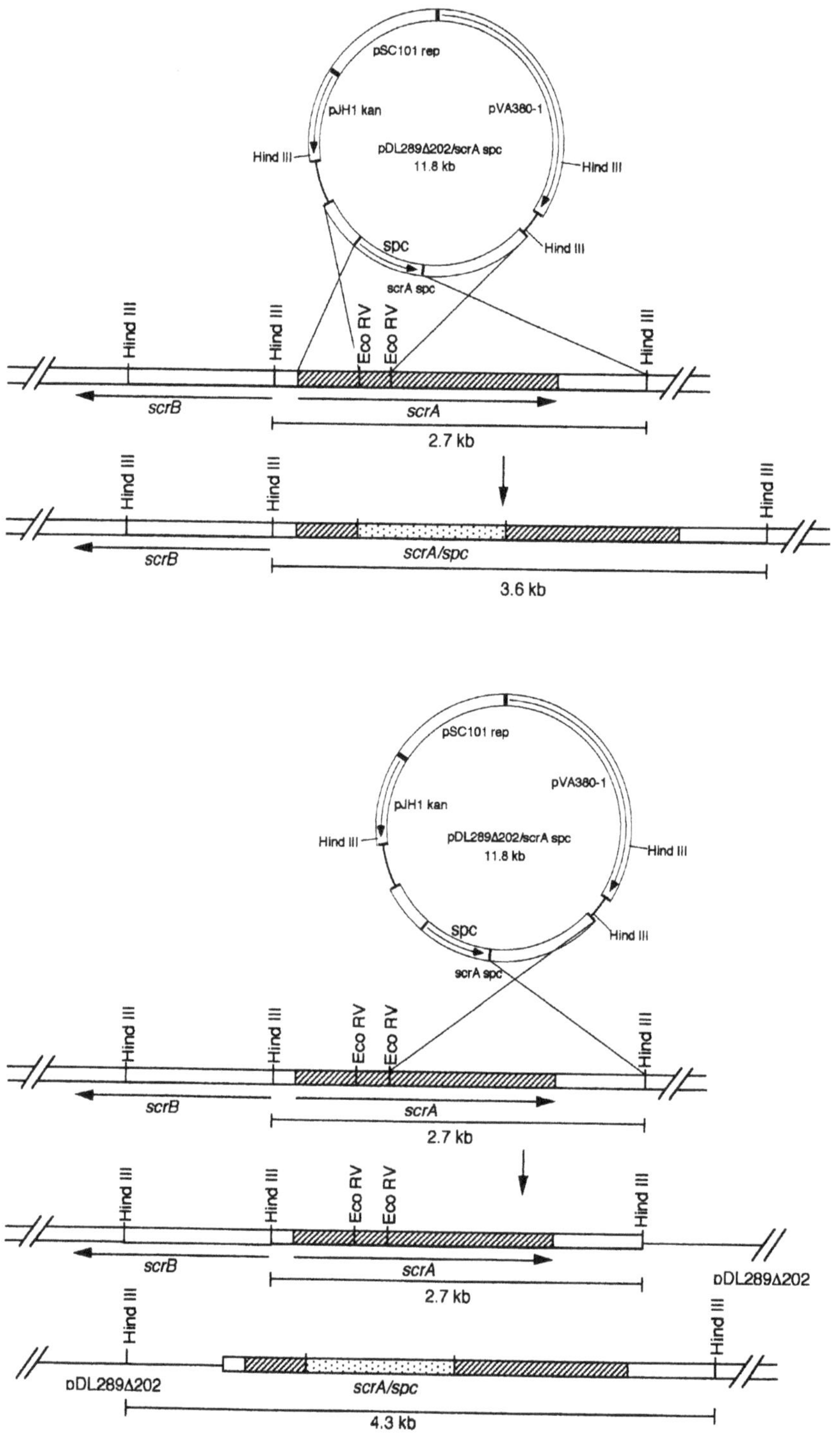

Figure 2. Diagrams of double and single crossover events between *scrA* in the chromosome of *S. sobrinus* and *scrA* interupted by *spc*, present on pDL289Δ202. Top segment depicts a double crossover event between the *S. sobrinus* chromosomal wild-type *scrA* gene within a 2.7 kb *Hin*dIII fragment of DNA, and the plasmid-borne *spc*-inactivated *scrA* gene in a 4.3 kb *Hin*dIII fragment, resulting in replacement of the wild-type gene on the chromosome with the inactivated gene in a 3.6 kb *Hin*dIII fragment. Lower segment depicts a single crossover event between the same chromosomal wild-typegene and the *spc*-inactivated *scrA* gene resulting in integration of the entire pDL289Δ202/*scrAspc* plasmid, and the presence of two copies of *scrA* in the *S. sobrinus* host chromosome, a wild-type allele within a 2.7 kb *Hin*dIII fragment, and an inactivated (by *spc* insertion) allele within a 4.3 kb *Hin*dIII fragment.

pAMβ1 into a representative transformant, *S. gordonii* (pDL289Δ/*scrAspc* + pAMβ1) was used as a donor in matings with *S. sobrinus* 6515-RF as the recipient. Transconjugants were obtained on media containing Rf, Fa, and Sp, and then tested for resistance to Km and to Em. Two different resistance phenotypes were observed, Sp^RKm^S, and Sp^RKm^R. None of the transconjugants tested were resistant to Em. Total cell DNA was prepared from representative transconjugant cultures and digested with *Hin*dIII. After separation of the DNA fragments by agarose gel electrophoresis, Southern blots were prepared and hybridized with a probe consisting of a DNA fragment internal to *scrA*, and including both sides of the two *Eco*RV sites. Two types of results were obtained, as predicted in Figure 2. Those transconjugants with the Sp^RKm^S phenotype yielded only a 3.6 kb *Hin*dIII fragment that hybridized to the *scrA*-specific probe. In these transconjugants, the 2.7 kb *Hin*dIII fragment containing the wild-type *scrA* gene was increased to 3.6 kb via a double crossover event in which the inactivated *scrA* gene replaced the wild-type gene (Figure 2, top segment). Greater than 50% of transconjugants selected for resistance to Sp only expressed the Sp^RKm^S phenotype, indicating that double-crossover events occur at relatively high frequencies in *S. sobrinus*. Transconjugants with the Sp^RKm^R phenotype yielded two *Hin*dIII fragments that hybridized to the *scrA*-specific specific probe, one of 2.7 kb and the other 4.3 kb in size. The size of the *Hin*dIII fragment on pDL289Δ202 containing the inactivated *scrA* gene is 4.3 kb in size. Although this latter group of transconjugants were not subjected to further analysis, three possibilities would explain these results. The first is that the plasmid, pDL289Δ202/*scrAspc* is replicating in the transconjugant strictly as an extrachromosomal element, and the wild-type *scrA* gene has not been affected. The second possibility is that the entire plasmid has integrated into the host chromosome via a single crossover event in which the wild-type and interrupted *scrA* genes each consist of DNA derived from the incoming plasmid and a portion derived from the recipient chromosome, as depicted in the bottom segment of Figure 2, and the plasmid has been lost. A third possibility is that integration of the plasmid has occurred by a single crossover event, but that it also continues to replicate in the host as an independent replicon. The three possibilities could be distinguished by conducting Southern hybridization experiments similar to those described above, but with DNA from the transconjugants with a Sp^RKm^R phenotype that has been digested with a different enzyme, e.g., *Eco*RI. In this case, transconjugants in which the plasmid is replicating solely as an independent replicon would yield a fragment that hybridized with the *scrA*-specific probe of approximately 10.6 kb in size. A second *Eco*RI fragment, 1.2 kb in size (see Figure 1), representing the DNA between the *Eco*RI site in the mcs and that in the *mob* gene, would not hybridize to the probe. The second possibility described above would yield a hybridizing *Eco*RI fragment that was much larger than 10.6 kb, since the *scrA* gene of *S. sobrinus* 6715 is in an *Eco*RI fragment of approximately 14 kb [4]. Transconjugants containing both an integrated and a replicating plasmid would yield two hybridizing *Eco*RI fragments, the 10.6 kb fragment, and the larger one.

The transformation system described under procedures has been used to transform *S. sobrinus* 6715 with the *E. coli/Streptococcus* shuttle vectors, pDL278 and pDL289, as well as with recombinant DNA derivatives of pDL278 and pDL289Δ202. Representative transformation frequencies obtained with each of these plasmids, all purified from *E. coli* DH5α, are shown in Table 1. At a concentration of 0.5 μg/ml, pDL278 transformed *S. sobrinus* at a frequency of $1.6\ \mu \times 10^3/\mu g$. This same plasmid, and with the same batch of competent cells, when present at a concentration of 1.0 μg/ml yielded a frequency of $5.4 \times 10^2/\mu g$ of DNA, suggesting that the transforming DNA reached saturation between 0.5 and 1.0 μg/ml. A fragment of *S. sobrinus* 6715 genomic DNA containing intact *scrB* plus the region 5′ of *scrB* encoding the putative promoter for *scrA* [5] (here designated *scrA*P), was cloned upstream of the promoterless *cat* gene of pCAT3-Basic (Promega). A DNA fragment containing *scrBscrA*P*cat* was then subcloned onto pDL278, and the recombinant DNA molecule was used to transform *S. sobrinus*. Transformation frequencies obtained with pDL278 were on average only approximately four times higher than those obtained with the recombinant pDL278 molecule (Table 1). Isolates of *S. sobrinus*} containing pDL278/*scrBscrA*P*cat* are resistant to Cm. Whether or not the plasmid is functioning as an independent replicon, or has integrated into the *S. sobrinus* chromosome via a single crossover event, or both, has yet to be determined. In general, frequencies of transformation of *S. sobrinus* with the mobilizable shuttle vector, pDL289, were at least 10-fold lower than the frequencies obtained with pDL278. A fragment of *S. sobrinus* chromosomal DNA from which most of *scrA* and *scrB* was deleted and replaced by the *spc* gene was subcloned into

Table 1. Frequencies of transformation of *S. sobrinus* 6715 with different plasmids

Transforming plasmid	Frequencies of transformation[a]
pDL278	$0.54–1.6 \times 10^3$
pDL278*scrBscrA*P*cat*	$2.30–3.0 \times 10^2$
pDL289	$0.15–1.5 \times 10^2$
pDL289*scrABspc*	$0.05–0.9 \times 10^2$

[a] Transformation frequencies expressed as transformants/μg DNA.

pDL289Δ202. This recombinant molecule, pDL289Δ202*scrABspc* also transformed *S. sobrinus*. One transformant with a Sp^RKm^R phenotype was divided into two cultures, A and B. Culture A was allowed to grow for 100 generations in the presence of Sp, while culture B was carried for the same number of generations in the absence of selection by any antibiotic. Dilutions of each culture were spread on agar media without added antibiotics, and 100 colonies from each were picked onto agar media containing either Sp or Km. Eighty-six percent of the colonies from culture A grew in the presence of Sp, but not Km, whereas the remainder grew in the presence of both antibiotics. Thirty-four percent of the colonies from culture B were resistant to Sp, but not Km, and the remaining 66 colonies were susceptible to both antibiotis. These results suggested that the plasmid was readily lost from the cells, and that, like the results obtained with pDL289Δ202*scrAspc* following its transfer to *S. sobrinus* via mobilization, double cross-over events due to homologous recombination were favored over single crossover events, following transfer by transformation.

Acknowledgment

This work was supported by DE08915 (DJL).

Notes on suppliers

1. Sigma Chemical Company, St. Louis, Missouri, USA
2. Difco Laboratories, Detroit, Michigan, USA
3. Becton Dickinson and Co., Cockeysville, Maryland, USA
4. Fisher Scientific, Pittsburgh, Pennsylvania, USA
5. Dupont, Wilmington, Delaware, USA
6. Oxford Labware, St. Louis, Missouri, USA
7. VWR Scientific, Bridgeport, New Jersey, USA
8. Heraeus Instruments, South Plainfield, New Jersey, USA
9. Scientific Products, McGaw Park, Illinois, USA
10. Bio-Rad Laboratories, Hercules, California, USA
11. PGC Scientifics, Gaithersburg, Maryland, USA
12. Life Technologies, Gaithersburg, Maryland, USA
13. Promega Corporation, Madison, Wisconsin, USA

References

1. Buckley ND, Lee LN, LeBlanc DJ (1995). Construction of a mobilizable vector for genetic analysis of *Streptococcus sobrinus*. Dev Biol Stand 85: 399–401.
2. Buckley ND, Lee LN, LeBlanc DJ (1995). Use of a novel mobilizable vector to inactivate the *scrA* gene of *Streptococcus sobrinus* by allelic replacement. J Bacteriol 177: 5028–5034.
3. Caparon MG, Scott JR (1991). Genetic manipulation of pathogenic streptococci. Meth Enzymol 204: 556–586.
4. Chen Y-YM, LeBlanc DJ (1992). Genetic analysis of *scrA* and *scrB* from *Streptococcus sobrinus* 6715. Infect Immun 60: 3739–3746.
5. Chen Y-YM, Lee LN, LeBlanc DJ (1993). Sequence analysis of *scrA* and *scrB* from *Streptococcus sobrinus* 6715. Infect Immun 61: 2602–2610.
6. Churchward G, Belin D, Nagamine Y (1984). A pSC101-derived plasmid which shows no sequency homology to other commonly used cloning vectors. Gene 31: 165–171.
7. Clewell DB, Yagi Y, Dunny GM, Schultz SK (1974). Characterization of three plasmid deoxyribonucleic acid molecules in a strain of *Streptococcus faecalis*: Identification of a plasmid determining erythromycin resistance. J Bacteriol 117: 28–289.
8. Dunny GM, Lee LN, LeBlanc DJ (1991). Improved electroporation and cloning vector system for gram-positive bacteria. Appl Environ Microbiol 57: 1194–1201.
9. Hershfield V (1979). Plasmids mediating multiple drug resistance in group B streptococcus: Transferability and molecular properties. Plasmid 2: 137–149.
10. Kuramitsu HK (1993). Virulence factors of mutans streptococci: Role of molecular genetics. Crit Rev Oral Biol Med 4: 159–176.
11. Lassiter MO, Doyle RJ (1994). Transpositional mutagenesis of *Streptococcus sobrinus* 6715 by electroporation with Tn*916*. In: Program and Abstracts of the ASM 4th International Conference on Streptococcal Genetics.
12. LeBlanc DJ (1994). The role of sucrose metabolism in the cariogenicity of the mutans streptococci. In: Miller VL, Kaper JB, Portnoy DA, Isberg RR (eds), The molecular biology of pathogenesis, pp 465–477. Washington, DC: American Society for Microbiology.
13. LeBlanc DJ, Chen Y-YM, Lee LN (1993). Identification and characterization of a mobilization gene in the streptococcal plasmid, pVA380-1. Plasmid 30: 296–302.
14. LeBlanc DJ, Hassell FP (1976). Transformation of *Streptococcus sanguis* Challis by plasmid deoxyribonucleic acid from *Streptococcus faecalis*. J Bacteriol 128: 347–355.
15. LeBlanc DJ, Hawley RJ, Lee LN, St. Martin EJ (1978). 'Conjugal' transfer of plasmid DNA among oral streptococci. Proc Natl Acad Sci USA 75: 3484–3487.
16. LeBlanc DJ, Inamine JM, Lee LN (1986). Broad geographical distribution of homologous erythromycin, kanamycin, and streptomycin resistance determinants among group D streptococci of human and animal origin. Antimicrob Agents Chemother 29: 549–555.
17. LeBlanc DJ, Lee LN, Abu-Al-Jaibat A (1992). Molecular, genetic, and functional analysis of the basic replicon on pVA380-1, a plasmid of oral streptococcal origin. Plasmid 28: 130–145.
18. LeBlanc DJ, Lee LN, Inamine JM (1991). Cloning and nucleotide base sequence analysis of a spectinomycin adenyltransferase (AAD(9) determinant from *Enterococcus faecalis*. Antimicrob Agents Chemother 16: 686–689.

19. Macrina FL, Keeler CL Jr, Jones KR, Wood PH (1980). Molecular characterization of unique deletion mutants of the streptococcal plasmid, pAMβ1. Plasmid 4: 8–16.
20. Macrina FL, Tobian JA, Jones KR, Evans RP, Clewell DB (1982). A cloning vector able to replicate in *Escherichia coli* and *Streptococcus sanguis*. Gene 19: 345–353.
21. Macrina FL, Wood PH, Jones KR (1980). Genetic transformation of *Streptococcus sanguis* (Challis) with cryptic plasmids from *Streptococcus ferus*. Infect Immun 28: 692–699.
22. Murchison HH, Barrett JF, Cardineau GA, Curtiss R 3rd (1986). Transformation of *Streptococcus mutans* with chromosomal and shuttle plasmid (pYA629) DNAs. Infect Immun 54: 273–282.
23. Perry D, Kuramitsu HK (1981). Genetic transformation of *Streptococcus mutans*. Infect Immun 32: 1295–1297.
24. Westergren G, Emilson CG (1983). Prevalence of transformable *Streptococcus mutans* in human dental plaque. Infect Immun 41: 1386–1388.

Address for correspondence: Donald J. LeBlanc, Lilly Research Laboratories, Eli Lilly and Company, Lilly Corporate Center, Drop Code 0438, Indianapolis, IN 46285, USA
Phone: (317)-433-3922; Fax: (317)-276-1743
E-mail: leblanc_donald_j@lilly.com

Methods in Cell Science **20**: 95–106 (1998).

Isolation of enterococcal antigen-encoding genes from genomic libraries

Yi Xu[1], Lingxia Jiang[1,*], Xiaomei Jin[3], Barbara E. Murray[2,3,4] & George M. Weinstock[1,2,4]

[1] *Department of Biochemistry and Molecular Biology;* [2] *Department of Microbiology and Molecular Genetics;* [3] *Division of Infectious Diseases, Department of Medicine;* [4] *Center for the Study of Emerging and Re-emerging Pathogens, University of Texas Medical School, Houston, Texas, USA (*Current address: The Institute for Genomic Research, Rockville, Maryland, USA)*

Abstract. We have devised a procedure using immune sera to identify antigen-encoding genes of strains of *Enterococcus faecalis*. First, genomic cosmid libraries containing large inserts were constructed and screened with sera from patients with enterococcal infectious endocarditis and with serum from a rabbit immunized with surface proteins of an enterococcal endocarditis isolate. Immunopositive cosmid clones were analyzed by restriction enzyme digestions and clones containing distinct inserts were chosen for subcloning. Sublibraries were screened with one of the five sera, and immunopositive subclones were subjected to DNA sequencing. BLASTX and BLASTN at NCBI were used to search for database similarities.

Key words: Antigen, Cosmid, Immunoscreening, Library construction

1. Introduction

Antigens of pathogenic bacteria are important in several ways: they may be good candidates for developing vaccines and serodiagnostic tools, they are potential targets for designing new drugs, and they can be cell surface associated or secreted and thus include molecules that interact with the host. Some antigens of pathogenic bacteria are in fact virulence factors which make them useful for studying the mechanisms of pathogenicity. Enterococci are Gram-positive organisms of intestinal origin with AT-rich DNA. They are among the leading causes of hospital-acquired infections in the United States and, prior to the emergence of vancomycin resistance, about 85–90% of the clinical enterococcal isolates were *Enterococcus faecalis* [18]. With the development of resistance to multiple antibiotics in the past several decades, especially the recent emergence of vancomycin-resistant strains, enterococci have posed a potential threat for effective therapy using currently available antimicrobial agents, particularly in patients with endocarditis caused by enterococci resistant to aminoglycosides. Thus, a better understanding of enterococcal infection mechanisms, which might lead to the development of new therapeutics or preventive modalities, is highly desirable.

We describe here a procedure for the construction of cosmid libraries to identify potential antigen-encoding genes of *E. faecalis* strains.

2. Materials

A. Chemicals
1. 4-chloro-1-naphthol, No. C-8890.[14]
2. Agarose DNA grade, No. BP164-100.[7]
3. Albumin, bovine serum, No. A-9647.[14]
4. Brain heart infusion (BHI), No. 0037-01-6.[6]
5. Gigapack III Gold Packaging Extract Kit, No. #200202.[15]
6. Hydrogen peroxide 30% (w/w) solution, No. H-1009.[14]
7. Klenow fragment of DNA polymerase I, No. PR-M2201.[7]
8. Lysozyme, No. L-6876.[14]
9. Non-fat dry milk (blotting grade), No. 170-6404.[2]
10. Phenol:chloroform, No. 0883-100ML.[1]
11. Protein A-peroxidase, No. P-8651.[14]
12. Proteinase K, No. BP1700-100.[7]
13. Qiagen Plasmid Mini Kit (100), No. 12125.[13]
14. RNase, No. R-9134.[14]
15. RQ1 RNase-free DNase, No. PR-M6101.[7]
16. *Sau*3AI, No. PR-R6191.[7]
17. SeeBlue™ Pre-stained Standards, No. LC5625.[12]
18. Shrimp alkaline phosphatase (SAP), No. 70092.[16]
19. T4 DNA ligase, No. 202S.[11]
20. Zwittergent 3-12, No. 610170.[3]

B. Equipment
1. 96-well microtiter dish, No. 25850.[5]
2. Boekel replicator, No. 05-450-9.[7]
3. Immobilon-P transfer membrane, No. IPVH 00010.[10]

4. Klett-summerson photoelectric colorimeter, No. 800-3.[8]
5. Microplate fluorometer, No. 7625.[4]
6. MilliBlot™ – SDE System, No. MBBD SDE 00.[10]
7. NitroPlus nitrocellulose transfer membrane, No. WO2HY00010.[9]

3. Procedure

A. Solutions

1. 1% skim milk: 1% non-fat dry milk (w/v) in 50 mm Tris-HCl (pH 7.4).
2. 10% sucrose solution: 10% (w/v) sucrose in 10 mM Tris-HCl (pH 8.0), 10 mM NaCl, 1 mM EDTA (pH 8.0); 40% sucrose solution: 40% (w/v) sucrose in 10 mM Tris-HCl (pH 8.0), 10 mM NaCl, 1 mM EDTA (pH 8.0).
3. 10× gel loading buffer (for agarose gel electrophoresis): 20% Ficoll 400, 0.1 M EDTA (pH 8.0), 1% SDS, and 0.25% bromophenol blue.
4. 1× DNase I reaction buffer: 50 mM Tris-HCl (pH 7.5), 10 mM $MnCl_2$.
5. 2× loading buffer (for SDS-PAGE): 100 mM Tris-HCl (pH 6.8), 200 mM dithiothreitol, 4% SDS, 0.2% bromophenol blue, and 20% glycerol.
6. 3% skim milk blocking buffer: 3% non-fat dry milk (w/v) in 50 mM Tris-HCl (pH 7.4).
7. Color development mix: 10 ml of 0.3% 4-chloro-1-naphthol (w/v in methanol), 40 ml of 50 mM Tris-HCl (pH 7.4), and 18 µl of hydrogen peroxide 30% solution.
8. J-buffer: 0.1 M Tris-HCl (pH 8.0), 0.1 M EDTA (pH 8.0), 0.15 M NaCl.
9. Lysis buffers: the stock is 100 mM Tris-HCl (pH 8.0), 150 mM NaCl, and 5 mM $MgCl_2$. Before use, to 100 ml of stock, add 1.5 g of bovine serum albumin, 10 µl of RQ1 RNase-free DNase, and 4 mg of lysozyme (lysis buffer I) or 40 mg of lysozyme (lysis buffer II).
10. Protein A-peroxidase solution: stock solution is 125 µg/ml in PBS (pH 7.7), dilute 1000 fold with 1% skim milk prior to use.
11. SM buffer (per liter): 5.8 g of NaCl, 2.0 g of $MgSO_4 \cdot 7H_2O$, 50.0 ml of 1 M Tris-HCl (pH 7.5), 5.0 ml of 2% (w/v) gelatin, add distilled H_2O to a final volume of 1 liter, autoclave.
12. TNT: 10 mM Tris-HCl (pH 8.0), 150 mM NaCl, and 0.05% (v/v) Tween 20.
13. TSS: LB broth, 10% PEG 3500 or 8000, 5% DMSO, 50 mM $MgCl_2$, pH 6.5, can be autoclaved together.

B. Construction of cosmid libraries

1. Preparation of genomic DNA from *Enterococcus faecalis* strain OG1RF.

Preparation of enterococcal genomic DNA was described by Murray et al. [19]. A simpler method, based on a procedure for isolating genomic DNA from *E. coli* [8], was also used and yielded sufficiently good quality DNA for the construction of cosmid libraries. In this procedure, OG1RF is grown overnight at 37 °C with shaking in 250 ml of Brain Heart Infusion (BHI) broth. The cells are pelleted by centrifugation at 4000 rpm for 10 minutes, washed once with 250 ml of J-buffer and resuspended in 8 ml of J-buffer. The suspension is then incubated at different temperatures after the addition of various reagents, as described:

a. Add 1 ml of freshly-made lysozyme solution (10 mg/ml in 0.25 m Tris-Cl, pH 8.0), incubate at 37 °C for 10 minutes.
b. Add 20 µl of boiled RNase solution, incubate at 37 °C for 10 minutes.
c. Incubate at 70 °C for 3 minutes, then add 0.8 ml of 30% sarcosyl (w/v in water), mix by gentle inversion several times (from this point on, occasional gentle inversions should be carried out to ensure good mixing), return to 70 °C for 20 minutes, and then shift to 37 °C for 1 hour.
d. Add 20 mg proteinase K, incubate at 37 °C for 2 to 4 hours.
e. Add another 20 mg proteinase K, transfer the solution to a dialysis tube and dialyze overnight at 37 °C against a buffer consisting of 0.01 M Tris (pH 8.0), 0.01 M EDTA (pH 8.0), and 0.15 M NaCl.
f. Transfer the solution to a plastic tube, extract twice with phenol:chloroform. Mix by gentle inversion for about 10 minutes. Extract with ether twice, and then dialyze the DNA for several hours against TE.
g. Transfer the DNA solution to a clean tube and measure the DNA concentration in a microplate fluorometer.

2. Preparation of *Sau*3AI partially digested genomic DNA.

Partial digestion with *Sau*3AI was based on the method describe by Sambrook et al. [22]. First, a pilot experiment in which a fixed amount of genomic DNA was used with a serial dilution of *Sau*3AI was carried out to determine the digestion conditions which give the best yield of DNA fragments in the desired size range (20–50 kb). The conditions determined in the pilot experiment were then applied to a large-scale digestion.

a. Pilot experiment
   1) Dilute 20 µg genomic DNA in 1× *Sau*3AI reaction buffer to a final concentration of 1 µg/20 µl.
   2) Prepare 10 eppendorf tubes. Add 40 µl

of the above diluted DNA to tube 1, 20 µl to tubes 2 through 10. Keep all tubes on ice.

3) Add 2 units of *Sau*3AI to tube 1, mix gently. Transfer 20 µl from tube 1 to tube 2, mix gently. Transfer 20 µl from tube 2 to tube 3, mix. Continue till tube 9. Discard 20 µl from tube 9. Do not add anything to tube 10. All the tubes should be kept on ice in this step.
4) Incubate all tubes at 37 °C for exactly 1 hour.
5) Heat the tubes at 70 °C for 15 minutes to inactivate the enzyme. Spin briefly. Add 2.2 µl 10× gel loading buffer, load onto a 0.4% agarose gel. Run the gel slowly (< 1V/cm). Take pictures.

b. Large-scale preparation of partially digested genomic DNA

1) Choose a condition in the pilot experiment that gives the highest yield of DNA fragments with sizes of 20–50 kb, in our case, it was about 0.7 units *Sau*3AI/mg DNA at 37 °C for 1 hour. Digest each 100 µg of high-molecular-weight genomic DNA with 0.07 and 0.14 units of *Sau*3AI, respectively, for 1 hour at 37 °C.

   Since the conditions in the large scale digestion are not absolutely identical to the conditions in the pilot, it is worthwhile to do an additional digestion using one of the neighboring conditions.
2) Add EDTA to a final concentration of 12 mM at the end of the reactions (for a 200 µl reaction, add 4.8 µl of 0.5 M EDTA, pH8.0). Load approximately 5 µg of DNA from each digestion reaction onto a 0.4% agarose gel to examine the extent of the digestions. Pool the digests and store at 4 °C.
3) Prepare a 38 ml 10–40% continuous sucrose density gradient in a Beckman SW 28 ultracentrifuge tube (or equivalent). Adjust the volume of the pooled digests to 2 ml with TE, and carefully load onto the gradient. Centrifuge at 24,000 rpm for 20 hours at 20 °C.
4) Prepare a rack of microfuge tubes. A 40 ml gradient will need about 80 tubes. Number all tubes and place them in appropriate racks in order. Collect sucrose fractions, ~0.5 ml/tube.
5) Dilute 20 µl of every fourth fraction with 20 µl water, add 4.4 µl of 10× gel loading buffer. Analyze by electrophoresis through a 0.4% agarose gel.
6) Pool the gradient fractions containing DNA fragments of the desired size (35 to 50 kb for pBeloBAC11, 30 to 40 kb for pTEX5176, and 20 to 35 kb for pLAFRx). Dialyze against TE at 4 °C, change buffer 3 to 4 times.
7) Transfer the DNA solutions to fresh microfuge tubes and precipitate with ethanol.
8) Dissolve the DNA in TE to an approximate concentration of 400 ng/µl, measure DNA concentration and check by electrophoresis through a 0.4% agarose gel. Store at 4 °C.

3. Preparation of vector DNA

a. Large-scale preparation of cosmid DNA uses the alkaline lysis procedure followed by CsCl-ethidium bromide gradient centrifugation [22].

b. Digest 5 µg clean DNA with 1 unit of *Bam*HI at 37 °C for 1 hour. Remove an aliquot and apply to a 0.7% agarose gel to examine if the digestion is complete. If it is not, add more enzyme and continue incubation for another hour.

c. When digestion is complete, extract the sample with phenol:chloroform and precipitate with ethanol. Dissolve in 50 µl TE. Remove 5 µl DNA solution and save for later use.

d. Add 10 µl 10× shrimp alkaline phosphatase (SAP) reaction buffer, 45 µl water and 1 unit SAP to the DNA solution, incubate at 37 °C for 1 hour. Heat at 65 °C for 20 min to inactivate the enzyme.

e. Extract twice with phenol:chloroform, precipitate with ethanol. Dissolve in 50 µl TE (approximately 100 ng/µl). Measure DNA concentration.

f. Test ligations of vector DNA

Set up six 5 µl ligation reactions containing the following:

1) 100 ng SAP treated vector + 50 ng *Bam*HI digested pBluescript SK (–) (as control insert DNA) + 1.5 units of T4 DNA ligase.
2) 100 ng SAP treated vector + 50 ng *Bam*HI digested pBluescript SK (–), no ligase.
3) 100 ng SAP treated vector +1.5 units T4 DNA ligase.
4) 100 ng SAP treated vector, no ligase.
5) 100 ng *Bam*HI digested, vector not treated with SAP + 1.5 units T4 DNA ligase.
6) 100 ng *Bam*HI digested, vector not treated with SAP, no ligase. Add 0.5 µl 10× ligation buffer, and water to each mixture. Incubate at room temperature for 4 hours or at 14 °C overnight. Add

0.6 μl gel loading buffer and load onto a 0.7% agarose gel to examine the extent of ligations.

4. Ligation of cosmid vector to partially digested genomic DNA, in vitro packaging and storage
   a. Set up two ligation reactions both containing 1 μl vector DNA, 0.5 μl 10× ligation buffer, 0.5 μl T4 ligase, and 1 μl water. Add 2 μl partially digested genomic DNA to the first tube, and 2 μl water to the second. Incubate at room temperature for 4 hours or 14 °C overnight.
   b. Preparation of host bacteria: inoculate 5 ml of LB with 0.2% (w/v) maltose-10 mM $MgSO_4$ with a fresh single colony of *E. coli* DH5α and the control host VCS257 (included in the Gigapack III Gold Packaging Extract kit), respectively, shake at 37 °C for 4–6 hours (do not grow past an OD_{600} of 1.0). Harvest the bacteria by centrifugation at 4000 rpm for 10 minutes. Gently resuspend the cells to an OD_{600} of 0.5 with sterile 10 mM $MgSO_4$ (should be used immediately after this).
   c. In vitro packaging and plating follow the instructions from the Gigapack III Gold Packaging Extract kit with slight modifications. A small-scale test packaging should be carried out first. Quickly thaw a packaging extract, carefully split into 4 aliquots. To each aliquot, add 0.5 μl of the first ligation mixture (vector + insert), 0.5 μl of the second ligation mixture (vector alone), 0.5 μl of wild-type lambda DNA (included in the kit), and 0.5 μl of water, respectively. Mix well by stirring each tube with a pipet tip. Incubate the tubes at room temperature for 2 hours (do not exceed 2 hours). Add 125 μl of SM buffer to each tube, add 5 μl of chloroform and mix gently. Store at 4 °C until ready for titering (1 to 2 days).
   d. Titering the test packaging reactions: prepare a 1:10 and a 1:50 dilution of each of the packaging reactions in SM buffer. Mix 25 μl of each dilution with 25 μl of the prepared host cells. For lambda DNA, VCS257 is used as plating cells, and for the others, DH5α is used. After incubation of the tubes at room temperature (RT) for 30 minutes, 200 μl of LB broth is added to each sample. Incubate the tubes at 37 °C for 1 hour with occasional shaking. Plate the cells on LB agar plates with appropriate antibiotics. Incubate overnight at 37 °C. Count the colonies and calculate the titer to colony forming units (cfu)/μg DNA. Plating of the lambda control follows the instruction manual and plaques are counted to calculate the efficiency as plaque forming units/μg DNA. Pick a number of colonies and inoculate LB broth with appropriate antibiotics to prepare DNA (will be described below). Digest the DNA with restriction enzymes to determine the sizes of the inserts.
   e. When the test packaging produces good results, a normal scale packaging can be performed following the instruction manual. Briefly, the remains of the first ligation mixture is split into two and added to two packaging extracts. After incubation at RT for 2 hours, add 500 μl of SM buffer followed by 25 μl of chloroform to each reaction. A positive control with lambda DNA and VCS257 is also carried out. Based on the colony numbers from the test packaging and plating, dilute the packaging reaction to a level that will yield approximately 100 to 200 colonies/plate. Use a freshly prepared batch of cells for plating.
   f. Storage of the cosmid libraries: add 200 μl LB broth with the appropriate antibiotics to the wells of 96-well microtiter plates. Pick with sterile toothpicks single colonies from the primary plating to each of the wells. Incubate the microtiter plates at 37 °C overnight with shaking. Add 30 μl of sterile glycerol to each well (final concentration 15%), transfer 115 μl culture from each well to a new microtiter plate as a back up. Store both at −80 °C. Store the remains of the packaging reaction mixture at 4 °C in the presence of chloroform in a clean glass tube.

C. Preparation of the antisera.
   1. Source of antisera
      Four sera were collected from patients diagnosed by their physicians as having *E. faecalis* endocarditis. A rabbit serum was obtained from a rabbit immunized with surface proteins of an *E. faecalis* endocarditis isolate [24].
   2. Testing the antisera against enterococcal strains and *E. coli* by Western blot analysis:
      a. Surface protein extracts were prepared using the detergent Zwittergent 3–12 [12]. Inoculate BHI broth with single colonies of *E. faecalis* strains and LB broth with *E. coli*, and shake overnight at 37 °C. Adjust to a density of about 160 Klett units (measured in a Klett-Summerson photoelectric colorimeter) with BHI or LB broth. Take an equal volume from each adjusted culture, harvest the cells by centrifugation at 4000 rpm for 10 minutes, and wash once with equal volume of phosphate buffered saline (PBS). Resuspend in 1% of the adjusted culture volume in PBS with 0.2%

Zwittergent 3–12. Place the suspension in a roller for 1 hour at RT, transfer to fresh microfuge tube and centrifuge for 5 minutes at 10,000 g. Dialyze the supernatant against 50 mm Tris-HCl, pH 7.5 overnight at 4 °C, and store at –20 °C.

b. SDS-polyacrylamide gel electrophoresis (PAGE) is performed based on the Laemmli system, using a 4% stacking and 10% separating gel in a Tris-glycine buffer system [22]. Add an equal volume of 2× loading buffer to each extract from about 2 ml of the initial adjusted culture, boil for 5 minutes and subjects to SDS-PAGE.

c. Transfer proteins from the gel to an Immobilon-P transfer membrane using the MilliBlot™-SDE System. Incubate the membrane in 3% skim milk blocking buffer for at least 1 hour at room temperature with gentle shaking, and then incubate with the primary antisera solution (1:500 dilution in 1% skim milk) overnight at 4 °C with gentle shaking. Wash the membrane with 1% skim milk 3 times (5 minutes each time), add the protein A-peroxidase solution, and incubate for at least 1 hour at RT. Wash the membranes thoroughly 3 times with 50 mm Tris, pH 7.4 (5 minutes each time), and then incubate with the color development mix for 30 minutes, covered with a sheet of aluminum foil. Rinse the membranes thoroughly with distilled water, and air dry at RT.

3. Absorption of antisera with *E. coli* lysates.

a. Preparation of lysates follows the method described [22]. Briefly, grow the *E. coli* host strain containing the cloning vector in 100 ml of LB broth with appropriate antibiotics. Harvest the cells by centrifugation at 4000 rpm for 10 minutes. Resuspend the cells in 3 ml of 50 mM Tris-HCl (pH 8.0) and 10 mM EDTA (pH 8.0). Freeze and thaw for a few times, and then sonicate to lyse the cells. Store the lysate at –20 °C.

b. Absorption: dilute 100 μl of the antiserum with 850 μl of 1% skim milk, add 50 μl of the lysate, and incubate at RT for 4 hours with gentle shaking. Store the absorbed serum at –20 °C.

D. Immunoscreening of the cosmid libraries for antigen-expressing cosmid clones

1. Preparation of filters. Label a number of sterile NitroPlus nitrocellulose transfer membrane filters which have been cut to the size of a 96-well microtiter plate (or 137 mm circular filters), and carefully place each on top of a LB agar plate with the appropriate antibiotics. Inoculate library clones stored in 96-well microtiter plates onto the filters using a multi-prong Boekel replicator. Incubate at 37 °C overnight. Lift the filters from the plates, and place in a chloroform chamber colony side up, for 15 minutes (this should be done in a chemical fume hood). Transfer the filters to lysis buffer I and shake at RT for 12–16 hours. Alternatively, transfer the filters to lysis buffer II which contains a higher concentration of lysozyme, incubate at RT for 1 hour, transfer to a fresh batch of lysis buffer II and incubate at RT for another hour. Transfer to TNT buffer, use a gloved finger to wipe off any residue of the colonies from the filters, incubate for 10 minutes at RT with shaking. Repeat twice. The filters are ready to use in immunoblotting, or can be wrapped in Saran Wrap and stored at 4 °C for 1–2 days.

2. Blocking, incubation with antiserum and protein A-peroxidase and color development are the same as in Western blot.

3. Re-screen. Immunopositive clones from the initial screening are streaked and picked to a new 96-well microtiter dish containing 200 ml of LB broth with appropriate antibiotics in each well, and incubated at 37 °C overnight with shaking. Add glycerol to each well, and split the contents into two dishes as in the storage of library clones. Inoculate these clones onto the filters and screen again as described above. All five sera are used in the second screen.

E. Subsequent analysis of antigen-expressing cosmid clones

1. DNA preparation and restriction enzyme digestion (RED) analysis

a. Grow up clones in 10 ml of LB broth with appropriate antibiotics, and carry out the standard SDS-alkaline lysis method [22]. Resuspend the cell pellet in 150 μl of solution I, add 300 μl of freshly made solution II, mix, incubate at RT for 5 minutes, add 225 μl of solution III, mix and chill on ice for 5 minutes. Centrifuge for 10 minutes in a desktop centrifuge (15,000 rpm), transfer the supernatants to clean microfuge tubes. Extract with phenol:chloroform, and precipitate with ethanol. Resuspend the DNA pellets in 50 μl of TE with RNase.

b. Set up restriction digestions in 50 μl reaction volumes. For pBeloBAC11 clones, use 25 μl of the miniprep DNA, and for pLAFRx and pTEX5176 clones, use 10 μl of DNA. Add 5 μl 10× reaction buffer, 1 μl of enzyme (*Eco*RI or *Hin*dIII), and water to bring the volume up to 50 μl. Incubate at 37 °C for 2–3 hours. Add 5.5 μl 10× gel loading buffer, apply to a thick 0.7% agarose gel and carry out electrophoresis.

c. Compare RED patterns to find clones with overlapping inserts.

2. Western blot analysis
   a. Preparation of protein extracts from recombinant clones involved the use of 3 methods. First, the detergent Zwittergent 3–12 is used as in the preparation of surface protein extracts from enterococcal strains. Second, 0.5 ml of an overnight culture of recombinant clones is centrifuged, resuspended in 40 µl of 1× SDS-sample buffer, boiled for 5 minutes and applied to SDS-PAGE. Third, 0.125 ml of ice-cold 100% trichloroacetic acid is added to the supernatant of 1.2 ml of overnight culture, the tubes are chilled on ice for 5 minutes and centrifuged for 15 minutes at the highest speed in a desktop centrifuge at 4 °C. After the centrifugation, 1 ml of ice-cold acetone was added to each pellet, the tubes were vortexed briefly and centrifuged again for 5 minutes. The pellets are then air dried, resuspended in 1× SDS sample buffer, boiled for 5 minutes and subjected to SDS-PAGE.
   b. SDS-PAGE and Western blot are as described above.

3. Subcloning of antigen-encoding genes from immunopositive cosmid clones.
   a. Preparation of cosmid DNA. The same method used to purify cosmid vectors (alkaline lysis and CsCl-ethidium bromide gradient centrifugation) is used.
   b. Pilot experiments to determine the conditions for DNase I treatment of cosmid DNA. Dilute 5 µg of cosmid DNA in 52.5 µl 1× DNase I reaction buffer, take a 5 µl aliquot and keep on ice. Dilute DNase I to 0.01 units/µl in 1× reaction buffer, add 2.5 µl of the diluted DNase I to the DNA solution, and incubate at RT. Take a 5 µl aliquot after every minute, add 1 µl of 0.1 M EDTA (pH 8.0) to each aliquot immediately and keep on ice. After the last aliquot is done, add gel loading buffer and apply to a 0.7% agarose gel, followed by electrophoresis. In our case, 4–5 minutes of incubation at RT produced DNA fragments largely concentrated at 0.5 to 2 kb.
   c. Incubate 5 µg of cosmid DNA with 2.5 units of DNase I in 100 µl reaction at RT for 4–5 minutes, stop the reaction by adding 12 µl of 0.1 m EDTA (pH 8.0). Extract twice with phenol:chloroform, precipitate with ethanol, and resuspend in 20 µl of TE. Take 1 µl for 0.7% agarose gel electrophoresis to check the extent of degradation.
   d. Fill-in the ends with Klenow fragment: to the remaining 19 µl DNase I treated cosmid DNA, add 5 µl 10× Klenow buffer, 1 µl 10× BSA, 1 µl dNTP mix (25 mM of each dNTP), 21 µl water and 3 µl Klenow fragment. Incubate at RT for 20 minutes, heat at 75 °C for 15 minutes to inactivate Klenow fragment. Extract with phenol:chloroform, precipitate with ethanol and resuspend in 20 µl of TE.
   e. Plasmid pBluescript SK (–) is used to clone fragments from cosmid DNA. Preparation of pBluescript SK (–) follows the procedure used for cosmid vectors, except that *Eco*RV is used instead of *Bam*HI. Test ligations are carried out using 50 ng of vector and 25 ng of a blunt ended DNA fragment as the control insert. Ligation mixtures are incubated overnight at 14 °C. At the end of the ligation, 1/25th of the ligation mix is used to transform DH5α competent cells (will be described below) to assess the quality of the prepared vector.
   f. Ligation of DNase I treated cosmid to pBluescript SK (–) vector. Set up a 5 µl ligation reaction with 1 µl of SAP treated pBluescript SK (–), 2.5 µl insert DNA, 0.5 µl 10× ligation buffer, 0.5 µl T4 DNA ligase, and 0.5 µl water. For ligation control, use 0.5 µl pBluescript SK (–) without SAP treatment, and 2.5 µl water instead of the insert DNA. Incubate overnight at 14 °C. Dilute the reactions to 50 µl with TE, use 2 µl to transform 100 µl of DH5α competent cells.
   g. Preparation of competent cells and transformation are based on the one-step TSS method [4].
   h. About 600 to 1000 transformants from each cosmid can be screened in a similar way as described for the immunoscreening of the cosmid library. The differences are that (a) after the colonies grow up on the primary transformation plates, they are lifted onto NitroPlus nitrocellulose transfer membrane filters directly; (b) immunopositive clones from the first screen are streaked and then spotted onto LB agar plates containing 100 µg of ampicillin and 0.5 mm IPTG; after overnight growth, the colonies are lifted and subjected to another round of screening; (c) tertiary screening is also carried out when necessary.

2. DNA sequencing of the immunopositive subclones and sequence analysis
   a. Templates for sequencing are prepared using the Qiagen plasmid mini kit, and primers are for the T3 and T7 promoter regions on pBluescript SK (–). DNA sequencing reactions are performed using

the *Taq* dye-deoxy terminator kit and run on either an ABI 373A or 377 DNA sequenator.

b. Delete the pBluescript SK (–) vector region sequence from the DNA sequences using a sequence editor such as Seqed in the GCG software package. The sequences are submitted through the Internet to the BLAST (BLASTX and BLASTN) network service at NCBI to look for homologous sequences in the databases.

4. Results and discussion

The advantage of cloning large fragments (20–45 kb) of chromosomal DNA into cosmid vectors is that a relatively small number of clones (two thousand at most) is needed to achieve high coverage of the genome. This number is small enough so that individual colonies from the primary plating can be picked and stored in 96-well microtiter dishes easily, thus avoiding amplification of the library which can result in over- or under-representation of certain sequences caused by unequal growth of the cosmid clones. The clones should be stable in the glycerol stock at –80 °C for years and can be used repeatedly for different studies. However, care should be taken to minimize contamination of clones by those of neighboring wells.

One concern in cloning large insert DNA in cosmid vectors, especially inserts of AT-rich DNA, is that inserts may be unstable in *E. coli*. In some cases, the instability was thought to be caused by promoter activity from AT-rich DNA [3]. This may result in selection for mutations that reduce promoter activity. Thus, the choice of vector may be important. One of our libraries was made in vector pTEX5176, a medium copy number expression cosmid vector derived from a broad-host-range plasmid RSF1010 (Figure 1, Table 1). RED analysis of library clones showed that small inserts (4–6 kb) were quite common, and the overall average insert size was 10 kb (Table 2). Since the insert DNA was size-fractionated and the ligation mix had been packaged with lambda extracts which should have the highest efficiency for recombinants from 38 to 52 kb, it seemed that enterococcal DNA in pTEX5176 was not very stable. Clones we generated in two other vectors, pBeloBAC11 (low copy number) and pLAFRx (medium copy number), did not show this unstability. Cosmid pBeloBAC11 is an expression vector derived from the bacterial F factor, a low copy number plasmid [24]. Other F factor-based vectors have been shown to maintain very large inserts in *E. coli* [23]. Cosmid pLAFRx was derived from the RK2 R-factor, has a copy number of about 10 copies/cell, and is not an expression vector (Table 1, Figure 2). The majority of clones in libraries constructed with these two vectors seemed to have the expected insert sizes (Table 2).

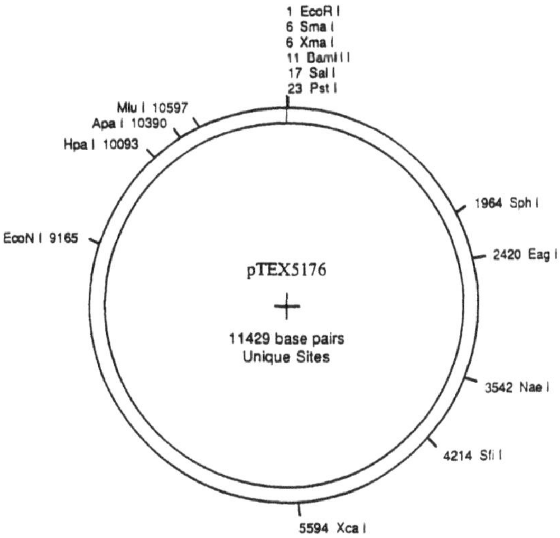

Figure 1. A *Hin*dIII fragment containing a spectinomycin resistance gene from plasmid pHP45Ω was made blunt-ended and inserted into the *Xmn*I site at position 1693 of pMMB190 to generate plasmid pXJ1. A 450 bp fragment containing the sequence for the lambda cos site was PCR-amplified from plasmid pGE240, digested with *Eag*I and *Sph*I, and ligated with *Eag*I and *Sph*I digested pXJ1. The resulting cosmid was designated pTEX5176. The unique restriction sites are shown in the map. Some features of pTEX5176 are as follows: *lacZ* (1–184 bp, 11415–11429 bp), ampicillin resistance gene (711–1571 bp), cos site (1970–2419 bp), spectinomycin resistance gene (2419–4061 bp), origin of replication (4374–4768 bp), and *oriT* (5031–5188 bp).

Table 1. Cosmid vectors

Cosmid vector	pTEX5176	pLAFRx	pBeloBAC11
Size (kb)	11	21.6	7.5
Derived from	RSF1010 (IncQ)	pLAFR (RK2)	*E. coli* F
Selection marker	Spectinomycin	Tetracycline	Chloramphenicol
Copy number/cell	~10	~10	1–2
Expression	*lac* promoter	Non-expression	*lac* promoter
Cloning site	*Bam*HI	*Bam*HI	*Bam*HI
Other features	*oriT*	*oriT*	

Table 2. Results of cosmid library constructions and immunoscreening

Vector Insert	pTEX5176 TX52	pTEX5176 OG1RF	pLAFRx OG1RF	pBeloBAC11 OG1RF	pBeloBAC11 TX52
Average insert size (kb)	6	11	24.5	34	34
Titer (cfu/μg DNA)	10^4	10^3	10^5	10^5	10^4
Background (%)[a]	0	4	10	0.5	0.5
No. of colonies picked	2400	0	1000	800	800
No. of clones screened[b]	2400	0	1000	800	800
Genome coverage screened[c]	5.1×		8.7×	9.7×	9.7×
No. of total immunopositive cosmid clones[d]	19		21	29	25
No. of total immunopositive clones/genome equivalent	3.7		2.4	3.0	2.6

[a] The background is calculated as the percentage number of colonies on the vector only plate versus colonies on the vector plus insert plate.
[b] Library pTEX5176-OG1RF was screened with a rabbit serum only, pTEX5176-TX52 was not screened, the others were screened with a patient serum first and the initial immunopositive clones were rescreened with five sera.
[c] Genome coverage is calculated as: the number of clones screened × average insert size/genome size.
[d] For pTEX5176-OG1RF, the number indicate clones reacted with the rabbit serum; for the others, the numbers indicate clones reacted with at least two of the five sera tested.

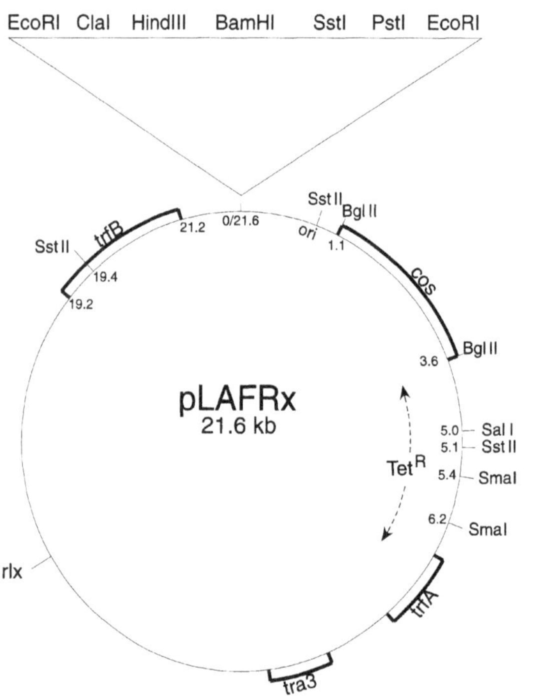

Figure 2. Two complementary oligonucleotides containing the recognition sites for the following enzymes: *Bam*HI, *Hin*dIII, *Pst*I and *Sst*I, were annealed and ligated with *Bam*HI and *Hin*dIII digested pLAFR1 [9] to form pLAFRx. The sequence for the oligonucleotides are: linker 1, 5′GATCTGCAGAGCTCGGATCCA, and linker 2, 5′AGCTTGGATCCGAGCTCTGCA. The unique restriction sites in the multiple cloning region are shown.

The procedure for the construction of libraries described here depends on the quality of insert DNA. The quality of DNA can vary from batch to batch, so it is important to perform the pilot experiments, test ligations and test packagings to determine the optimal conditions. The pilot experiment for DNase I treatment is especially critical since it is a highly active enzyme and the Mn^{+2} in the reaction buffer is very sensitive to oxidizing agents. The test ligations and packagings are also important when large quantities of genomic or cosmid DNA are not available. The method has been successfully applied to the construction of libraries from two enterococcal strains, OG1RF and TX52, in two cosmid vectors pBeloBAC11 and pLAFRx (Table 2).

Western blots showed that common antigen bands, as well as variable bands, were present in most of the enterococcal strains (Figure 3). The five sera also showed varying degrees of reaction with *E. coli* proteins (data not shown). The reaction was reduced after absorption with *E. coli* lysates. One patient serum was used in the primary screening of the cosmid libraries since it had the lowest background reaction with *E. coli* among the patient sera tested. The use of multiple antisera in the screening of cosmid libraries minimizes the possibility of selecting immunopositive clones that were biased for antigens from one specific strain or one specific patient. Furthermore, clones that reacted with all the sera could be strong candidates for ‘common antigens’, while clones reacting with only the patient sera but not the rabbit serum may contain antigens that are only expressed during infection. Results of the immunoscreening are summarized in Figure 4.

Several observations were made from the results

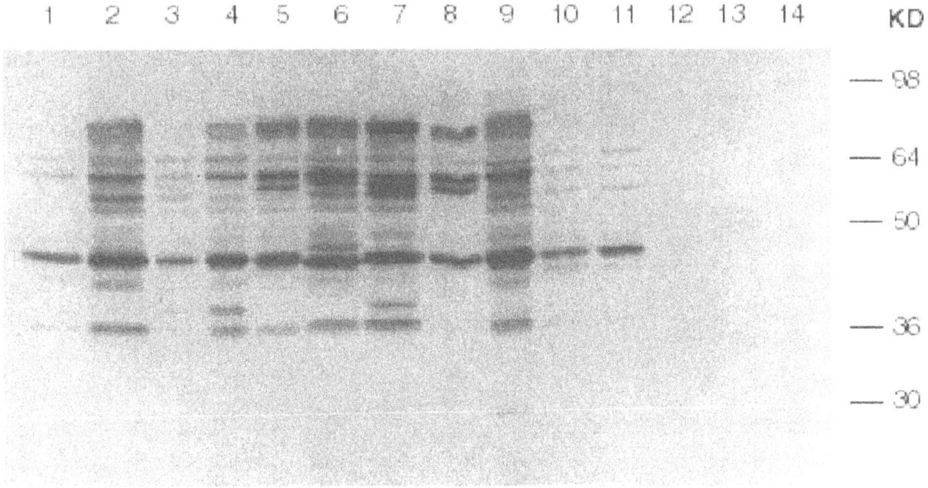

Figure 3. Western blot analysis of *E. faecalis* strains with one of the patient sera. All samples were prepared using the Zwittergent method. From lane 1 to lane 14 are: TX11, TX65, TX 20, TX52, TX44, TX12, TX17, TX1, TX4, TX5, OG1RF, an *E. coli* recombinant clone, and DH5α/pBeloBAC11, and the molecular weight standard (see Blue™ Prestained Standards, Novex, San Diego, CA). The patient serum was diluted 1000 fold.

of Western blot analysis of immunopositive cosmid clones (data not shown): 1) there were differences in the antigen bands with protein samples prepared by different methods, e.g., some antigen bands only appeared in Western blots using the Zwittergent extract; 2) some cosmid clones were obviously positive in the immunoscreens, but did not show any clear antigen bands in Western blots, possibly because the antigen co-migrated with a host band that had cross reactivity with the antisera, or the antigen was aggregated and did not dissolve easily in the gel loading buffer. In addition, one clone turned out to produce a polysaccharide antigen in *E. coli*, explaining why it behaved differently than protein antigens.

The procedure of subcloning antigen-encoding genes from immunopositive cosmid clones has been applied successfully to at least 38 clones, and several thousand transformants can be easily obtained from each sublibrary. Twenty-five clones gave 53 immunopositive subclones while 13 did not produce any immunopositive subclones, possibly because the antigen-encoding genes were unstable in a high-copy number vector or the production of antigen required more than one gene.

BLAST searches of the DNA sequences from the immunopositive subclones showed that a wide range of genes were cloned (Table 3), including those encoding transporters, potential virulence factors, cell surface proteins, metabolic enzymes and hypothetical proteins. The majority of these were predicted to be surface or membrane proteins, consistent with the hypothesis that these proteins could be seen by the immune system in infection. In addition, we also identified a pBeloBAC11 clone (45 kb insert) containing a putative polysaccharide biosynthesis gene cluster. We do not believe it would have been possible to isolate this clone with the standard small insert expression library approach.

The possibility that a sequence of interest is present in a random library can be estimated using the following equation based on Poisson distribution: $N = \ln(1 - P)/\ln[1 - (I/G)]$ where I is the average size of the inserts, G is the size of the genome, and N is the number of independent clones that must be screened to isolate a particular sequence with probability P (10). For a 99% probability of isolating a sequence of interest, the equation predicts that the total number of base pairs screened ($I \times N$) be 4.6-fold in excess over the total number of base pairs in the genome (G). According to this calculation, N should be 380 for pBeloBAC11 clones (I = 34 kb), and 530 for pLAFRx clones (I = 24.5 kb). However, it is known that certain regions of a genome are more difficult to clone, possibly due to unique DNA structure or toxic functions encoded in the region, and consequently the library is not completely random. Therefore we screened a larger number of clones to increase the chance of isolating the antigen genes. Critical evaluation as to whether we have isolated all the antigen-encoding genes is difficult to make because: (1) it is not known how many antigens the enterococcal strains produce, and (2) it is not clear how many antibodies against enterococci are present in each serum at high enough titer for immunodetection. However, a comparison of the results obtained by Western blot analysis to those obtained by immunoscreening, assuming that each antigen band on the Western blot represents a different antigen, can provide an estimate. There were 10 to

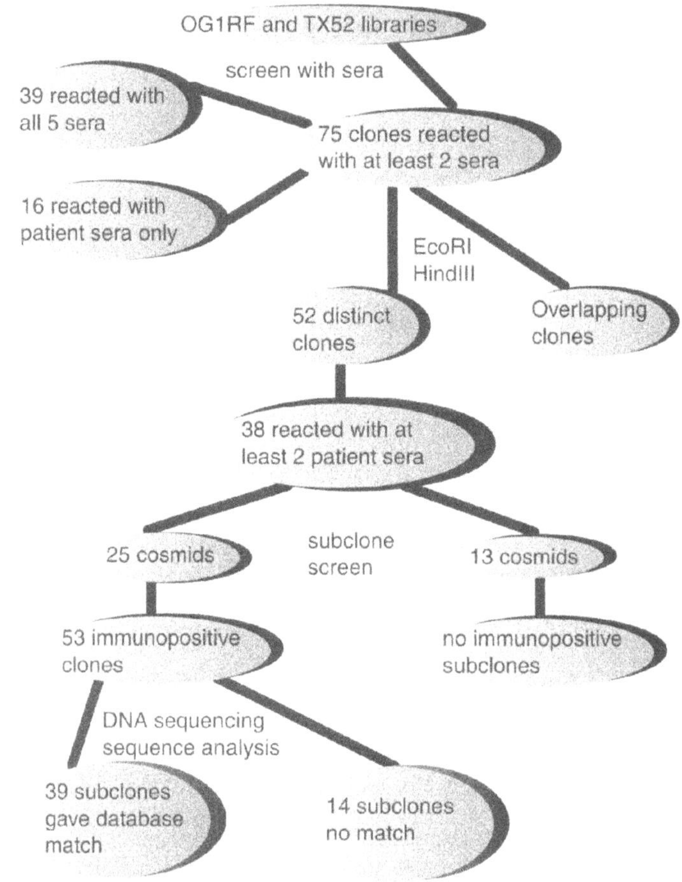

Figure 4. A flow chart of the immunoscreening, cloning and DNA sequencing results.

Table 3. BLAST search results

Function	Blast hit	Organism	Ref.
Transport/binding	Glutamine transport ATP-binding protein Q (*glnQ*)	*Methanococcus jannaschii*	
	Transmembrane protein LPLB (*lplB*)	*Bacillus subtilis*	
	Hyaluronate synthase (*has*)	*Streptococcus equisimilis*	[16]
	Probable amino-acid ABC transporter (*orf1*)	*Bacillus subtilis*	[21]
	High-affinity periplasmic glutamine binding protein	*Salmonella typhimurium*	[15]
	FliY (*fliY*)	*Escherichia coli*	[20]
Virulence	*E. faecalis* endocarditis antigen EfaA (*efaA*)	*Enterococcus faecalis*	[17]
	20KDa protein in SSAB 3′ region (*orf3*)*	*Streptococcus sanguis*	[11]
	PepM49	*Streptococcus pyogenes*	[13]
Cell envelope/surface	Envelope protein EnvC (*envC*)	*Escherichia coli*	[14]
	Autolysin	*Enterococcus faecalis*	[2]
	P54	*Enterococcus faecium*	[10]
Regulators	PfeR (*pfeR*)	*Pseudomonas aeruginosa*	[6]
Metabolism	Phosphoribosylaminoimidazole carboxylase (*purK*)	*Bacillus subtilis*	[7]
	Dihydrolipoamide acetyltransferase (*pdhC*)	*Enterococcus faecalis*	[9]
Hypothetical proteins	N15OR, N15NR	*Bacillus subtilis*	
	Hypothetical protein 2 (*orf2*)	*Lactobacillus leichmannii*	

* SSAB, *Streptococcus sanguis* adhesin B.

20 antigen bands in most of the strains tested (Figure 3), and at least 18 unique putative antigen-encoding genes that showed sequence similarity to genes in the database (Table 3) were found by immunoscreening. Taking into consideration the 14 sequences from immunopositive subclones that did not have any database similarity (Figure 4), the number of potential antigen-encoding genes isolated by immunoscreening is of the same order as the number of antigen bands observed in the Western blots. Two previously identified *E. faecalis* genes, one of which encodes an antigen associated with endocarditis infections (EfaA) [17], and the other encoding a surface protein (autolysin) [2], were detected by our immunoscreening. These results suggest that our libraries are representative, and that immunoscreening is reasonably thorough. However, we note that each library only identified a few antigen-encoding genes per genome equivalent cloned (Table 2), clearly lower than the abundance of antigen-encoding genes. This is likely due in part to the non-randomness of libraries, discussed above, but may also relate to the low copy number of the vectors. Since inserts are large, expression in *E. coli* relies on endogenous signals, and at low copy number, this may be inefficient. It is possible that expression is affected by the location of the gene in the insert, i.e., inserts in pBeloBAC11 with an antigen encoding gene near the *lac* promoter are more likely to be immunopositive. Thus, high coverage is important. It is likely that libraries made in high copy number expression vectors, such as pBluescript SK (–), may facilitate detection of antigens with lower titer antibodies in serum. However, antigens dependent on large inserts, such as polysaccharides, would be missed.

Acknowledgment

This work was supported by grant AI 33516 from the PHS.

We are very grateful to Ronald Christopher Mackenzie for his technical support in constructing the genomic and DNase I random libraries, performing DNA sequencing and advice on sequence analysis, Kavindra V. Singh and Monjula Chidambaram for general technical support, H. Shizuya and M. Simon at the California Institute of Technology, Pasadena, CA, for providing us with the cosmid vector pBeloBAC11, Laura Bankey and Beto Zuniga for technical support.

Notes on suppliers

1. AMRESCO, Solon, Ohio 44139, USA
2. Bio-Rad Laboratories, 2000 Alfred Nobel Dr., Hercules, CA 94547, USA
3. Calbiochem Co., La Jolla, CA 92037, USA
4. Cambridge Technology, 23 Elm Street, Watertown, MA 02172, USA
5. Corning Glass Works, Corning, New York 14831, USA
6. Difco Laboratories Inc., Detroit, MI 48232-7058, USA
7. Fisher Scientific, Fair Lawn, New Jersey 07410, USA
8. Klett MFG. CO. INC., New York, USA
9. Micro Separations Inc., Westboro, MA 01581, USA
10. Millipore Co., Bedford, MA 01730, USA
11. New Englsnd Biolabs Inc., 32 Tozer Road, Beverly, MA 01915-5599, USA
12. Novex, 11040 Roselle St., San Diego, CA 92121, USA
13. Qiagen Inc., 9600 De Soto Avenue, Chatsworth, CA 91311, USA
14. Sigma, PO Box 14508, St. Louis, MO 63178, USA
15. Stratagene, 11011 North Torrey Pines Road, La Jolla, CA 92037, USA
16. United States Biochemical Co., PO Box 22400, Cleveland, Ohio 44122, USA

References

1. Allen AG, Perham RN (1991). Two lipoyl domains in the dihydrolipoamide acetyltransferase chain of the pyruvate dehydrogenase multienzyme complex of *Streptococcus faecalis*. FEBS Lett 287: 206–210.
2. Beliveau C, Potvin C, Trudel J, Asselin A, Bellemare G (1991). Cloning, sequencing, and expression in *Escherichia coli* of a *Streptococcus faecalis* autolysin. J Bacteriol 173: 5619–5623.
3. Chen J-D, Morrison DA (1988). Construction and properties of a new insertion vector, pJDC9, that is protected by transcriptional terminators and useful for cloning of DNA from *Streptococcus pneumoniae*. Gene 64: 155–164.
4. Chung CT, Niemela SZ, Miller RH (1989). One-step preparation of competent *Escherichia coli*: Transformation and storage of bacterial cells in the same solution. Proc Natl Acad Sci USA 86: 2172–2175.
5. Clarke L, Carbon J (1976). A colony bank containing synthetic *Col*E1 hybrids representative of the entire *E. coli* genome. Cell 9: 91–99.
6. Dean CR, Poole K (1993). Expression of the ferric enterobactin receptor (PfeA) of *Pseudomonas aeruginosa*: Involvement of a two-component regulatory system. Mol Microbiol 8: 1095–1103.
7. Ebbole DJ, Zalkin H (1987). Cloning and characterization of a 12 gene cluster from *Bacillus subtilis* encoding nine enzymes for *de novo* purine nucleotide synthesis. J Biol Chem 262: 8274–8287.
8. Ebel-Tsipis J, Botstein D, Fox MS (1972). Generalized transduction by phage P22 in *Salmonella typhimurium*. I. Molecular origin of transducing DNA. J Mol Biol 71: 433–448.
9. Friedman AM, Long SR, Brown SE, Buikema WJ, Ausubel FM (1982). Construction of a broad host range cosmid cloning vector and its use in the genetic analysis of *Rhizobium* mutants. Gene 18: 289–296.
10. Furst P, Mosch H-U, Solioz M (1989). A protein of unusual composition from *Enterococcus faecium*. Nucleic Acids Res 17: 6724.

11. Ganeshkumar N, Hannam PM, Kolenbrander PE, McBride BC (1991). Nucleotide sequence of a gene coding for a saliva-binding protein (SsaB) from *Streptococcus sanguis* 12 and possible role of the protein in coaggregation with actinomyces. Infect Immun 59: 1093–1099.
12. Kessler RE, Yagi Y (1983). Identification and partial characterization of a pheromone-induced adhesive surface antigen of *Streptococcus faecalis*. J Bacteriol 155: 714–721.
13. Khandke KM, Fairwell T, Acharya AS, Trus BL, Manjula BN (1988). Complete amino acid sequence of streptococcal PepM49 protein, a nephritis-associated serotype. Conserved conformational design among sequentially distinct m protein serotypes. J Biol Chem 263: 5075–5082.
14. Klein JR, Henrich B, Plapp R (1991). Molecular analysis and nucleotide sequence of the *envCD* operon of *Escherichia coli*. Mol Gen Genet 230: 230–240.
15. Klose KE, Mekalanos JJ (1997). Simultaneous prevention of glutamine synthesis and high-affinity transport attenuates *Salmonella typhimurium* virulence. Infect Immun 65: 587–596.
16. Lansing M, Lellig S, Mansolf A, Martini I, Crescenzi F, O'Regan M, Prehm P (1993). Hyaluronate synthase: Cloning and sequencing of the gene from *Streptococcus* sp. Biochem J 289: 179–184.
17. Lowe AM, Lambert PA, Smith AW (1995). Cloning of an *Enterococcus faecalis* endocarditis antigen: Homology with adhesins from some oral streptococci. Infect Immun 63: 703–706.
18. Murray BE (1990). The life and times of the enterococcus. Clin Microbiol Rev 3: 46–65.
19. Murray BE, Singh KV, Ross RP, Heath JD, Dunny GM, Weinstock GM (1993). Generation of restriction map of *Enterococcus faecalis* OG1 and investigation of growth requirements and regions encoding biosynthetic function. J Bacteriol 175: 5216–5223.
20. Mytelka DS, Chamberlin MJ (1996). *Escherichia coli fliAZY* operon. J Bacteriol 178: 24–34.
21. Rodriguez F, Grandi G (1995). An operon encoding a novel ABC-type transport system in *Bacillus subtilis*. Microbiology 141: 1781–1784.
22. Sambrook J, Fritsch EF, Maniatis T (1989). Molecular cloning, a laboratory manual. Cold Spring Harbor, NY: Cold Spring Harbor Laboratory Press.
23. Shizuya H, Birren B, Kim U-J, Mancino V, Slepak T, Tachiiri Y, Simon M (1992). Cloning and stable maintenance of 300-kilobase-pair fragments of human DNA in *Escherichia coli* using an F-factor-based vector. Proc Natl Acad Sci USA 89: 8794–8797.
24. Xu Y, Jiang L, Murray BE, Weinstock GM (1997). *Enterococcus faecalis* antigens in human infections. Infect Immun 65: 4207–4215.

Address for correspondence: George M. Weinstock, Department of Microbiology and Molecular Genetics, University of Texas Medical School, 6431 Fannin St., Houston, TX 77030, USA
Phone: (713)-500-6083; Fax: (713)-500-0652
E-mail: georgew@utmmg.med.uth.tmc.edu

Methods in Cell Science **20**: 107–111 (1998)

A simple microtiter plate screening assay for bacterial invasion or adherence

Victor Nizet[1], Arnold L. Smith[2], Paul M. Sullam[3] & Craig E. Rubens[4]

[1] *Department of Pediatrics, University of California, San Diego, La Jolla, California, USA;* [2] *Department of Molecular Microbiology & Immunology, University of Missouri-Columbia Medical School, Columbia, Missouri, USA;* [3] *Department of Medicine, Veteran Affairs Medical Center and University of California – San Francisco, San Francisco, California, USA;* [4] *Division of Infectious Diseases, Children's Hospital & Medical Center, University of Washington, Seattle, Washington, USA*

Abstract. Many bacteriologic studies, including cellular invasion and adherence assays, require enumeration of viable organisms or colony forming units. When attempting to screen large numbers of clinical or environmental isolates or laboratory-derived mutants for differences in invasion or adherence phenotype, standard plating methods can be cumbersome and severely limit the number of organisms or conditions which can be tested. As a potential alternative, we describe a simple, rapid and inexpensive soft agar-based technique for semi-quantitative determination of bacterial colony counts directly within the wells of a 96-well microtiter plate.

Key words: Bacterial adherence, Bacterial invasion, Bacteriological techniques, Invasion assay, Microbial colony count, Soft agar

1. Introduction

Many bacteriologic studies, including cellular invasion and adherence assays, depend at some point on data derived from enumeration of viable organisms or colony forming units. When screening large numbers of clinical or environmental isolates or laboratory-derived mutants in such quantitative assays, standard plating methods can be time-consuming and limit the number of organisms or conditions which can be tested. Alternative methods to agar plating for bacterial quantification include: direct microscopic enumeration in counting chambers [6], epifluorescence microscopy using acridine orange or other fluorochromes [3, 4], flow cytometry [8], impedance bacteriometry [5], tetrazolium dye reduction [7], measurement of adenosine triphosphate bioluminescence [10], and enzyme-labeled, rRNA-targeted oligonucleotide probes [1]. None of these methods, however, allows the investigator to assay bacterial viability, antibiotic susceptibility and metabolic phenotype simultaneously, as can be done with growth on specialized agar media.

In the course of our studies of cellular invasion and adherence by group B streptococci (GBS) and other bacterial pathogens, we have employed a simple, rapid and cost-saving soft agar technique for the semi-quantitative determination of bacterial colony counts directly within the wells of 96-well microtiter plates. Although we hesitate to claim that a similar technique has never been described, the current literature in the field of bacterial pathogenesis indicates that standard labor-intensive plating techniques are widely used, even when great numbers of isolates or laboratory-derived mutants are being screened. Therefore, we offer a brief description of our screening invasion and adherence assays, in hope that others may find a useful application for the soft agar technique in their own experiments.

The screening cellular invasion assay is adapted from the antibiotic protection procedure first described by Devenish and Schiemann [2]. Certain antibiotics (e.g., gentamicin) will effectively kill extracellular or surface-adherent bacteria, but inefficiently penetrate eukaryotic cells so as not to reach microbicidal levels in the intracellular compartment. The differential antibiotic killing can be exploited to quantify the degree of invasion of viable bacteria into the host cell. Previously, we had quantified the number of internalized bacteria by plating small aliquots of an epithelial cell lysate onto standard agar petri dishes containing the appropriate growth media. When screening large numbers of transposon mutants for modifications of invasion phenotype, however, we found plating samples from each individual well to be quite tedious. A single person could screen comfortably no more than 100 mutants in a single work day. Moreover, the inherent time delay between plating of the first and last samples introduced an undesired variable into determination of viable counts, and replicate samples often demonstrated differences attributable to sampling or plating error. For screening purposes, semi-quantitative discrimination of colony forming units would suffice. We therefore modified our protocol to eliminate the labor-intensive plating steps, and employed soft agar to enumerate the bacteria directly within wells of a 96-well microtiter plate. We subsequently adapted this methodology to develop a screening assay for streptococcal adherence to human platelets.

2. Materials

1. 96 well tissue culture plates, flat bottom, sterile with lid (Costar #3598).
2. Todd Hewitt Broth (Difco #DF0492-17-6).
3. Bacteriologic grade agar (Difco DF0140-01-0).
4. A549 human lung carcinoma cell line (ATCC #CCL-185).
5. RPMI 1640 tissue culture media, with HEPES (Sigma #R 6504).
6. Fetal calf serum, heat inactivated (Sigma #F 4135).
7. 12-channel pipettors (Labsystems #4510040 and 4510050).
8. Boekel 96-pin stainless steel replicator (Fisher #05-450-9).
9. Reagent reservoir (Sigma #R 1936).
10. Bent stainless steel manifold (Drummond #3-00-094).
11. Gentamicin sulfate (Sigma #G 1264).
12. Penicillin G (Sigma #P 7794).
13. 0.25% Trypsin/EDTA solution (Sigma #T 4049).
14. Triton X-100 (Sigma #X-100).
15. Poly-L-Lysine (Sigma #P 4832).
16. Brain-Heart Infusion (BHI) Broth (Difco DF0037-07-0).
17. Tyrode's salt solution (Sigma #T 2397).

3. Procedures

A. *Cellular Invasion Assay*

1. *Preparation of bacteria and tissue culture cells*
   A549 lung epithelial cells are seeded and grown in tissue culture media (RPMI media + 10% fetal calf serum) in 96-well tissue culture plates until confluent monolayers are formed. The day before the screening assay, single colonies of GBS mutants are picked with a sterile toothpick and used to inoculate 200 μl of Todd Hewitt broth (THB) in individual wells of 96-well microtiter plates. Control wells on each plate are inoculated with either the wild-type GBS strain or a noninvasive bacterial isolate (e.g., *Escherichia coli* strain DH5 or *Streptococcus gordonii* Challis). The plates are incubated overnight at 37 °C to allow the bacteria in each well to reach stationary growth phase. Cellular invasion by certain bacterial species may require factors expressed only during exponential growth phase. In such cases, a replica plate with 200 μl of fresh media per well may be inoculated with 10 μl of the overnight cultures and growth monitored by optical density measurement using a microplate reader.
2. *Inoculation of cell monolayers*
   Immediately prior to the invasion assay, each plate is vortexed gently to resuspend settled bacteria. Optical density measurement can be obtained to identify any strains or mutants which grew poorly, such that any spuriously low input inoculum can be considered in interpretation of final invasion assay results. Using an 8- or 12-channel pipettor, 10 μl of each bacterial culture is transferred as an inoculum to a corresponding well in a 96-well plate containing 200 μl antibiotic-free tissue culture media and a confluent A549 cell monolayer. Alternatively, the plates may be inoculated using a Boekel 99-pin stainless steel replicator which transfers a standard volume of the liquid culture by capillary action.
3. *Antibiotic protection method for determination of cellular invasion*
   Following inoculation, the 96-well tissue culture plates are centrifuged at 800 ×*g* for 10 min to place GBS at the surface of the cell monolayer, then incubated for 2 h at 37 °C with 5% CO_2 to allow cellular invasion by the GBS. After the incubation, medium is removed from the monolayers by gentle aspiration of the wells using a bent-hub 8-channel stainless steel manifold attached to a vacuum source. The subsequent wash and treatment steps all involve liquids dispensed from a sterile reagent reservoir. Monolayers are washed ×3 by adding 200 μl of phosphate buffered saline (PBS) via a multichannel pipettor, followed by gentle aspiration of the wash buffer with the vacuum manifold. Alternatively, removal of medium or wash buffer can be accomplished by rapid inversion and brief shaking of the plate over a sink or disposal container. After the primary wash steps, 200 μl of tissue culture medium containing 100 μg/ml gentamicin and 5 mg/ml penicillin G is added to each well, and the plates incubated for 2 h at 37 °C with 5% CO_2 to kill extracellular and surface-adherent bacteria. The monolayers are once again washed ×3 with PBS. Finally, 50 μl of a 1:4 mixture of 0.25% trypsin/EDTA solution and 0.025% Triton X-100 is added to each well, the plates incubated for 10 min at 37 °C in order to disrupt the epithelial cell monolayers and liberate intracellular bacteria.
4. *Semiquantitative determination of bacterial colony counts using soft agar*
   Todd Hewitt soft agar (THSA) medium is prepared in advance by using 0.7% Bacto

agar (Difco) rather than our standard 1.5%, maintained in liquid form by incubation in a 45–50 °C water bath, and dispensed into a reagent reservoir just prior to use. Addition of 150 μl of THSA to each well using the multichannel pipettor results in uniform dispersion of the 50 μl lysate and solidification within a few minutes at room temperature. The microtiter plates are incubated overnight at 37 °C, and the next day bacterial colonies are observed growing within the THSA. Identification of isolates or mutants that are invasive (some colonies present), hyperinvasive (many colonies present) or noninvasive (no colonies present) is easy and reproducible. Mutants with suspected alterations in invasion phenotype are subsequently confirmed by means of a standard 24-well plate quantitative cellular invasion assay [9].

B. *Platelet adherence assay*

1. *Preparation of bacteria and platelet monolayers*

   To identify transposon mutants of *Streptococcus sanguis* deficient in binding to human platelets, we have performed a similar 96-well microtiter plate screening assay. Briefly, washed human platelets are obtained from fresh blood donated by healthy volunteers, isolated by centrifugation (100 ×*g*, 15 min), followed by washing, fixation in 0.8% formaldehyde (30 min at 37 °C), additional washing, and suspension in Tyrode's solution. Platelet monolayers are prepared by placing 10^7 fixed platelets into each well of 96-well tissue culture plates pretreated with poly-L-lysine (0.01% solution).

2. *Semiquantitative determination of platelet binding using soft agar*

   Overnight cultures of *S. sanguis* mutants are grown in BHI broth, washed and resuspended in Tyrode's salt solution. One hundred μl of each bacterial suspension is transferred (~10^7 cfu) onto the immobilized platelets in a corresponding microtiter well and rocked gently at 4 °C for 1 h. The unbound organisms are then removed by washing three times with Tyrode's solution. The wells are treated with trypsin (1 mg/ml, 10 min) to release the adherent bacteria. Brain-Heart Infusion soft agar (BHISA) medium is prepared in advance by using 0.7% Bacto agar (Difco), maintained in liquid form by incubation in a 45–50 °C water bath, and dispensed into a reagent reservoir just prior to use. Using a multichannel pipettor, 175 μl of BHISA is added to each well. After overnight incubation at 37 °C, the number of organisms per well is assessed qualitatively by visual inspection. Those wells with markedly decreased quantities of colony forming units represent possible low-platelet-binding mutants, whose phenotype is subsequently confirmed by a quantitative platelet binding assay [11].

4. Results and discussion

Figure 1 demonstrates GBS recovered within microtiter wells by the soft agar technique following invasion of lung epithelial cell monolayers. Serial twofold dilutions of the organism were used as initial inoculums over a range at which invasion is approximately linear, in order to demonstrate the observed colonial morphology and reproducibility of the assay. Each dilution was performed in triplicate, and three replica wells were assayed by spread plating on Todd-Hewitt agar in standard petri dishes. After overnight incubation, high colony counts imparted a fine granular appearance to the solidified soft agar, whereas at low colony counts, larger discrete colonies were evident. Below approximately 1,000 colonies per well, even two-fold differences in colony number were easily discernable. For quantitative comparison, the colony forming units counted on the spread plates for each dilution were as follows (+ st dev): 1260 + 80, 635 + 32, 301 + 34, 164 + 20, 79 + 4, 42 + 6, 19 + 6, 12 + 8, and 4 + 2.

Figure 2 illustrates how a low platelet binding mutant of *S. sanguis* can be readily identified by visual inspection following the semiquantitative soft agar-based screening platelet adherence assay. A similar screening assay allowed identification of low platelet binding variants of *Staphylococcus aureus* [11]. In additional experiments using appropriate soft agar media, we have successfully adapted our method to assay the invasion potential of *Escherichia coli*, *Burkholderia cepacia* and *Haemophilus influenzae*. The soft agar microtiter plate technique should be generalizable to a host of cellular adherence and invasion assays, microbicidal assays and numerous other bacteriologic experiments which can be performed in a microtiter well and require only semi-quantitative interpretation. Once a particular assay is developed, 'standard curves' from known serial dilutions of the organism grown in soft agar media can be constructed and used to assign approximate numerical values to matching experimental wells.

We have found several advantages to the use of the soft agar microtiter well assay as compared to standard bacteriologic plating methods. First, the entire well is assayed for colony forming units, eliminating both sampling error in obtaining a small aliquot and plating error associated with 'hockey-sticking' or other dispersion techniques. Second, the assay is extremely rapid, requiring but seconds for delivery of the soft agar with a multichannel pipettor. Severalfold more samples can be screened efficiently in a given experiment, and intersample variation,

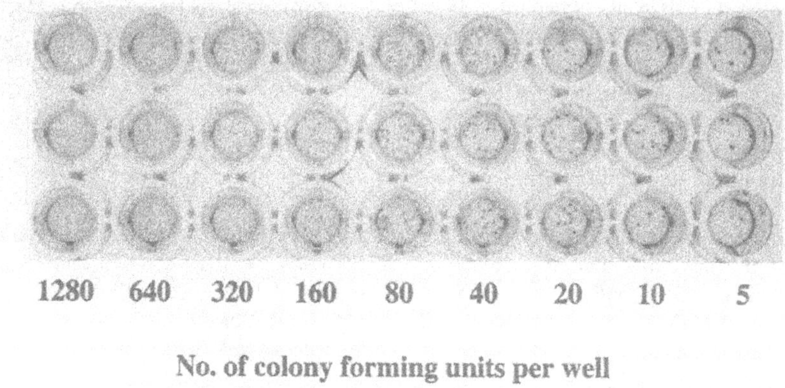

Figure 1. Semi-quantitative invasion assay in which serial dilutions of group B streptococci are added to lung epithelial cell monolayers in microtiter wells, extracellular bacteria eliminated by washing and gentamicin treatment, and intracellular bacteria recovered by addition of soft agar and overnight incubation at 37 °C. The experiment was performed in triplicate and colony counts confirmed by standard plating methods.

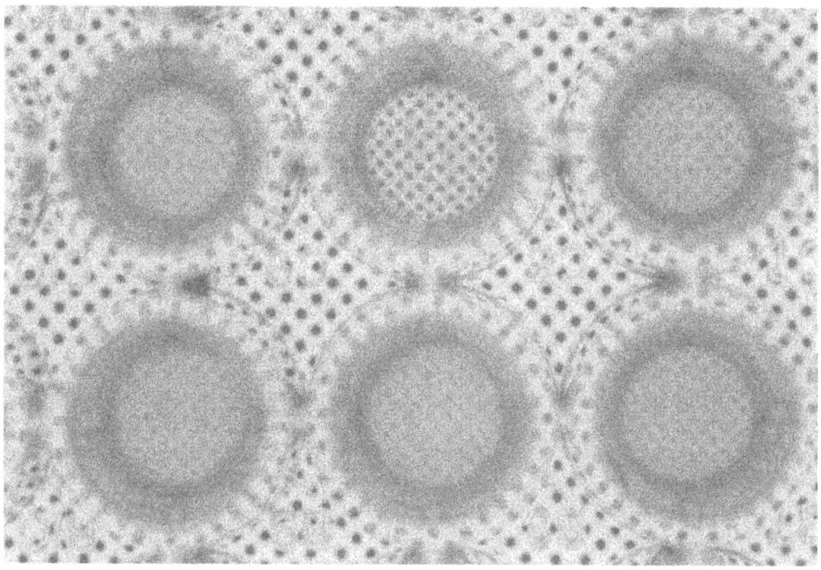

Figure 2. Use of soft agar to identify low platelet binding variants of Streptococcus sanguis M99. One of the center wells is much clearer, indicating greatly decreased colony forming units associated with low binding to platelets immobilized on the microtiter well surface.

due to time elapsed between individual platings, is avoided. Third, the assay is easy to interpret, since positive and negative controls can be placed on the same plate for direct visual comparison with sample wells. Lastly, the assay requires almost no additional expense, in contrast to the considerable cost and/or labor involved in preparing large numbers of standard agar petri dishes. Stocks of soft agar medium can be made in advance, allowed to solidify, and the required amount melted using a microwave on the day of the assay.

All soft agar assays should be calibrated such that <1,000 colony forming units are typically recovered in a single well; alternatively, appropriate dilution into a replica microtiter plate can be performed prior to addition of the soft agar. We recognize that a soft agar microtiter well assay may not be applicable to organisms requiring high oxygen tensions and those acutely sensitive to the initial temperature elevation required to maintain the agar in liquid form. Formation of large or confluent surface colonies would also tend to obscure organisms embedded within the agar. Nevertheless, we encourage other investigators interested in large scale screening of microorganisms for viable colony counts to consider this semi-quantitative technique. By delivering the agar to the organism (rather than the other way around), substantial laboratory time, effort and expense may be saved.

Acknowledgments

This work was supported in part by grants HD 07233 and AI 01451 (VN), AI 29549 (ALS), AI 32506 (PMS) and AI 30068 (CER) from the National Institutes of Health.

Notes on suppliers

1. Corning Costar Corporation, One Alewife Center, Cambridge, MA 02140, USA
2. Difco Laborotories LTD, P.O. Box 14B, Central Avenue, West Molesey Surrey, England, KT8 2SE
3. Drummond Scientific Co., 500 Parkway Blvd., Broomal, PA 19008, USA
4. Fisher Scientific, 711 Forbes Avenue, Pittsburgh, PA 15219, USA
5. Labsystems Oy, P.O. Box 8, FIN-00881 Helsinki, Finland
6. American Type Culture Collection (ATTC), 12301 Parklawn Drive, Rockville, MD 20852, USA
7. Sigma, 3050 Spruce Street, St. Louis, MO 63103, USA

References

1. Amann RI, Zarda B, Stahl DA, et al. (1992). Identification of individual prokaryotic cells by using enzyme-labeled, rRNA-targeted oligonucleotide probes. Appl Environ Microbiol 58: 3007–3011.
2. Devenish JA, Schiemann DA (1981). HeLa cell infection by *Yersinia enterocolitica*: evidence for lack of intracellular multiplication and development of a new procedure for quantitative expression of infectivity. Infect Immun 32: 48–55.
3. Francisco DE, Mah RA, Rabin AC (1973). Acridine orange epifluorescent technique for counting bacteria in natural waters. Trans Am Microsc Soc 92: 416–421.
4. Lundgren B (1981). Fluorescein diacetate as a stain of metabolically active bacteria in soil. Oikos 36: 17–22.
5. Noble PA, Ashton E, Davidson CA, et al. (1991). Heterotrophic plate counts of surface water samples by using impedance methods. Appl Environ Microbiol 57: 3287–3291.
6. Norris KP, Powell EO (1961). Improvements in determining total counts of bacteria. J R Micros Soc 80: 107–119.
7. Peck R (1985). A one-plate assay for macrophage bactericidal activity. J Immunol Methods 82: 131–140.
8. Pinder AC, Purdy PW, Poulter SA, et al. (1990). Validation of flow cytometry for rapid enumeration of bacterial concentrations in pure cultures. J Appl Bacteriol 69: 92–100.
9. Rubens CE, Smith S, Hulse M, et al. (1992). Respiratory epithelial cell invasion by group B streptococci. Infect Immun 60: 5157–5163.
10. Selan LE, Berlutti FN, Passariello CE, Thaller MC, Renzini G (1992). Reliability of a bioluminiscence ATP assay for detection of bacteria. J Clin Microbiol 30: 1739–1742.
11. Sullam PM, Bayer AS, Foss WM, et al. (1996). Diminished platelet binding in vitro by Staphylococcus aureus is associated with reduced virulence in a rabbit model of infective endocarditis. Infect Immun 64: 4915–4921.

Address for correspondence: Victor Nizet, M.D., Assistant Professor of Pediatrics, Division of Infectious Diseases, Mail Code 0672, University of California, San Diego, 9500 Gilman Drive, La Jolla, CA 92093, USA
Phone: (619) 534-7408; Fax: (619) 534-7411
E-mail: vnizet@ucsd.edu

Methods in Cell Science **20**: 113–118 (1998).

End-probing: A non-radioactive approach to mapping transposon insertions

Martin H. Lee, Aphakorn Nittayajarn & Craig E. Rubens
Children's Hospital Medical Center CH-32, University Of Washington, Department of Pediatrics, 4800 Sand Point Way NE, Seattle, Washington, USA

Abstract. Two pieces of data are needed to fully map the location of a transposon inserted in a plasmid, the site of insertion and the transposon's orientation. Both of these parameters can be determined from a map of restriction sites, which can be derived by end-probing. Like restriction mapping, end-probing reveals the distance between restriction sites on a plasmid. In contrast to restriction mapping, end-probing unambiguously reveals the order of those restriction sites. End-probing is similar to end-labeling, except that the uncertainties inherent in the radiolabeling reaction (and the problems of working with radionucleotides) are avoided. In this paper we discuss the use of end-probing as a means to map insertions of transposon Tn*917* into the *Streptococcus agalactiae* plasmid pGB354.

Key words: End-labeling, GBS, Mapping, Restriction enzyme mapping, Tn*917*

Abbreviations: bp = Base pair; cm = Chloramphenicol; EDTA = Ethylenediaminetetraacetic acid; em = Erythromycin; GBS = Group B Streptococcus; HRP = Horseradish Peroxidase; kan = Kanamycin; LTR = Left Terminal Repeat; MSI = Micron Separations Inc.; NEB = New England Biolabs; PBS = Phosphate Buffered Solution; SDS = Sodium dodecyl sulfate; TBS = Tris Buffered Saline; THB = Todd Hewitt Broth; USB = United States Biochemical

1. Introduction

An efficient mapping procedure was needed to map Tn*917* insertions in plasmid pGB354, as part of characterizing Tn*917* insertion-site bias. Site bias is of interest because we are developing a Tn*phoA*-like tool for Gram-positive organisms. Tn*phoA* utilizes Tn*5*, which exhibits relatively little insertion site bias [1]. End-probing was devised to perform this mapping.

Plasmid pGB354 was selected as a target vector because it accumulated to high copy number in Group B Streptococcus (GBS), restriction maps were available [2], partial sequences were published [3, 4], and pGB354 was compatible with plasmid pTV10K [5], the source of Tn*917*. Tn*917* was chosen because we have already shown that transposition into the GBS chromosome is random [6], the sequence of the transposon has been reported [7], and (in other organisms) Tn*917* was known to transpose into plasmid targets [8, 9]. With these tools it was possible to derive large numbers of Tn*917* insertions in pGB354. We devised end-probing to map the insertion site and orientation of these Tn*917* insertions.

2. Materials

A. Equipment
1. Incubator-Shaker, model 25, New Brunswick Scientific.[28]
2. Tabletop centrifuge, model MC-12C, Sorvall.[23]
3. Spectrophotometer, model DU-6, Beckman.[24]
4. Heatblock, #13529-005, VWR Scientific.[8]
5. Electrophoresis Apparatus, model Horizon 11·44, #1068BD, Life Technologies.[17]
6. Power Supply, model 250, #11066-016, Life Technologies.[17]
7. UVP Dual Intensity Transilluminator, UVP.[25]
8. UV crosslinker, Hoefer.[7]
9. Orbital shaker, model Red Rotor, Hoefer.[7]
10. Hybridization Incubator, model 1000, Robbins Scientific.[27]
11. X-ray film processor, model M35A X-OMAT, Kodak.[18]
12. Oligonucleotide synthesizer, ABI.[16]
13. Concentrator, model SVC100H Speedvac, Savant.[21]

B. Culture medium and reagents
1. Antibiotics
   - Kanamycin, #K-4000, Sigma.[2]
   - Chloramphenicol, #C-0378, Sigma.[2]
   - Erythromycin, #E-6376, Sigma.[2]

2. Todd Hewitt Broth
   - THB, #0492, Difco.[1]
   - Select Agar, #5054, Sigma.[2]
3. Protoplast buffer
   - Sucrose, #S-9378, Sigma.[2]
   - $MgCl_2$, #M-0250, Sigma.[2]
   - Trizma, #T-1503, Sigma.[2]
   - TritonX-100, #T-9284, Sigma.[2]
   - Mutanolysin (#M-9901), Sigma.[2]
4. Restriction enzymes and DNA markers
   - *Pvu*II #140S, New England Biolabs.[4]
   - *Bcl*I # 160S, New England Biolabs.[4]
   - biotinylated Lambda/*Bst*EII size markers, #301-4BTS, New England Biolabs.[4]
   - NEB Buffer 2, #007-2, New England Biolabs.[4]
5. 6× Sample buffer
   - Bromophenol blue #B-7021, Sigma.[2]
   - Xylene Cyanole FF X-4126, Sigma.[2]
   - Glycerol G-7757, Sigma.[2]
6. SSC
   - NaCl, #S-9888, Sigma.[2]
   - Na_3Citrate, #C-7254, Sigma.[2]
7. PBS
   - NaCl, #S-9888, Sigma.[2]
   - KCl, #6845, Mallinckrodt.[9]
   - Na_2HPO_4, #7917, Mallinckrodt.[9]
   - KH_2PO_4, #P-0662, Sigma.[2]
8. Kits and miscellaneous reagents
   - Qiaprep Spin Kit, #27104, Qiagen.[3]
   - PCR Non-radioactive Labeling System, #10200-012, Life Technologies.[17]
   - ABI Prism Dye Terminator Cycle Sequencing Core Kit, #402111, Perkin Elmer.[19]
   - Agarose (Seakem, LE), #50004, FMC.[5]
   - Boric Acid, #0084-05, J. T. Baker.[6]
   - EDTA, #EDS, Sigma.[2]
   - Ethidium bromide, #E-2515, Sigma.[2]
   - HCl, #VW3110-3, VWR.[8]
   - NaOH, #S-5881, Sigma.[2]
   - Casein, #C-0376, Sigma.[2]
   - Tween-20, #P-1379, Sigma.[2]
   - Denhardt's solution, #70468, United States Biochemical.[10]
   - Deionized formamide, #423-502, Curtin Matheson Scientific Inc.[11]
   - SDS, #L-4509, Sigma.[2]
   - Dextran sulfate, #54030, Oncor.[12]
   - Supersignal, #34080SE, Pierce.[13]
   - Streptavidin-peroxidase conjugate, #14-30-00, Kirkegaard & Perry Laboratories.[14]
   - Nylon membrane, Magnagraph, #NJOHY00010, MSI.[15]
   - X-ray film, X-Omat RP, #195-2126, Kodak.[18]
   - ABI Prism Dye Terminator Cycle Sequencing Core Kit, #402111, Perkin Elmer.[19]
   - Sequencing analysis performed at the FHCRC.[22]
   - Desalting columns, Centri-sep, #PSR0015, Princeton Separations.[20]
   - Type 57 Polaroid film, #0083-170, Glazers.[26]
9. Strains and plasmids
   - COHI [10].
   - pMHL201 (this report) contains a Tn*917* insertion into pGB354.

3. Procedures

A. Preparation of materials

1. THB (in dH_2O)
   - 3% THB.
   - Autoclave for 45 minutes.
   - Cool to ~50 °C and add antibiotics as required.
2. THA (in dH_2O)
   - 3% THB.
   - 1.5% agar.
   - Autoclave for 45 minutes.
   - Cool to ~50 °C and add antibiotics as required.
3. Protoplast Buffer (in dH_2O)
   - 20% sucrose.
   - 10 mM $MgCl_2$.
   - 20 mM Trizma pH-7.0.
   - 0.05% TritonX-100.
   - (filter sterilize).
4. 6× Sample Buffer (in dH_2O)
   - 0.25% bromophenol blue.
   - 0.25% Xylene Cyanole FF.
   - 30% glycerol (G-7757).
5. 10× TBE (in dH_2O)
   - 890 mM Tris pH 8.
   - 890 mM boric acid.
   - 20 mM EDTA pH 8.0.
6. 20× SSC (in dH_2O)
   - 3 M NaCl.
   - 0.3 M Na_3Citrate pH 7.0.
7. Denaturation buffer (in dH_2O)
   - 1.5 M NaCl.
   - 0.5 M NaOH.
8. Neutralization buffer (in dH_2O)
   - 1 M Trizma, pH 7.4.
   - 1.5 M NaCl.
9. PBS (in dH_2O)
   - 137 mM NaCl.
   - 2.7 mM KCl.
   - 4.3 mM Na_2HPO_4.
   - 1.4 mM KH_2PO_4.
   - (autoclave).
10. TBS (in dH_2O)
    - 0.1 M Trizma pH 7.5.
    - 0.5 M NaCl.
    - (autoclave).

11. Prehybridization buffer (in TBS)
   - 1% Casein.
   - 0.05% Tween-20.
   - Dissolve the casein by heating gently (below boiling) with stirring for 2 hours.
12. Hybridization buffer (in dH_2O)
   - 1× Denhardt's solution.
   - 45% deionized formamide.
   - 0.1% SDS.
   - 5× SSC.
   - 20 mM sodium phosphate buffer pH 7.0.
   - 10% dextran sulfate.
13. Prewash Buffer (in dH_2O)
   - 2× SSC.
   - 1% SDS.
14. Wash buffer (in dH_2O)
   - 0.16× SSC.
   - 0.1% SDS.
15. Buffer 1 (in dH_2O)
   - 2 mM $MgCl_2$.
   - 0.1 M Trizma pH 7.5.
   - 0.1 M NaCl.
   - 0.05% TritonX-100.
16. Blocking buffer (in PBS)
   - 1% Casein.
   - 1% Tween-20.
   - Dissolve the casein by heating gently (below boiling) with stirring for 2 hours.
17. Binding buffer (in PBS)
   - 5% Tween-20.
18. Detection buffer I (in PBS)
   - 1% Tween-20.
19. Detection buffer II (in dH_2O)
   - 0.1 M Trizma pH 9.5.
   - 0.1 M NaCl.
   - Autoclave.
   - 50 mM $MgCl_2$ (added fresh with each use from a 1.0 M stock).
20. SSPE (in dH_2O)
   - 175.3 g NaCl.
   - 27.6 g $NaH_2PO_4 \cdot H_2O$.
   - 7.4 g EDTA.
   - Raise to 800 ml with dH_2O and pH to 7.4 with 10 M NaOH.
   - QS to 1 L and autoclave.

B. Primers
1. oML104
   AAT ATC CCT TTT GTT GTA GAA AC
2. oML105
   CAG ATT TAA ATT CTG ATT TTG AAG
3. Tn*917* L
   AGA GAG ATG TCA CCG TCA AG
4. Tn*917* R-ext
   TAG GCC TTG AAA CAT TGG TT

C. Plasmid preparation
1. Streak COHI [pMHL201] onto THA plate supplemented with cm to 10 μg/ml and em to 10 μg/ml.
2. Incubate overnight at 37 °C.
3. Inoculate a single colony into 3 ml THB-10 μg/ml cm and 10 μg/ml em.
4. Incubate overnight at 37 °C.
5. Pellet culture in a benchtop centrifuge at 13,600 g.
6. Resuspend the pellet in 1.5 ml PBS, repellet as before.
7. Resuspend the pellet in 40 μl protoplast buffer.
8. Add 2 μl of 10 U/μl mutanolysin, incubate at 37 °C for 1 hour.
9. Pellet protoplasts as before, resuspend pellet in 250 μl of buffer P1 from Qiaprep Spin Kit. Process the plasmids, thereafter, exactly according to the manufacturor's instructions. Note that, after lysis, the solution should clear. If the culture remains turbid it is very unlikely that the cells lysed sufficiently.
10. Elute the plasmid DNA off of the spin column in 50 μl 10 mM Tris pH 8.5.
11. Estimate the plasmid concentration assuming an absorbance of 1.0 equates to a 50 μg/ml DNA concentration (given a 1 cm light path) [11].

D. PCR amplification of biotinylated end-probe
1. Dilute pMHL201 to 1 ng/μl and dilute primers oML104 and oML105 to 5 μM in 10 mM Trizma pH 8.0. (The primers were designed amplify a 487 bp fragment adjacent to the unique *Pvu*II site in pMHL201, see Figure 1.)
2. Amplify the biotinylated probe using PCR Non-radioactive Labeling System, according to the manufacturer's recommendations. Set up four control reactions that omit one of the DNA components or $MgCl_2$. Set up trial reactions having 1, 2, 4, or 6 mM $MgCl_2$. (A faint 560 bp contaminating band is seen with 2–6 mM $MgCl_2$.)
3. Separate the fragments from the biotinylated cytosine with a Centri-sep column, according to the manufacturer's instructions.

E. Test the labeling reaction
   - Dilute the probe through 8 twofold dilutions in 10 mM Tris pH 8.0.
   - Spot 1 μl of each dilution onto a nylon membrane.
   - Crosslink with UV at 120 mJ/cm^2.
   - Block the membrane by incubating with blocking buffer for 2 hours at room temperature.
   - Wash three times in binding buffer for 3 minutes at room temperature.
   - Incubate the membrane with streptavidin-peroxidase diluted 5000-fold in binding buffer for 1 hour at room temperature.
   - Wash the membrane three times in blocking buffer for 15 minutes.
   - Wash the membrane twice in detection buffer I for 5 minutes.

- Dry the membrane on Whatman paper.
- Detect the probe using the Supersignal substrate according to the manufacturer's instructions.

F. Restriction digests
- Set up 4 digests of pMHL201 using 250 to 500 ng of plasmid per digest in 1× NEB 2 buffer.
- Add 10 units of *Pvu*II and digest for 1 hour at 37 °C.
- Dilute *Bcl*I in 1× NEB 2 buffer to produce 10, 3.33, 1, and 0.4 units/μl.
- When the *Pvu*II digest is complete, add 1 μl from one of the *Bcl*I dilutions to each of the complete *Pvu*II digests, and digest for 5 minutes at 50 °C.
- Add 4 μl of 6× sample buffer to each 20 μl digest, and resolve the fragments in a 0.7% agarose gel in 1× TBE at 2–3 V/cm.
- Stain the digested fragments for 10 minutes in TBE/ethidium bromide.
- Destain in TBE for 10 minutes, and photograph with UV transillumination.

G. Southern blotting. Blot the DNA onto nylon membrane using standard Southern blotting methods [12], with the following modifications
- Soak the agarose gel in 10× SSC for 30 minutes.
- Depurinate in 0.25M HCl for 30 minutes.
- Denature the DNA with two 30 minute incubations in denaturing buffer.
- Prepare the gel for transfer with two 30 minute incubations in neutralization buffer.
- Use overnight capillary transfer to move the DNA onto the membrane.
- After transfer, wash the membrane in 5× SSPE for 5 minutes at 60 °C. Dry the membrane on the bench for 30 minutes (it should remain slightly damp) and UV crosslink the DNA to the filter at 120 mJ/cm.
- Incubate the membrane for 1 hour at 42 °C in prewash buffer using a rotating hybridization.
- Block in prehybridization buffer for 30–40 minutes at room temperature.
- Convert the probe to single stranded form by adding an equal volume (usually 50 μl) of 0.2 M NaOH and incubate the probe at 37 °C for 30 minutes.
- Remove the prehybridization buffer and add 10 ml hybridization buffer with the denatured probe, then incubate overnight at 42 °C.
- Wash the membrane twice in wash buffer (to cover the membrane) for 3 minutes at room temperature on an orbital shaker at room temperature.
- Wash twice more at 65 °C.
- Rinse the membrane twice in 2× SSC for 3 minutes at room temperature in the hybridization oven.
- Block the membrane in Buffer 1 for 1 minute at 65 °C.
- Incubate in blocking buffer for 1 hour at room temperature, exchange for fresh buffer, and incubate 30 minutes further.
- Wash twice for 5 minutes in binding buffer.
- Cover the membrane in binding buffer with a 5000:1 dilution of streptavidin-peroxidase (it helps to have a small, flat-bottomed container to minimize the volume of peroxidase required, we found the tops of pipette tip racks to work well) and incubate for 1 hour at room temperature.
- Wash the membrane three times for 5 minutes at room temperature in blocking buffer.
- Wash three more times in detection buffer II for 5 minutes at room temperature.
- Perform chemilumenescent detection using Supersignal in accordance with the manufacturer's instructions. Exposure times vary between 2 and 20 minutes.

H. Sequencing
- Sequence from the left and right terminal repeats of Tn*917* with primers Tn*917*-L and Tn*917*-R-ext (respectively), using the ABI Prism Dye Terminator Cycle Sequencing Core Kit.
- Desalt with Centri-sep columns.
- Dry samples in a concentrator.
- Fluorescent sequencing was performed at the Fred Hutchinson Cancer Research Center facility (FHCRC).[22]

4. Results and conclusion

The approach is diagrammed in Figure 1a. First the plasmid was digested to completion at the unique *Pvu*II site. The linearized plasmid was then partially digested at the 5 *Bcl*I sites. This produces a complicated mixture of fragments, which would ordinarily be difficult to interpret. However, there is an informative subset of fragments (thin gray lines), all of which have one end at the *Pvu*II site and run clockwise to one of the *Bcl*I sites. Resolving the fragments on a gel and hybridizing a probe (thick gray line) to a Southern blot identified this subset. The probe was generated by PCR amplification of sequences adjacent to the *Pvu*II site.

Figure 1b shows such a blot. Plasmid pMHL201 was digested to completion with *Pvu*II, then the linear fragment was digested for 5 minutes with various amounts of *Bcl*I (varying *Bcl*I concentrations altered the relative abundance of short vs. long DNA fragments). The blotted fragments were probed with the biotinylated PCR fragment and the bands visualized using a streptavidin-HRP conjugate in the presence of a chemilumenescent substrate.

Six bands hybridized with the probe; the 12 kb band was produced by *Pvu*II cleavage (no *Bcl*I cleavage) and the 5 smaller bands were each

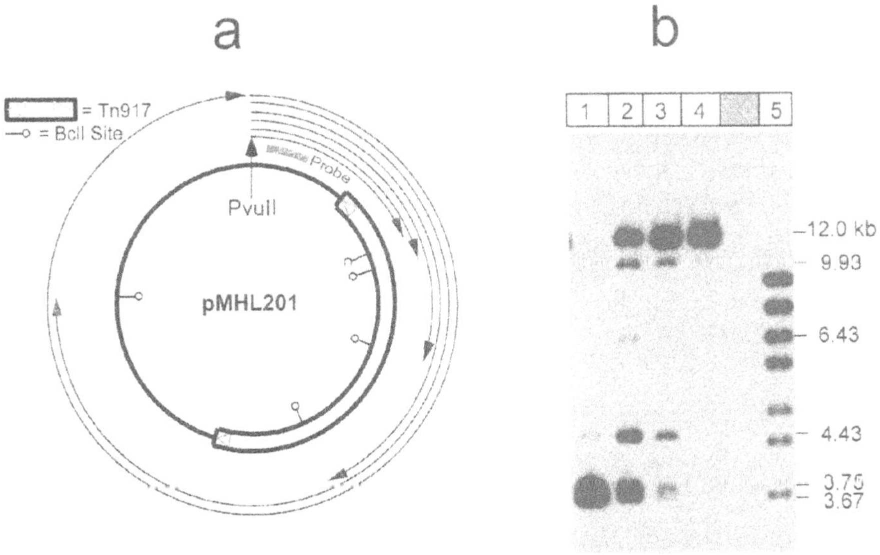

Figure 1. End-probing schematic and results. **(a)** Schematic of end-probing method. Plasmid pMHL201, composed of pGB354 with a Tn*917* insertion, was digested to completion with *Pvu*II and then partially digested with *Bcl*I (which cleaves 4 times in Tn*917* and once in pGB354). The fragments are separated on an agarose gel, blotted, and hybridized with biotin-labeled probe (thick gray line). The probe hybridizes to all fragments that extend clockwise from the *Pvu*II restriction site (thin gray lines). **(b)** Results of end-probing. *Bcl*I concentrations used were 10, 3.3, 1.1, and 0.4 units in lanes 1 to 4, respectively. Biotinylated lambda/*Bst*EII fragments were used as size markers in lane 5. Bands were detected using a streptavidin-HRP conjugate and a chemilumenescent substrate for HRP.

produced by *Pvu*II cleavage at one end and *Bcl*I cleavage at the other. The size of these bands, therefore, maps the distance from *Pvu*II (clockwise) to a *Bcl*I site. Reading from the bottom to the top, there are *Bcl*I sites at 3.67 kb, 3.75 kb, 4.43 kb, 6.43 kb, and 9.93 kb from the *Pvu*II site. The pattern of *Bcl*I restriction sites showed that Tn*917* is inserted with its left terminal repeat (LTR) closest to the probe. Within Tn*917*, the *Bcl*I site nearest the LTR is 1.83 kb from the LTR end, so the insertion site was at 1.84 kb (3.67–1.83) in pGB354.

How accurate is end-probing? The transposon's orientation and its exact insertion site could be identified by sequencing. Twelve Tn*917* insertions were first mapped by end-probing and then sequenced. All twelve had been assigned the correct orientation. The insertion site assignments made by end-probing were misplaced, on average, by 173 base pairs. The size of the average error was greatly increased by a small number of assays in which the smallest bands were quite large (~6 kb). We note that smaller and more accurately sized bands could have been derived by stripping the probe off of the blot and reprobing with sequences from the opposite flank of the *Pvu*II site.

Problems with end-probing can occur when the enzyme used for the partial digest cleaves different locations at significantly different rates (e.g., [13]). If a given site is cleaved too slowly then the expected fragment may not accumulate to detectable levels during the short incubations used in these experiments. A complete digest is a useful control in determining the number of expected fragments.

End-probing proved to be an efficient means of mapping a large number of transposon insertions. The problematic aspects of end-labeling were avoided, while the traditional advantage of end-labeling (the unambiguous ordering of restriction sites on the plasmid) was maintained. End-probing is based on robust methods, and investigators with PCR and Southern blotting experience can expect to obtain interpretable results immediately. End probing is time consuming compared to a simple restriction digest (probe production and verification, as well as the washing and hybridizing steps, make it a two-day process at a minimum). Consequently, it may be best exploited in projects where a single probe can be used to analyze many plasmids simultaneously, e.g., to identify overlapping DNA fragments in a physical-mapping project.

Acknowledgments

This work was supported by NIH Grant AI25152 The Group B Streptococcus Initiative. Thanks to Paul Framson and Harry Yim for help in creating pMHL201 and to Amanda Jones for a critical reading of this manuscript.

Notes on suppliers

1. Difco, PO Box 331058, Detroit, MI 48232-7058, USA
2. Sigma, PO Box 14508, St. Louis, MO 63178, USA
3. Qiagen, 9600 DesSoto Avenue, Chatsworth CA 91311, USA
4. NEB, 32 Tozer Rd., Beverly, MA 01915-5599, USA
5. FMC BioProducts, 191 Thomaston Str., Rockland, ME 04841, USA
6. J. T. Baker, 222 Red School Lane, Phillipsburg, NJ 08865, USA
7. Hoefer Scientific Instruments, Box 77387, San Francisco, CA 94107-0387, USA
8. VWR Scientific, PO Box 1380, Piscataway, NJ 08855, USA
9. Mallinckrodt, 222 Red School Lane, Phillipsburg, NJ 08865, USA
10. USB, PO Box 22400, Cleveland, OH 44128, USA
11. Curtin Matheson Scientific Inc., Box 1546, Houston, TX 77038, USA
12. Oncor, 209 Perry Parkway, Gaithersburg, MD 20877, USA
13. Pierce, PO 117, Rockford, IL 61105, USA
14. Kirkegaard & Perry Laboratories, 2 Cessna Court, Gaithersburg, MD 20879, USA
15. Micron Separations, Inc, 135 Flanders Road, Westborough, MA 01581, USA
16. ABI (division Perkin-Elmer) 850 Lincoln Center Drive, Foster City, CA 94404, USA
17. Life Technologies, 1 Kendall Square, Grand Island, NY 14072-0068, USA
18. Eastman Kodak, 1001 Lee Road, Rochester, NY 14652-4115, USA
19. Perkin Elmer, 850 Lincoln Center Drive, Foster City, CA 94404, USA
20. Princeton Separations Inc, PO Box 300, Adelphia, NJ 07710, USA
21. Savant Industries, 110-103 Bi-County Boulevard, Farmingdale, NY 11735, USA
22. FHCRC, PO Box 19024, Seattle, WA 98109-1024, USA
23. Sorvall, PO Box 5509, Newtown, CT 06470-5509, USA
24. Beckman Instruments Inc., 8920 Route 108, Columbia, MD 21045, USA
25. UVP (Ultraviolet Products), 2066 W. 11th Street, Upland, CA 91786, USA
26. Glazers Camera Supply, 430 8th Street N., Seattle, WA 98109, USA
27. Robbins Scientific, 814 San Aleso Avenue, Sunnyvale, CA 94086-1411, USA
28. New Brunswick Scientific Co. Inc., 44 Talmadge Road, Edison, NJ 08818-4005, USA

References

1. Berg DE, Schmandt MA, Lowe JB (1983). Specificity of transposon Tn*5* insertion. Genetics 105: 813–828.
2. Behnke D, Gilmore MS, Ferretti JJ (1981). Plasmid pGB301, a new multiple resistance streptococcal cloning vehicle and its use in cloning of a gentamicin/kanamycin resistance determinant. Mol Gen Genet 182: 414–421.
3. Trieu-Cuot P, de Cespedes G, Horaud T (1992). Nucleotide sequence of the chloramphenicol resistance determinant of the streptoccal plasmid pIP501. Plasmid 28: 272–276.
4. Brantl S, Behnke D, Alonso JC (1990). Molecular analysis of the replication region of the conjugative *Streptococcus agalactiae* plasmid pIP501 in *Bacillus subtilis*. Comparison with plasmids pAM beta 1 and pSM19035. Nucleic Acids Res 18: 4783–4790.
5. Gutierrez JA, Crowley PJ, Brown DP, et al. (1996). Insertional mutagenesis and recovery of interrupted genes of *Streptococcus mutans* by using transposon Tn*917*: preliminary characterization of mutants displaying acid sensitivity and nutritional requirements. J Bacteriol 178: 4166–4175.
6. Framson P, Nittayajarn A, Merry J, et al. (1997). New genetic techniques for Group B Streptococci: High-efficiency transformation, maintenance of temperature-sensitive pWV01 plasmids, and mutagenesis with Tn*917*. Appl Environ Microbiol 1997; 63: 3539–3547.
7. Shaw JH, Clewell DB (1985). Complete nucleotide sequence of macrolide-lincosamide-streptogramin B-resistance transposon Tn*917* in *Streptococcus faecalis*. J Bacteriol 164: 782–796.
8. Tomich PK, An FY, Clewell DB (1980). Properties of erythromycin-inducible transposon Tn*917* in *Streptococcus faecalis*. J Bacteriol 141: 1366–1374.
9. Cosby WM, Axelsson LT, Dobrogosz WJ (1989). Tn*917* transposition in Lactobacillus plantarum using the highly temperature-sensitive plasmid pTV1Ts as a vector. Plasmid 22: 236–243.
10. Wessels MR, Haft RF, Heggen LM, et al. (1992). Identification of a genetic locus essential for capsule sialylation in type III group B streptococci. Infect Immun 60: 392–400.
11. Sambrook J, Fritsch EF, Maniatis T (1989). Molecular cloning: A laboratory manual, 2nd edn. Plainview, NY: Cold Spring Harbor Laboratory Press.
12. Ausubel FM, Brent R, Kingston RE, et al. (1995). Current protocols in molecular biology, in: K. Janssen (ed), New York: John Wiley & Sons, Inc.
13. Gingeras TR, Brooks JE (1983). Cloned restriction/modification system from *Pseudomonas aeruginosa*. Proc Natl Acad Sci USA 80: 402–406.

Address for correspondence: Craig E. Rubens, Children's Hospital Medical Center CH-32, University Of Washington, Department of Pediatrics, 4800 Sand Point Way NE, Seattle, WA 98105, USA
Phone: (206) 528-2771; Fax: (206) 527-3890
E-mail: cruben@chmc.org

Methods in Cell Science **20**: 119–126 (1998)

A method for mapping phage-inducible promoters for use in bacteriophage-triggered defense systems

G. M. Djordjevic[1,*] & T. R. Klaenhammer[2]

[1] *Departments of Microbiology and* [2] *Food Science, North Carolina State University, Raleigh, North Carolina, USA (*Current Address: University of California, San Diego, Department of Biology, 0116, 9500 Gilman Drive, La Jolla, California, USA)*

Abstract. We have recently developed a novel bacteriophage-protection system for *Lactococcus lactis*, based on a two-component genetic 'trap'. An inducible promoter from a lytic bacteriophage is used to activate a lethal gene after infection, killing the host cell and halting phage proliferation. To expand the potential use of this novel defense strategy, promoters specific to any particular phage of interest must be available for fusion to a universal death gene. A method to localize regulated promoters within the context of the total phage genome was evaluated. The 'capping' activity of the vaccinia virus guanylyltransferase was exploited to label newly synthesized mRNA extracted from infected cells at sequential time points over the course of a phage infection. The labeled mRNAs were then used as probes in Southern hybridization reactions to identify restriction fragments in the phage genome where new transcripts were initiated during progression of the phage lytic cycle. This method has been used successfully in our laboratory to map the general location of a number of inducible promoters on the genomes of bacteriophages attacking lactic acid bacteria. Once identified and cloned, small fragments encoding inducible promoters can be partially sequenced and primer extension reactions carried out on phage RNA, isolated over the course of an infection, to pinpoint the precise location of the promoter. In this study, we illustrate the use of the capping method to map the phage-inducible promoter on the genome of the lactococcal bacteriophage sk1. This approach provides a rapid and efficient means to identify promoter regions on the genomes of relatively uncharacterized phages. These promoters can then be used in a variety of applications, including phage-triggered defenses and inducible gene expression systems.

Key words: 'Capping', *Lactococcus lactis*, Phage-triggered promoters

1. Introduction

Lactococcus lactis, used widely as a starter culture in dairy product fermentations, remains susceptible to periodic bacteriophage attacks. Repeated and long term use of pure starter cultures provides numerous opportunities for bacteriophage contamination. Our group has developed a unique defense system that is designed to trap infecting phages and prevent their proliferation in fermentation environments [6]. The system, first constructed in *L. lactis*, uses a phage-specific promoter to trigger expression of a bacterial suicide system following phage infection [20, see Figure 1]. The lethal component of the model suicide system is the restriction endonuclease cassette (*Lla*IR$^+$), composed of three genes (*llaI.1, llaI.2, llaI.3*) from the *Lla*I operon which are essential for restriction activity [4, 11, 18, 19; Figure 2]. When *L. lactis* harbored the ϕ31P/*Lla*IR$^+$ suicide cassette on a high-copy plasmid (pTRK414H), the plaquing efficiency (EOP) of phage ϕ31 was limited to 10^{-4}. Moreover, liquid cultures, challenged at multiplicities of infection (MOI) of 0.1, developed normally with no inhibition of growth or acid production. The efficiency of the suicide cassette was further improved by increasing promoter strength, enhancing restriction activity, and stacking this phage-triggered suicide system with other defense mechanisms [5]. In the later example, phage present initially in broth cultures at levels of 10^6 pfu/ml, were eliminated within 4–5 hours of growth ($<10^2$ pfu/ml).

The phage-triggered suicide system is a promising approach for phage protection of lactococci, or potentially any other prokaryote under siege by a virulent bacteriophage. However, due to the high intrinsic specificity of phage transcription, promoters are fired only by a few closely-related phages [6]. Broader application of triggered suicide systems against different phages will require isolation of tightly-regulated inducible promoters from the phage of interest or, ideally, promoters which are more widely recognized by several different bacteriophages. Information about phage-specific inducible promoters is scarce. Except for ϕ31P [20], all the phage-specific promoters identified thus far in lactococcal phages bIL41, c2, and sk1 have resulted

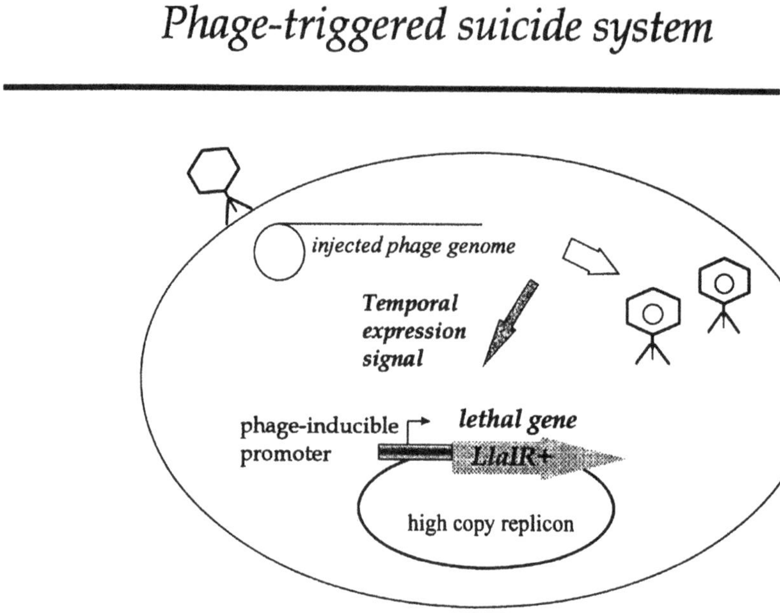

Figure 1. Representation of the phage-inducible suicide system constructed in *Lactococcus lactis*.

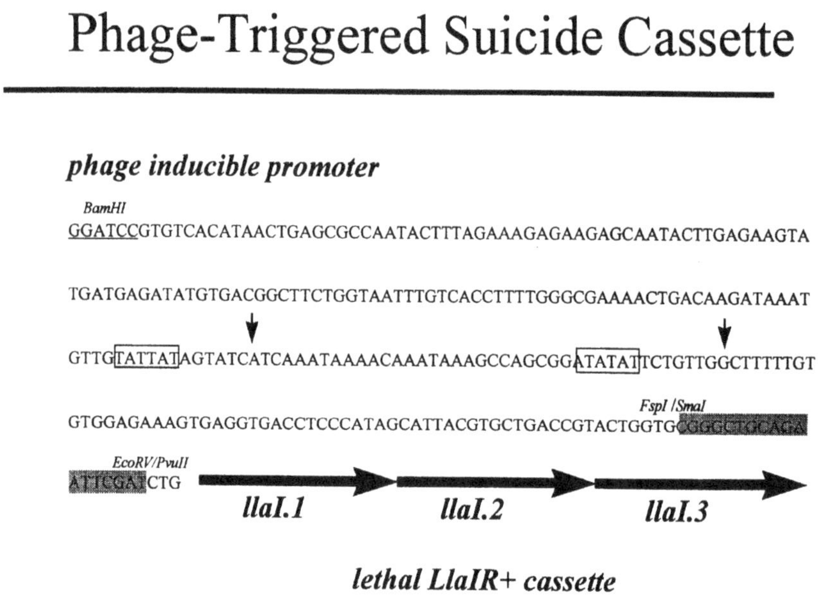

Figure 2. Nucleotide sequence of the 239-bp φ31P promoter fused to the *Lla*IR⁺ restriction cassette of *llaI.1*, *llaI.2*, and *llaI.3* (φ31P/*Lla*IR⁺). The transcription start sites at nucleotides 703 and 744 [16] are represented by vertical arrows. The −10 consensus promoter regions are boxed. The shaded region represents a portion of the multiple cloning site (MCS) from pBluescript II KS+. Junctions between the MCS and φ31P and *llaI.1* are represented by *Fsp*I/*Sma*I and *Eco*RV/*Pvu*II, respectively. The *Bam*HI site was introduced by the oligonucleotide primer used for PCR subcloning of the φ31P. Adapted [5].

from partial or complete sequencing of phage genomes [2, 21, 28]. Neither sequencing nor shotgun cloning promise to be effective methods to detect and isolate phage-specific promoters from lytic phages that appear and destroy starter cultures in industrial fermentations. A rapid, efficient, and accurate method is needed which could be universally applied to different bacteriophages to identify and characterize inducible promoters. Ideally, the approach would not require detailed genetic information about the phage or its genomic sequences.

Transcription initiation sites used by the

Escherichia coli bacteriophage N4 RNA polymerase (RNAP) were mapped previously using the enzyme guanylyltransferase [3, 10]. Stoddard and Howe [26] isolated RNA at varying time points after induction of Mu, radiolabeled the RNA by an in vitro capping assay, and used it in Southern hybridization experiments against DNA fragments of Mu. The capping enzyme, guanylyltransferase has been isolated and purified from vaccinia virions [14], rat nuclei [13], and *Saccharomyces cerevisiae* [12]. The enzyme exclusively catalyzes the transfer of the guanosine-mono-phosphate (GMP) moiety ('cap') from the guanosine-tri-phosphate (GTP) to RNA transcripts possessing di- or tri-phosphate 5′ termini [15]. Therefore, the product of the exchange reaction catalyzed by guanylyltransferase is a 'capped' RNA molecule, with the $G^{5'}ppp^{5'}N$ structure on it's 5′ terminus:

$$\overset{\gamma\beta\alpha}{pp\mathbf{pG}} + \overset{\gamma'\beta'\alpha'}{p\mathbf{pp}N(pN)n} \xrightarrow[]{\text{Guanylyl transferase}} \underset{\text{'capped' RNA}}{\overset{\alpha\beta'\alpha'}{G\mathbf{ppp}N(pN)n}} + \overset{\gamma'}{p_i} + \overset{\gamma\beta}{pp_i}\text{GTP RNA}$$

When $^{\alpha 32}$P-labeled GTP is used in the capping reaction, the RNA molecules are specifically labeled at their 5′ ends.

In this study, the capping activity of guanylyltransferase was used to identify the phage-specific inducible promoter in the *L. lactis* bacteriophage sk1. Phage transcripts synthesized *de novo* at different stages of phage infection cycle were first labeled in vitro with vaccinia virus guanylyltransferase and then used as hybridization probes to identify genomic restriction fragments in phage sk1 where phage-specific transcripts were initiated.

2. Materials

A. Bacteria and bacteriophage
   1. *L. lactis* MG1363 [12], sensitive host for bacteriophage sk1
   2. Bacteriophage sk1, a small isometric-headed, 936-species, cohesive-ended, lactococcal bacteriophage with a double-stranded DNA genome of 28.45-kb [2, 22]

B. Culture growth medium and supplements
   1. M17 broth, No. 1856-17.[1]
   2. Glucose, No. G 8270.[2]

C. Reagents
   1. *Taq* DNA polymerase, No. 1 146 165.[3]
   2. PCR purification kit, No. 28104.[5]
   3. TRIzol™, No. 15596-018.[6]
   4. Glass beads (<106 μm, 0.5 g), No. G 4649.[2]
   5. Tris-HCl, No. T 3253.[2]
   6. $MgCl_2$, No. M 2670.[2]
   7. KCl, No. P 9541.[2]
   8. DTT, No. D 5545.[2]
   9. Guanylyltransferase, No. 18024-018.[6]
   10. $^{\alpha 32}$P[GTP], No. NEG-006H.[8]
   11. Ethylenediaminetetraacetic acid, disodium salt (EDTA), No. ED 2SS.[2]
   12. Sodium dodecyl sulfate (SDS), No. L 5750.[2]
   13. NaCl, No. S 9625.[2]
   14. Sodium hydroxide (NaOH), No. S 8045.[2]
   15. Hepes, No. H 2393.[2]
   16. *Eco*RV, No. 667145.[3]
   17. *Hin*dIII, No. 656313.[3]
   18. *Sau*3AI, No. 709743.[3]
   19. *Ssp*I, No. 972967.[3]
   20. RNAse inhibitor, No. 799017.[3]
   21. SeaKem GTG agarose gel, No. 50072.[10]
   22. MSI Magnacharge nylon membranes, No. NBOHY BOO10.[11]
   24. 20× SSC (3 M NaCl, No. S 9625.[2], 0.3 M Na-citrate, No. S 4641.[2]; pH = 7.0)
   25. Hybridization buffer (50% formamide, No. 100144.[3], 5× SSPE, 20% SDS, No. L 5750.[2], 10× Denhardt)
   26. 20× SSPE (NaCl, No. S 9625.[2]/175.3 g/L, $NaH_2PO_4{\cdot}H_2O$, No. S 9638.[2]/27.6 g/L, EDTA, No. ED 2SS.[2],/7.4 g/L, pH = 7.4)
   27. Denhardt (50×; Ficol Type 400, No. 17-0400-01[12],/1 g/100 ml, polyvinylpyrrolidone, No. P 6755.[2]/1 g/100 ml, BSA (Pentex Fraction V), No. A 2153.[2],/1 g/100 ml)
   28. Whatman paper, No. 3030 917.[13]
   29. $^{\alpha 32}$P[dCTP], No. BIVO13H.[8]
   30. Multiprime DNA labeling kit, No. RPN.1601Z.[14]
   31. 1× STE buffer (0.1 M NaCl, No. S 9625.[2], 20 mM Tris-HCl (pH 7.5), No. T 3253.[2], 10 mM EDTA (pH 8.0), No. ED 2SS.[2])

D. Major equipment
   1. Thermal cycler (PowerBlock™ System).[4]
   2. NucTrap® probe purification push columns, No. #400701.[9]
   3. NucTrap® push column device, No. #400700.[9]
   4. UV-Stratalinker-Stratagene, Model 1800.[9]
   5. Mini bead beater-8™ cell disrupter.[7]
   6. KODAK BIOMAX MS film, No. 116 5893.[15]
   7. Intensifying screen, No. KP 105763.[15]
   8. KODAK X-omatic cassette, No. KP 69065-B.[15]

3. Procedure

A. Culture conditions
   1. Propagate *L. lactis* MG1363 at 30 °C in M17 broth supplemented with 0.5% glucose (GM17). L. lactis MG1363 was maintained as frozen stock (GM17 with 20% glycerol) at −20 °C. A 5 ml overnight culture was inoculated from the single colony. The 5 ml culture was used to inoculate 50 ml of GM17 (10% inoculum, log. culture) and propagated at

30 °C until OD600 reached 0.4 (approx. 3–4 hours).

B. Bacteriophage infection and phage DNA isolation
   1. Conduct bacteriophage infection as described by Terzaghi and Sandine [27].
   2. Isolate phage DNA using the large scale protocol described by Raya et al. [23].

C. RNA Isolation
   1. All solutions and glass/plasticware must be RNAse free. Plastic/glassware was soaked in 3% hydrogen peroxide, several hours or overnight, rinsed well with picopure water (double deionized water), and autoclaved, dry cycle, 40 minutes. Water (picopure) was autoclaved first in RNAse-free glassware and then used to make desired solutions. Measuring spatulas were soaked in 3% peroxide and autoclaved.
   2. Incubate *L. lactis* MG1363 cells at 30 °C until OD_{600} = 0.4 is reached and then infect with phage sk1 at multiplicity of infection (MOI) of 5 (five ml of phage stock @ 1×10^{10} pfu/ml was added to the 50 ml culture).
   3. At different time points over the course of the infection, remove 9 ml of cells from the culture, centrifuge, and freeze the cell pellets in an ethanol:dry ice bath. Isolate RNA from the frozen pellets using 1 ml of TRIzol™ reagent according to the manufacturers instructions. Add glass beads and homogenize the cells for 90 seconds in a Mini Bead Beater.[7] Determine RNA concentrations spectrophotometrically, 1 OD_{260} = 40 mcg/ml RNA. For pure RNA the OD_{260}/OD_{280} ration should be 2 (> 1.8 is acceptable).

D. 5′-end labeling of RNA samples (all solutions and glass/plasticware must be RNAse free)
   1. Use the procedure originally described by Haynes and Rothman-Denes [10] with the following modifications: Incubate RNA (100 μg) for 30 minutes at 37 °C in a reaction buffer containing 50 mM Tris-HCl (pH 7.9), 1.2 mM $MgCl_2$, 6 mM KCl, and 2.5 mM DTT, with 10 units of guanylyltransferase and 3.5 nmol $^{\alpha 32}P$[GTP] (3000 Ci/mmol), in a total reaction volume of 50 μl.
   2. Terminate the reaction by adding 1 mM EDTA (pH 8.0) and 0.2% SDS. Five microliters of 10× stock solution (10 mM EDTA, 2% SDS, pH = 8) per 50 microliters reaction (see item D1).
   3. Add 20 μl of 1× STE buffer and separate the RNA from the unincorporated nucleotides using NucTrap[R] probe purification columns according to the manufacturers instructions.
   4. When the purification is completed, wash the columns with 70 μl of 1× STE buffer and combine eluates with the originally eluted RNA samples (total volume per sample 140 μl).
   5. To generate short end-labeled RNA species, hydrolyze the purified RNA samples with 0.2 N NaOH (30 min, on ice), and neutralize with 0.5 vol 1M Hepes acid as described previously by Haynes and Rothman-Denes [10].
   6. Use short RNA species as hybridization probes to locate phage sk1 fragments which encode phage-inducible transcription signals.

E. RNA/DNA hybridization (all solutions and glass/plasticware must be RNAse free)
   1. Digest phage sk1 genomic DNA (3–5 μg) with restriction enzymes *Eco*RV, *Hin*dIII, *Sau*3AI, and *Ssp*I using the standard methods [24].
   2. Add 10 units of RNAse inhibitor per sample and incubate the tubes at 37 °C for 30 minutes.
   3. Heat the samples for 30 minutes at 65 °C (to open phage cohesive ends) and place in an ice bath until loading on a 1.2% SeaKem GTG agarose gel.
   4. Gel-separate restriction fragments and transfer to MSI Magnacharge nylon membranes as described by Southern [25]. Perform transfer overnight at room temperature. Use 20× SSC buffer.
   5. When transfer is completed, cross-link the DNA to the membrane in UV-Stratalinker.
   6. Pre-wet the filters in 5× SSPE and prehybridize for 1 hour at 42 °C in prehybridization buffer, 30 ml (0.2 ml/cm^2 of membrane 150 cm^2) containing 50% formamide [8]. 20 ml of hyb. buffer (approx. 0.13 ml/cm^2 of membrane). Hybridize at high-stringency (at 42 °C, over-night in 20 ml of hybridization buffer, identical to prehybridization buffer) to 5′-end labeled RNA probes. The labeled RNA probe (100 mcg) was in a total volume of 213 microliters (213 mcl. = 140 mcl. (see step D4) + 70 mcl. of Hepes + 2.8 mcl of 10 N NaOH (see step D5). The probe was boiled for 5 min, mixed with 0.5 ml of hybridization buffer and added to a bottle containing 19.5 ml of the same buffer (0.13 ml/cm^2 membrane). Hybridization was performed in roller oven.
   7. Perform post-hybridization washes at room temperature:
      - Wash filters once for 30 sec in 30 ml of 2× SSPE, 0.1% SDS to remove unbound probe
      - Wash filters twice, for 15 min each time, in 200 ml 2× SSPE, 0.1% SDS
      - Wash filters twice, for 15 min each time, in 200 ml 0.1× SSPE, 0.1% SDS
   8. Rinse filters in 2× SSPE and blot excess buffer by placing membrane on a piece of Whatman paper.
   9. Wrap the filters in plastic wrap and set up autoradiography. Use KODAK BIOMAX MS film, with adequate intensifying screen.

F. DNA/DNA hybridization

1. Digest phage DNA with restriction enzymes *Eco*RV, *Hin*dIII, *Sau*3AI, and *Ssp*I, transfer to MSI Magnacharge nylon membranes, and UV-crosslinked as described above.
2. Generate the hybridization probe by PCR amplification of a phage sk1 genomic DNA template over the region encoding the middle promoter, at nucleotide postions 26,331 to 26,487 (10 & A. Hillier and B. Davidson, personal communication). The primers used were 5′ GTCTCTAGCCATTGTTACC (forward) and 5′ CTGTTCTTCTGTCATTTGC (reverse) corresponding to positions 26,331–26349 and 26,487–26,469, respectively in the phage sk1 genome [2]. Use the Taq polymerase and the following amplification conditions: 1 cycle of 8 min at 95 °C (hot start), 30 cycles of 1 min at 95 °C, 1 min at 55 °C, 2 min at 68 °C, 1 cycle of 10 min at 68 °C in an Ericomp[4] thermal cycler. Purify PCR products using the QIAGEN[5] PCR purification kit following the manufacturer's instructions.
3. Hybridize to $^{\alpha 32}$P[dCTP]-labeled DNA probes, prepared with a Multiprime DNA labeling kit according to the manufacturers instructions (See Amersham Multiprime DNA Labeling Instruction booklet, p. 8. Rapid Protocol). Use the same hybridization and post-hybridization wash conditions as described for RNA/DNA hybridization.

4. Results and discussion

In this study, we describe the rapid and efficient method to identify phage genomic regions encoding inducible promoters. The 'capping' activity of vaccinia virus guanylyltransferase was exploited to in vitro label RNA transcripts generated after infection of *L. lactis* with bacteriophage sk1. In eukaryotic cells, the 'cap' structure, consisting of 7-methylguanosine residue, is naturally added to the initial 5′ nucleoside of RNA transcripts via a 5′-5′ triphosphate bridge by a specific guanylyltransferase. Absence of the 'cap' at 5′-ends of prokaryotic transcripts represents the principal difference between the prokaryotic and eukaryotic mRNAs. Because guanylyltransferase catalyzes transfer of the GMP moiety ('cap') from GTP to any RNA transcript possessing a di- or tri-phosphate 5′ termini, the newly initiated prokaryotic transcripts, which have tri-phosphate groups at their 5′-termini, are 'capped' in vitro.

The capping method was used to identify precisely the sk1 genomic regions where phage-specific transcription is initiated during the timely-regulated expression of phage genes. *L. lactis* MG1363 cells were infected with phage sk1 at MOI of 5 and total RNA isolated from lactococcal cells at 3, 5, 8, and 15 minutes after infection, and labeled with vaccinia virus guanylyltransferase and [$^{\alpha 32}$P]GTP. Because phage-encoded transcripts usually encompass a large genomic regions capped RNA molecules were hydrolyzed with 0.2 M NaOH to obtain shorter end-labeled probes, suitable for hybridization.

A partial restriction map of sk1 was constructed recently [1, Figure 3A]. Phage sk1 genomic DNA was isolated and digested with *Eco*RV, *Hin*dIII, *Sau*3AI, and *Ssp*I (Figure 3B). The restriction fragments were then immobilized on the nylon membranes and hybridized with RNA probes generated from the RNA samples isolated at 3, 5, 8, and 15 minutes after the infection. Comparison of the hybridization patterns at 3 and 8 minutes (Figure 3C) identified DNA fragments where new phage-specific transcription was initiated 8 minutes after the infection. Within the sk1 genome, these were the overlapping DNA fragments *Hin*dIII 'D' and *Ssp*I 'I' (Figure 3A), as designated by Chandry et al. [see 1, Figure 1]. Probes generated from RNA samples isolated at 5 minutes and 15 minutes after the infection showed hybridization patterns identical to those obtained with the 3 minute and 8 minute probes, respectively (data not shown). Strong hybridization signals were obtained only with single restriction fragments, regardless of the restriction enzyme used to digest phage DNA (Figure 3C). The results indicated that phage-specific transcription was initiated within a single phage sk1 genomic region 8 minutes after the phage infection.

Some of the early phage sk1 transcripts, generated at 3 minutes after the infection, were more abundant during the later stages of lytic development (Figure 3C, lanes 2 & 6, 4 & 8). These transcripts correspond to sk1 genomic regions where early and middle-late transcription overlaps [1]. However, the sk1 transcripts which hybridize to phage genomic fragments with the same intensity at 3 minutes and 8 minutes after the infection, represent early phage transcripts, exclusively (Figure 3C, lanes 5 & 9).

The entire phage sk1 genome has been sequenced by Chandry et al. [2] and a phage-inducible, middle promoter mapped by primer-extension. For our study, a short segment of the sk1 genome (157-bp), encoding the middle phage-specific promoter (P_M, 10 & A. Hillier & B. Davidson, personal communication), was amplified by PCR and labeled with $^{\alpha 32}$P[dCTP]. The labeled fragment was then used as a probe for Southern hybridization [25] to restriction fragments of phage genomic DNA immobilized on the nylon membrane (Figure 3D). Strong hybridization signals were obtained only with the phage sk1 genomic fragments which also hybridized to transcripts generated during the middle-late stage of lytic development (Figure 3C). Therefore, the *Hin*dIII fragment 'D' and the *Ssp*I fragment 'I', mapped previously with guanylyltransferase, were found to encode a single phage-inducible transcrip-

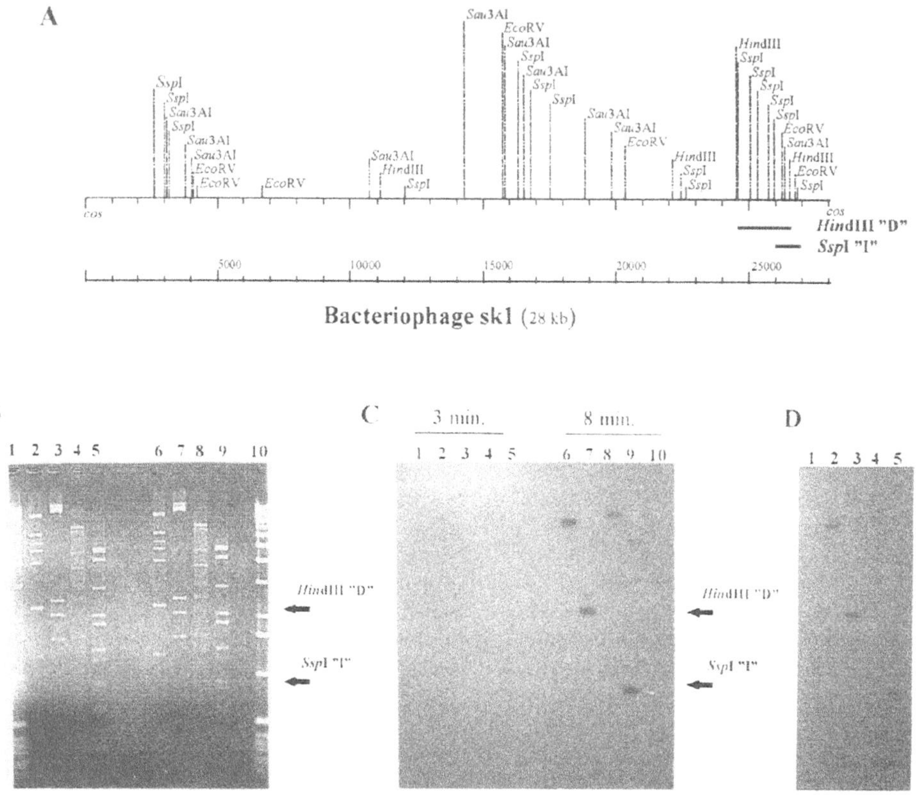

Figure 3. Location of phage sk1 genomic fragments encoding new transcripts initiated after infection of *Lactococcus lactis* MG1363. (A) Restriction map of bacteriophage sk1 adapted from [1]. Only relevant restriction sites are shown. Overlapping DNA fragments *Hin*dIII 'D' and *Ssp*I 'I' are indicated. (B) Restriction enzyme patterns of phage sk1 restricted with *Eco*RV (lanes 2, 6), *Hin*dIII (lanes 3, 7), *Sau*3AI (lanes 4, 8), and *Ssp*I (lanes 5, 9). The molecular weight standard was 1-kb ladder (GIBCO-BRL, lanes 1, 10). The *Hin*dIII fragment 'D' and *Ssp*I fragment 'I' are indicated by arrows. (C) Hybridization profiles of Southern transfer containing phage sk1 genomic fragments restricted with *Eco*RV (lanes 2, 6), *Hin*dIII (lanes 3, 7), *Sau*3AI (lanes 4, 8), and *Ssp*I (lanes 5, 9) probed with 5′-end labeled RNAs isolated at 3 minutes and 8 minutes after the infection. Hybridization signals indicated by arrows represent newly initiated transcripts corresponding to the DNA fragments *Hin*dIII 'D' and *Ssp*I 'I'. (D) Hybridization profiles of Southern transfer containing phage sk1 genomic fragments restricted with *Eco*RV (lane 2), *Hin*dIII (lane 3), *Sau*3AI (lane 4), and *Ssp*I (lane 5) probed with the $^{\alpha32}$P[dCTP]-labeled sk1 DNA fragment encoding the phage-specific promoter. Hybridization signals corresponding to the *Hin*dIII fragment 'D' and the *Ssp*I fragment 'I' are indicated by arrows.

tion signal. Only one such promoter was found to be encoded within the entire phage sk1 genome. In fact, phage-specific promoters are not abundant in the genomes of lactococcal bacteriophages; ϕ31 [20], bIL41 [21], and c2 [28] each encode only a single phage-inducible promoter. Therefore, use of the 'capping' method to localize these rare genetic elements in the future will have an important advantage over shot-gun cloning or sequencing of phage genomes, two labor intensive approaches used previously to identify fragments encoding phage-specific promoters. Once identified, these DNA fragments can be cloned and sequenced. The precise location of a phage-inducible promoter within a small DNA fragment could be identified using primer extension reactions, carried out on RNA isolated over the course of a phage infection.

A temporal transcription map of phage sk1 was determined previously [1]. Northern analysis revealed that early transcription of the sk1 genome was initiated within 5 minutes after infection, followed by middle (7 to 10 minutes), and late (15 minutes) transcription. The positions and lengths of early and middle phage transcripts were determined by Northern hybridization. The transcriptional starts and exact lengths of late transcripts could not be determined due to RNA processing or degradation. The authors speculated that late sk1 transcripts could be generated from either a late promoter located within the middle region of the phage genome, or by antitermination or processing of transcripts initiated from a middle phage-specific promoter. Since processed transcripts are not labeled by the capping method, the late sk1 transcripts detected by Chandry et al. [1] were most likely generated from the sole middle phage-specific promoter. Therefore, one major advantage of the 'capping' method over Northern analysis is precise mapping of temporal transcription within phage genomes.

Molecular analysis of phage-specific inducible promoters isolated from lactococcal phages to date (ϕ31, [20]; bIL41, [21]; c2, [28]; and sk1, [2] revealed that these promoters contain a –10 consensus hexamer, but no –35 promoter region. Several phage-specific late promoters isolated from the *E. coli* bacteriophage T4 share the common –10 consensus sequence, but all lack those seqeuences conserved around position –35 [7]. Several late promoters isolated from the *Bacillus subtilis* bacteriophages ϕ29 and ϕNf show conserved –10 regions, but no –35 promoter region [16, 17]. Interestingly, all these bacteriophages depend on a host-encoded RNA polymerase for their lytic development and each encodes unique transcriptional activators capable of altering the specificity of the enzyme. Therefore, the modified RNAP apparently does not require a –35 promoter region to initiate transcription from phage-specific promoters. Because of the high specificity of phage-inducible promoters, definition of possible consensus promoter sequences which could potentially be recognized by several different phages will require isolation of a large number of phage-specific promoter elements. Rapid screening procedures, similar to the capping method described herein, will greatly accelerate progress along these lines. Moreover, definition of promoter elements and characterization of factors regulating their expression will be extremely valuable for dissecting, and potentially interfering with the temporal pathways exploited by bacteriophages.

One great advantage of the 'capping' method over other approaches used to isolate phage-specific promoters is that little information about the genome of the targeted phage is needed. Knowledge of several restriction enzymes capable of cutting the phage DNA into short genomic fragments is all that is required. Therefore, this approach could be readily employed to identify phage-inducible promoters from industrial bacteriophages for any bioprocessing bacterium, which evolve rapidly and for which little genetic information is available. Phage-triggered bacterial suicide systems have great potential for future use if universal and effective lethal genes can be directed against quickly-emerging industrial phages. Rapid isolation and characterization of inducible promoters encoded by these bacteriophages is a critical prerequisite for a prompt implementation of this potentially powerful defense into starter cultures used throughout industrial fermentations. Moreover, these phage-specific inducible promoters could be widely exploited in tightly regulated gene expression systems now being developed in a variety of prokaryotes important in bioprocessing and biotechnology.

Acknowledgments

This study was supported by the NRICGP project No. 92-37500-8018, and the Marschall Dairy Business, Rhone-Poulenc, Inc. During this study, G. Djordjevic was a Ph.D student at NCSU, on educational leave from the Institute of Molecular Genetics and Genetic Engineering, Vojvode Stepe 283, P.O. Box 794, Belgrade 11000, Yugoslavia.

Notes on suppliers

1. Difco Laboratories, Detroit, MI, USA
2. Sigma Chemical, St. Louis, MO, USA
3. Boehringer Mannheim Biochemicals, Indianapolis, IN, USA
4. Ericomp, San Diego, CA, USA
5. Qiagen Inc, Chatsworth, CA, USA
6. GIBCO-BRL, Gaithersburg, MD, USA
7. Biospec Products, Bartlesville, OK, USA
8. NEN[R] Research Products, Boston, MA, USA
9. Stratagene, La Jolla, CA, USA
10. FMC BioProducts, Rockland, ME, USA
11. Micron Separations, Inc., Westboro, MA, USA
12. Pharmacia Biotech Inc., Piscataway, NJ, USA
13. Whatman Inc., Clifton, NJ, USA
14. Amersham, Arlington Heights, IL, USA
15. Eastman Kodak Company, New Haven, CT, USA

References

1. Chandry PS, Davidson BE, Hillier AJ (1994). Temporal transcription map of *Lactococcus lactis* bacteriophage sk1. Microbiology 140: 2251–2261.
2. Chandry, PS, Moore, SC, Boyce, JD, Davidson BE, Hillier AJ (1997). Analysis of the DNA sequence, gene expression, origin of replication, and modular structure of the *Lactococcus lactis* lytic bacteriophage sk1. Mol Microbiol 26: 49–64.
3. Cho N-Y, Choi M, Rothman-Denes LB (1995). The bacteriophage N4-coded single-stranded DNA-binding protein (N4SSB) is the transcriptional activator of *Escherichia coli* RNA polymerase at N4 late promoters. J Mol Biol 246: 461–471.
4. Djordjevic GM, Klaenhammer, TR (1996). Positive selection, cloning vectors for Gram-positive bacteria based on a restriction endonuclease cassette. Plasmid 35: 37–45.
5. Djordjevic GM, Klaenhammer TR (1997). Bacteriophage-triggered defense systems: phage adaptation and design improvements. Appl Environ Microbiol 63: 4370–4376.
6. Djordjevic GM, O'Sullivan DJ, Walker SA, Conkling MA, Klaenhammer TR (1997). A triggered-suicide system designed as a defense against bacteriophages. J Bacteriol 179: 6741–6748.
7. Elliott T, Geiduschek EP (1984). Defining a bacteriophage T4 late promoter: absence of '–35' region. Cell 36: 211–219.
8. Farrell RE Jr (1993). Practical nucleic acid hybridiza-

tion. In: RNA methodologies, pp 191–203. Academic Press, Inc.

9. Gasson MJ (1983). Plasmid complements of *Streptococcus lactis* NCDO712 and other lactic streptococci after protoplast-induced curing. J Bacteriol 154: 1–9.
10. Haynes LL, Rothman-Denes LB (1985). N4 virion RNA polymerase sites of transcription initiation. Cell 41: 597–605.
11. Hill C, Pierce K, Klaenhammer TR (1989). The conjugative plasmid pTR2030 encodes two bacteriophage defense mechanisms in lactococci, restriction modification (R^+/M^+) and abortive infection (Hsp^+). Appl Environ Microbiol 55: 2416–2419.
12. Itoh N, Mizumoto K, Kaziro Y (1984). Messenger RNA guanylyltransferase from *Saccharomyces cerevisiae*. J Biol Chem 259: 13923–13929.
13. Mizumoto K, Lipman F (1979). Transmethylation and transguanylylation in 5′-RNA capping system isolated from rat liver nuclei. Proc Natl Acad Sci USA 76: 4961–4965.
14. Monroy G, Spencer E, Hurwitz J (1978a). Purification of mRNA guanylyltransferase from vaccinia virions. J Biol Chem 253: 4481–4489.
15. Monroy G, Spencer E, Hurwitz J (1978b). Characteristics of reactions catalyzed by purified guanylyltransferase from vaccinia virus. J Biol Chem 253: 4490–4498.
16. Monsalve M, Mencia M, Rojo F, Salas M (1995). Transcription regulation in *Bacillus subtilis* phage ϕ29: expression of the viral promoters throughout the infection cycle. Virology 207: 23–31.
17. Nuez B, Salas M (1993). Bacteriophage Nf DNA region controlling late transcription: structural and functional homology with bacteriophage ϕ29. Nucleic Acid Res 21: 2861–2865.
18. O'Sullivan DJ, Klaenhammer TR (1995). C *Lla*I is a bifunctional regulatory protein of the *llaI* restriction modification operon from *Lactococcus lactis*. In: Feretti JJ, Gilmore MS, Klaenhammer TR, Brown F (eds), Genetics of the streptococci, enterococci, and lactococci, vol 85: Dev Biol Stand, pp 591–595. Basel: Karger.
19. O'Sullivan DJ, Zagula K, Klaenhammer TR (1995). In vivo restriction by *Lla*I is encoded by three genes, arranged in an operon with *llaI*M, on the conjugative *Lactococcus* plasmid pTR2030. J Bacteriol 177: 134–143.
20. O'Sullivan DJ, Walker SA, West SG, Klaenhammer TR (1996). Development of an expression strategy using a lytic phage to trigger explosive plasmid amplification and gene expression. Bio/Technology 14: 82–87.
21. Parreira R, Valyasevi R, Lerayer ALS, Ehrlich SD, Chopin M-C (1996). Gene organization and transcription of a late-expressed region of *Lactococcus lactis* phage. J Bacteriol 178: 6158–6165.
22. Powell IB, Arnold PM, Hillier AJ, Davidson BE (1989). Molecular comparison of prolate- and isometric-headed bacteriophages of lactococci. Can J Microbiol 35: 860–866.
23. Raya RR, Kleeman EG, Luchansky JB, Klaenhammer, TR (1989). Characterization of the temperate bacteriophage ϕadh and plasmid transduction in *Lactobacillus acidophilus* ADH. Appl Environ Microbiol 55: 2206–2213.
24. Sambrook J, Fritsch EF, Maniatis, T (1989). Molecular cloning: a laboratory manual, Second edition. Cold Spring Harbor, NY: Cold Spring Harbor Laboratory.
25. Southern EM (1975). Detection of specific sequences among DNA fragments separated by gel electrophoresis. J Mol Biol 98: 503–517.
26. Stoddard SF, Howe MM (1989). Localization and regulation of bacteriophage Mu promoters. J. Bacteriol 171: 3440–3448.
27. Terzaghi BE, Sandine WE (1975). Improved medium for lactic streptococci and their bacteriophages. Appl Microbiol 29: 807–813.
28. Waterfield NR, Lubbers MW, Polzin KM, Le Page RWF, Jarvis AW (1996). An origin of DNA replication from *Lactococcus lactis* bacteriophage c2. Appl Environ Microbiol 62: 1452–1453.

Address for correspondence: T. R. Klaenhammer, Department of Food Science, North Carolina State University, Raleigh, NC 27695-7624, USA
Phone: (919) 515-2971; Fax (919) 515-7124
E-mail: klaenhammer@ncsu.edu

Methods in Cell Science **20**: 127–136 (1998).

Secretion of heterologous proteins by genetically engineered *Streptococcus gordonii*

Teruaki Shiroza[1] & Howard Kuramitsu[2]

[1] *Department of Biochemistry, Nihon University School of Dentistry, Matsudo, Japan;* [2] *Department of Oral Biology, State University of New York, Buffalo, New York, USA*

Abstract. The secretion of heterologous proteins by *Streptococcus gordonii* from genes present on plasmids or on the chromosome has been achieved with a novel recombination system. We have constructed an integration-mediated transformation system in oral streptococci to clone foreign DNA into either resident plasmids or the host chromosome. In this system, *Escherichia coli* plasmid pACYC184 was extensively modified and utilized to construct anchor, integration, and heterodimer plasmids for introduction of the foreign genes. In order to evaluate this system, we cloned the gene for cycloisomaltooligosaccharide glucanotransferase (CITase) isolated from *Bacillus circulans* T3040 into *S. gordonii*. A portion of the CITase gene devoid of its signal sequence was fused inframe to the signal sequence of the *Streptococcus sobrinus gtf*I gene. The CITase fused gene was then introduced into either a resident plasmid or the *S. gordonii* chromosome. The presence of the enzymatically active CITase in the culture fluids from plasmid-borne transformants was confirmed. These results indicate that the integration-mediated transformation system described is an effective means of engineering recombinant streptococci capable of secreting heterologous proteins.

Key words: Cycloisomaltooligosaccharide glucanotransferase, Glucosyltransferases, Plasmids, Secretion, *Streptococcus gordonii*

Abbreviations: CITase = cycloisomaltooligosaccharide glucanotransferase; GTF = glucosyltransferase; MPIS = multi-copy plasmid integration system; SCIS = single-copy chromosome integration system; dNTP = deoxynucleotide triphosphates

1. Introduction

One approach toward controlling pathogenic microorganisms could be to colonize the human host with commensal organisms expressing proteins which would antagonize the virulence of the former organisms. These proteins could be antigens which induce antibodies against the pathogen, enzymes which inhibit their colonization, or proteins which might interfere with adhesin or virulence properties. In this regard, expression of such proteins on the cell surface of the human commensal bacterium *Streptococcus gordonii* has been achieved as fusions with the cell surface M-protein [10]. Another variation of this approach would be to genetically engineer transformable commensal streptococcal strains so that they might secrete such proteins following colonization of specific host sites. Recently, we [13, 14] and others [5] have genetically engineered strains of *S. gordonii* to actively secrete heterologous proteins. In the later case, the secreted enzyme dextranase was demonstrated to inhibit the colonization of cariogenic *Streptococcus sobrinus* in vitro. In these systems, the promoter and signal sequence regions of mutans streptococcal glucosyltransferases (GTF)s were fused inframe to heterologous protein genes. This general approach has proven successful in secreting proteins not only from streptococci but also from other bacteria as well [11, 16].

Secretion of foreign genes could be obtained in the *S. gordonii* system from either the chromosome [14] or from shuttle plasmids [5, 13, 14]. Since some fusion genes could not be stably maintained in *Escherichia coli*-streptococcal shuttle plasmids when cloned into *E. coli*, a novel heterodimer plasmid system was developed in order to circumvent this problem [14]. In this approach, fusion genes were integrated into resident plasmids following transformation and homologous recombination. A similar approach was also used to integrate the fusion genes into the chromosome of *S. gordonii*. The present communication describes an improved version of this integration system which might also prove useful in secreting heterologous proteins from other transformable bacteria.

The mutans streptococci are strongly correlated with the development of human dental caries [7]. Among these organisms, *Streptococcus mutans* and *S. sobrinus* secrete extracellular enzymes, GTFs, which catalyze water-insoluble and water-soluble glucan synthesis. The expression of these molecules has been recognized as an important virulence factor in dental decay. One possible approach for reducing cariogenicity would be to construct recombinant oral commensal bacteria which produce enzymes leading to inhibition of GTF activity. One of the objectives of the present study was to construct *S. gordonii* strains able to secrete the enzyme cycloisomaltooligosaccharide glucanotransferase (CITase) which has been demonstrated to inhibit glucan formation by mutans streptococci [4]. We have now utilized the integration strategy to secrete CITase from *S. gordonii*.

Another objective of the present study was to expand the versatility of the previous system. In the newer version, the anchor plasmid (containing target DNA sequences for integration of heterologous DNA) was introduced into either a resident plasmid or the host chromosome and acts as the common integration site. This strategy allowed us to integrate heterologous genes originally present on a single *E. coli* plasmid into either multi-copy plasmids or the single copy chromosome.

2. Materials

A. Equipment
   1. PCR amplifier.[1]

B. Chemicals and reagents
   2. T4 DNA polymerase.[2]
   3. Brain heart infusion broth.[3]
   4. Erythromycin.[4]
   5. Spectinomycin.[4]
   6. Blue dextran.[5]
   7. Q-Sepharose.[5]
   8. Horse serum.[4]
   9. Agarose, SeaKem.[6]

3. Procedures

A. Recombinant DNA procedures

The CITase gene (GenBank accession number D61382) isolated from *Bacillus circulans* T3040 was originally cloned into the pUC derivative, pCI429 [9]. DNA purification, endonuclease restriction, ligation, and *E. coli* transformation were carried out under standard conditions as previously described [1]. Nucleotide sequencing was carried out as outlined earlier [15].

B. PCR amplification

Since the signal peptides of both the CITase and the GTF-I enzymes consist of the first 38 amino acids [3, 9], a set of two PCR primers complementary to the respective regions were designed: 5′-CGCACTAGCTACTGAAGCACC-3′ for the *gtf*I (GenBank D90213) and 5′-TCAGGCTCTG-GCGGCATCGAG-3′ for the CITase structural gene. To fuse the *gtf*I secretion domain encoding 38 amino acids and the CITase structural gene, PCR amplification was carried out using approximately 200 ng of pNS01 (see Figure 4D) as a template with 10 pmoles of each primer in a final volume of 25 μL in a GeneAmp PCR System 9600 thermocycler[1] under the following conditions: (i) initial denaturation, 94 °C for 3 min; (ii) 30 cycles of amplification, denaturation at 94 °C for one min, primer annealing at 66 °C for 2 min, and extension at 72 °C for 3 min with AmpliTaq DNA polymerase[1]; and (iii) final extension, 72 °C for 10 min. The PCR product was blunt-ended with T4 DNA polymerase[2] in the presence of dNTP, and pNS03 (see Figure 4E) was isolated following standard kinase[2] treatment, ligation, and transformation [1].

C. Transformation of *S. gordonii*

Transformation of *S. gordonii* Challis can be carried out as previously described [13] with slight modifications. Add the host bacterium (1:20 dilution) to 1.0 ml of brain heart infusion[3] (BHI) broth containing 10% heat-inactivated horse serum[4] (HIHS-BHI) and maintain the culture at 37 °C. When bacteria harboring plasmids are transformed, no antibiotics need to be added to the media for positive selective pressure. After 16 h incubation, transfer a 50 μl aliquot of this overnight culture into 1.0 ml of fresh HIHS-BHI and incubate at 37 °C for 30 min. Transfer the competent cells (100 μl) into fresh tubes containing 100–200 ng of appropriate DNA and incubate at 37 °C. After 30 min, add 100 μl of fresh HIHS-BHI and continue the incubation for an additional 2 h. Select transformants on BHI agar plates supplemented with either erythromycin[4] (Em, 10 μg/ml), or spectinomycin[4] (Sp, 200 μg/ml), following 2 days of incubation at 37 °C under anaerobic conditions.

D. Enzyme assays

Since the CITase exhibited dextranase activity [9], blue dextran was used to monitor CITase activity in this study. To screen for *S. gordonii* CITase positive transformants, use BHI plates containing 0.5% blue dextran[5] and examine each for clearing zones around the colonies after 48 h of anaerobic growth at 37 °C. To detect CITase activity in cell samples, zymography was also carried out. Dialyze culture fluids (50 ml) from CITase positive *S. gordonii* cells against 10 mM Tris-HCl buffer (pH 8.0) and apply onto a Q-Sepharose[5] column (0.6 × 15 cm) at 4 °C. Elute the proteins with the same buffer containing sodium chloride (0 to 0.5 M linear gradient) at approximately 0.1 ml per min and subject the eluates to 7.5% polyacrylamide gel electrophoresis under non-denaturing conditions [13]. After electrophoresis, overlay the gel with 0.7% agarose6 containing 1.0% blue dextran and incubate at 37 °C. After several hours of incubation, clear zones can be observed where enzymatically active protein was present in the corresponding polyacrylamide gel.

4. Results

Overview of this approach

Figure 1 summarizes an overview of the present study. To facilitate the cloning of foreign genes into *E. coli* plasmids, the medium copy-number plasmid,

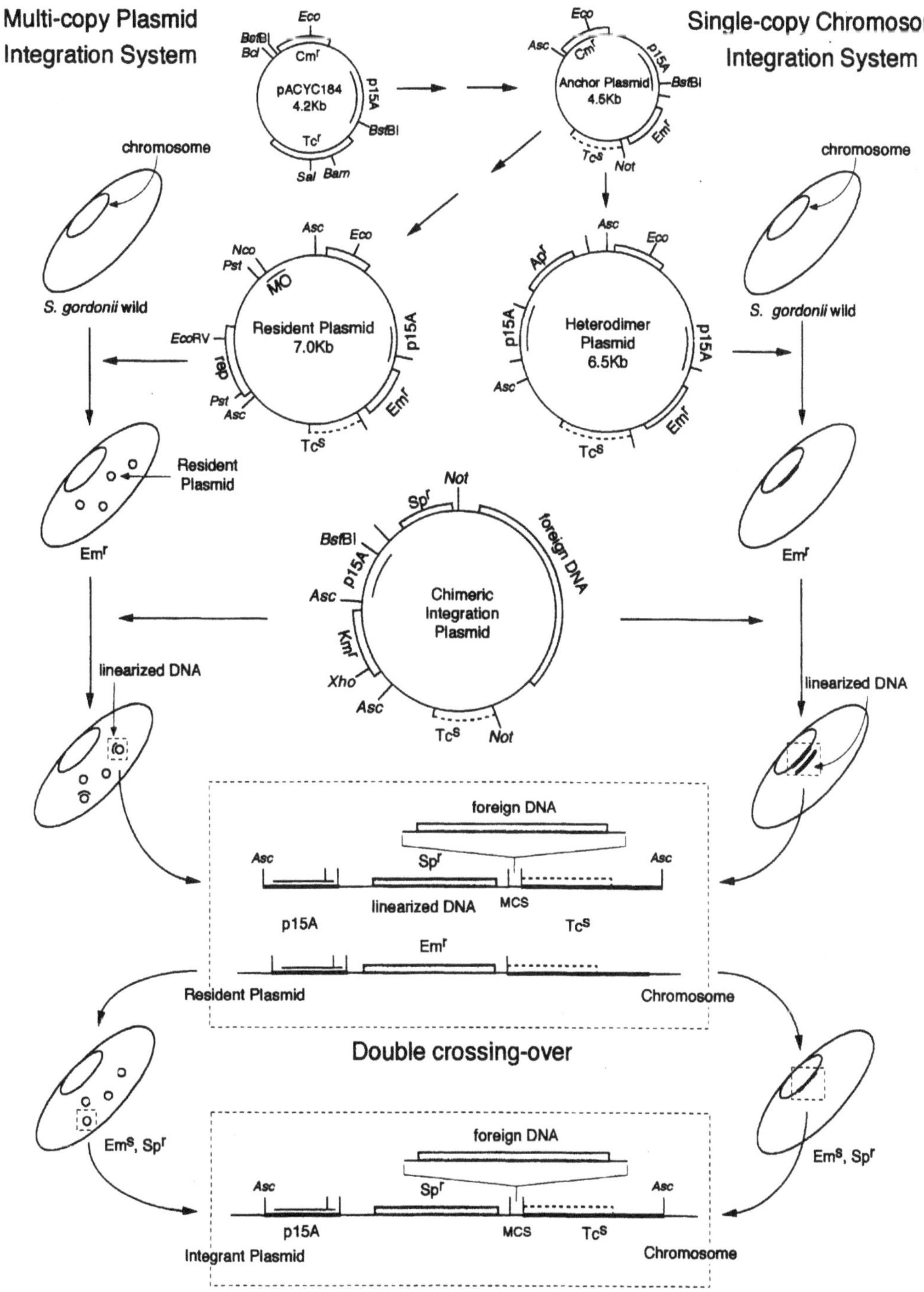

Figure 1. Overview of the SCIS and MPIS strategies. See text for details.

pACYC184 [2], and its replication origin, p15A, were used throughout this study. Initially, pACYC184 was extensively modified and a portion of the Tc^r gene was replaced with either an Em^r or Sp^r gene (in this example, only the Em^r version is depicted). In the anchor plasmid, the remaining 3′-portion of the Tc gene was designated Tc^s. Using this anchor plasmid, both resident and heterodimer plasmids were constructed. The former plasmid was transformed into *S. gordonii* cells to establish the Multi-copy Plasmid Integration System (MPIS), while the latter plasmid was integrated into *S. gordonii* chromosome as a basis for the Single-copy Chromosome Integration System (SCIS). Next, foreign DNA was cloned into the integration plasmid (Sp^r version) which was a derivative of the anchor plasmid. Following the linearization of the chimeric integration plasmid with the restriction enzyme *Asc*I, competent *S. gordonii* cells were transformed. Since the two regions, p15A and Tc^s, present within the acceptor DNA and linearized integration plasmid are identical, double cross-over integration resulted in replacement of the Em^r gene with the Sp^r gene as well as the heterologous DNA fragment.

Construction of resident, anchor, heterodimer, and integration plasmids

To construct resident, anchor, heterodimer, and integration plasmids, *E. coli* plasmid pACYC184 was extensively modified (Figure 2). Both *Bst*BI (nucleotide position 3716) and *Bcl*I (3541) sites were converted to *Asc*I and a 170-bp DNA fragment was removed. This manipulation resulted in the rendering of a unique *Bst*BI site (1318). Next, the *Sal*I site (2145) positioned in the middle of the Tc^r gene was converted into a NotI site following linker ligation (A). This modified pACYC plasmid and pResEmNot

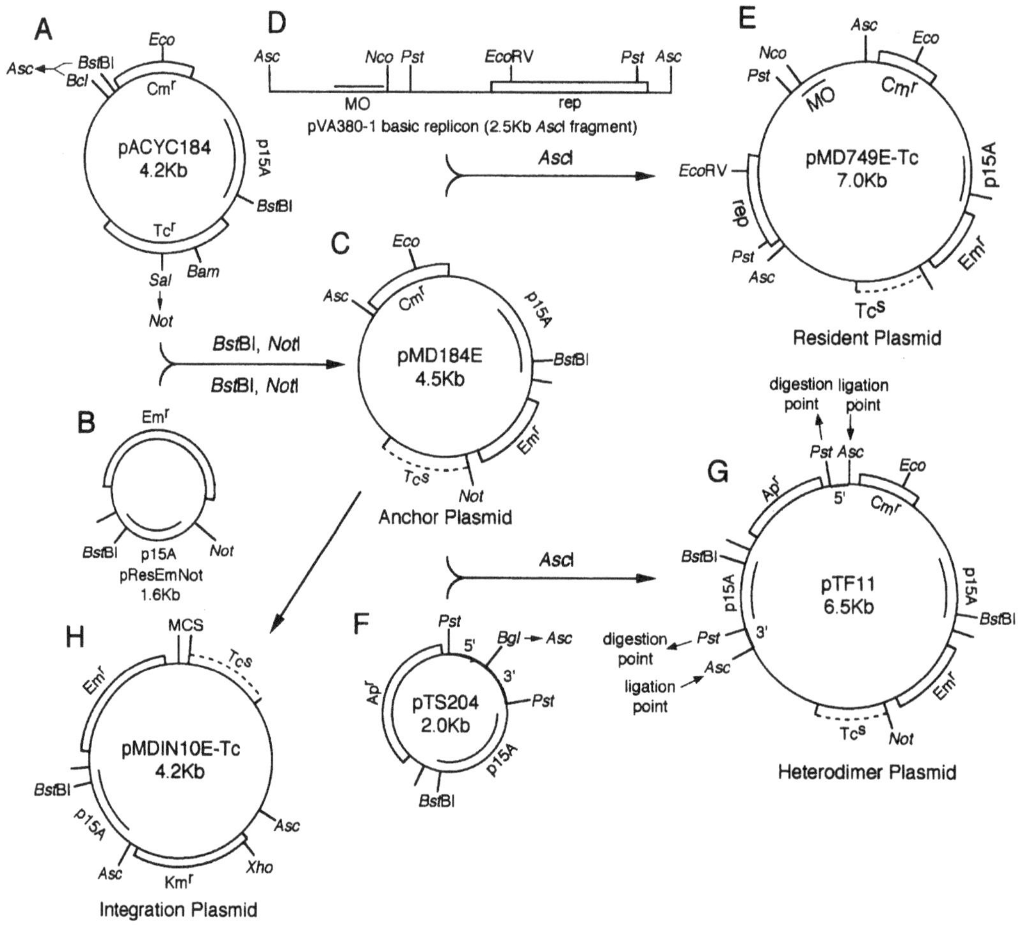

Figure 2. Construction of resident, anchor, heterodimer, and integration plasmids. For details, see the text. In heterodimer plasmid (G), both ligation and digestion sites are such that the orientation of the two coupler fragments is the same as is present on the chromosome. The structure of the MCS of the integration plasmid pMDIN10-Tc (H) is: *Not*I-*Bgl*II-*Eco*RI–*Bam*HI–*Hin*dIII-*Bcl*I-*Not*I in the clockwise orientation. The MCS structure between *Eco*RI and *Hin*dIII is the same as that of pUC vectors. The MCS structure of the other integration plasmid pMDIN11-Tc is the same except that it is in the opposite orientation relative to the plasmid.

(B) were digested with *Bst*BI and *Not*I and the 800-bp *Bst*BI-*Not*I fragment corresponding to the 5′ end of the Tc^r gene in the modified pACYC was replaced with the gel-purified Em^r gene containing a portion of the p15A ori to yield the anchor plasmid pMD184E (C). To the unique *Asc*I site of the anchor plasmid, the pVA380-1 [14] replicon (D) or pTS204 (F, for construction of this plasmid, see below) was introduced and the resident plasmid pMD749E-Tc (E) and heterodimer plasmid pTF11 (G) were isolated. The anchor plasmid was also further modified and the integration plasmids, pMDIN10E-Tc and pMDIN11E-Tc (H), were constructed. In Figure 2, only the Em^r versions are depicted. Using pResSpNot, a counterpart plasmid of pResEmNot (B), Sp^r versions of resident (pMD749S-Tc), anchor (pMD184S), and integration plasmids (pMDIN10S-Tc and pMDIN11S-Tc) were also constructed (data not shown).

Construction of pTS204 and related plasmids; introduction of 'coupler fragments'

In order to develop a SCIS not only for *S. gordonii* but also for use in other transformable bacteria, we introduced 'coupler fragments' originating from the *S. mutans* GS-5 chromosome as outlined in Figure 3. The *S. mutans* GS-5 *gtf*B and *gtf*C genes are arranged tandemly on the chromosome (A). The 250 bp *Pst*I-*Bgl*II fragments flanking the *gtf*B and *gtf*C

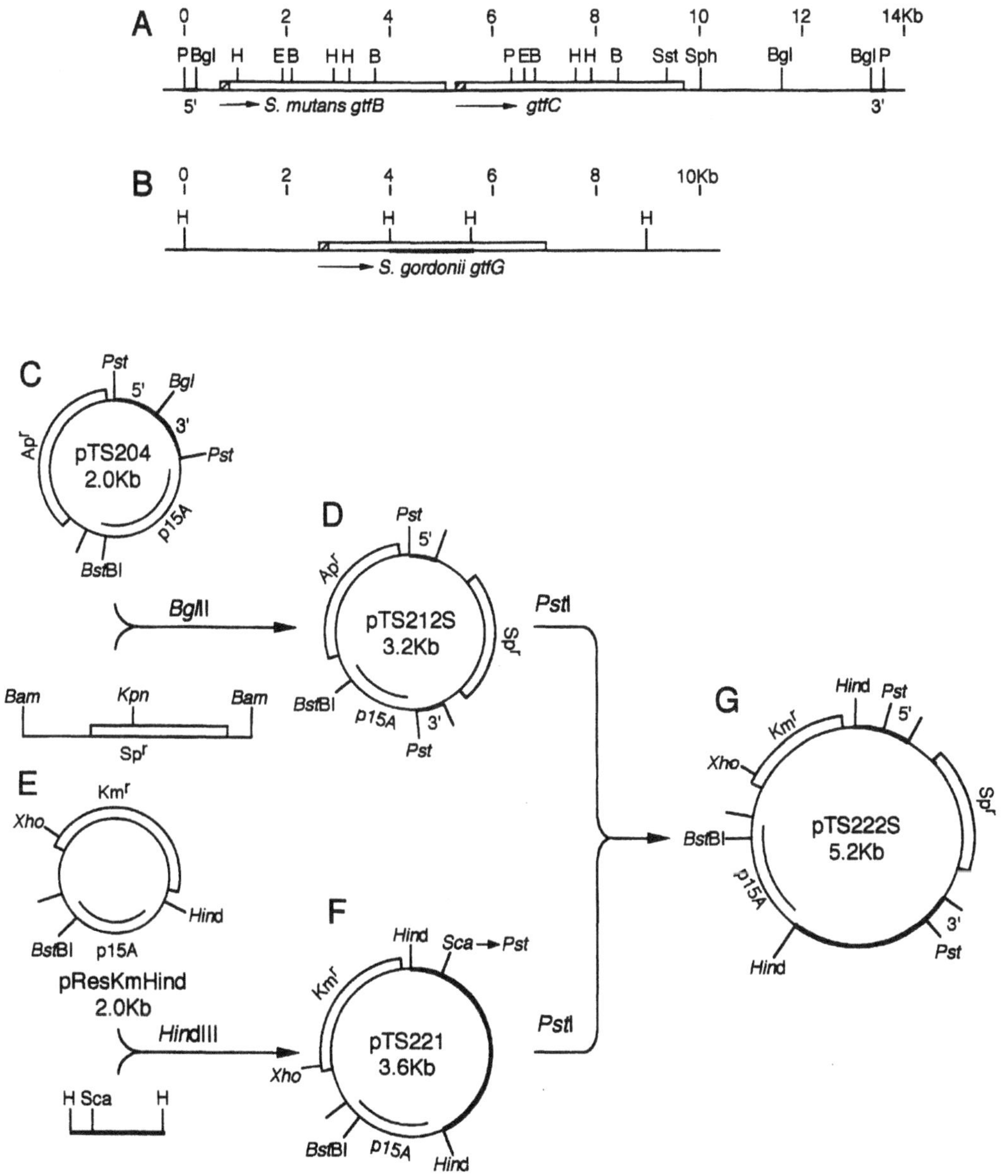

Figure 3. Construction of pTS204 and related plasmids as well as the strategy for introduction of the coupler fragments.

genes, referred to as 5′ and 3′, respectively, were cloned into the *Pst*I site of the plasmid pResAmpPst (not shown), and pTS204 (C) was constructed. Into the unique *Bgl*II site of the plasmid was ligated the Sp^r gene and pTS212S (D) was isolated. To introduce the anchor plasmid into the *S. gordonii* chromosome, the 1.5 kb *Hin*dIII fragment present within the *S. gordonii gtf*G gene (B) was cloned into the intermediate plasmid, pResKmHind (E), and pTS221 (F) was constructed. The unique *Sca*I site found in the 1.5 kb fragment was next converted into a *Pst*I site using linker DNA. Following *Pst*I digestion and ligation of these two plasmids, pTS222S (G) was obtained.

Cloning of a portion of the CITase gene and replacment of its secretion domain with that from the S. sobrinus gtfI *gene*

The gene coding for the CITase was first isolated on a 5 kb *Sau*3AI fragment and the non-essential 3′-region (1.2 kb) was removed by *Pst*I digestion to produce 429ΔPst (Figure 4A, 3.8-kb *Bam*HI-*Pst*I fragment). Our preliminary experiments indicated that although this gene was expressed in *E. coli*, *S. gordonii* harboring this gene secreted only very small amounts of CITase into the culture fluids. This suggested the possibility that the secretion domain originating from *B. circulans* was inactive in streptococci. A similar observation was also made for the secretion of a *S. mutans* glucan-binding domain from *S. gordonii* [14]. Therefore, we initially isolated a portion of the CITase gene devoid of the 700 bp 5′-region (429ΔPstΔEco) and then fused the remainder of the gene inframe with the 5′-end of the *S. sobrinus gtf*I gene coding for the signal sequence.

To accomplish this, the 429ΔPstΔEco fragment was cloned into the integration plasmid pMDIN10S-Tc and pTF01 (C) was isolated. To construct the secretion domain, the 1.1 kb *Bam*HI-*Sal*I fragment of the CITase gene and the 1.2 kb *Dra*I-*Bcl*I fragment from the *S. sobrinus gtf*I gene (B) were cloned into pResEmNotMCS10, a derivative of pResEmNot (see Figure 2B), and pNS01 (D) was constructed. Two PCR primers were designed to anneal with the 3′-end of the signal peptide coding region of the *gtf*I gene and 5′-end of the structural gene of the CITase gene. Following PCR amplification, self-ligation, and transformation, pNS03 was isolated (E). As shown in Figure 5, DNA sequencing of the junction region of pNS03 confirmed that the secretion domain of the GTF from *S. sobrinus* was fused with the CITase structural gene in the correct reading frame.

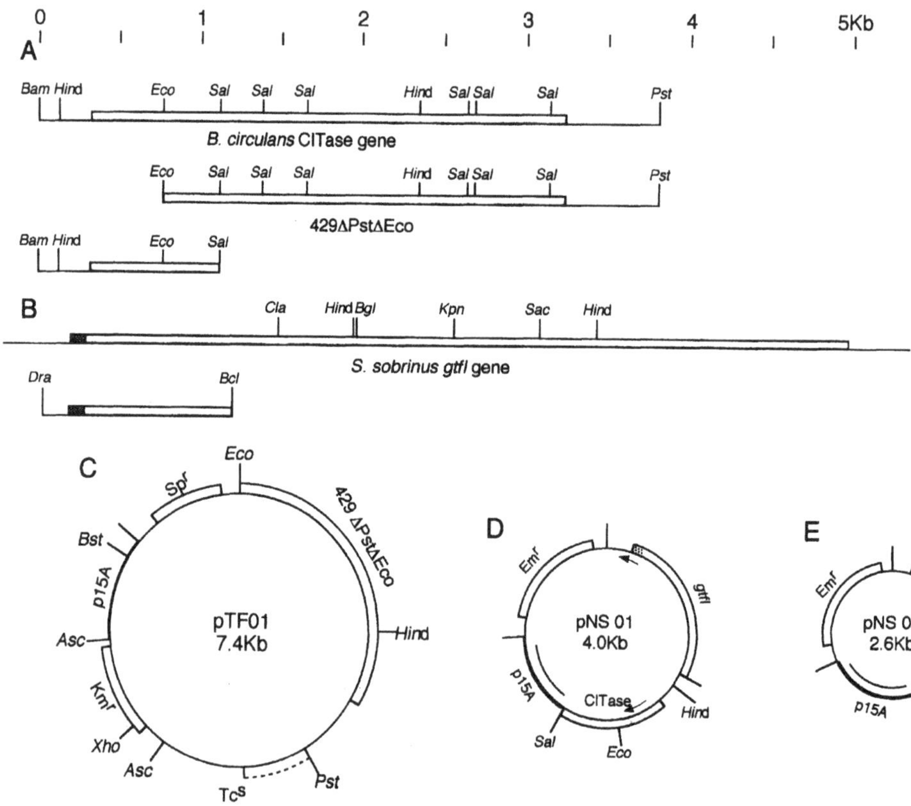

Figure 4. Cloning of a portion of the CITase gene and replacement of its secretion domain with that from the *S. sobrinus gtf*I gene. The hatched region at the 5′-end of the *gtf*I open reading frame (B) corresponds to the signal sequence coding region. Small inverted arrows in (D) denote the direction of PCR amplification.

Sequential integration of the CITase gene

The strategy for integrating the CITase gene into the resident plasmid (upper) or host chromosome (lower) is depicted in Figure 6. This strategy involved initially integrating the CITase gene without a functional signal sequence into the target region via the acceptor sequences. Subsequently, the GTF signal sequence was fused to the CITase gene following homologous recombination. *S. gordonii* harboring the resident plasmid pMD749E-Tc (Em^r-version, top, line A) was made competent and *Asc*I-digested pTF01 (top, line B) was transformed into these cells. The expected genotype of the initial plasmid-integrants would be Em^s and Sp^r. However, most of the transformants were resistant to both Em and Sp. *S. gordonii* secondary transformants generated using the plasmids isolated from the initial transformation introduced into parental *S. gordonii* did exhibit the Em^s and Sp^r phenotypes and transformants harboring pTF02-Tc were isolated (top, line C). These results suggested that integration of the linearized plasmid occurred in only a portion of the multi-copy plasmids in each cell in the initial transformation. Therefore, each cell would contain a mixture of Em^r and Em^s plasmids. Next, pNS03 was linearized by digesting with *Asc*I (top, line D) and *S. gordonii* harboring pTF02-Tc was transformed. Transformants were screened as described above and *S. gordonii* harboring pNS04-Tc was isolated (top, line E).

The target site for chromosomal integration used was within the *S. gordonii gtf*G gene (1.5 kb *Hin*dIII fragment, bottom, line A). pTS222S was digested with *Asc*I (bottom, line B) and transformed into *S. gordonii*. An Em^s and Sp^r integrant designated dBCSp^r was isolated (lower line C) and 'coupler fragments' were successfully introduced into the host chromosome. This strain was further transformed with *Pst*I-digested heterodimer plasmid, pTF11, (bottom, line D) and integrant I was constructed (bottom, line E). Following sequential integration using the linearized pTF01 (bottom, line F) and pSN03 (bottom, line H), integrants II and III were isolated and the CITase gene was successfully integrated into the chromosome of integrant III (bottom, line I).

Secretion of CITase from S. gordonii

Figure 7 shows the expression and secretion of CITase from plasmid-borne *S. gordonii* transformants. When *S. gordonii* harboring pTF02-Tc and pNS04-Tc were spread on BHI agar plates containing 0.5% blue dextran, only the latter transformant exhibited clear halos around the colonies suggesting that enzymatically active CITase proteins were secreted (Figure 7A, right). Plasmids were isolated from these cells and subjected to agarose gel electrophoresis. Figure 7B clearly indicated that following the sequential integration of linearized DNA, the fused CITase gene was successfully introduced into the resident plasmid. The presence of the active CITase enzyme in the culture fluids was also confirmed by zymography (Figure 7C).

On the other hand, the single copy chromosome integrant did not exhibit halos around the colonies on blue dextran plates. Therefore, chromosomal DNA was isolated from integrants II and III, and the presence of the CITase gene in the chromosome was examined by PCR amplification (Figure 7D). Using a set of primers corresponding to the p15A and CITase sequences (see Figure 6), 1.8 kb and 2.4 kb bands were obtained from integrants II and III (lanes 1 and 2, respectively). After *Not*I digestion of these two fragments, the former band yielded 1.4-kb and 0.4-kb bands, while the latter fragment produced 1.3-kb and 1.1-kb bands (lanes 3 and 4, respectively). Since the 1.4-kb and 0.4-kb bands corresponded to the Sp^r gene sequences containing a portion of p15A ori and with the 5′-end of the 429ΔPstΔEco while the latter 1.3-kb and 1.1-kb bands coincided with the Em^r gene containing a portion of p15A ori and the 5′-end

	32		34		36		38 ↓		40		42		44		
B. circulans CITase	TCC	CCC	CCG	CAA	GCG	CTA	GCA	TCA	GGC	TCT	GGC	GGC	ATC	GAG	A
	Ser	Pro	Pro	Gln	Ala	Leu	Ala	Ser	Gly	Ser	Gly	Gly	Ile	Asp	
S. sobrinus gtfI	GGT	GCT	TCA	GTA	GCT	AGT	GCG	GAT	ACA	GAC	ACT	GCT	AGT	GAT	B
	Gly	Ala	Ser	Val	Ala	Ser	Ala	Asp	Thr	Asp	Thr	Ala	Ser	Asp	
pNS03	GGT	GCT	TCA	GTA	GCT	AGT	GCG	TCA	GGC	TCT	GGC	GGC	ATC	GAG	C
	Gly	Ala	Ser	Val	Ala	Ser	Ala	Ser	Gly	Ser	Gly	Gly	Ile	Asp	

Figure 5. Nucleotide and corresponding amino acid sequences around the signal peptide cleavage sites. The amino acid sequence of the first 32 to 45 residues from the initiator Met of the *B. circulans* CITase (A), *S. sobrinus* GTF-I (B) and pNS03 fusion protein (C) as well as the corresponding nucleotide sequences are shown. The nucleotide sequences of the two PCR primers utilized to fuse the *gtf*I secretion domain and the CITase structural genes are underlined. The arrow denotes the signal peptide processing site of both the CITase and the GTF-I proteins.

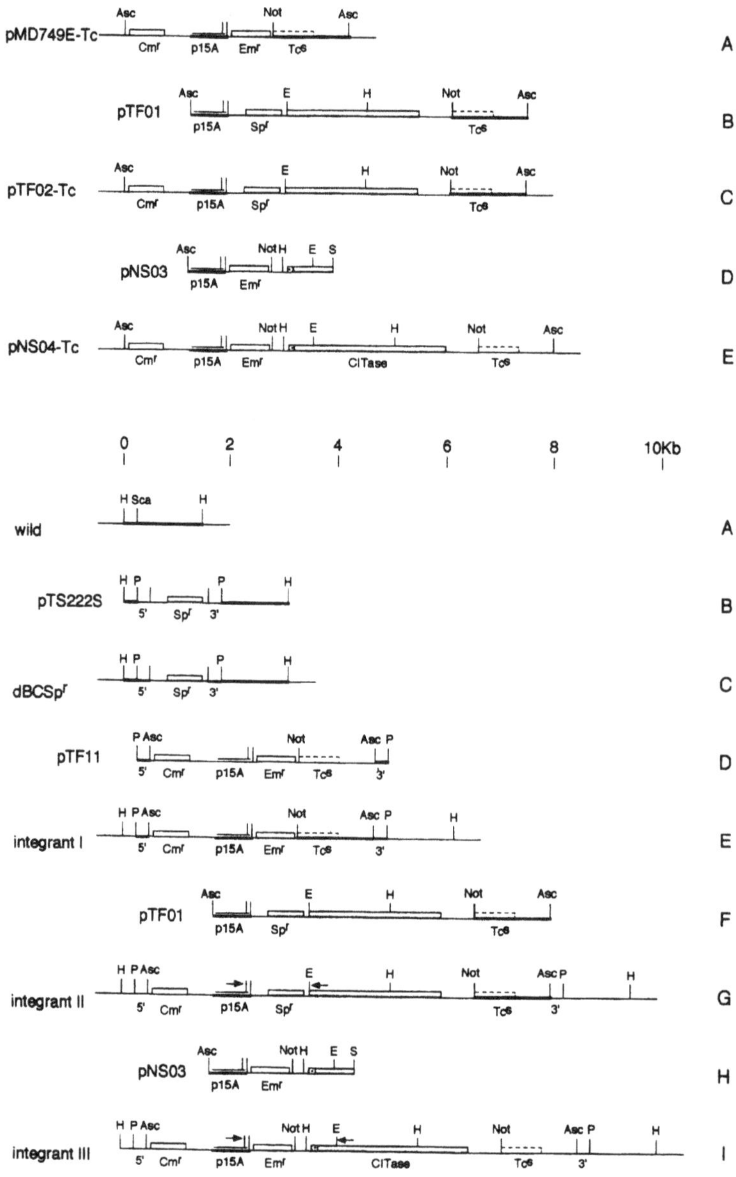

Figure 6. Sequential integration of the CITase gene. All constructs depicted were made in *S. gordonii*. For details, see the text. The inverted arrows in *S. gordonii* integrants II and III (bottom, lines G and H) denote the primer annealing sites and direction of PCR amplification for detection of the CITase gene integrated into the chromosome.

of the fused CITase gene, respectively, it was confirmed that the CITase gene was successfully integrated into the *S. gordonii* chromosome. Therefore, it is likely that the single copy of the CITase gene on the chromosome secreted only low levels of the enzyme which could not be readily detected with the blue dextran assay systems.

5. Discussion

We have described an integration-mediated transformation system in *S. gordonii* using an extensively modified *E. coli* plasmid pACYC184 as the anchor plasmid. In this system, several DNA fragments including anchor plasmids, coupler fragments, integration plasmids, and cloning plasmids were prepared and these DNA fragments were systematically manipulated. To introduce the CITase gene into *S. gordonii*, each of two 250-bp coupler fragments originating from the *S. mutans* GS-5 chromosome were initially integrated into the host chromosome. To subsequently integrate an anchor plasmid, a heterodimer plasmid was constructed since this construct could readily provide the linearized DNA fragment in which the orientation of the two terminal

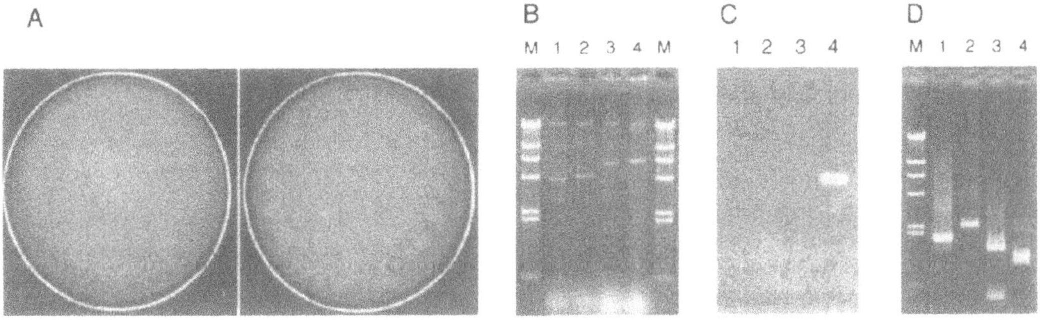

Figure 7. Secretion of CITase from *S. gordonii*. A, BHI agar plates containing 0.5% blue dextran spread with *S. gordonii* transformants harboring pTF02-Tc (left) and pSN04-Tc(B). B, agarose gel electrophoresis of plasmids isolated from *S. gordonii* transformants. Lanes 1 and 2, integration plasmids pMDIN10E-Tc; lanes 3 and 4, pTF02-Tc and pSN04-Tc. M represents the molecular size marker (λphage *Hin*dIII digest). C, zymogram of CITase activity. Samples in lanes 1 and 4 are the same as those in panel (B). For preparation of samples, see Materials and Procedures. D, agarose gel electrophoresis of PCR products to detect the integrated CITase gene. Lanes 1 and 3, PCR product from integrant II; lanes 2 and 4, integrant III. Lanes 1 and 2, intact products; lanes 3 and 4, *Not*I digests.

regions (coupler fragments) are exactly the same as that present in the host chromosome. The integration of the coupler fragments into any host chromosome prior to the introduction of the anchor plasmid would allow for the construction of a similar system in bacteria other than *S. gordonii*. This is an improvement of the original system developed in our laboratory [14] and the utility of coupler fragments as universal integration anchor sites for many bacteria is suggested. Indeed, these coupler fragments were recently integrated into the *S. milleri* chromosome and the *S. sobrinus gtf*I gene was inserted into the chromosome and secreted GTF activity (data not shown). Therefore, once DNA fragments of interest are cloned into an *E. coli* integration plasmid, it is possible to introduce heterologous genes into either plasmids or chromosomes in *S. gordonii*, *S. milleri* and possibly other gram-positive transformable bacteria

Although the CITase gene was integrated into the *S. gordonii* chromosome, detection of CITase activity in the culture fluids was unsuccessful. It is possible that the promoter activity of the *S. sobrinus gtf*I gene may not be sufficient to secrete detectable amounts of enzyme protein. Additional approaches will be required to determine why the fusion protein was not highly expressed from the *S. gordonii* chromosome. In the SCIS, utilization of the secretion domain from the *S. mutans* GS-5 *gtf*B gene or the corresponding regions from other secreted proteins might be necessary to accomplish efficient secretion of heterologous proteins.

At first glance, the strategy depicted here would appear to be excessively laborious since multiple genetic engineering steps are required. Indeed, it has been possible to construct secretion systems in *S. gordonii* using a simpler approach by cloning a fusion gene composed of a heterologous gene fused inframe to a streptococcal secretion domain directly into a *E. coli*-streptococcal shuttle plasmid [5, 13]. However, we have demonstrated that such an approach does not appear to be suitable for all heterologous genes. For example, it was not possible to stably clone a fusion gene composed of the *S. mutans gtf*B secretory domain with the glucan-binding domain DNA fragment from the *S. mutans gtf*D gene into the shuttle plasmid pTS749 in *E. coli* [14]. For this reason an earlier version of the MPIS and SCIS was developed, without the coupling fragment strategy, which was successful in secreting the glucan binding domain from *S. gordonii* [14]. The present strategy expands the potential usefulness of this system to other transformable streptococci and potentially other gram-positive organisms as well. Whether or not this strategy could also prove useful for strains transformable only by electroporation awaits further evaluation.

In attempting to secrete heterologous proteins from streptococci, it would be practical to first utilize the simplest strategy for constructing a secretable derivative of the heterologous gene directly in a shuttle plasmid in *E. coli* followed by transformation into the gram-positive bacteria. If difficulty in isolating such a stable chimeric shuttle plasmid is encountered, either the MPIS or SCIS could prove to be viable alternatives. Since the MPIS expresses higher levels of the heterologous proteins, this system would likely be preferable for most purposes. However, maintenance of the relevant plasmids would likely require positive selection with antibiotics. For utilization in in vivo systems, this may prove to be impractical. Nevertheless, non-antibiotic selection strategies are available for plasmid maintenance [8].

An alternate strategy for secreting heterologous proteins and peptides from *S. gordonii* has recently been developed [12]. This procedure utilizes a variation of the surface expression system based on the localization of the M-protein on the cell surface. By genetically engineering a stop codon between the chimeric M-protein-heterologous protein sequences and the cell surface anchor sequences it has been possible to secrete peptides comprising the *Porphyromonas gingivalis* FimA fimbrillin subunit protein from *S. gordonii*. However, since only relatively small peptides have been secreted in this system, it is not known if larger proteins can also be secreted by this strategy. In this regard, a 98 kDa protein has been secreted using the *S. mutans gtf*B secretory domain on a shuttle plasmid [13].

The ability to genetically engineer the oral commensal *S. gordonii* to secrete heterologous peptides or proteins may have potential therapeutic value. These organisms have been demonstrated to colonize the oral cavity of rodents [6] and may therefore be implantable into the human oral cavity. Strains which stably secrete peptide antigens in the oral cavity may elicit mucosal immunity against bacterial, fungal, or viral pathogens. In addition, the secretion of peptides which interfere with the colonization of the pathogens might also prove to be beneficial. Likewise, the secretion of bacteriocins or enzymes which interfere with the growth or colonization of oral pathogens might also prove useful. Therefore, it will be of interest to pursue these possibilities not only for oral streptococci but for other organisms which could be engineered with the MPIS or SCIS to antagonize microbial virulence in other areas of the human host (skin, reproductive tract, throat). Other genes besides the *gtf*G gene of *S. gordonii* or alternate promoter fragments may prove to be more suitable for such constructs.

Acknowledgments

This study was supported in part by a Grant-in-Aid for General Individual Research Grants of the Suzuki Memorial Grant of Nihon University School of Dentistry at Matsudo (TS) and by grant DE03258 from the National Institutes of Health (HK).

Notes on suppliers

1. Perkin-Elmer Corp., Norwalk, CT, USA
2. Takara Syuzo, Kyoto, Japan
3. Difco Laboratories, Detroit, MI, USA
4. Sigma Chemical Co., St. Louis, MO, USA
5. Pharmacia Biotech., Piscataway, NJ, USA
6. BioProducts, Rockland, ME, USA

References

1. Aoki H, Shiroza T, Hayakawa M, et al. (1986). Cloning of a *Streptococcus mutans* gene coding for insoluble glucan synthesis. Infect Immun 53: 587–594.
2. Chang ACYC, Cohen SN (1978). Construction and characterization of amplifiable multicopy DNA cloning vehicles derived from the p15A cryptic miniplasmid. J Bacteriol 134: 1141–1156.
3. Ferretti JJ, Gilpin ML, Russell RRB (1987). Nucleotide sequence of a glucosyltransferase gene from *Streptococcus sobrinus* MFe28. J Bacteriol 169: 4271–4278.
4. Kobayashi M, Funane K, Oguma T (1995). Inhibition of dextran and mutan synthesis by cycloisomaltooligosaccharides. Biosci Biotechnol Biochem 59: 1861–1865.
5. Kubo S, Kubota H, Ohnishi Y, et al. (1993). Expression and secretion of an *Arthrobacter dextranase* in the oral bacterium *Streptococcus gordonii*. Infect Imun 61: 4375–4381.
6. Loach DM, Jenkinson HF, Tannock GW (1994). Colonization of the murine oral cavity by *Streptococcus gordonii*. Infect Immun 62: 2129–2131.
7. Loesche WJ (1986). Role of *Streptococcus mutans* in human dental decay. Microbiol Rev 50: 353–380.
8. Nakayama K, Kelly SM, Curtiss III R (1988). Construction of an *asd*$^+$ expression cloning vector: Stable maintenance and high level expression of cloned genes in a *Salmonella* vaccine strain. Biotechnology 6: 693–697.
9. Oguma T, Kurokawa T, Toke K, et al. (1995). Cloning and sequence analysis of the cycloisomaltooligosaccharide glucanotransferase gene from *Bacillus circulans* T-3040 and expression in *Escherichia coli*. Appl Carb Sci 42: 415–419.
10. Pozzi G, Contorni M, Oggioni MR, et al. (1992). Delivery and expression of heterologous antigens on the surface of streptococci. Infect Immun 60; 1902–1907.
11. Savijoki K, Kahala M, Palva A (1997). High level heterologous protein production in *Lactococcus* and *Lactobacillus* using a new secretion system, based on the *Lactobacillus brevis* S-layer signals. Gene 186: 255–262.
12. Sharma A, Sojar HT, Hruby DE, et al. (1997). Secretion of *Porphyromonas gingivalis* fimbrillin polypeptides by recombinant *Streptococcus gordonii*. Biochem. Biophys Res Commun (in press).
13. Shiroza T, Kuramitsu HK (1993). Construction of a model secretion system for oral streptococci. Infect Immun 61: 3745–3755.
14. Shiroza T, Kuramitsu HK (1995). Development of a heterodimer plasmid system for introduction of heterologous genes into streptococci. Plasmid 34: 85–95.
15. Shiroza T, Ueda S, Kuramitsu HK (1987). Sequence analysis of the *gtf*B gene from *Streptococcus mutans*. J Bacteriol 169: 4263–4270.
16. Udaka S, Yamagata H (1993). High level secretion of heterologous proteins by *Bacillus brevis*. Methods Enzymol 217: 23–33.

Address for correspondence: Dr Howard Kuramitsu, Department of Oral Biology, SUNY, 3435 Main Street, Buffalo, NY 14214-3092, USA
Phone: (716) 829-2068; Fax: (716) 829-3942
E-mail: kuramits@acsu.buffalo.edu

Methods in Cell Science **20**: 137–142 (1998).

Examination of streptococcal gene expression in the mammalian environment

W. Todd Grey[1], Joshua D. Lasker[1], Roy Curtiss III[2] & Michael C. Hudson[1]
[1] *The University of North Carolina at Charlotte, Department of Biology, Charlotte, North Carolina, USA;*
[2] *Washington University, Department of Biology, St. Louis, Missouri, USA*

Abstract. Laboratory studies of bacterial gene expression have not allowed the investigation of in vivo factors crucial to pathogenesis. Recently, genetic tools have been used to identify in vivo-activated genes. These tools, however, have not allowed the measurement of specific expression levels of virulence-associated bacterial genes in vivo. This report describes a system, previously tested utilizing the fructosyltransferase (*ftf*) gene of *Streptococcus mutans*, that allows the measurement of activity of specific virulence-associated streptococcal genes while bacteria are living in the animal environment.

Key words: Gene expression, Mammalian environment, *Streptococcus*

1. Introduction

Recently, Mahan et al. have utilized *Salmonella typhimurium* cells containing both *lacZY* and *cat* fusions to identify genes which are activated during host infection [12]. Other studies have described the development of a technique which allows the detection of *Vibrio cholerae* infection-associated genes which are transiently or constitutively expressed in vivo [2, 3]. These techniques have allowed the identification of bacterial genes whose expression is involved in the causation of disease. These investigations, however, have not resulted in the measurement of expression of specific virulence-associated genes while bacteria are living in the animal host. Such techniques are critical to understanding bacterial pathogenesis and could facilitate successful treatment and prevention of disease.

The current report describes a procedure to examine the expression of specific streptococcal genes when bacteria are present in the animal environment. The method has been utilized successfully to determine the influence of the mammalian oral cavity on expression of the *Streptococcus mutans* fructosyltransferase (*ftf*) gene [4]; however, the approach presented should be useful to examine gene expression as other streptococcal species interact with their eukaryotic hosts. The system relies on the use of chloramphenicol acetyltransferase (*cat*) fusions as a tool to study gene expression. The in vitro essay for CAT is both rapid and simple, and the *cat* gene product can be assayed with specificity and great sensitivity [14]. Importantly, very few organisms produce CAT, which alleviates the problem of background enzymatic activities [15]. *S. mutans* strain SMS101 [6] (Figure 1) contains a fusion of the *ftf* promoter to a formerly promoterless *cat* gene. This *ftf-cat* operon fusion approach maintained the integrity of the complete *ftf* gene and also resulted in duplication of the *ftf* promoter such that it is, in addition, directing *cat* expression. Thus, for strain SMS101, any factors involved in the regulation of transcription of the *cat* gene are also necessarily involved in regulation of transcription of the

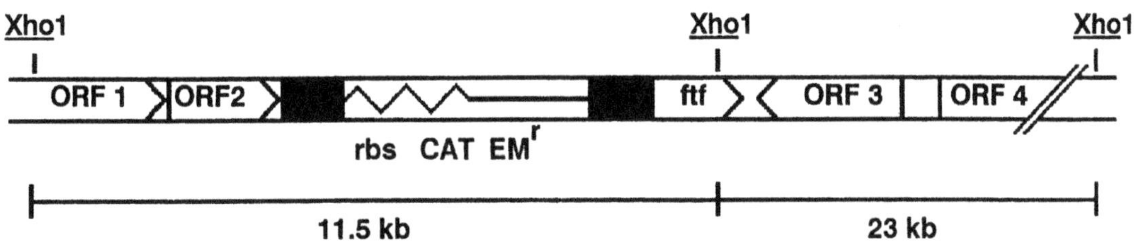

Figure 1. Region of the genome of *S. mutans* SMS101 (*ftf-cat*) [6]. ORF, open reading frame; rbs, *cat* gene ribosome binding site; CAT, promoterless chloramphenicol acetyltransferase gene; EM^r, erythromycin resistance gene; ftf, fructosyltransferase gene; ■, *ftf* promoter (duplicated); kb, kilobase pairs. The arrowheads depict the direction of transcription.

ftf gene. Compared to fructosyltransferase, the intracellular CAT molecule is relatively easy to easy for, making the enzyme an extremely effective tool for the study of factors influencing transcription of bacterial genes. Understanding the role of various environmental factors in influencing the expression of bacterial virulence-associated genes is extremely important in the study of the pathogenesis of any bacterium. Because the mammalian environment is so complex, it cannot be successfully duplicated in vitro. With regard to the oral streptococci, host-dependent factors such as carbohydrate consumption amounts and animal age, for example, could influence bacteria present in the animal environment. Because the complex mammalian environment may influence streptococcal gene expression in ways that cannot be observed in the laboratory, we developed a method where the expression of *cat*, controlled by the *ftf* regulatory region, was used as a reflection of *ftf* expression while *S. mutans* was present in the mammalian oral cavity.

It has been demonstrated that members of the rat normal oral flora are sensitive to the antibiotic streptomycin [1, 8, 9, 13]. Therefore, a stable, spontaneous streptomycin-resistant (str^r) version of *S. mutans* fusion strain SMS101 [6], designated NCH106, was isolated to facilitate strain enumeration.

2. Materials

A. Equipment
- High speed centrifuge (RC5C, Sorvall Instruments).[1]
- Micropipetter (Oxford 40–200 μl range, Fisher Scientific #2100231).[2]
- Vortex mixer (Vortex Genie 2, Fisher #12-812).
- Microscope (Micromaster HL, Fisher #12-576-2S).
- 37 °C incubator (Isotemp, Fisher #11-690-650D).
- Mini Beadbeater apparatus (Biospec Products).[3]
- Microcentrifuge (235C, Fisher).
- Water bath (Isotemp, Fisher #15-461-2).
- Liquid scintillation spectrophotometer (LS 6000 SC, Beckman Instruments, Inc.).[4]

B. Supplies
- Brain Heart Infusion (BHI) broth and agar (Difco Laboratories #DF0037-17-8, DF0418-17-7).[5]
- Todd Hewitt Broth (THB) (Difco #DF0492-17-9).
- 50 ml polypropylene tubes (Fisher #05-538-55).
- GasPak anaerobic culture system (Fisher #11-814-21).
- Streptomycin sulfate (Sigma Chemical Co. #S6501).[6]
- Sprague-Dawley rats (Charles River Laboratories).[7]
- Standard rodent chow (Laboratory Rodent Diet #5001, Purina Mills).[8]
- Feeding bowls (5 cm deep with a 4 cm opening, custom made).
- Pipet tips (Fisher #21-244-1).
- Cotton-tipped applicators (Fisher #14-959-90).
- Trizma base (Sigma #T6791).
- Scalpel (Fisher #08-920A).
- Forceps (Fisher #13-812-41).
- Microcentrifuge tubes (Fisher #05-407-10).
- 0.1 mm zirconium beads (Biospec).
- Chloramphenicol (Sigma #C0378).
- ^{14}C-Acetyl CoA solution (ICN Pharmaceuticals, Inc. #13070L).[9]
- Ethyl acetate (Fisher #BP1125-1).
- Polyethylene scintillation vials (Fisher #03-337-11B).
- Bio-Safe II scintillation cocktail (Research Products International Corp. #111195).[10]
- 100% ethanol (McCormick Distilling Co. Inc.).[11]

C. Animal diets
- Dextrin (Sigma #D2006).
- Polypeptone (Difco #DF0123-17-3).
- L-methionine (Sigma #M6039).
- Calcium phosphate (dibasic, Sigma #C7263).
- Sodium chloride (Sigma #S7653).
- D-Sucrose (Sigma #S7903).
- D-Glucose (Sigma #G7528).
- D-Fructose (Sigma #F2543).
- Vitamin B_1 (thiamine, Sigma #T4625).
- Vitamin B_6 (pyridoxine, Sigma #P9755).
- Vitamin C (ascorbic acid, Sigma #A5960).
- Vitamin B_3 (niacinamide, Sigma #N3376).
- Vitamin B_5 (calcium pantothenate, Sigma #P5710).
- Vitamin B_2 (riboflavin, Sigma #R4500).
- Vitamin E (tocopherol acetate, Sigma #T3001).
- Vitamin D_2 (ergocalciferol, Sigma #E5750).
- Vitamin A (retinol acetate, Sigma #R4632).
- Vegetable oil (Proctor and Gamble).[12]
- Whole wheat flour (The Pillsbury Co.).[13]

3. Procedures

A. Str^r bacterial strain

Transfer an isolated colony of the streptococcal *cat* fusion strain from a brain heart infusion (BHI) agar plate into 45 ml of BHI broth (Difco) in a 50 ml polypropylene tube. Grow anaerobically (GasPak Plus Anaerobic System) at 37 °C for 3 days, standing, to achieve a dense culture (approximately 10^{10} colony forming units (cfu)/ml). Harvest bacteria by centrifugation at 4300 ×*g* for 10 minutes at 4 °C to concentrate cells. Spread plate approximately 10^{11} cfu on BHI agar containing 200 μg/ml streptomycin sulfate (Sigma) and grow anaerobically at 37 °C for 4 days. Streak resistant clones for isolation on fresh media containing streptomycin. Uniform colony size will be indicative of stability of the mutation responsible for resistance. Determine the maximal level of expressed streptomycin resistance by streaking isolates on BHI agar containing increasing amounts of streptomycin. This will dictate the amount of streptomycin to be used to select for *cat* fusion strains once removed from the animal environment.

B. Animal inflections

Maintain 18- to 40-day-old Sprague-Dawley rats (Charles River) on indirect bedding in groups of 4 throughout the experiment. Feed the animals standard rodent chow (Laboratory Rodent Diet #5001, Purina) for at least 4 days prior to the inoculation and experimental periods (allowing time for acclimation to the animal facility), and supply them with distilled water lacking fluoride. To initiate the experiments, feed the rats soft defined diet M2000 (containing 56% sucrose) [8] (Table 1) for 4 days (inoculation period). The soft diet will facilitate retention of bacteria on the teeth once infection has occurred. Feeding dishes will either need to be purchased or made that will allow the rats to feed on the soft sticky diet without getting the food on their bodies and in particular on their feet (animals actually will become stuck to the bottom of their cages). We had pottery made that were 5 cm deep with a 4 cm opening for this purpose. Add soft diet anytime animals ingest all provided and replace any remaining diet with fresh food once each day.

Table 1. Diet M2000

Component	Amount
Dextrin	115 g
Polypeptone	270 g
L-methionine	10 g
Calcium phosphate	8 g
Sodium chloride	20 g
Sucrose	560 g
Vitamin B_1 (thiamine)	25 μg
Vitamin B_6 (pyridoxine)	25 μg
Vitamin C (ascorbic acid)	750 μg
Vitamin B_3 (niacinamide)	250 μg
Vitamin B_5 (calcium pantothenate)	100 μg
Vitamin B_2 (riboflavin)	25 μg in 2.5 ml 0.1 N NaOH
Vitamin E (tocopherol acetate)	1 ml
Vitamin D_2 (ergocalciferol)	0.063 μg in 2 μl ethanol
Vitamin A (retinol acetate, 2.9×10^6 U/g)	8.62 μg in 148 μl ethanol
Vegetable oil	6 ml
Distilled water	400 ml

Transfer an isolated colony of the streptococcal str^r *cat* fusion strain from a BHI agar plate containing streptomycin into Todd Hewitt broth (THB) (Difco) containing 1% sucrose and grow anaerobically at 37 °C overnight. Concentrate bacteria by centrifugation at 4300 ×*g* and suspend in fresh THB containing 1% sucrose to a density of 1×10^{10} cfu/ml. Inoculate cell suspensions into the rat oral cavities on days 2 and 3 of the 4 day inoculation period. Inoculation of animals is best performed by 2 individuals. To accomplish animal inoculations on days 2 and 3 of the 4 day high sucrose diet, remove animals from their cages singly. Cradle the animal, using both hands, while not applying any pressure. Hold hands with thumbs up, knuckles forward, and 1 set of fingers enclosed by the other so that a small hole will be created that the rat can sit in with head protruding. Mix the bacterial inoculum by pipetting in and out several times with a micropipetter followed by withdrawal of 100 μl of suspension into the pipette tip. Gently steady the animal's head and insert the pipet tip into 1 side of the rat's mouth. Expel the solution slowly and observe that the animal is drinking. If the animal is hesitant to drink the suspension, gently touch the pipet tip to the tongue or mouth roof of the animal. This usually will cause the animal to begin drinking. Administer the bacterial suspension until the pipet tip is empty. Reinoculate the animal if less than 75% of the suspension is retained. Do not provide antibiotics to animals so that streptococcal fusion strains can interact with members of the rat normal oral flora.

C. Experimental treatment

Following the 4 day inoculation period, place animals on diet 2000 (which lacks sucrose) [8, 10] (Table 2) containing 5% glucose for 1 to 2 days. After this 1 to 2 day period, remove diet from the animals so that they fast overnight. Once fasted, separate animals into individual cages and feed them diet 2000 containing a particular carbohydrate at a 5% final concentration for 30 minutes, and during this time confirm that the animals actually ingest the provided diet (confirmation facilitated by use of individual cages). The approach utilizing feeding following fasting will allow for precise control of exposure of streptococcal cells to experimental carbohydrates.

D. Collection of bacteria and preparation of bacterial lysates

Following the experimental treatment of animals, collect bacteria from the animals' teeth by swabbing the tooth surfaces with sterile cotton-tipped applicators soaked in 0.05 M Tris-HCI buffer (pH 7.8). Collection of bacteria is best performed by 2 individuals. Place the animal into

Table 2. Diet 2000

Component	Amount
Dextrin	115 g
Polypeptone	270 g
L-methionine	10 g
Calcium phosphate	8 g
Sodium chloride	20 g
Whole wheat flour	510 g
Experimental carbohydrate (D-glucose, D-fructose, or D-sucrose)	50 g
Vitamin B_1	25 μg
Vitamin B_6	25 μg
Vitamin C	750 μg
Vitamin B_3	250 μg
Vitamin B_5	100 μg
Vitamin B_2	25 μg in 2.5 ml 0.1 N NaOH
Vitamin E	1 ml
Vitamin D_2	0.063 μg in 2 μl ethanol
Vitamin A	8.62 μg in 148 μl ethanol
Vegetable oil	6 ml
Distilled water	400 ml

the left hand, head protruding between the thumb and forefinger. Close the right hand over top of the rat, and apply a minimal amount of pressure with the right thumb and forefinger on the animal's shoulder blades and sides of the head. This will prevent the animal from withdrawing into or jumping out of the hand. This holding technique does not apply pressure to the throat or neck and thus in no way interferes with breathing or swallowing by the animal. Insert the sterile moistened swab into the animal's mouth and slowly twirl the applicator while moving it back and forth along the outer and inner surfaces of the upper and lower teeth of the left and right side.

Aseptically remove the cotton from the applicator by first loosening it with an alcohol-flamed scalpel, then pulling it from the end of the stick with alcohol-flamed forceps. Place the removed cotton in 900 μl of 0.05 M Tris-HCI buffer (pH 7.8) in a microcentrifuge tube on ice. Remove bacteria from the cotton and disrupt streptococcal chains by 30 seconds of vigorous vortexing. Confirm chain disruption by microscopy. Dilute aliquots of bacterial suspensions and enumerate streptococcal *cat* fusion strains following plating on BHI agar containing streptomycin.

Use remaining solutions (held on ice) to prepare cell-free extracts. Add 350 mg of 0.1 mm zirconium beads (Biospec Products) to each microcentrifuge tube containing cells in 0.05 M Tris-HCI (pH 7.8) solution. Disrupt bacteria by a 3 minute treatment in a Mini Beadbeater apparatus (Biospec Products) and incubate lysates on ice for 2 minutes. Remove cellular debris and zirconium beads by centrifugation at 13,600 ×*g* for 2 minutes at 4 °C.

E. Chloramphenicol Acetyltransferase (CAT) assay Measure the activity of the CAT enzyme in the cell-free extracts [16] as a reflection of expression from the streptococcal promoter directing *cat*. Following lysis and centrifugation, remove the supernatant and dispense a 55 μl aliquot into a new microcentrifuge tube. Add 5 μl of 3 M Tris-HCI (pH 7.8) to each 55 μl aliquot, and incubate the solutions for 10 minutes at 65 °C. Add 25 μl of 8 mM chloramphenicol (Sigma) and 40 μl of a 0.5 mM ^{14}C-Acetyl CoA solution (0.1 μCi) (ICN). Mix solutions by repeated inversion and incubate at room temperature for 30 minutes. Following incubation, extract each sample by the addition of 230 μl of cold ethyl acetate (Fisher) (4 °C). Vortex samples vigorously for 30 seconds, then centrifuge at 13,600 ×*g* for 5 minutes at 4 °C. Transfer 150 μl of the organic phase, containing molecules of ^{14}C-acetylated chloramphenicol, to a polyethylene scintillation vial containing 10 ml of Bio-Safe II scintillation cocktail (Research Products International). Quantify radioactivity by liquid scintillation spectrometry and report as disintegrations per minute (dpm). Recount samples demonstrating a coefficient of variance of 5 or greater, as well as samples showing a greater that 5% contribution to total counts due to luminescence as determined by a scintillation spectrophotometer (Beckman). Generate CAT specific activities by dividing dpm values by the number of streptococcal fusion strain cfu present in the sample used in the CAT assay. Express specific activities as dpm/1000 cfu. Analyze data using single factor analysis of variance and with determination of 95% confidence intervals.

4. Results and discussion

Most genes encode products for which no easy assay exists. Examination of gene expression is further complicated if one is interested in changes in expression as bacteria interact with their animals hosts. A method is described here utilizing *cat* fusions as a tool to study bacterial gene expression in a mammalian environment. It has been used successfully to examine *S. mutans ftf* expression [4] but the approach should be applicable to any system once fusions between a promoter of interest and a promoterless *cat* gene are available. *S. mutans ftf* expression was consistent between experiments and only varied as a function of specific carbohydrate consumed and the age of the host animal. No significant variation was observed due to the amount of carbohydrate consumed or the sex of the rats used in the studies.

The method described was utilized to demonstrate that *S. mutans ftf* expression is induced by sucrose in the mammalian oral cavity. This finding is consistent with earlier in vitro studies [6, 7, 11, 17] and at least partially provides a clear genetic explanation for the increased virulence of this oral bacterium in the presence of table sugar.

Young animals are more susceptible to dental caries than older animals [5]. The method described allowed us to determine that bacterial *ftf* expression in glucose-fed rats averaging 27 days of age was almost five-fold that seen in animals averaging 36 days of age. Expression of *ftf* in rats fed diets containing fructose, with a mean age of 29 days, was almost three-fold that seen in 37 day-old animals. For sucrose-fed rats, *ftf* expression in 27 day-old animals was nearly five-fold that seen in 36 day-old animals. These differences were statistically significant ($p < 0.05$) for bacteria living within animals fed diets containing either sucrose or glucose. Weyrauch et al. reported that caries-induced tooth loss among military personnel declines with increasing age [18]. The existence of age-dependent phenomena relative to caries warranted a study of whether animals age directly affects *S. mutans* gene expression. The protocol described here was used to demonstrate that there is an animal age-dependent induction of a *S. mutans* virulence-associated gene whose product plays an important role in the causation of dental caries. The results suggest a molecular basis for age-dependent dental caries incidence. It is unclear whether animal age has a generalized effect on *S. mutans* gene expression or a specific influence on *ftf*. Factors responsible for causing this age-dependent expression might be host-specific, or might be due to age-dependent changes in the members of the normal bacterial flora.

The method described will allow measurement of the expression of specific virulence-associated bacterial genes while bacteria are living in the mammalian oral cavity. Furthermore, it provides a model to study virulence-associated genes in other pathogenic streptococcal species.

Acknowledgments

This research was funded by National Institutes of Health Award DE10179 (to M. C. H.). Appreciation is due Vickey L. Morin, Brian E. Grey, Stephen B. King, and Chris S. Leslie for technical assistance with this project.

Notes on suppliers

1. Sorvall Instruments, Wilmington, DE 19898, USA
2. Fisher Scientific, 711 Forbes Avenue, Pittsburgh, PA 15219, USA
3. Biospec Products, Box 722, Bartlesville, OK 74005, USA
4. Beckman Instruments, Inc., Fullerton, CA 92634, USA
5. Difco Laboratories, Detroit, MI 48232, USA
6. Sigma Chemical Company, P.O. Box 14508, St. Louis, MO 63178, USA
7. Charles River Laboratories, 251 Ballardvalle Street, Wilmington, MA 01887, USA
8. Purina Mills, 1710 N. Tryon Street, Charlotte, NC 28206, USA
9. ICN Pharmaceuticals, Inc., 3300 Hyland Avenue, Costa Mesa, CA 92626, USA
10. Research Products International Corp., 410 N. Business Center Drive, Mount Prospect, IL 60056, USA
11. McCormick Distilling Co. Inc., DSP-MO-5, Weston, MO 64098, USA
12. Proctor and Gamble, Box 5558, Cincinnati, OH 45201, USA
13. The Pillsbury Co., 2866 Pillsbury Center, Minneapolis, MN 55402, USA

References

1. Bowen WH, Madison KM, Pearson SK (1988). Influence of desalivation in rats on incidence of caries in intact cagemates. J Dent Res 67: 1316–1318.
2. Camilli A, Beattie DT, Mekalanos JJ (1991). Use of genetic recombination as a reporter of gene expression. Proc Natl Acad Sci USA 91: 2634–2638.
3. Camilli A, Mekalanos JJ (1995). Use of recombinase gene fusions to identify *Vibrio cholerae* genes induced during infection. Mol Microbiol 18: 671–683.
4. Grey, WT, Curtiss R III, Hudson MC (1997). Expression of the *Streptococcus Mutans* fructosyltransferase gene within a mammalian host. Infect Immun 65: 2488–2490.
5. Hamada S, Slade HD (1980). Biology, immunology, and cariogenicity of *Streptococcus mutans*. Microbiol Rev 44: 331–384.
6. Hudson MC, Curtiss R III (1990). Regulation of expression of *Streptococcus mutans* genes important to virulence. Infect Immun 58: 464–470.
7. Hudson MC, and Curtiss R III (1991). Use of operon fusions to study gene expression in *Streptococcus mutans*. In: Dunny GM, Cleary PP, McKay LL (eds), Genetics and molecular biology of streptococci, lactococci, and enterococci, pp 261–266. Washington, DC: American Society for Microbiology.
8. Ikeda T, Koulourides T, Kurita T, Housch T, Hirasawa M (1985). Anti-dental caries effect in rats and man of a bacteriocin purified from the oral bacterium *Streptococcus mutans* C3603. Archs Oral Biol 30: 381–384.
9. Kawanabe J, Hirasawa M, Takeuchi T, Oda T, Ikeda T (1992). Noncariogenicity of erythritol as a substrate. Caries Res 26: 358–362.

10. Keyes, PH, Jordan HV (1964). Periodontal lesions in the Syrian hamster-III Findings related to an infectious and transmissible component. Archs Oral Biol 9: 377–400.
11. Kiska, DL, Macrina FL (1994). Genetic regulation of fructosyltransferase in *Streptococcus Mutans*. Infect Immun 62: 1241–1251.
12. Mahan, MJ, Tobias JW, Slauch JM, Hanna PC, Collier RJ, Mekalanos JJ (1995). Antibiotic-based selection for bacterial genes that are specifically induced during infection of a host. Proc Natl Acad Sci USA 92: 669–673.
13. Ooshima T, Sumi N, Izumitani A, Sobue S (1988). Effect of inoculum size and frequency on the establishment of *Streptococcus mutans* in the oral cavities of experimental animals. J Dent Res 67: 964–968.
14. Shaw WV (1979). Chloramphenicol acetyltransferase from chloramphenicol-resistant bacteria. Methods Enzymol 43: 737–755.
15. Shaw WV, Packman LC, Burleigh BD, Dell A, Morris HR, Hartley BS (1979). Primary structure of a chloramphenicol acetyltransferase specified by R plasmids. Nature (London) 282: 872–872.
16. Sleigh, MJ (1986). A nonchromatographic assay for expression of the chloramphenicol acetyltransferase gene in eucaryotic cells. Anal Biochem 156: 251–256.
17. Wexler DL, Hudson MC, Burne RA (1993). *Streptococcus mutans* fructosyltransferase (*ftf*) and glucosyltransferase (*gtfBC*) operon fusion strains in continuous culture. Infect Immun 61: 1259–1267.
18. Weyrauch CD, Burgess JO, Reagan SE (1995). Tooth loss during compulsory dental care. Military Med 160: 194–197.

Address for correspondence: Michael C. Hudson, Ph.D., Department of Biology, The University of North Carolina at Charlotte, 9201 University City Boulevard, Charlotte, NC 28223, USA
Phone: 704/547-4048; Fax: 704/547-3457
E-mail: mchudson@email.uncc.edu

Methods in Cell Science **20**: 143–151 (1998)

Analysis of adherence-associated gene expression in *Streptococcus parasangusis*: A method for RNA isolation

Eunice H. Froeliger & Paula Fives-Taylor
University of Vermont, Department of Microbiology and Molecular Genetics, Burlington, Vermont, USA

Abstract. The objective of the current research was to define a consistent and reliable method for the preparation of high quality, intact total RNA from Gram-positive oral bacteria. The method described, adapted from previously published protocols [1, 8, 17], employs mechanical means to disrupt cells and release RNA. Bacterial cells are subjected to vigorous vortexing in the presence of high density glass beads and the denaturing agents, phenol and sodium dodecyl sulfate. Under these conditions, disruption of the bacterial cell walls is rapid enough to expose intracellular contents to the denaturing agents before endogenous ribonucleases have a chance to hydrolyze RNA. Based on the ethidium bromide straining pattern of 23S and 16S rRNAs, either denaturing (formaldehyde/agarose) or nondenaturing (TAE/agarose) gel electrophoresis could be used to determine the general integrity of the RNA preparation. Analysis by either method revealed sharp rRNA bands indicative of high quality RNA. Expression of adherence-associated transcripts could be detected using formaldehyde/agarose gel electrophoresis and northern hybridization analysis. This method of RNA preparation proved to be useful for isolation of intact RNA from the recalcitrant bacterium, *Streptococcus parasanguis*, and may prove useful for other such oral bacteria.

Key words: Adherence, Gene expression, RNA isolation, Streptococci, Transcriptional regulation

Abbreviations: cpm = counts per minute; DEPC = diethyl pyrocarbonate; EDTA = ethylenediaminetetraacetic acid, EtBr = ethidium bromide; kb = kilobases; mRNA = messenger RNA; MOPS = 3-(N-morpholino)propanesulfonic acid; RNA = ribonucleic acid; RNase = ribonuclease; rRNA = ribosomal RNA; SDS = sodium dodecyl sulfate; THB = Todd Hewitt Broth; Tris = Tris[hydroxymethyl]aminomethane

1. Introduction

The sanguis streptococci (including *Streptococcus parasanguis, S. sanguis, S. gordonii* and *S. oralis*) are the primary colonizers of the tooth surface and are found abundantly in human dental plaque [6], the forerunner of dental caries and inflammatory periodontal disease. In addition, when the oral streptococci gain access to the bloodstream, they can successfully colonize heart tissue, becoming a major cause of subacute bacterial endocarditis [21]. Successful colonization and the persistence of these organisms within the constantly changing conditions of the oral cavity or within heart tissue implies that they have evolved mechanisms adaptive to their survival in these unique ecological niches. A critical first step in the colonization process is the ability of the bacterium to attach firmly to the host surface. Adherence is particularly important in areas such as the oral cavity, where mucosal surfaces are constantly being washed by host fluids. Numerous studies suggest that oral streptococci bind to a variety of substrates in the mammalian host, including salivary components, extracellular matrix or serum components, host cells and other microbial cells present in the oral cavity [11, 12, 15]. Adherence of oral streptococci to this wide range of substrates is facilitated by multiple adhesins exposed on the surface of the bacterial cell [12]. These adhesins are considered to be important factors in determining the success of oral streptococcal colonization and survival within the human host.

Given the constantly changing environmental conditions micro-organisms face when attempting to colonize the human host, it is not surprising that they have evolved the ability to regulate the expression of adherence- and other virulence-associated genes. The differential expression of these genes enables the bacteria to selectively produce factors necessary for successful adaptation to changing host conditions. For example, the types of adhesins expressed by the oral streptococci at the mucosal surfaces of the oral cavity may be very different from those required for adherence to heart tissue. The regulation of bacterial virulence factors is often associated with environmental cues such as growth phase, carbon source, pH, temperature, osmolarity, iron or oxygen levels and several of these changes in gene expression appear to be regulated at the transcriptional level [2, 5, 9, 10, 14, 16, 22].

If environmental conditions are critical in controlling pathogen-related gene expression, it will be necessary to determine which signals affect the expression of specific genes. Historically, northern hybridization analysis has been one of the most common methods used in the analysis of gene expression. It is used to measure the size of a message, as well as to determine the steady-state level of message produced from a specific gene. The first, and often the most critical step in performing northern analysis, is the isolation of high quality, intact RNA from cells. This is often hindered by the presence of active RNases that are released immediately upon cell disruption. While nucleolytic degradation of RNA can sometimes be eliminated by denaturation of these enzymes, one of the principal complications in obtaining undegraded RNA is the difficulty in disrupting cells quickly and efficiently.

Isolation of high quality RNA becomes even more challenging when working with cell types such as Gram-positive bacteria, particularly those prominent in the resident microflora of the oral cavity and nasopharynx. These bacteria have thick cell walls that are resistant to disruption or degradation. Methods of cell lysis that employ enzymatic digestion of cell wall peptidoglycans by lysozyme or proteinase K, have not proven to be effective with the oral bacteria. Thus, with these organisms, rapid cellular disruption in the presence of denaturing agents becomes the most critical factor in the preparation of intact RNA.

Previously, our laboratory had developed a method for obtaining total RNA from the oral bacteria, *S. parasanguis*, using a combination of the lytic enzymes, lysozyme and mutanolysin [4]. This method, while somewhat successful, still resulted in partially degraded RNA. Because of our interest in studying the expression of adherence-associated genes from *S. parasanguis* using northern hybridization analysis, we attempted to define a method that would yield high quality RNA from *S. parasanguis* consistently and reliably. A number of procedures have been published that release RNA from cells by vigorous vortexing in the presence of glass beads and denaturing agents. This method has been used to prepare total RNA from yeast [1], *Bacillus subtilis* [17] and from a member of the 'salivarius group' of human oral streptococci, *Streptococcus salivarius* [8]. We have adapted this method of mechanical disruption of cells into our protocol for isolation of RNA from the sanguis streptococci.

2. Materials

A. Equipment
  1. SPECTRONIC 20D spectrophotometer, Cat. No. 22348-110.[1]
  2. Barnstead MEGA-PURE water still MP-6A, Cat. No. 26295-001.[1]
  3. Anaerobic system, model 1025.[2]
  4. CO_2 incubator, model 3157.[2]
  5. Sorvall high speed centrifuge, model RC5B with SS34 rotor.[3]
  6. Vortex-Genie 2 Mixer, Cat. No. 58815-178.[1]
  7. Mini-sub cell system, Cat. No.170-4307.[4]
  8. Power supply, model VWR105, Cat. No. 27370-265.[4]
  9. Polaroid gel documentation system, Cat. No. 170-3742.[4]
  10. Hybridization oven, model 2710, Cat. No. 35923-454.[1]
  11. Vacuum oven, model 1400E, Cat. No. 52201-218.[1]
  12. Exposure cassette/intensifying screen, Cat. No. E 5138.[5]

B. Glassware and plastics
  1. Pyrex 15 ml culture tubes with closures, Cat. Nos. 60820-206 and 60825-872.[1]
  2. Pyrex culture flasks, 250 ml, Cat. No. 29174-149.[1]
  3. Corning disposable polypropylene centrifuge tubes, 15 ml, 50 ml, Cat. Nos. 21008-678, 21008-736.[1]

C. Culture medium
  1. Todd Hewitt Broth, Cat. No. BB11736.[1]

D. Reagents for RNA isolation and analysis
  1. Diethyl pyrocarbonate (DEPC), Cat. No. D 5758.[5]
  2. Glass beads, acid-washed (150–212 microns), Cat. No. G 1145.[5]
  3. Chloroform, Cat. No. C 2432.[5]
  4. Isoamyl alcohol, Cat. No. I 9392.[5]
  5. Phenol, saturated with 0.1 M citrate buffer, Cat. No. P 4682.[5]
  6. TRIZMA-HCl (Tris[hydroxymethyl]aminomethane hydrochloride), Cat. No. T 7149.[5]
  7. EDTA (ethylenediaminetetraacetic acid), disodium salt, dihydrate, Cat. No. E 5134.[5]
  8. SDS (sodium dodecyl sulfate), Cat. No. L 4390.[5]
  9. Lithium chloride, Cat. No. L 9650.[5]
  10. Acetic acid, glacial, Cat. No. A 6283.[5]
  11. Ethidium bromide, tablets, Cat. No. E 4391.[5]
  12. Agarose (Ultra Pure), Cat. No. 15510-019.[6]
  13. MOPS (3-(N-morpholino)propanesulfonic acid), Cat. No. M 8899.[5]
  14. Sodium acetate, Cat. No. S 2889.[5]
  15. Formaldehyde, Cat. No. F 8775.[5]
  16. Glycerol, Cat. No. G 5516.[5]
  17. Bromophenol blue, Cat. No. B 5525.[5]
  18. Xylene cyanol FF, Cat. No. X 4126.[5]
  19. Formamide, deionized, Cat. No. F 9037.[5]
  20. RNA molecular size standards, 0.24-9.5 kb ladder, Cat. No. 15620-016.[6]
  21. RNA molecular weight markers, 461-4061 bases, Cat. No. 361.[7]
  22. Citric acid, trisodium salt, dihydrate (sodium citrate) Cat. No. C 8532.[5]

23. Sodium chloride, Cat. No. S 3014.[5]
24. Denhardt's solution, 50x concentrate, lyophilized, Cat. No. D 9905.[5]
25. DNA for hybridization, sonicated from salmon testes, Cat. No. D 7656.[5]
26. S&S Optitran BA-S supported nitrocellulose membrane, Cat. No. 28152-363.[1]
27. [α-^{32}P]-deoxycytidine 5′-triphosphate, 3000 Ci/mmol, Cat. No. NEG013H.[9]
28. Nick translation system, Cat. No. 18160-010.[6]
29. TE SELECT-D, G25 spin columns, Cat. No. 1-232195.[8]
30. Kodak X-OMAT XAR-2 autoradiography film, Cat. No. IB1651579.[1]

3. Procedures

A. Preparation of labware and stock solutions

Sterile, disposable plasticware is free of RNases and can be used without pretreatment for preparation of stock solutions or for the preparation and storage of RNA. Treat all other glassware, labware and solutions to be used in RNA preparation with diethylpyrocarbonate (DEPC) to inactivate ribonucleases [18]. **Caution**: DEPC is a suspected carcinogen and should be handled with care. Wear gloves and use in a chemical fume hood (DEPC is inactivated by autoclaving which causes its hydrolysis into CO_2 and ethanol). Use water that has been purified by reverse osmosis and distillation (dH_2O). A major source of RNase contamination is the hands, so wear gloves when handling DEPC-treated labware, preparing solutions or preparing RNA.

1. Preparation of labware.
   a) Fill bottles to be used for storage of stock solutions half full with water and add DEPC to a final concentration of 0.1%. Shake vigorously. In a closed container, soak beakers, graduated cylinders, stir bars and spatulas in a 0.1% solution of DEPC. Allow the labware to stand for at least 1 h at 37 °C. Remove the DEPC-treated labware from the solution, shaking off excess moisture. Cover beakers and graduated cylinders with aluminum foil and package stir bars and spatulas in autoclavable pouches. Autoclave these items for 30 min at 15 lb/ square inch using a dry cycle. Use these sterile, DEPC-treated items for the preparation of stock solutions. Without removing the water, autoclave the DEPC-treated bottles for 20 min at 15 lb/square inch on liquid cycle. Use the DEPC-treated water in the bottles to prepare stock solutions and the bottles for the storage of stock solutions.
2. Preparation of solutions
   a) 0.1 M EDTA, pH 8.0: To a 250 ml beaker add 75 ml DEPC-treated water and 3.72 g of Na_2EDTA. Adjust the pH to 8.0 and bring up the volume to 100 ml. Add 100 μl DEPC of and shake vigorously. Allow the solution to stand for at least 1 h at 37 °C and then autoclave for 20 min at 15 lb/square inch on liquid cycle.
   b) 0.1 M Tris, pH 8.0: Solutions containing a primary amine (such as Tris and ammonium acetate) should not be treated with DEPC. Instead, use Tris-HCl from a bottle reserved for the preparation of RNase-free solutions. To a 250 ml beaker add 75 ml DEPC-treated water and 1.576 g of Tris-HCl. Dissolve the Tris and adjust the solution to pH 8.0 and bring up the volume to 100 ml. Transfer the solution to a DEPC-treated bottle and autoclave for 20 min at 15 lb/square inch on liquid cycle. Store at room temperature.
   c) TE, pH 8.0: To a glass bottle, add 89 ml DEPC-treated water, 10 ml of 0.1 M Tris (pH 8.0) and 1.0 ml of 0.1 M EDTA (pH 8.0) and mix. Autoclave for 20 min at 15 lb/square inch on liquid cycle. Store at 4 °C.
   d) 10% SDS: To a 50 ml disposable polypropylene centrifuge tube, add 40 ml DEPC-treated water and 5.0 g SDS. Mix gently until the SDS dissolves and then bring up the volume to 50 ml. Store at room temperature.
   e) 10 M LiCl: Using LiCl from a bottle reserved for the preparation of RNase-free solutions, add 21.2 g of LiCl to a 50 ml disposable polypropylene centrifuge tube. Add DEPC-treated water to a final volume of 50 ml. Store at 4 °C.
   f) 70% ethanol: To a 50 ml disposable polypropylene centrifuge tube, and 15 ml DEPC-treated water and 35 ml 100% ethanol and mix. Store at 4 °C.

B. Growth conditions of *S. parasanguis*

Although the nutritional requirements of *S. parasanguis* are complex, there are several media, including a chemically defined medium, FMC [19, 20] that are suitable for culture.

1. Start overnight cultures of *S. parasanguis* under defined growth conditions. For example, if oxygen level is the environmental cue under investigation, inoculate 10 ml cultures of THB with a single colony taken from a fresh plate culture of *S. parasanguis*. Incubate the culture overnight at 37 °C with 5% CO_2 or in an anaerobic chamber at the same temperature. Have two 250 ml flasks containing 125 ml each of THB prepared for both the aerobic and

anaerobic environmental conditions. Include this medium in the growth chambers so that the medium has equilibrated with the environmental conditions of each growth chambers prior to use.

2. After overnight incubation, the cultures will be in stationary phase (Do not incubate the cultures longer than 18 h. After this point cells enter a logarithmic death phase and will go through a long lag phase when resuspended in fresh medium). Add two mls of the aerobic culture to each of the two flasks containing 125 ml of THB equilibrated to aerobic conditions. Incubate at 37 °C, 5% CO_2 until mid-log growth phase (about 3.5 h). Do the same for the culture growing under anaerobic conditions.
3. Pour the cultures from each growth conditions into five 50 ml disposable polypropylene centrifuge tubes and incubate on ice for 10 min.

C. Preparation of total RNA

For preparation of total RNA, perform all steps at 4 °C and, with the exception of 10% SDS, keep all solutions and tubes on ice. Chill the glass beads on ice before use. Centrifuge larger volumes in a refrigerated Sorvall RC5B centrifuge using a pre-chilled SS34 rotor. The 50 ml and 15 ml polypropylene tubes will fit into a SS34 rotor if a nut, rather than the rotor cover, is used to secure the rotor. Alternatively, other types of refrigerated centrifuges can be used. For smaller volumes use a refrigerated microcentrifuge or a microcentrifuge kept in a cold room. The procedure below is for 250 ml of culture grown under one environmental condition. More than one culture can be processed simultaneously.

1. Centrifuge tubes for 10 min, 6K, 4 °C. Aspirate or carefully pour off supernatant and resuspend bacteria in a total volume of 4.0 ml of ice cold TE, pH 8.0. Transfer the bacteria to a 15 ml disposable polypropylene centrifuge tube.
2. Add 3.5 g chilled acid-washed glass beads, 3 ml of 100:24:1 (v/v/v) acid-saturated phenol/chloroform/isoamyl alcohol and 0.25 ml 10 % SDS to the 15 ml tube and mix well to ensure that beads are suspended. Vortex immediately for a total of 3 min at highest speed. After each min, place on ice for 1 min. Repeat two more times.
3. Centrifuge for 1 min at 6K. Transfer top aqueous phase to a clean 15 ml polypropylene centrifuge tube, making sure not to take any of the interface. Add an equal volume of 100:24:1 (v/v/v) phenol/chloroform/isoamyl alcohol and vortex at top speed for 10 sec. Centrifuge 1 min at 6K. Repeat extraction 3 more times.
4. To precipitate RNA, add 0.1 volume of 10 M LiCl and 2 volumes of 100% ethanol. Mix well and place at –20 °C for at least 30 min.
5. Centrifuge 10K for 15 min. Pour off supernatant and wash pellet gently with ice cold 70% ethanol. Centrifuge for 5 min. Remove as much of the ethanol as possible and then allow the pellet to air dry by inverting the tube onto a clean KimWipe for 5 min at room temperature. Resuspend the pellet in 200 μl DEPC-treated water. Freeze RNA in aliquots at –70 °C.

D. Analysis of RNA integrity

Before proceeding to northern hybridization analysis or to other applications, it is necessary to first assess the integrity of the isolated RNA. Most often, RNA is analyzed by electrophoresis through a denaturing agarose gel in the presence of ethidium bromide. Several denaturing agents, such as formaldehyde, glyoxal and dimethyl sulfoxide (DMSO) or methyl mercuric hydroxide, can be used [1]. However, formaldehyde is used most frequently because it is less toxic than mercuric hydroxide and formaldehyde gels are less difficult to run than those containing glyoxal/DMSO [1]. Integrity of RNA can also be assessed rapidly by electrophoresis on a nondenaturing agarose gel. This latter method eliminates problems associated with toxic chemicals. Both methods are presented below. The protocol for analysis of RNA using a denaturing agarose gel is adapted from Ausubel et al. [1]. Both methods, as well as northern hybridization analysis, are performed on 7 × 10 cm horizontal submarine gels. Longer gel rigs can be used if additional resolution is required. Use distilled water and sterile Pipetman tips and microfuge tubes that have been reserved for use with RNA applications.

1. Analysis of RNA using a denaturing agarose gel
   a) Preparation of 10X MOPS buffer: 10X MOPS buffer is 0.2 M 3-(N-morpholino)propanesulfonic acid (MOPS), 50 mM sodium acetate, 10 mM EDTA, pH 8.0. To prepare one l, add 41.9 g of MOPS and 4.1 g of sodium acetate to 800 ml of water and adjust the pH to 7.0. Add 20 ml of a 0.5 M stock solution of Na_2EDTA, pH 8.0 and adjust the volume to 1 l with water. Store at 4 °C, protected from light.
   b) Preparation of a 0.44 M formaldehyde denaturing gel: To make 50 ml of a 1% agarose/formaldehyde gel, melt 0.5 g of agarose in 43.2 ml of water and cool to about 65 °C. Add 1.8 ml of formaldehyde and 5 ml of 10× MOPS buffer to the melted agarose, swirl to mix and pour into gel tray. **Caution:** Formaldehyde vapors are toxic. Prepare solutions containing

formaldehyde in a fume hood. Pour and run agarose/formaldehyde gels in a fume hood.

c) Preparation of 1× MOPS running buffer: To make 1 l, add 100 ml 10× MOPS buffer and 36 ml of formaldehyde to 864 ml water.

d) Preparation of 5× RNA loading dye: 5× RNA loading dye contains 50% glycerol, 1 mM EDTA, pH 8.0, 0.25% bromophenol blue, and 0.25% xylene cyanol FF [18]. Use DEPC-treated water and DEPC-treated 0.1 M EDTA, pH 8.0 and prepare in a 15 ml disposable, polypropylene centrifuge tube. Store at 4 °C.

e) Sample preparation: Prepare 20 μl RNA samples by combining on ice in the following order:

10× MOPS buffer	2.0 μl
Formaldehyde	0.7 μl
formamide	10.0 μl
5× loading dye	2.0 μl
RNA	1.0 μl
DEPC-treated water	4.2 μl
Ethidium bromide (10 mg/ml)	0.1 μl

f) Mix the samples and centrifuge briefly. Denature samples for 10 min at 65 °C, chill on ice, and load immediately onto gel. **Caution:** Ethidium bromide is a mutagen and should be handled with care.

g) Gel electrophoresis: Run the gel in 1× MOPS running buffer at 40 to 50 V until the leading dye has migrated approximately 2/3 the length of the gel. After electrophoresis, visualize RNA under UV transillumination, and record the image using a camera and Type 55 Polaroid film.

2. Analysis of RNA using a native agarose gel

a) Preparation of 10× TAE buffer: Add 48.4 g TRIZMA base, 11.1 ml glacial acetic acid, 7.44 g $Na_2EDTA \cdot 2H_2O$ to water to make one l. Approximate pH is 8.5.

b) RNA sample preparation: Combine on ice:

RNA	1.0 μl
5× loading dye	2.0 μl
DEPC-treated water	4.0 μl

c) Gel electrophoresis: Run the RNA samples on a 1.0% agarose, 1× TAE, 0.5 μg/ml ethidium bromide mini-gel at 50–75 V until the leading dye has migrated approximately 2/3 the length of the gel. After electrophoresis, visualize RNA under UV transillumination, and record the image using a camera and Type 55 Polaroid film.

d) If RNA integrity is good, freeze in aliquots at –70 °C. If RNA concentration needs adjustment, EtOH precipitate, resuspend the RNA and then freeze in aliquots.

E. Northern hybridization analysis

The following procedure for the analysis of RNA by northern hybridization is adapted from a previously published protocol [1] that can be consulted for additional details.

1. Prepare an agarose/formaldehyde gel and RNA samples as described above under 'Analysis of RNA using a denaturing agarose gel'. When comparing gene expression under varying environmental conditions, be sure to load equivalent amounts of RNA for each sample. Amounts of RNA can be estimated from gels used to determine RNA integrity. Compare intensity of staining of one of the rRNAs with a known concentration of a molecular weight marker of a similar molecular weight. Based on the level of gene expression, between 1–20 μg of RNA will need to be analyzed for each sample. Start with approximately 10–15 μg of RNA for each sample. Run RNA molecular weight markers on outside lanes leaving a lane between the markers and the RNA samples. For a mini-gel load 2 μg (500 μg/ml) New England BioLabs 0.46–4.061 kb markers or 3 μg of Gibco BRL 0.24–9.5 kb RNA ladder.
2. When electrophoresis is complete, remove the gel and photograph it on a short wave transilluminator. Include a clear or fluorescent ruler on the gel so that mobility of the RNA molecular weight markers can be used to construct a standard curve. Carefully cut out the portion of the gel to be transferred to nitrocellulose.
3. Place the gel in a large tray and rinse 30–60 min with shaking in several changes of dH_2O to remove the formaldehyde.
4. Soak the gel for 15–30 min with shaking in 20× SSC and then transfer the RNA from the gel to nitrocellulose overnight (12–20 h) as described in detail by Ausubel et al. [1]. 20× SSC is 3 M NaCl, 0.3 M sodium citrate, pH 7.0. When transfer is complete, remove the filter and rinse briefly in 5× SSC and then allow the filter to air dry at room temperature. Place the filter between 2 pieces of filter paper and bake at 80 °C under vacuum for 2 h.
5. Prehybridize and hybridize the filter according to conventional methods [1, 18] in a hybridization oven. If a hybridization oven is not available, prehybridization and hybridization can be done in a heat-sealable bag [1, 18]. Prehybridize for 1–2 h and hybridize overnight at 42 °C. We routinely use a prehybridization and hybridization solution of 5× SSC, 5× Denhardt's reagent, 0.5% SDS (w/v), 50% formamide (v/v), 100 μg/ml denatured, fragmented salmon sperm DNA. We use double-

stranded DNA probes labeled by nick translation to a specific activity of at least 1×10^8 cpm/μg using [α-^{32}P]dCTP using the GibcoBRL Nick Translation System according to the manufacturer's directions. Unincorporated label is removed from the probe using TE SELECT-D, G25 spin columns according to the manufacturer's directions. Add the denatured probe directly into the prehybridization solution (about 500,000 cpm of the probe/ml hybridization solution).

6. Following hybridization, wash the filter for 10 min at room temperature in 2× SSC, 0.5% (w/v) SDS. Repeat. Wash at 65 °C for 15 min in 1× SSC, 0.1% SDS. Repeat. Wash 15 min at 65 °C in 0.2× SSC, 0.1% (w/v) SDS. Repeat.
7. Expose the filter to Kodak film at –70 °C with an intensifying screen. Develop the film after 24 h.

4. Results and discussion

Our interest in studying the expression of specific streptococcal adherence-associated genes by northern hybridization has led us to explore alternative methods for the isolation of high quality, intact RNA from *S. parasanguis*. Although useful information about the steady-state levels of RNA produced from a specific gene can be garnered with lower quality RNA preparations using slot or dot hybridization of RNA [18], coupling gel electrophoresis with northern hybridization offers several advantages. First of all, transcript size can be determined using gel electrophoresis. More importantly, however, it is possible to detect the presence of multiple transcripts produced from a single gene or operon or to detect changes in the relative size of a transcript produced from a single gene or operon. This could become extremely important when comparing gene expression under different environmental conditions when, not only message abundance but also message size or number, may vary in response to environmental signals. In addition, gel electrophoresis allows for the direct comparison of relative message abundance and transcript sizes between samples on the same blot. The blot can be stripped and rehybridized several times with probes to other virulence-associated genes or to other genes of interest, greatly facilitating analysis of gene expression.

This approach, however, requires high quality, intact RNA. The objective of the current research was to improve upon methods developed in our laboratory for the preparation of RNA [4] by defining a protocol that would consistently and reliably yield high quality RNA from *S. parasanguis*. Although methods based on enzymatic digestion of cells walls have not been optimal with the oral streptococci, we first investigated the utility of a RNA isolation kit for yeast and Gram-positive bacteria commercially available from Gentra Systems, Inc. (15200 25th Ave. N., Suite 104, Minneapolis, MN 55447), for the isolation of RNA from *S. parasanguis*. This kit (PURE*script*, Cat. No. R-6000) employs a proprietary 'Lytic Enzyme Solution' to generate spheroplasts. Using this kit, RNA was isolated from *S. parasanguis*, following the manufacturer's protocol, and then RNA integrity was evaluated using native and denaturing agarose gel electrophoresis and northern hybridization analysis. Some degradation of the RNA was evident and the quality of the Northern blot, using this RNA, was not satisfactory (data not shown). Since all of the solutions in the kit were RNase free and were handled carefully, the most likely explanation for degradation of the RNA was that, for *S. parasanguis* cells, the enzymatic disruption was too slow and inefficient. Good lysis of *S. parasanguis* cells was not observed following incubation with the 'Lytic enzyme solution'. We concluded that a strong physical force may be necessary to disrupt the cell walls of the oral streptococci.

A number of reports have indicated that RNA can be released from cells that are difficult to disrupt by vigorously vortexing cells in the presence of glass beads and denaturing agents, phenol and SDS [1, 8,17]. We adapted this method for use with *S. parasanguis* and found that the streptococcal cells could be disrupted quickly and effectively by this mechanical means to yield high quality RNA. The procedure requires few manipulations and is done in the cold room, or on ice, with prechilled reagents to minimize RNase activity. Another advantage of the method is that it prevents RNA degradation by RNases during lysis since samples are always in the presence of the denaturing agents. The use of phenol in methods for RNA preparations was first described by Kirby [13] and phenol extraction is an effective means of inactivating and removing RNases. In the past, however, the preparation of phenol for use in molecular biology experiments was difficult and often hazardous, with the phenol being subject to varying impurities. Now, the use of phenol is safer and more convenient due to the introduction of prepared high quality, RNase-free, saturated liquid phenols by numerous biotechnology companies. In addition, the method does not require hot phenol for the extraction of RNA as does our previous method, thus safety risks associated with the handling of hot phenol are minimized.

The general integrity of the RNA released from cells was assessed based on the typical ethidium bromide staining pattern of total RNA after electrophoresis on both a simple native agarose gel (Figure 1A) or on a denaturing formaldehyde/agarose gel (Figure 1B). Total RNA preparations contain primarily rRNA. If the two bands, representing the 23S and 16S, in either lane were not sharp, it would be indicative of RNA degradation during sample preparation. This was not the case. Two sharp bands appear in each lane on both gels, indicating high quality, intact RNA had been isolated from both sample preparations. Chromosomal DNA is also apparent in the RNA preparations and is particularly evident when the RNA is analyzed on native agarose gels (Figure 1A). The method releases both RNA and DNA, so an A_{260} reading will be a measure of total nucleic acid. Lithium chloride precipitation is incorporated into the protocol as this salt preferentially precipitates RNA. While, the RNA preparation still contains some contaminating DNA, this usually will not interfere with most studies, including S1, primer extension and northern hybridization analysis. If further purification of the RNA is required, DNase treatment can also be used [1, 18]. Expected yields from a 250 ml mid-log culture of *S. parasanguis* are approximately 500 μg of RNA.

Both native and denaturing agarose gels gave comparable results (Figure 1A and 1B), indicating that either type of gel could be used to assess RNA quality. Native gels offer an advantage over denaturing gels in that sample and gel preparation is easier and therefore somewhat quicker. Toxic chemicals are avoided and gels can be run on the laboratory bench rather than in a fume hood.

To analyze the integrity of the total RNA preparations further, we employed northern hybridization analysis using an adherence-associated gene, *fimA*, from *S. parasanguis* FW213 [3, 4, 6] as a hybridization probe. RNA was denatured with formaldehyde/formamide and separated on a 0.44 M formaldehyde/1% agarose gel. This procedure gave good separation of RNAs over a range of molecular weights as indicated by the distinct RNA markers ranging from 461 to 4061 bases (Figure 2A). Addition of EtBr directly to RNA samples prior to loading on the formaldehyde/agarose gel gave excellent visualization of RNA with ultraviolet light immediately after electrophoresis (Figure 2A). It was also advantageous to be able to view the RNA at any point during electrophoresis. If the RNA has not been separated sufficiently, electrophoresis could be continued. After transfer to a nitrocellulose membrane by capillary action, the gel retained undetectable amounts of RNA indicating that the stained RNA had been transferred efficiently (Figure 2B).

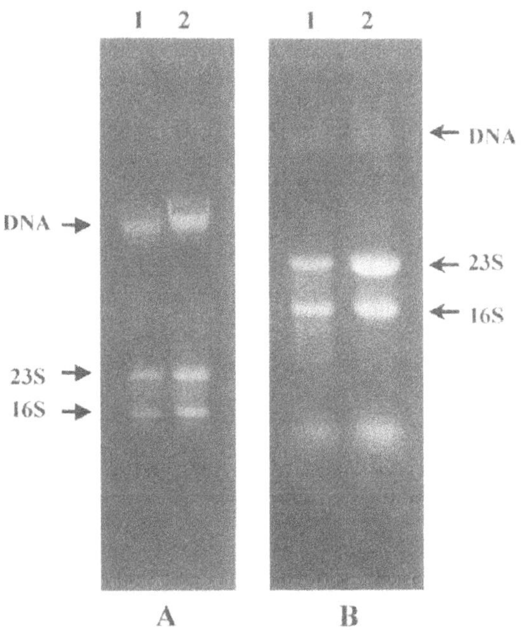

Figure 1. Electrophoretic pattern of RNA isolated from *S. parasanguis* FW213 using vigorous vortexing of cells in the presence of glass beads and denaturing agents. EtBr staining of rRNA reveals RNA integrity. Samples were prepared and gels electrophoresed as described in Procedures. (A) Native 1% agarose gel in TAE buffer. Lanes 1 and 2 are RNA samples prepared from cultures of *S. parasanguis* FW213 grown anaerobically or aerobically in 5% CO_2, respectively. (B) Denaturing 1% agarose gel in formaldehyde/MOPS buffer with the same samples.

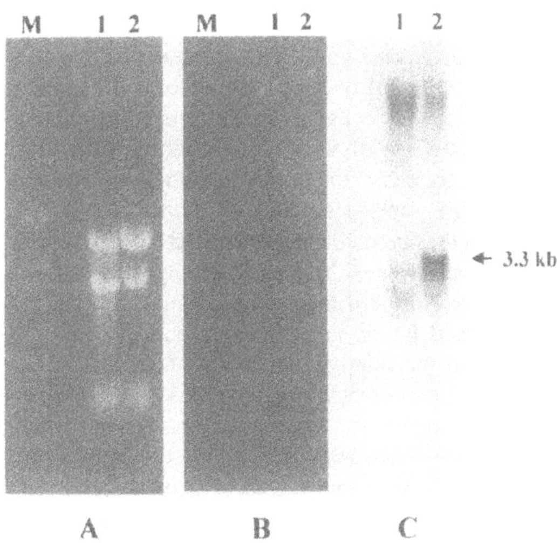

Figure 2. Northern hybridization analysis of RNA isolated from *S. parasanguis* FW213. RNA samples were prepared and separated on a 1% agarose/formaldehyde gel as described in Procedures. (A) Electrophoretic pattern of EtBr-stained RNA before transfer to nitrocellulose. Lane M, RNA molecular weight markers (2 μg) are, from top to bottom, 4061, 2193, 1701, 1280, 754, 461 bp; lanes 1 and 2, RNA samples (10 μg) prepared from cultures grown anaerobically or aerobically in 5% CO_2, respectively. (B) Gel depicted in (A) after transfer of RNA to nitrocellulose by capillary action as described in Procedures. (C) Northern blot of RNA samples transferred to nitrocellulose and analyzed for the presence of FimA mRNA using a nick-translated vector probe containing internal *fimA* sequences as described in Procedures. Lanes 1 and 2, total RNA prepared from cultures grown anaerobically or aerobically in 5% CO_2, respectively.

Our laboratory has shown previously that the *fimA* operon produces a 3.3 kb transcript when grown aerobically in THB with 5% CO_2 [4]. The results of northern blotting of total RNA isolated from *S. parasanguis* FW213 grown either anaerobically or aerobically in 5% CO_2 are shown in Figure 2C. The appearance of nondegraded FimA mRNA (Figure 2C, Lane 2) further demonstrates that the described method for preparing total RNA from *S. parasanguis* yields high quality, intact RNA. The ability to isolate high quality RNA is the first and often most critical step in performing many fundamental molecular biology experiments and it is essential to analyzing gene expression. For our work, it has made it possible to begin to analyze the regulation of adherence-associated genes in *S. parasanguis* FW213. Northern hybridization analysis suggets that expression of FimA is stimulated at the transcriptional level during aerobic growth (Figure 2C, Lane 2) but not during anaerobic growth (Figure 2C, Lane 1). Further analysis of environmental signaling may reveal other pathways that may be involved in the regulation of adherence-associated gene expression.

In summary, we describe here a simple, reliable and relatively rapid method for preparing high quality, intact total RNA from the Gram-positive bacteria, *S. parasanguis*. This procedure may have general applicability for use with oral bacteria or other cells with tough walls that are difficult to disrupt. The use of vigorous vortexing with glass beads greatly assists the initial disruption of cells and, because the sample is always in the presence of phenol/SDS, this method prevents RNA degradation during lysis. The resulting RNA preparation is useful for routine Northern hybridization analyses and can also be used for primer extension and S1 analyses. Because of the relative ease with which this procedure can be performed, it is possible to process several samples at the same time. This feature is particularly important when comparing the steady-state level of a message under various growth conditions or over time. This method of RNA preparation will facilitate analyzes of the expression of virulence-associated genes in *S. parasanguis* and in other oral streptococci.

Acknowledgments

We are grateful for very helpful experimental suggestions from Dr Michel Frenette (University of Laval, Quebec City, Quebec, Canada), Dr Dominique Galli (Indiana University, Indianapolis, IN) and Dr Mary Tierney (University of Vermont, Burlington, VT). We would also like to thank Dr Tierney for a critical review of this manuscript. This research is support by Public Health Service Grant R37-DE11000 from the National Institutes of Health.

Notes on suppliers

1. VWR Scientific Products Corp., 501 Heron Dr., Bridgeport, NJ 08014, USA
2. Forma Scientific, Box 649, Marietta Ohio 45750-0649 USA
3. DuPont, 31 Pecks Lane, P.O. Box 5509, Newtown, CT 06470-5509 USA
4. Bio-Rad Laboratories, 2000 Alfred Nobel Drive, Hercules, CA 94547 USA
5. Sigma Chemical Company, P.O. Box 14508, St. Louis, MO 63178 USA
6. Gibco BRL, P.O. Box 68, Grand Island, NY 14072-0068 USA
7. New England Biolabs, 32 Tozer Road, Beverly, MA 01915-5599 USA
8. 5 Prime-3 Prime, Inc., 5603 Arapahoe Road, Boulder, CO 80303 USA
9. NEN Life Science Products, 549 Albany Street, Boston, MA 02118-2521 USA

References

1. Ausubel FM, Brent R, Kingston RE, Moore DD, Seidman JG, Smith JA, Struhl K, eds (1989). Current protocols in molecular biology. New York: John Wiley and Sons.
2. Caparon MG, Geist RT, Perez-Casal J, Scott JR (1992). Environmental regulation of virulence in group A streptococci: transcription of the gene encoding M protein is stimulated by carbon dioxide. J Bacteriol 174: 5693–5701.
3. Fenno JC, LeBlanc DJ, Fives-Taylor PM (1989). Nucleotide sequence of a type 1 fimbrial gene of *Streptococcus sanguis* FW213. Infect Immun 57: 3527–3533.
4. Fenno JC, Shaikh A, Spatafora G, Fives-Taylor P (1995). The *fimA* locus of *Streptococcus parasanguis* encodes an ATP-binding membrane transport system. Mol Microbiol 15: 849–863.
5. Finlay BB, Falkow S (1997). Common themes in microbial pathogenicity revisited. Microbiol Mol Biol Rev 61: 136–169.
6. Fives-Taylor PM, Macrina FL, Pritchard TJ, Peene SS (1987). Expression of *Streptococcus sanguis* antigens in *Escherichia coli*: cloning of a structural gene for adhesion fimbriae. Infect Immun 55: 123–128.
7. Frandsen EVG, Pedrazzoli V, Kilian M (1991). Ecology of viridans streptococci in the oral cavity and pharynx. Oral Microbiol Immunol 6: 129–133.
8. Gagnon G, Vadeboncoeur C, Gauthier L, Frenette M (1995). Regulation of *ptsH* and *ptsI* gene expression in *Streptococcus salivarius* ATCC 25975. Mol Microbiol 16: 1111–1121.
9. Gibson CM, Caparon MG (1996). Insertional inactivation of *Streptococcus pyogenes sod* suggests that *prtF* is regulated in response to a superoxide signal. J Bacteriol 178: 4688–4695.
10. Hudson MC, Curtiss R III (1990). Regulation of expression of *Streptococcus mutans* genes important to virulence. Infect Immun 58: 464–470.
11. Jenkinson HF (1995). Genetic analysis of adherence by oral streptococci. J Ind Microbiol 15: 186–192

12. Jenkinson HF, Lamont RJ (1997). Streptococcal adhesion and colonization. Crit Rev Oral Biol Med 8: 175–200.
13. Kirby KS (1968). Isolation of nucleic acids with phenolic solvents. Meth Enzymol 12B: 87–99.
14. Kiska DL, Macrina FL (1994). Genetic regulation of fructosyltransferase in *Streptococcus mutans*. Infect Immun 62: 1241–1251.
15. Kolenbrander PE, London J (1993). Adhere today, here tomorrow: oral bacterial adherence. J Bacteriol 175: 3247–3252.
16. Mekalanos JJ (1992). Environmental signals controlling expression off virulence gene determinants in bacteria. J Bacteriol 174: 1–7.
17. Putzer H, Gendron N, Grunberg-Manago M (1992). Co-ordinate expression of the two threonyl-tRNA synthetase genes in *Bacillus subtilis*: control by transcriptional antitermination involving a conserved regulatory sequence. EMBO J 11: 3117–3127.
18. Sambrook J, Fritsch EF, Maniatis T (1989). Molecular cloning: a laboratory manual 7.0–7.87. Cold Spring Harbor: Cold Spring Harbor Laboratory Press.
19. Terleckyj B, Shockman GD (1975). Amino acid requirements of *Streptococcus mutans* and other oral streptococci. Infect Immun 11: 656–664.
20. Terleckyj B, Willett NP, Shockman GD (1975). Growth of several cariogenic strains of oral streptococci in a chemically defined medium. Infect Immun 11: 649–655.
21. Weinberger I, Rotenberg Z, Zacharovitch D, Fuchs J, Davidson E, Agmon J (1990). Native valve infective endocarditis in the 1970's versus the 1980's: underlying cardiac lesions and infecting organisms. Clin Cardiol 13: 94–98.
22. Wexler DL, Hudson MC, Burne RA (1993). *Streptococcus mutans* fructosyltransferase (*ftf*) and glucosyltransferase (*gtfBC*) operon fusion strains in continuous culture. Infect Immun 61: 1259–1267.

Address for correspondence: Dr Paula Fives-Taylor, University of Vermont, Department of Microbiology & Molecular Genetics, Stafford Hall, Burlington, VT 05405-0084, USA
Phone: (802) 656-1121; Fax: (802) 656-8749
E-mail: pfivesta@zoo.uvm.edu

Methods in Cell Science **20**: 153–163 (1998).

Development of an integrative, *lacZ* transcriptional-fusion plasmid vector for *Streptococcus mutans* and its use to isolate expressed genes

Francesca Peruzzi, Patrick J. Piggot & Lolita Daneo-Moore
Department of Microbiology and Immunology, Temple University School of Medicine, Philadelphia, Pennsylvania, USA

Abstract. A new integrational vector, pFP1, containing a promoterless *lacZ* gene has been constructed for use with *Streptococcus mutans*. The vector can be grown in *Escherichia coli*, but cannot replicate in *S. mutans*. pFP1 can be transformed into *S. mutans* with selection for kanamycin resistance. It integrates into the chromosome when homologous DNA is present in the vector. pFP1 provides a way for cloning and identifying new genes of *S. mutans*. When a number of *S. mutans* transformants were tested on agar containing different sugars, some 19 distinct clones with sucrose- and/or glucose-responsive promoters were isolated. Sequence analysis indicated that the cloned DNA encoded all or part of 29 putative proteins with 52% to 100% similarity to known proteins. When assayed for β-galactosidase activity in BTR medium containing 2% sucrose, glucose or maltose, several genes showed some evidence of sugar regulation, including *gtfBK*, *ftf*, *scrA*, and most dramatically for sucrose regulation, *fruA*.

Key words: Integrational vector, Promoters, *S. mutans*, Sugar regulation

1. Introduction

The study of gene expression is greatly facilitated by the use of promoter probes. The probes have been used both to analyze expression of particular genes and to identify and characterize groups of genes responding to particular signals [42]. Transposons have been widely used as promoter-probe vehicles in searches for genes responding to a variety of signals. In *Streptococcus mutans*, Tn*917* insertions have been obtained, allowing the characterization of mutants with acid sensitivity [17]. However, transposon vectors suffer from two disadvantages: insertions are not obtained in essential genes, and transposon insertion may not be random. The latter reservation is particularly important when considering a new transposon or a new host species.

For transformable species, integrative plasmid vectors provide a very good alternative to transposons as vehicles for the promoter probe [32]. Such plasmids can readily be used as vectors for essential genes, and the randomness of the DNA clone bank constructed in the vector largely determines the randomness of the promoter screen. Integrative plasmids have been used successfully for *S. mutans* [22, 39]. The integrative plasmids replicate in *Escherichia coli*, but not in *S. mutans*, although they carry a drug-resistance determinant that can be expressed in *S. mutans*. If the plasmid contains cloned *S. mutans* DNA, then, when selection is made for the plasmid-encoded resistance, *S. mutans* transformants can be obtained in which the plasmid has integrated into the homologous region of the chromosome by a single crossover (Campbell-like) recombination [39, 46].

S. mutans is found in the human oral cavity and is responsible for the development of dental caries [23]. The response of *S. mutans* to sugars and the metabolism of those sugars are major factors in the cariogenicity. There have been several studies of the effect of sugars on the expression of particular genes already known to be associated with sugar metabolism [20, 21, 39]. Moreover a translational *lacZ* fusion to *scrA* has been used to study the effects of sugars on expression of *scrA*, which encodes enzyme II of the phosphoenolpyruvate-dependent sucrose transport system [39]. However, there has not been a general search specifically designed to find genes whose expression responds to specific sugars. Nor has extensive use been made of *lacZ* as promoter probe for *S. mutans*.

We report here the construction of an integrative promoter-probe vector, pFP1, for *S. mutans*. This plasmid can express kanamycin (KM) resistance both in *E. coli* and in streptococci. The plasmid contains a promoterless *lacZ* gene positioned downstream of a multicloning site. (*S. mutans* GS-5 produces little, if any, β-galactosidase.) The plasmid provides a convenient way to identify and study genes that are expressed in *S. mutans*. We have tested the general utility of pFP1 as a promoter-probe vector by identifying a number of *S. mutans* genes whose expression may be regulated by sugars.

2. Materials

Bacterial strains and media
For DNA preparation, *E. coli* was grown in Luria-Bertani (LB) broth (Difco).[1] Todd-Hewitt (TH) broth (Difco) supplemented with 0.1% cysteine and 0.2% glucose and 2% agar for plates were used to grow *S. mutans*. Alternately, the BTR medium described by Sato et al. [39] was used with addition of various sugars, and of 2% agar for plates. In some instances, we used a chemically defined medium described by Hudson and Curtiss [20]. Horse serum was obtained from Sigma.[2] Restriction enzymes were obtained from Promega.[3]

The *S. mutans* GS-5 strain (GS-5.Ku) was supplied by H. K. Kuramitsu. In some experiments, a GS-5 transformant of a spontaneous derivative of *S. mutans* GS-5.Ku (L7) that contains a deletion in the *gtfB/C* region was used [44, 48]. Cells of the L7 derivative do not clump in sucrose-containing media. Km (Sigma) was used for GS-5 at 300 μg/ml, when appropriate. Chromosomal DNA was extracted using a glycine-enhanced procedure [36].

3. Procedures

For the preparation of plasmid DNA, *E. coli* DH5-α is grown in LB broth. LB broth is solidified with 2% agar (Bacto Agar, Difco) when required. Km is used at 50 μg/ml.

Precompetent *S. mutans* bacteria are made by diluting 1/1000 of an overnight culture in TH broth containing 10% horse serum. The culture is grown at 37 °C for 4 hours and then centrifuged at 4 °C. The bacterial pellet is resuspended in 1/10 of the initial volume in TH broth plus 30% glycerol and stored at –70 °C until use.

Transformation of *S. mutans* GS-5 or its L7 derivative is carried out following a modification of the Perry and Kuramitsu protocol [34]. A 1/10 dilution of GS-5 precompetent cells is made in TH broth plus 10% horse serum and the cells are grown for 2 hours at 37 °C (to about 5×10^7 colony-forming units per ml). Portions of the culture are then removed and exposed to 0.5 to 2 μg of plasmid DNA. After an additional 2 hours of incubation, an appropriate amount of culture is spread on TH agar (TH broth plus 0.1% cysteine solidified with 2% agar) containing 300 μg Km per ml, 2% sucrose and 60 μg X-gal (5-bromo-4-chloro-3-indolyl-β-d-galactoside [Sigma]) per ml. The frequencies of transformation range from 10^2 to 10^3 transformants per μg DNA, depending on the size of the insert. β-galactosidase-positive transformants are detected as blue colonies, and they are screened on three different TH-X-gal or on BTR-X-gal agar plates containing, respectively, no sugar added, 2% sucrose, and 2% glucose. Some of the transformants are also plated on X-gal plates containing 2% of other sugars.

The *E. coli-Streptococcus* integration plasmid pSF151 [46] is digested with *Pst*I and *Pvu*II, and the larger fragment is ligated to the *Pst*I-*Bal*I fragment of pPP207 [49] that contains a promoterless *lacZ* gene (Figure 1). The resulting plasmid, pFP1, replicates in *E. coli* but not in streptococci, because of the absence of a streptococcal origin of replication. Additionally, it has a Km marker which can be expressed in both *E. coli* and streptococci (respectively, at the concentrations of 50 μg/ml, and 250 to 500 μg/ml). pFP1 contains no *bla* gene.

A library of GS-S DNA is constructed in plasmid pFP1 and is maintained in *E. coli*. DNA for strain GS-5 is completely digested with *Sau*3A, and four different series of fragments are isolated from a 1.0% low-melting-point agarose gel. Fragments of class #1 range between 0.05 and 0.9 kb, of class #2 between 1.0 and 1.3 kb, of class #3 between 1.4 and 4.8 kb, and of the last class #4, between 4.9 and 18 kb. After purification (midiprep using a purification kit [Wizard, Promega]), the resulting fragments are ligated to the vector pFP1 previously linearized with *Bam*HI, and treated with calf intestinal phosphatase. *Sau3*A, T4 DNA ligase and calf intestinal phosphatase are from Boehringer Mannheim.[4] *Bam*HI is from Promega. This ligation positions the fragments upstream from the promoterless *lacZ* gene. The ligation mixture is introduced into *E. coli* DH5-α-competent cells by electroporation using a Bio-Rad[5] gene pulser, and transformants are selected on LB agar plates containing 50 μg Km per ml. Each colony is transferred to a well of a 96-well plate containing 150 μl of LB agar plus 15% glycerol, and grown overnight at 37 °C with gentle shaking. The *E. coli* stocks are stored at –70 °C. The numbers of recombinant plasmid obtained from the different size classes range from 216 to 288.

From each glycerol well 10 μl is used to inoculate 150 μl of fresh LB broth containing Km, and the culture is grown overnight at 37 °C with shaking, always in the 96-well plates. Pooled or single plasmid extractions are performed using Qiagen miniprep kits (QIAGEN).[6] Pools are made using 100-μl lots from 12 cultures.

Sequence determination and analysis
All the sequences are performed in both directions by automated sequencing (DNA Sequencing Facility, University of Pennsylvania). Comparison with protein sequences is by the BLAST program [3].

Two sets of primers, SCRA1-SCRA2 and FTF1-FTF2, are used to amplify, respectively, a 1758-bp fragment of the *scrA* gene and a 1420-bp fragment of the *ftf* gene. The sequence of the primers is: SCRA1 5′-GACTAGTCTGTTGTCTGATATCTTTGTA, SCRA2 5′-CGGGATCCCGGCTTTAATACTATTGTTACT, FTF15′-GACTAGTCAGCTTAATCTAATATGTGAAT, FTF2 5-′CGGGATCCCGTTCACCCACTTCAGATATT. The *scrA* primers

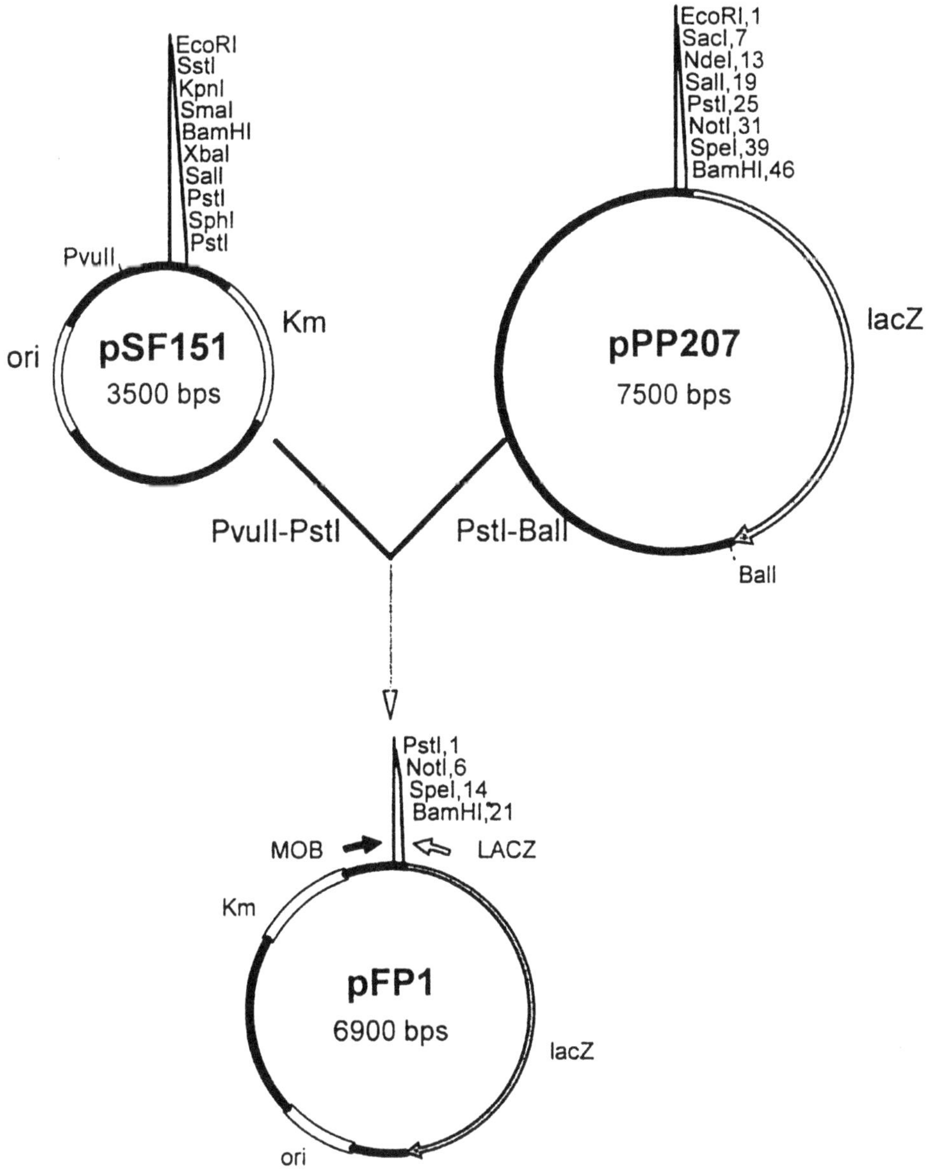

Figure 1. Plasmid pFP1 was constructed in two steps. The integrational vector pSF151 (kindly provided by Dr L. Tao, Oklahoma City, Oka.) was digested with *Pst*I and *Pvu*II, and ligated to the promoterless *lacZ* fragment (*Pst*I-*Bal*I) from pPP207 [49]. The smaller arrows flanking the multicloning site in pFP1 indicate the direction of the two oligonucleotides (MOB and LACZ) used both for PCR and sequencing. Abbreviations: *Km*, kanamycin resistance; *Ori, E. coli* origin of replication.

are designed on the sequence published by Sato et al. [38], and the *ftf* primers are designed on the sequence published by Shiroza et al. [40]. The *Spe*I restriction site is added to the 5′ of the sequence of the primers SCRA1 and FTF1, while the *Bam*HI restriction site is added to the 5′ of SCRA2 and FTF2 primers. These restriction sites are used to clone the *scrA* and the *fif* genes upstream of the *lacZ* promoterless fragment in pFP1.

For o-nitrophenyl-β-D-galactopyranoside (ONPG) analysis, cells of GS-5 or of its L7 derivative are grown in BTR containing 2% glucose. The next day the cells are washed once with 1 volume of BTR containing no sugars and then diluted 1/10 in the same medium. The cells are grown in BTR containing 2% sucrose, 2% glucose, or 2% maltose. One ml of the culture is harvested at an OD at 675 nm of ~0.8. (or at various other OD when specified) and used for the ONPG analysis. Assays for β-galactosidase activities are performed on toluene-permeabilized cells, and ONPG is used as the substrate as described by Nicholson and Setlow [28]. One ml of the culture is transferred to a 1.5-ml Eppendorf tube and centrifuged for 5 min at 4,000 rpm. To the pellet are added sequentially 1 ml of Z buffer (60 mM Na_2HPO_4, 40 mM NaH_2PO_4, 1 mM $MgSO_4$, 10 mM KC1, 50 mM β-mercaptoethanol) and one drop (about 10 µl) of toluene. The tubes are then vortexed for 15 s and are left 10 min at room temperature. The tubes are then opened and transferred to a 28 °C waterbath. After 10 min 200 µl of the ONPG solution (4 mg/ml ONPG in Z buffer) is added to the tubes. The time is recorded at this point. When the solution has become clearly yellow, the reaction is stopped by adding 300 µl of 1 M Na_2CO_3, and the time is recorded again. The cells are spun down at 13,000 rpm for 4 min, and the supernatant is transferred to a cuvette and the OD_{420} measured with a spectrophotometer. The β-galactosidase activity is calculated according to the formula: Miller units = $A_{420} \times 1000/\text{reaction time (min)} \times OD_{675}$.

4. Results

S. mutans GS-5 was transformed with 94 distinct pooled-plasmid preparations (12 plasmids per pool), with selection for Km^r on TH agar containing 2% sucrose. Transformants were screened for β-galactosidase activity. Of the 94 plasmid pools tested, 64 gave some transformants with detectable β-galactosidase activity. All these transformants were then patched onto three TH-X-Gal agars containing, respectively, 2% sucrose, 2% glucose, and no added sugar. From 23 pools, transformants were obtained that exhibited a potentially sugar-regulated phenotype, displaying substantially higher β-galactosidase activity on TH agar with no added sugar than on TH agar with 2% sucrose or 2% glucose. Individual plasmid preparations of each of the 12 members of the 23 pools were then used to transform *S. mutans*, and transformants were again selected on TH-Km plates containing X-/Gal and 2% sucrose. Seventy-seven of the 276 plasmids were found to give transformants that exhibited a possible sugar-regulated phenotype. Thus, several plasmid pools contained more than one plasmid that led to a possible sucrose-regulated phenotype. In a first analysis, 26 of 77 were chosen for sequencing. The inserts analyzed ranged from 112 to 1600 base pairs.

To begin characterization of the inserts fused to the *lacZ* reporter gene, an oligonucleotide was used to obtain single-stranded sequence, using a primer that hybridized to *lacZ*, 5′-GTTGGGTAACGCCAGGG. In addition, single-stranded sequence was obtained for the other end of the insert, using a primer that hybridized to *mob*, 5′-CACGACCAGTCTTTCTGCA (Figure 1, arrows). Sequences were performed using the entire plasmids prepared from *E. coli* and also polymerase chain reaction (PCR) products obtained from the plasmids. Appropriate oligonucleotides were synthesized and used to sequence regions of interest (based on suggested open reading frames [ORFs]) in the opposite direction. Potential ORFs were analyzed using the BLAST program for proteins. The results of this analysis are shown in Table 1. All plasmids yielded ORFs showing similarities to proteins or ORFs in GenBank. In most plasmids analyzed, the cloned inserts included portions of single genes. In four instances, the amino terminus of one putative gene product or an entire gene and the carboxy terminus of another (*dagK* and *sgp*, *tuf* and *tpi*, *acmA* and *dac*, and *glnB* and *serC*) were present in the cloned DNA. In one instance the insert contained the amino domains of divergently transcribed genes (*ORF1* and *pepX*). In each case, β-galactosidase expression may indicate expression of the gene closer to, and in the same direction as, *lacZ*; there are no data to indicate if the more distal gene is coregulated with the more proximal gene. Two clones were recovered coding for different portions of the same gene, *mfd*.

Integration of the plasmids into the *S. mutans* chromosome is presumed to be by a Campbell-like mechanism. This presumption was confirmed in the case of transformants obtained with the plasmids pFP11 and pFP19, which contained, respectively, a portion of *gtfB/C* and a portion of *fruA* (see Table 1). *gtfB/C* analysis by Southern blots of appropriately restricted DNA indicated that the plasmid pFP11 had indeed integrated into the region of homology on the chromosome by a single cross-over (Campbell-like recombination) (data not shown). In the case of *fruA*, two primers were designed based on the sequence of Burne et al. [7]: FRUA1, 5′-CACTGGAATAACCAATGGT; and FRUA2, 5′-GACTTGTCCTAACAAGACT, both external to the region included in the plasmid. Analysis by PCR using two sets of primers, FRUA1-LACZ and FRUA2-MOB, confirmed the integration of the plasmid into the region of homology on the chromosome.

In 13 of the 26 clones the insert was entirely internal to the putative gene. In these cases the gene would be insertionally inactivated by Campbell-like integration. As a result, it can be inferred that the following genes are not essential: *adhE*, *bglH*, *fruA*, *gcp*, *gtfB/C*, *gtfD*, *leuD*, *ilvG*, *mfd*, *sapR*, *uvrA*, and *yieG*; it is already known that *fruA*, *gtfB/C*, and *gtfD* are not essential genes for *S. mutans*.

Data obtained from plates containing TH and 2% sucrose or 2% glucose were difficult to interpret, presumably due to the presence of 0.2% glucose added to the TH medium. Consequently, we assayed the transformants on BTR X-gal plates containing 2% glucose or 2% sucrose [39]. The results of this screening were easier to interpret. Specifically, one *purB* transformant gave white colonies on glucose-

Table 1. *S. mutans* sequences cloned in pFP1

Clone	Accession number	Homologous gene and function (reference)	% identity/ similarity[a]	Number of amino acids compared
pFP1-1	U75476	*purB*; adenylosuccinate lyase (*Bacillus subtilis*) [10, 11, 25]	75/86	53[b]
pFP1-2	U75471	*livG; liv-I* from *E. coli* [2, 43] (high similarity also with the *livG* gene product from *Salmonella typhimurium*, with the *lktB* gene product of *Pasteurella* with the *gltL* gene product of *E. coli*, and with the *rtx* gene product of *Actinobacillus pleuropneumoniae*)	54/68	90[c]
pFP1-3	U75473	*mfd*; transcription repair coupling factor (*B. subtilis*) [30]	64/83	98[c]
pFP1-4	U75474	*mfd*; transcription repair coupling factor (*B. subtilis*) [30]	73/87	72[c]
pFP1-5	U75472	*bglH*; β-glucosidase (*B. subtilis*) (D. Le Coq et al., direct submission to the EMBL/GenBank/DDBJ data banks, 1994)	58/76	++[c]
pFP1-6[d]	U78605[d]	*ORF1*; hypot. ORFI in *pepX* 5′ region (*Lactococcus lactis*) [24, 25]	74/86	78[b]
pFP1-6[e]	U78605	*pepX*; X-prolyl-dipeptidyl-peptidase (*L. lactis*, *L. lactis* subsp. *cremoris*) [26, 27]	55/70	127[b]
pFP1-7[e]	U78599	*acmA*, N-acetylmuramidase (*Lactococcus lactis*) [5] (high similarity also with the autolysin of *Enterococcus faecalis* and *Enterococcus hirae*)	34/60	164[b]
pFP1-7[d]	U78599[d]	*dac*; D-D-carboxypeptidase (*E. faecalis*) (S. Evers & P. Courvalin, unpublished)	38/65	136[f]
pFP1-8	U78602	*ycaJ*; hypot. Protein HI1590 (*Haemophilus influenzae*) [12] (W. Oswald, Thesis, Justus Liebig Universitatet, Frankfurt, Germany 1994)	41/64	211[b]
pFP1-9	U75478	*adhE*; alcohol dehydrogenase (*E. coli*) [16] (high similarity also with *adhE* gene product of *Entoemeba histolytica*, *Clostridium acetobutylicum*, *Citrobacter freundii*, *Saccharomyces cerevisiae*, *Zymomonas mobilis*, *Schizosaccharomyces pombe*, and *Bacillus methanolicus*)	44/73	52[e]
pFP1-10	U75477	*oppF*; oligopeptide transport ATP-binding protein (*B. subtilis*) [33, 37]	53/71	109[b]
pFP1-11	M17361	*gtfB/C*; glucosyltransferase B and C (*S. mutans*) [41, 47]	100/100	46[c]
pFP1-12	U75475	*lauD*; hypothetical ABA transporter in *leuD* 3′ region (*L. lactis*) [15]	65/82	58[c]
pFP1-13	U78604	Membrane protein; (*Streptococcus pneumoniae*) (J. Krauss & R. Hakenbeck, unpublished sequence in GenBank)	34/62	129[c]
pFP1-14[c]		*dagK*; diacylglycerol kinase (*S. mutans*) [48]	99/100	137[g]
pFP1-14[e]	L03428	*sgp*; GTP-binding protein ERA homolog (*S. mutans*) [48}	100/100	128[b]
pFP1-15	U78601	*gcp*; sialoglycoprotease (*P. haemolitica*) [1] (M. D. Potter & R. Y. C. Lo, unpublished)	47/73	157[c]
pFP1-16	M29296	*gtfD*; glucosyltransferase D (*S. mutans*) [19]	100/100	146[c]
pFP1-17[3]	U78606	*glnB*; nitrogen regulatory protein (*Azospirillum brasilense*) [9]	63/77	57[f]
pFP1-17[e]	U78606[d]	*serC*; phosphoserine aminotransferase (*Yersinia enterocolitica*) [29]	57/75	84[b]
pFP1-18	U75479	*uvrA*; excinuclease ABC subunit A (*Mycoplasma genitalium*) [13] (high similarity also with the *uvrA* gene product of *Mycoplasma capricolum*, *Micrococcus luteus*, *Neisseria gonorrhoeae*, *E. Coli*, *S. typhimurium*, *H. influenzae*, and *Pseudomonas aeruginosa*)	69/81	134[c]
pFP1-19	L03358	*fruA*; exo-β-D-fructosidase (*S. mutans*) [7]	99/99	79[c]

Table 1. *(Continued)*

Clone	Accession number	Homologous gene and function (reference)	% identity/ similarity[a]	Number of amino acids compared
pFP1-20	U75470	*pksC*; polyketide synthase (*Mycobacterium leprae*) (K. Robinson, direct submission to the EMBL/GenBank/DDBJ data banks, 1994; D. R. Smith, unpublished data) (high similarity also with the *eryA* gene product of *Saccharopolyspora erythracea* and with the *pksC* gene product from *B. subtilis, Streptococcus hygroscopicus*, and *Mycobacterium tuberculosis*)	49/67	110[b]
pFP1-21	U78603	*yieG*; hypothetical 46.9-kilodalton protein and *tnaB-bglB* intergenic region (*E. coli*) [6]	31/52	74[c]
pFP1-22	U75480	*lgt*; prolipoprotein diacylglyceryl transferase (*Staphylococcus aureus*) [35] (high similarity also with the *lgt* gene product of *M. genitalium, S. typhimurium, E. coli*, and *H. influenzae*)	47/65	187f
pFP1-23	U78600	*ptsG*; glucose permease (*S. pneumoniae*) [31] (high similarity also with *ptsG* gene product of *B. subtilis* and *B. stearothermophilus* and the *ptsA* gene product of *E. coli, Staphylococcus carnosus* and *Klebsiella pneumoniae*)	38/55	415[c]
pFP1-24	U78607	P54 protein precursor (*E. faecium*) [14] (high similarity also with the *usp45* gene product of *L. lactis*)	39/65	182[b]
pFP1-25[e]	U75482[d]	*tpi*; triosephosphate isomerase (*L. lactis*) [8] (high similarity also with the *tpi* gene product from *Bacillus megaterium, B. subtilis, B. stearothermophilus, E. coli*, and *H. influenzae*)	68/85	106[6]
pFP1-25[e]	U75481	*tuf*; elongation factor-Tu (*Streptococcus oralis*) [24]	92/99	148[f]
pFP1-26	U75483	*sapR*; response regulator in the sakacin A gene cluster [4] (*Lactobacillus sake*)	32/52	170[c]

[a] Identical amino acids plus conservative substitutions.
[b] Insert contains the amino terminus.
[c] Insert contains the central portion.
[d] Sequence upstream of *lacZ* (see Figure 1).
[e] Insert contains two genes.
[f] Insert contains the carboxy terminus.
[g] Insert contains the entire putative gene.

and sucrose-containing BTR. Transformants from pFP-2 to pFP-20 had blue colonies on sucrose, but white colonies on glucose-containing X-gal plates. Lastly, six transformants had blue colonies on glucose- and on sucrose-containing X-gal plates. Among the colonies that were blue on sucrose- but white on glucose-containing X-gal plates were transformants with a lacZ insert in the gtfB/C and in the *fruA* genes (pFP-11 and pFP-19 in Table 1).

As an additional control, we cloned the *ftf* gene and the *scrA* gene in the pFP1 vector and transformed *S. mutans*, selecting for Campbell-like insertions. Both transformants gave blue colonies in sucrose- and white colonies in glucose-containing media. A transformant containing the region downstream of the 3′ terminus of the *scrA* gene gave white colonies in both sucrose- and glucose-containing media (data not shown).

In order to analyze our results in greater detail, we grew each transformant of *S. mutans* GS-5 L7 in BTR medium to an OD 675 of 0.8. BTR medium containing either 2% or 0.2% sucrose, or 2.0 or 0.2% glucose, and the specific activity of β-galactosidase was determined and is expressed as Miller units. Table 2 provides the results obtained using the L7 derivative of *S. mutans* GS-5. However, we found no difference qualitatively between results obtained with the GS-5 parent, or with the L7 derivative, and media containing 2.0% or 0.2% sugar.

The *purB* strain which had white colonies on X-gal plates supplemented with sucrose or with glucose had low levels of β-galactosidase (Table 2). The six transformants (numbers 21–26) that gave blue colonies on both glucose and on sucrose had specific activities of β-galactosidase that ranged from less than 16 to 524 Miller units in both sugars (Table 2). However, the most interesting results were obtained from transformants that had blue colonies in sucrose

Table 2. β-galactosidase activity of *S. mutans* transformants

GS-517::pFP1 derivatives	β-galactosidase specific activity (in Miller units)		
	2% sucrose	2% glucose	2% maltose
1. *purB*	27	25.5	35.5
2. *livG*	50	47.5	44.6
3. *mfd*	36.9	25.2	27
4. *mfd*	ND	ND	ND
5. *bglH*	25.9	35.4	25
6. *pepX*	23.6	22.5	20.7
7. *dac*	95.7	95.7	65.5
8. *ycaJ*	42	39.8	31.7
9. *adhE*	991	959	143
10. *oppF*	69.5	65	59.2
11. *gtfB/C*	19.5	18.9	19.7
12. *leuD*	58.7	53.5	38.8
13. Membrane protein	52.6	41.9	35.8
14. *sgp*[a]	69.7	58.4	39.5
15. *gcp*	0.53	0.72	0.56
16. *gtfD*	45	55.7	ND
17. *serC*[a]	85.3	85	ND
18. *uvrA*	37.3	39	ND
19. *fruA*	148.3	9.4	5.4
20. *pksC*	43	37	ND
21. *yieG*	88.5	106	136
22. *lgt*	99	91	ND
23. *ptsG*	157.6	210.7	ND
24. *P54*	453.3	477.7	ND
25. *tpi*[a]	480.3	524	563.4
26. *sapR*	15.7	17.4	20
27. *scrA*	20.8	16.4	5.7
28. *ftf*	13	12.4	6.8

ND: not done.

and white colonies in glucose. In liquid media and in late exponential growth at an Abs 675 of about 0.8, these colonies gave specific activities in sucrose from 0.53 to 991 units of β-galactosidase. However, the transformant containing *fruA*, had a distinctly higher specific activity for β-galactosidase in sucrose than in glucose (Table 2).

We examined the specific activity of β-galactosidase at various times after resuspension of selected cultures in sucrose, glucose or other sugars. These data are shown in Figure 2A, 2B, 2C and 2D. Figure 2A and 2B shows that in these conditions the specific activities of *lacZ* for plasmids inserted at the promoter for the *gtfB/C*, are essentially the same in glucose, sucrose, or maltose (Figure 2A). For the *ftf* gene there no increase in β-galactosidase specific activity in maltose (Figure 2B). The *scrA* gene is marginally more active in sucrose than in glucose. Both sucrose and glucose increase the enzyme-specific activity (Figure 2C). However, the most dramatic effect of sucrose is on the *fruA* gene (Figure 2D), where the *lacZ* specific activity is much greater in sucrose than in glucose, and glucose does not increase the specific activity of the enzyme. In each case shown in Figure 2, the specific activity of *lacZ* is affected by the time of the analysis.

5. Discussion

We report here the construction and utilization of a promoter-probe, integrative-plasmid vector for *S. mutans*. The vector has enabled us to identify 31 putative *S. mutans* genes, and the study has focused on those genes whose expression was regulated by sugars. The extensive protein data bases now available have made possible a reasonable assessment of function for most of those genes, even though complete sequences were not obtained for the genes. Five of the ORFs identified had 99% to 100% identity to previously report *S. mutans* proteins (Table 1), and we presume that we have reisolated the previously cloned genes: *fruA*, *gtfB/C*, *gtfD*, *sgp*, and *dagK*, encoding, respectively, exo-β-D-fructosidase [7], glucosyltransferase B/C [47], glucosyltransferase D [19], GTP-binding protein ERA homolog [48], and diacylglycerol kinase [48]. Twenty of the ORFs had extensive similarity to proteins encoded by genes that had been identified and characterized for other species. A further five ORFs had similarity to ORFs of unknown function from other species. Thus, the use of pFP1 has provided a rapid way to identify, at least tentatively, several new genes for *S. mutans*.

The similarities in sequence of the *S. mutans* inserts were particularly apparent for proteins of gram-positive species. Five of the ORFs showed similarities to proteins encoded by *L. lactis*: an N-acetylmuramidase encoded by *acmA* [5], a hypothetical ABC transporter encoded in the *leuD* 3′ region [15], a triosephosphate isomerase encoded by *tpi* [8], *ORF1* encoded in the *pepX* operon of *L. lactis* [26, 27], as well as the peptidase encoded by the *pepX* gene of *L. lactis* (and the *pepX* of *Streptococcus cremoris*). One ORF gave 92% identity to the *tuf* gene from *S. oralis*. As might be expected, the putative Tuf Protein showed strong similarity to the proteins encoded by *tuf* genes from a range of species; for example, 92% identity over 148 amino acids to the *tuf* gene product from *S. oralis* [24]; 81% to 86% identity to the *tuf* gene products from *E. coli*, *S. typhimurium*, *M. leprae*, *M. tuberculosis*, *H. influenzae*, *Pseudomonas cepacia*, *Thermus aquaticus*, and many others.

Four ORFs had 53% to 75% identity to products of genes identified previously in *B. subtilis*. The genes were: *bglH*, which encodes β-glucosidase (Le Coq et al., direct submission to the EMBL/GenBank, DDBJ data banks, 1994); *mfd*, which encodes a transcription-repair coupling factor necessary for strand-specific repair [30]; *oppF*, which encodes an oligopeptide transport ATP-binding protein [33];

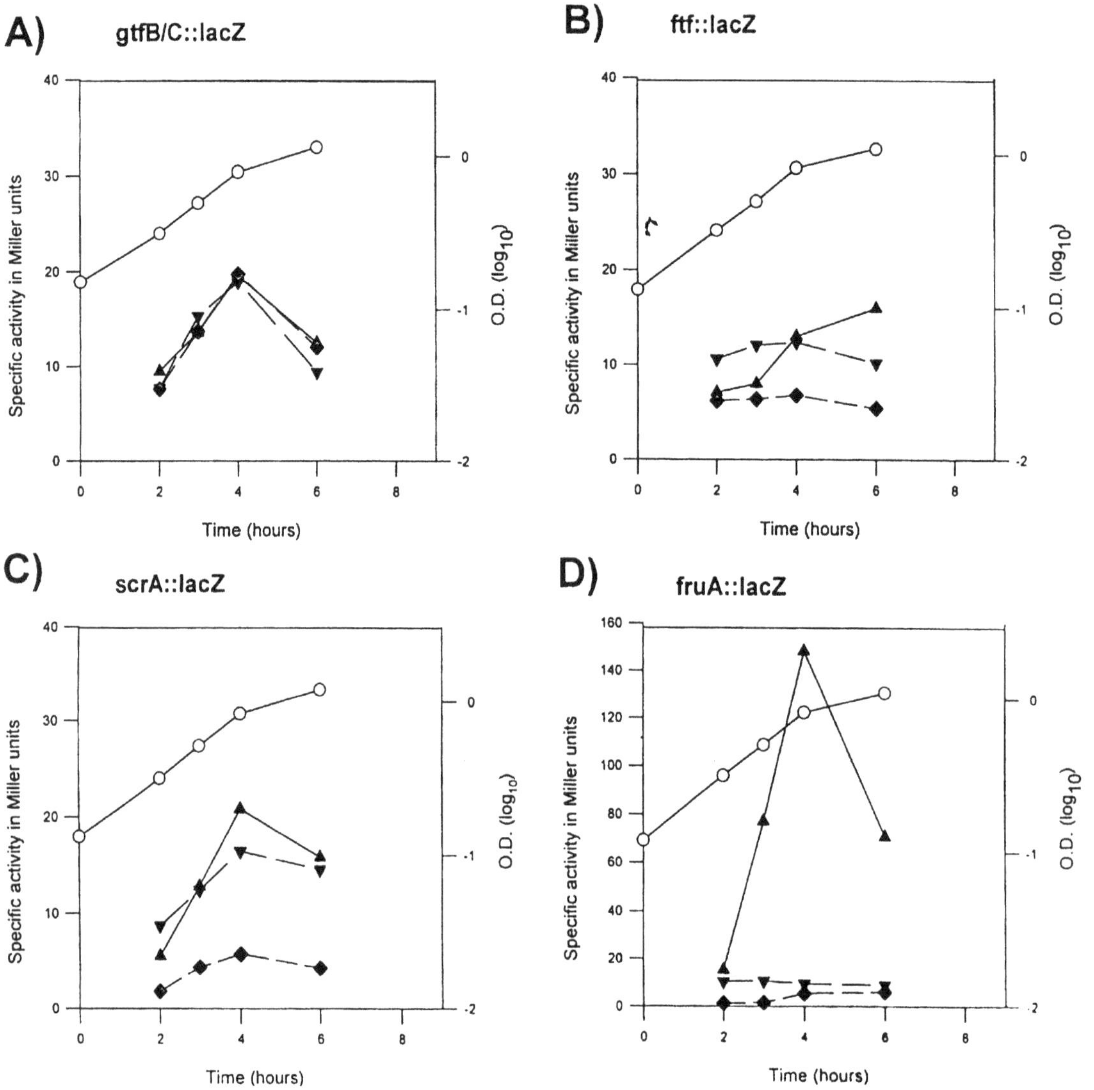

Figure 2. Specific activity of β-galactosidase at various times after growth in sucrose (▲), glucose (▼) or maltose (◆). **A.** *gtfB/C::lacZ.* **B.** *ftf::lacZ.* **C.** *scrA::lacA.* **D.** *fruA::lacZ* (note that the scale for *fruA::lacZ* differs from that for the others). (○) Represents growth in O.D. units.

purB, which is one of the 12 genes belonging to the *pur* operon encoding purine biosynthesis enzymes [25]. The *S. mutans* insert, sharing high similarity with the *oppF* gene product of *B. subtilis*, also gave a high degree of identity to ATP-binding proteins of other organisms, such as *L. lactis*, *S. pneumoniae*, *S. typhimurium*, *H. influenzae*, *E. coli* and *Bordetella pertussis*.

The *S. mutans* clones were isolated as showing sugar-responsive gene expression when grown on agar. some of the genes identified seem to be clearly involved in sugar metabolism (*adhE*, *bglH*, *fruA*, *gtfB/C*, *gtfD*, and *tpi*). A putative glucose permease encoded by the ptsG gene was found whose expression appeared relatively insensitive to sugars. This protein is part of the PTS system, and it has been identified in *E. coli* as an Enzyme III-independent Enzyme II^{Glc}; as in *E. coli* [summarized in ref. 45], the *ptsG* in *S. mutans* maps to a chromosomal locus far from the *pts* operon (M.G. Cappiello & F. Peruzzi, unpublished observations).

For many of the clones identified, it is difficult to find a relation between their function and sugar metabolism. Among those genes, five belong to the family of the ABC (ATP-binding cassette) transporters: *leuD*, *livG*, *oppF*, *sapR*, and *uvrA*. The substrates for the ABC transporters include oligopeptides, proteins, sugars, amino acids, and polysaccharides. With the exception of the *uvrA* gene product that is cytoplasmic, the ABC proteins are usually transmembrane proteins, and they are involved in many biological processes in bacteria [18]. Many of the ABC transporters are known to be associated with membrane functions and are regulated in response to specific growth conditions [18]; it is possible to speculate that the addition of sugars or sugar-alcohols affects those growth conditions.

The survey of the effects of glucose and sucrose on gene expression relied on plate assays of β-galactosidase. It seems interesting that our initial screening identified *gtfB/C* and *fruA*. The latter also appears to be clearly induced by sucrose, but not by glucose in liquid culture. Previous reports suggested that this gene appeared to be induced by sucrose [7]. In contrast to the results with *fruA*, our screen of several other genes, including *gtfB/C*, *scrA* and *ftf* did not exhibit a striking induction by sucrose or glucose in liquid culture. One possibility for the difference in the results obtained with plates and with liquid cultures is that both glucose and sucrose induced β-galactosidase to approximately the same level during the exponential phase of growth in liquid culture. Such a possibility was reported by Sato et al. for the scrA::lacZ translational fusions [39]. Hudson and Curtiss reported little difference for *ftf* or *gtfB/C* expression in transcriptional promoters fusion with the CAT gene. Our plate assay would be in the stationary phase, since the colonies are examined at 24 hours, and so may indicate gene expression during stationary phase. Nevertheless, the plate screening did identify four of the genes thought to be regulated by sucrose (Table 2 and Figure 2).

Our library of *S. mutans* DNA in the integrational plasmid pFP1 was screened for genes regulated by both glucose and sucrose. The analysis was not exhaustive. Nevertheless, the use of pFP1 described here clearly provide a rapid way to identfy and characterize *S. mutans* genes, and to identify genes that respond to particular environmental signals. The genes identified here substantially add to the number of known *S. mutans* genes, and will greatly expand the *S. mutans* genetic and physical map. They suggest that the effects of sugars on *S. mutans* metabolism may be greater than previously appreciated.

Acknowledgments

This work was supported by Public Health Service grant DE-08942 from the National Institutes of Health.

We thank Dr L. Tao for providing pSF151, Dr R. Burne for providing *fruA* clones, Dr G. Shockman, Dr O. Massidda and Dr M. G. Cappiello for their helpful advice, Dr M. Prisco for his assistance in the experiments for Table 2, and G. Harvey for help in preparation of the manuscript.

Notes on suppliers

1. Difco Laboratories, Detroit, MI, USA
2. Sigma Chemical Co., St. Louis, MO, USA
3. Promega Corp., Madison, WI, USA
4. Boehringer Mannheim, Indianapolis, IN, USA
5. Bio-Rad Laboratories, Richmond, CA, USA
6. QIAGEN, Inc., Chartsworth, CA, USA

References

1. Abdullah KM, Lo RYC, Mellors A (1991). Cloning, nucleotide sequence, and expression of the *Pasteurella haemolytica* A1 glycoprotease gene. J Bacteriol 173: 5597–5603.
2. Adams MD, Wagner LM, Graddis JJ, Landick R, Antonucci TK, Gibson AL, Oxender DL (1990). Nucleotide sequence and genetic characterization reveal six essential genes for the Liv-I and LS transport systems of *Escherichia coli*. J Biol Chem 265: 11436–1443.
3. Atshul SF, Gish W, Myers EW, Lipman DJ (1990). Basic local alignment tool. J Mol Biol 215: 403–410.
4. Axelsson L, Holck A (1995). The genes involved in production of and immunity to sakacin A, a bacteriocin from *Lactobacillus sake* Lb706. J Bacteriol 177: 2125–2137.
5. Buist G, Kok J, Leenhouts KJ, Dabrowska M, Venema G, Haandrikman AJ (1995). Molecular cloning and nucleotide sequence of the gene encoding the major peptidoglycan hydrolase of *Lactococcus lactis*, a muramidase needed for cell separation. J Bacteriol 177: 1554–1563.
6. Burland V, Plunkett G III, Daniels DL, Blattner FR (1993). DNA sequence and analysis of 136 kilobases of the *Escherichia coli* genome: Organizational symmetry around the origin of replication. Genomics 16: 551–561.
7. Burne RA, Penders JE (1992). Characterization of the *Streptococcus mutans* GS-5 *fruA* gene encoding exo-beta-D-fructosidase. Infect Immun 60: 4621–4632.
8. Cancilla MR, Davidson BE, Hillier AJ, Nguyen NY, Thompson J (1995). The *Lactococcus lactis* triosephosphate isomerase gene, *tpi*, is monocistronic. Microbiology 141: 229–238.
9. de Zamaroczy M, Delorme F, Elmerich C (1990). Characterization of three different nitrogen-regulated promoter regions for the expression of *glnB* and *glnA* in *Azospirillum brasilense*. Mol Gen Genet 224: 421–430.
10. Ebbole DJ, Zalkin H (1987). Cloning and characterization of a 12-gene cluster from *Bacillus subtilis* encoding nine enzymes for de novo purine nucleotide synthesis. J Biol Chem 262: 8274–8287.
11. Ebbole DJ, Zalkin H (1988). Detection of *pur* operon-attenuated mRNA and accumulated degradation intermediates in *Bacillus subtilis*. J Biol Chem 263: 10894–10902.
12. Fleischmann RD, Adams MD, White JM, McKenney K, Sutton G, FitzHugh W, Fields C, Gocayne JD, Scott J, Shirley R, Liu L-I, Glodek A, Kelley JM, Weidman JF, Phillips CA, Spriggs T, Hedblom E,

Cotton MD, Utterback TR, Hanna MC, Nguyen DT, Saudek DM, Brandon RC, Fine LD, Fritchman JL, Fuhrmann JL, Geoghagen NSM, Gnehm CL, McDonald LA, Small KV, Fraser CM, Smith HO, Venter JC (1995). Whole-genome random sequencing and assembly of *Haemophilus influenzae* Rd. Science 269: 496–512.

13. Fraser CM, Gocayne JD, White O, Adams MD, Clayton RA, Fleischmann RD, Bult CJ, Kerlavage AR, Sutton G, Kelley JM, Fritchman JL, Weidman JF, Small KV, Sandusky M, Fuhrmann J, Nguyen D, Utterback TR, Saudek DM, Phillips CA, Merrick JM, Tomb J-F, Dougherty BA, Bott KF, Hu P-C, Lucier TS, Peterson SN, Smith HO, Hutchison CA III, Venter JC (1994). The minimal gene complement of *Mycoplasma genitalium*. Science 270: 397–403.
14. Fuerst P, Moesch HU, Solioz M (1989). A protein of unusual composition from *Enterococcus faecium*. Nucleic Acids Res 17: 6724.
15. Godon J-J, Chopin M-C, Ehrlich SD (1992). Branched-chain amino acid biosynthesis genes in *Lactococcus* subsp. *lactis*. J Bacteriol 174: 6580–6589.
16. Goodlove PE, Cunningham PR, Parker J, Clark DP (1989). Cloning and sequence analysis of the fermentative alcohol-dehydrogenase-encoding gene of *Escherichia coli*. Gene 85: 209–214.
17. Gutierrez JA, Crowley PJ, Brown DP, Hillman JD, Youngman P, Bleiweis AS (1996). Insertional mutagenesis and recovery of interrupted genes of *Streptococcus mutans* by using transposon Tn*917*: Preliminary characterization of mutants displaying acid sensitivity and nutritional requirements. J Bacteriol 178: 4166–4175.
18. Higgins CF, Hiles ID, Salmond GPC, Gill DR, Downie JA, Evans IJ, Holland IB, Gray L, Buckel SD, Bell AW, Hermodson MA (1986). A family of related ATP-binding subunits coupled to many distinct biological processes in bacteria. Nature 323: 448–450.
19. Honda O, Kato C, Kuramitsu HK (1990). Nucleotide sequence of the *Streptococcus mutans gtfD* gene encoding the glucosyltransferase-S enzyme. J Gen Microbiol 136: 2099–2150.
20. Hudson MC, Curtiss R III (1990). Regulation of expression of *Streptococcus mutans* genes important for virulence. Infect Immun 58: 464–470.
21. Kiska DL, Macrina F (1994). Genetic regulation of fructosyltransferase in *Streptococcus mutans*. Infect Immun 62: 1241–1251.
22. Lane MA, Bayles KW, Yasbin RE (1991). Identification and initial characterization of glucose-repressible promoters of *Streptococcus mutans*. Gene 100: 225–229.
23. Loeshe WJ (1986). Role of *Streptococcus mutans* in human dental decay. Microbiol Rev 50: 353–380.
24. Ludwig W, Weizenegger M, Beltz D, Leidel E, Lenz T, Ludvigsen A, Moellenhoff D, Wenzig P, Schleifer KH (1990). Complete nucleotide sequences of seven eubacterial genes coding for the elongation factor Tu: Functional, structural and phylogenetic evaluations. Arch Microbiol 153: 241–247.
25. Makaroff CA, Zalkin H, Switzer RL, Vollmer SJ (1983). Cloning of the *Bacillus subtilis* glutamine phosphoribosylpyrophosphate amidotransferase gene in *Escherichia coli*: nucleotide sequence determination and properties of the plasmid-encoded enzyme. J Biol Chem 258: 10586–10593.
26. Mayo B, Kok J, Venema K, Bockelmann W, Teuber M, Reinke H, Venema G (1991). Molecular cloning and sequence analysis of the X-prolyldipeptidylaminopeptidase gene from *Lactococcus lactis* subsp. *cremoris*. Appl Environ Microbiol 57: 38–44.
27. Nardi M, Chopin MC, Chopin A, Cals MM, Gripon JC (1991). Cloning and DNA sequence analysis of an X-prolyl dipeptidyl aminopeptidase gene from *Lactococcus lactis* subp. *lactis* NCDO763. Appl Environ Microbiol 57: 45–50.
28. Nicholson WL, Setlow P (1990). Sporulation, germination and outgrowth. In: Harwood CR, Cutting SM (eds), Molecular biological methods for Bacillus, pp 391–429. West Sussex, England: John Wiley and Sons, Ltd.
29. O'Gaora P, Maskell D, Coleman D, Cafferkey M, Dougan D (1989). Cloning and characterization of the *serC* and *aroA* genes of *Yersinia enterocolitica*, and construction of an *aroA* mutant. Gene 84: 23–30.
30. Ogasawara N, Nakai S, Yoshikawa H (1994). Systematic sequencing of the 180 kilobase region of the *Bacillus subtilis* chromosome containing the replication origin. DNA Res 1: 1–14.
31. Pearce BJ, Yin YB, Masure HR (1993). Genetic identification of exported proteins in *Streptococcus pneumoniae*. Mol Microbiol 9: 1037–1050.
32. Perego M (1993). Integrational vectors for genetic manipulation in *Bacillus subtilis*. In: Sonenshein AL, Losick R, Hoch JA (eds), *Bacillus subtilis* and other gram-positive bacteria, pp 615–624. Washington, DC: American Society for Microbiology.
33. Perego M, Higgins CF, Pearce SR, Gallagher MP, Hoch JA (1991). The oligopeptide transport system of *Bacillus subtilis* plays a role in the initiation of sporulation. Mol Microbiol 5: 173–185.
34. Perry D. Kuramitsu HK (1981). Genetic transformation of *Streptococcus mutans*. Infect Immun 32: 1295–1297.
35. Qi HK, Sankaran K, Gan K, Wu HC (1995). Structure-function relationship of bacterial prolipoprotein diacylglyceryl transferase: Functionally significant conserved regions. J Bacteriol 177: 6820–6824.
36. Reider JL, Macrina F (1976). Plasmid DNA isolation in *Streptococcus mutans*: Glycine-enhanced cell lysis. In: Stiles HM, Loesche WJ, O'Brien TC (eds), Proceedings: Microbial aspects of dental caries (a special supplement to Microbial Abstracts, Vol 3), pp 725–736. Washington, DC: Information Retrieval.
37. Rudner DZ, Ledeaux JR, Ireton K, Grossman AD (1991). The *spoOK* locus of *Bacillus subtilis* is homologous to the oligopeptide permease locus and is required for sporulation and competence. J Bacteriol 173: 1388–1398.
38. Sato Y, Poy F, Jacobson GR, Kuramitsu H (1989). Characterization and sequence analysis of the *scrA* gene encoded enzyme IIScr of the *Streptococcus mutans* phosphoenolpyruvate-dependent sucrose phosphotransferase system. J. Bacteriol 171: 263–271.
39. Sato Y, Yamamoto Y, Suzuki R, Kizaki H, Kuramitsu HK (1991). Construction of scrA::laxZ gene fusions to investigate regulation of the sucrose PTS of

Streptococcus mutans. FEMS Microbiol Letts 79: 339–346.

40. Shiroza T, Kuramitsu H (1988). Sequence analysis of the *Streptococcus mutans* fructo-syltransferase gene by flanking regions. J Bacteriol 170: 810–816.
41. Shiroza T, Ueda S, Kuramitsu HK (1987). Sequence analysis of the *gtfb* gene from *Streptococcus mutans*. J Bacteriol 169: 4263–4270.
42. Silhavy TJ, Berman ML, Enquist LW (1984). Experiments with gene fusions. Cold Spring Harbor, NY: Cold Spring Harbor Laboratory.
43. Sofia HJ, Burland V, Daniels DL, Plunkett G III, Blattner RF (1994). Analysis of the *Escherichia coli* genome, V. DNA sequence of the region from 76.0 to 81.5 minutes. Nucleic Acids Res 22: 2576–2586.
44. Sun S, Daneo-Moore L, Marri L, Hantman M, Shockman GD (1991). Genetic and physiological studies of variants of *Streptococcus mutans* GS-5 that produce nonmucoid colonies on sucrose-containing media. In: Dunny GM, Cleary PP, McKay LL (eds), Genetics and molecular biology of streptococci, lactococci, and enterococci, pp 256–260. Washington, DC: ASM Publications.
45. Sutrina SL, Reddy P, Saier MH Jr, Reizer J (1990). The glucose permease of *Bacillus subtilis* is a single polypeptide chain that functions to energize the sucrose permease. J Biol Chem 265: 18581–18589.
46. Tao L, Le Blanc DJ, Ferretti J (1992). Novel streptococcal-integration shuttle vectors for gene cloning and inactivation. Gene 120: 105–110.
47. Ueda S, Shiroza T, Kuramitsu HK (1988). Sequence analysis of the *gtfC* gene from *Streptococcus mutans* GS-5. Gene 69: 101–109.
48. Yamashita Y, Takehara T, Kuramitsu HK (1993). Molecular characterization of a *Streptococcus mutans* mutant altered in environmental stress responses. J Bacteriol 175: 6220–6228.
49. Yu J, Hederstedt L, Piggot PJ (1995). The cytochrome *bc* complex (menaquinone: cytochrome *c*reductase) in *Bacillus subtilis* has a nontraditional subunit organization. J Bacteriol 177: 6751–6760.

Address for correspondence: Dr Lolita Daneo-Moore, Department of Microbiology and Immunology, Temple University School of Medicine, 3400 N. Broad Street, Philadelphia, PA 19140, USA
Phone: 215-707-3229; Fax: 215-707-7788
E-mail: mooreld@astro.ocis.temple.edu

Methods in Cell Science **20**: 165–179 (1998)

Use of proteomics and PCR to elucidate changes in protein expression in oral streptococci

Robert G. Quivey, Jr[1, 2], Wendi L. Kuhnert[2] & Roberta C. Faustoferri[1]

[1] *Department of Dental Research and* [2]*Department of Microbiology and Immunology, University of Rochester, Rochester, New York, USA*

Abstract. Several techniques are available for the exploration of changes in gene expression and protein repertoire in bacteria due to changes in environment. We have been interested in the changes that *S. mutans*, a cariogenic oral micro organism, makes in response to growth at low pH values. Two approaches that we have utilized to evaluate alterations in gene expression are the polymerase chain reaction and two-dimensional gel electrophoresis. The use of degenerate primers in PCR amplification of genes by homology-based approaches is included as well as a method for the construction of genetic fusions by splice-overlap extension PCR (SOE-PCR). As genome nucleotide sequences become available for the streptococci, PCR will likely be used more in combination with protein databases, or proteomes, to explore the expression of genes under specific environmental conditions. Therefore, we include a description of two-dimensional gels prepared using protein extracts of cells grown in steady-state under different environmental conditions.

Key words: Dental caries, Oral streptococci, Polymerase chain reaction, Proteomics, *S. mutans*

Abbreviations: BHI = Brain heart infusion medium; bp = base pairs; kbp = kilobase pairs; PMSF = phenyl methyl sulfonyl fluoride; rpm = revolutions per minute; rcf = relative centrifugal force; TE = Tris EDTA buffer; TAE = Tris acetate buffer; X-Gal = 5-bromo-4-chloro-3-indolyl-β-D-galactopyranoside; A, C, T, G = common deoxyribonucleotides; R = A or G deoxyribonucleotides; Y = C or T deoxyribonucleotides; D = A or G or T deoxyribonucleotides; N = A or C or T or G deoxyribonucleotides

1. Introduction

As a member of the oral microflora, *Streptococcus mutans* experiences wide fluctuations in the availability of carbohydrates, which it requires for growth. As sugars become available and are subsequently fermented by *S. mutans* to acidic end-products, its microbiological niche, dental plaque, acidifies. If plaque pH values remain below a critical level, generally regarded as 5.5, the tooth enamel underlying plaque will begin to demineralize. Over prolonged periods, demineralization will progress to the level of caries formation. The ability of *S. mutans* to survive in these acidic environments is predicated by its ability to adapt to changes, not only in nutrient supply, but also to changes in environmental pH.

Previous results have shown that a central element of the microbe's aciduance is attributable to changes in proton-translocating ATPase activity. At low pH values, ATPase activity increases, apparently to remove protons thereby maintaining a ΔpH across the organism's membrane [7]. The dependence of the ATPase for the organism's ability to acid-adapt is apparently a key feature in many bacteria, if not a universal bacterial phenomena [2, 3, 11, 13, 20, 24].

Our research has focused on the mechanisms by which *S. mutans* adapts to life at low pH values; including the way in which it regulates its ATPase levels, as well as other proteins potentially involved with adaptation and the organism's pathogenic potential. The investigation of bacterial virulence can be approached using a variety of techniques. Since its development, the technique of polymerase chain reaction (PCR [35]) has been used in virtually all biological systems to obtain genes, gene segments, construct genetic mutations, and to form hybrid genes [21]. More recently, proteomics, the study of the repertoire of proteins expressed under a given set of environmental conditions, has been used to evaluate the impact of stress on bacterial gene expression [11, 42–44]. We will describe how the PCR technique has been used to begin the genetic dissection of acid-adaptation in *S. mutans* and how the methodology is likely to be coupled with proteomics in the future.

2. Materials

Use of the PCR technique requires familiarity with general molecular biological methods and pertinent apparatus. For the purpose of the presentation here, we refer the reader to well-established treatments of methods for propagating DNA fragments by cloning methodologies [1, 36]. Included here are specific descriptions of experimental conditions that we have found useful in the amplification and recovery of streptococcal DNA templates [30, 31, 40].

A. Computers and software
   1. Desktop computer (e.g., Power Macintosh) with the ability to access either a DEC VAX running GCG software or the National Center for Biotechnology Information database containing genetic sequences
   2. MacVector software for DNA and protein sequence analysis
   3. Genetics Computer Group software for DNA and protein sequence analysis
   4. NIH Image v. 1.6 for the Macintosh (available as a public domain resource on the Internet at http://rsb.info.nih.gov/nih-image/)
   5. Apple Color One Scanner
B. Equipment
   1. Gene pulser electroporation apparatus.[5]
   2. Homogenizer, model mini-Bead Beater.[6]
   3. Micro-centrifuge, refrigerated, Eppendorf.[30]
   4. Investigator 2-D gel electrophoresis system.[12]
   5. Adjustable pipettes: nos. P-2; P-20; P-200; P-1000.[26]
   6. Thermal cycler.[23]
   7. Thermal cycler, model OmniGene.[14]
   8. Centrifuge, refrigerated-preparative, no. RC5C.[28]
   9. Rotor locking screw for RC5C centrifuge, no. 13017.[28]
   10. SS-34 rotor for RC5C centrifuge.[28]
   11. GSA rotor for RC5C centrifuge.[28]
   12. Shaking water bath.[4]
   13. Gel apparatus, model Horizon 58.[18]
   14. Pipette-aid, Drummond, VWR. no. 53498-001.[30]
   15. UV transilluminator, model no. 3-3000.[31]
C. Culture medium
   - Brain heart infusion medium, no. 0037-07-0.[10]
   - Todd Hewitt medium, no. 0492-17-6.[10]
   - Luria-Bertani (LB) agar, as follows: in 1 l of de-ionized water, 10 g tryptone, no. 0123-07-5[10]; 5 g yeast extract, no. 0127-07-1[10]; 10 g NaCl, no. 3624-15[16]; 15 g bacteriological grade agar, no. 00-5778KA[3].
D. Reagents for polymerase chain reaction
   - Deoxyribonucleotides, no. 1-277-049.[7]
   - 10× BRL Taq DNA Polymerase Buffer, included with no. 18038-042.[18]
   - 50 mmol/l $MgCl_2$, included with no. 18038-042.[18]
   - custom oligodeoxyribonucleotides.[18]
   - Taq DNA polymerase, no. 18038-042.[18]
   - 10× modified PCR buffer: made in distilled, de-ionized water with 200 mmol/l $(NH_4)_2SO_4$; 1 mol/l Tris, pH, 8.3; 100 mmol/l $MgCl_2$; 5% NP-40, no. I-3021[27] and filter-sterilized.
   - mineral oil, no. 2705-01.[13]
   - *E. coli* strain DH10B, no. 18290-015.[18]
E. Solutions
   - 20 mmol/l Trizma base (Tris), no. T-1530[27], adjusted to a pH value of 8.0 with HCl.
   - 10 mmol/l Tris adjusted to a value of pH 8.0 with HCl.
   - Lysis buffer: 10 mmol/l Tris pH 8.0; 0.4 M sucrose, no. S-5016[27]; 50 mg/ml lysozyme, no. L-6876.[27]
   - TE buffer (10 mmol/l Trizma base, pH 8.0; 1 mmol/l EDTA, no. E-5134[27])
   - TAE buffer (0.04 mol/l Trizma base, adjusted to pH 8.0 with glacial acetic acid; 1 mmol/l EDTA)
   - Glacial acetic acid, no. JT-9508-33.[30]
   - Phenol, buffer saturated, no. 9712.[2]
   - $CHCl_3$: isoamyl alcohol (mixed in a 24:1 ratio), nos. 9180-22 and 9038-01, respectively.[16]
   - 95% EtOH, no. 9401-25.[16]
   - 5 mol/l NaCl, no. 3624-15.[16]
   - 10 mg/ml RNAse A in TE buffer, no. R-4875.[27]
   - 10% (w/v) sodium dodecyl sulfate, no. 15525-017.[18]
   - PMSF, 10 mg/ml stock solution, in 100% ethanol, no. P-7626.[27]
   - Bio-Rad dye-binding reagent, 500-0006.[5]
   - Quanti-Gold kit, no. QG-250.[11]
   - tracking dye: prepared in distilled, de-ionized water as 0.25% (w/v) bromphenol blue, no. 161-0404[5]; 0.25% (w/v) xylene cyanol FF, no 161-0423[5]; 40% sucrose, no. S-0389.[27]
   - ethidium bromide stock solution (10 mg/ml in water), no. E-8751.[27]
   - ampicillin, 100 mg/ml stock solution in water, no. A-9518.[27]
   - chloramphenicol, 20 mg/ml stock solution in 95% ethanol, no. C-0378.[27]
   - X-Gal, no B-8465.[19] 40 mg/ml stock solution in dimethyl formamide, no. D-8654.[27]
   - molecular weight markers, 123 bp ladder, no. 15613-011[18] and 1 kbp ladder, no. 15615-016.[18]
F. DNA recovery reagents
   - agarose, 15510-027.[18]
   - QIAEX II gel extraction kit, no. 20051.[25]
   - glass spin columns, assembled as described [15].
G. DNA cloning reagents
   - pGEM-T kit, no. A-3600.[24]
   - TA Cloning kit, no. K-2000-01.[15]
   - CAT GenBlock, no. 27-4895.[31]
H. Glassware and plastics
   - 250 ml centrifuge bottles, no. 3120-0250.[20]
   - Oakridge tubes, no. 3119-0050.[20]

- 50 ml conical Corning tube, no. 25330-50.[9]
- 15 ml conical Corning tube, no. 25319-15.[9]
- 10 ml disposable plastic pipette, Falcon, no. 7530.[30]
- P-20 Aerosol resistant pipette tips (ART), no. MBP-2149P.[17]
- P-200 Aerosol resistant pipette tips (ART), no. MBP-2069.[17]
- 0.5 ml micro-centrifuge tubes, no. 8508-FT.[17]
- 0.2 ml thin-wall micro-centrifuge tubes, no. 435.[17]
- caps for 0.2 ml thin-wall micro-centrifuge tubes, no. 440.[17]
- glass beads, no. G-9018.[27]
- cryovials, with O-rings and screw-caps, no. 4204-S.[17]

3. Procedures

The procedure for amplifying any fragment of DNA requires a number of preparative steps, including the following: preparation of genomic DNA, to act as a template supporting the replication of a target gene sequence; selection and synthesis of oligodeoxyribonucleotide primers; amplification of the target sequences via PCR; recovery of the fragments by genetic cloning; and verification of the cloned fragment by nucleotide sequence determination. Each of these discrete steps, as performed in our laboratory, will be described below.

A. Preparation of template DNAs: streptococcal DNA isolation (based on [6])
   1. Day one – Inoculate, in the morning, a 25 ml culture of Brain Heart Infusion medium with *S. mutans*. The strains used in these studies were UA159 [25] or UR100 [30]. Late in the afternoon, transfer the 25 ml culture to a 1 litre flask containing fresh medium, pre-warmed to 37 °C. Allow the culture to grow overnight.
   2. Day two – Harvest the cells by centrifugation in the Sorvall centrifuge using the centrifuge bottles at 8000 rpm (10,400 rcf) for 10 min at 4 °C. Wash the collected cell pellets in 150 ml of chilled 20 mmol/l Tris pH 8.0.
   3. Resuspend the cells in 20 ml of lysis buffer. Resuspend the cells uniformly by vortexing. Transfer the suspension to a sterile flask and incubate the mixture with gentle shaking at 37 °C for two hours.
   4. Collect the cells by centrifugation in an SS-34 rotor at 5000 rpm (2,988 rcf) for 15 min at 4 °C. Resuspend the pellet in 20 ml of TE buffer and add 4 ml of 5 mol/l, NaCl solution. Add 2 ml 10% (w/v) sodium dodecyl sulfate solution. The suspensions should become viscous almost immediately as the cells lyse and release DNA into solution. If the suspension does not become viscous, place the flask in a 60 °C water bath for 5–30 min.
   5. Transfer the solution to a 50 ml conical Corning tube and add buffered phenol. Gently invert the tube several times to extract the proteins into the organic (phenol) layer. Separate the organic and aqueous phases by centrifugation in the Sorvall centrifuge at 2000 rpm (478 rcf) for 15 min using the SS-34 rotor, without the lid. Note: when using the SS-34 rotor without the lid, it is imperative to use the rotor locking screw to fasten the rotor to the drive spindle. Using a 10 ml plastic pipette and the pipette-aid, transfer the aqueous phase to fresh tube. Add 20 ml TE buffer to the first phenol layer and re-mix the solution. This 'back' – extraction generally results in improved recovery of DNA left at the interface after the first extraction. Repeat the centrifugation and combine the first and second aqueous layers.
   6. Using the $CHCl_3$: isoamyl alcohol solution, extract the aqueous phase recovered in step 5 (above).
   7. Add 2 1/2 volumes of cold 95% EtOH plus 1/6 volume 5 M NaCl to the aqueous phase. Mix well and allow the DNA to precipitate on ice. Collect the precipitate by centrifugation at 19,000 rpm (43,152 rcf) for 30 min at 4 °C. Carefully invert the centrifuge tubes to allow the liquid to pour off. Blot the inverted tubes on paper towel to remove as much buffer as possible. Resuspend the nucleic acid-containing pellet in 10 ml of TE buffer.
   8. Remove contaminating RNA by the addition of 0.2 ml of the RNAse A solution. Incubate the suspension for 30 min at room temperature.
   9. Add 10 ml of buffered phenol and repeat the extraction process described in step 5. Repeat the back extraction on the aqueous phase. Repeat Step 6. To the final aqueous phase, add 2 1/2 volumes of ice-cold 95% EtOH plus 1/6 volume of the 5 M NaCl solution. Collect the precipitate by centrifugation at 19,000 rpm (43,152 rcf) for 30 min at 4 °C. Carefully pour off the liquid phase and wash the DNA by resuspending the pellet in 10 ml of 70% ethanol. Re-collect the pellet by centrifugation, as before. Gently pour off the liquid and allow the pellet to air-dry at room temperature. The pellet will become clear as it dries. The recovered DNA can be spectrophotometrically quantitated by resuspending the pellet in 1–2 ml of TE buffer and measuring the absorbance of the material at 260 nm [18]. Typically, we recover approximately 20 mg/ml of DNA from this procedure.

B. Preparation of DNA templates from *E. coli*

In addition to PCR-based amplifications of DNA sequences from genomic templates, we have also found it useful to prepare streptococcal DNA templates from clones maintained in strains of *E. coli* [31]. A large number of protocols are available in the literature for the preparation of plasmid DNAs from *E. coli*. The protocol accompanying the QIA-prep mini-prep kit is reliable and cost-effective when used according to the manufacturer's instructions. We have also found it useful to use fragments of DNA isolated from agarose gels. In those circumstances, the preparation of DNAs has been accomplished by separating restriction enzyme digested DNA fragments by agarose gel electrophoresis in TAE buffer. The position of specific DNA is identified following ethidium bromide staining of the gel and comparison with molecular weight standards during illumination of the stained gel on a UV transilluminator. Note: UV light can cause skin irritation and long-term exposure may lead to eye damage. We invariably use a UV-protective face shield when working over the transilluminator. Appropriately sized fragments are excised from the gel and further purified using the QIAEX II gel extraction kit to recover DNA from agarose gels.

C. Selection of primer pairs for degenerate PCR

1. The goal of primer sequence selection is to achieve both the lowest possible degeneracy and the amplification of a useful fragment length. Operationally, there is no minimum for fragment size, though we have typically attempted to generate PCR products between 500 and 1000 bp. Fragments of this size can be amplified with highly degenerate primer mixes and have sufficient length as to be useful in subsequent insertional mutation studies [30].
2. Homologous DNA or protein sequences are collected using the BLAST or FASTA algorithms in the GCG software package.
3. In the case of related DNA sequences, the nucleotide sequences are translated into protein sequence using either the GCG or MacVector algorithms.
4. Deduced amino acid sequences are then aligned using either the multiple sequence alignment and PileUp algorithms available in GCG or using the ClustalW alignment algorithm in MacVector.
5. Deduced amino acid sequences aligned by Clustal are formatted to show conserved amino acid sequences. Typically, it is useful to focus on regions of at least five contiguous amino acids, though longer regions of conservation are desirable.
6. Regions fitting the criteria of conserved region length and distance from each other can be reverse translated, either manually or by computer, to create a degenerate nucleotide sequence. Typically, we have attempted to maintain a 50% G:C proportion in degenerate sequences and the 3′ base must not be degenerate. For example, the codons for glycine are GGN, where N represents all four possible bases. if glycine is the last amino acid in the region of conservation, then the N would be deleted from consideration and the last two bases in the oligodeoxyribo nucleotide sequence would be GG.

D. Synthesis of primers

1. In the past, we have synthesized primers using Model 381A and 391 instruments from Applied Biosystems [31]. However, a number of commercial suppliers have become available to provide rapid turnaround times for the custom synthesis of oligonucleotides containing common and uncommon bases. Mixtures of oligos are readily prepared by including all four of the common bases in the synthesis step at any point in the construction of a custom DNA sequence. Thus, highly degenerate primer mixes can be obtained representing all possible combinations of codons for a given amino acid sequence. The arrows in Figure 1 indicate the amino acid sequences that were used as the basis for synthesis of ATPase-specific oligodeoxyribonucleotide primers.

E. Degenerate PCR conditions

1. The mixture first synthesized for the forward primer consisted of the following sequence: 5′-GAATTCGTNTTYGCNGGNGTNGGNGAR-CGTNACNCGNG-3′ and accounted for all 262,144 possible sequence variations in the amino acid sequence. The reverse primer consisted of the sequence: 5′-GTCGACCATNC-CNARDATNGCDATDATRTCYTGNA-3′, which represented 55,296 possible sequence variations in that segment of ATPase β-subunit. In later experiments, we also used much shorter regions of the ATPase sequence. In those primers, we included restriction enzyme sites in both primers at the 5′ end of the sequences and the amount of material homologous to the ATPase was reduced, permitting lower degeneracy in the primer and a higher fidelity to the template. The forward primer was as follows: 5′- CTATGACCAT-GAATTCGGNGTNGGNGARCGNAC-NCGNG-3′ (8192 possible combinations), where the underlined sequence represents an engineered EcoRI site. The reverse primer sequence was as follows: 5′-CTGCCG-GTCAGTCGACARDATNGCDATDATRT-CYTGNA-3′ (3456 possible combinations),

where the double-underlined sequence represented an engineered SalI restriction endonuclease site.

2. Transfer into a 500 μl micro-centrifuge tube the following: 10 μl of 10X modified PCR buffer; 16 μl of 1.25 mmol/l dNTPs, final concentration of 200 μmol/l; 2 μl of sense-strand primer (20 μmol/l stock) final concentration of 1 μM; 2 μl of complementary strand primer (20 μmol/l stock); 1–10 μl of template DNA (range from 130 ng to 130 μg); adjust to 100 μl with distilled water; add 100 μl of mineral oil; initiate the thermal cycling, pausing the instrument when 94 °C is reached to add 0.5 μl of Taq polymerase; resume the cycle at 94 °C for 2 min, 37 °C for 2 min, 72 °C for 3 min; continue for 30–35 cycles. Remove a 10 μl aliquot and separate the products by electrophoresis in 1–2% agarose in TAE buffer. Note: the difference between 10× modified PCR buffer and the PCR buffer which comes with commercial preparations of Taq DNA polymerase is that the modified form contains magnesium at a defined concentration; whereas some commercial buffers are magnesium-free, permitting the investigator to vary the magnesium concentration. The effect of varying the magnesium concentration is to increase the likelihood that base substitutions will occur. This is often template and temperature dependent. For that reason, we generally hold the concentration to 1 mmol/l.
3. We have found that these conditions yield products visible in stained agarose gels. However, we have found more consistently, that re-amplification of the material results in significantly greater quantities of product, simplifying recovery of the DNA for cloning.
4. Re-amplify the initial PCR products by placing into a 500 μl micro-centrifuge tube the following: 10 μl of 10× modified PCR buffer; 16 μl of 1.25 mmol/l dNTPs, for a final concentration of 200 μmol/l; 2 μl of sense-strand primer (20 μmol/l), final concentration of 1 μmol/l; 2 μl of complementary strand primer (20 μmol/l); 1–10 μl of the initial PCR reaction described above; adjust to 100 μl with distilled water; add 100 μl of mineral oil; start the thermal cycler, pausing the instrument when 94 °C is reached, to add polymerase, then resume the cycling program at: 94 °C for 2 min, 50 °C for 2 min, 72 °C for 3 min.

F. Cloning of PCR fragments

1. PCR products are generally propagated as clones carried in *E. coli* cloning vectors. Cloning of PCR products is accomplished either by directly cloning fragments from the reaction mixture or by first purifying a fragment by gel electrophoresis.
2. Commercial kits are available for cloning fragments directly from reactions (e.g., the TA-cloning kit or the pGEM-T kit). The kits are supplied with instructions, which in our hands have proven reliable.
3. PCR products can also be separated by agarose gel electrophoresis, typically in 1–2% agarose in TAE buffer. 10 μl aliquots of reaction are mixed with 1 μl of loading dye and transferred to a well in the agarose gel apparatus. Molecular weight markers are either 123 bp ladder or 1 kbp ladder. Voltage is then applied to the gel, usually at 5–10 V/cm of gel, depending on the time-frame of your investigation. The length of electrophoresis time is dependent on the size of the desired PCR fragment. However, 1–2 hours is usually sufficient time to resolve fragments.
4. Gels are removed from the apparatus and placed in a glass tray with approximately 200 ml of TAE buffer plus 25 μl of ethidium bromide dye stock solution (sufficient to cover the gel). Allow several minutes for the gel to soak up the stain. Ethidium bromide is a known carcinogen and care, including the use of latex gloves and a laboratory coat, should be exercised with its use.
5. Following the staining interval, the dye solution is poured off, and the gel is illuminated on a long-wavelength UV transilluminator to reveal the position of the fragments with respect to the molecular weight markers.
6. The DNA band with the desired molecular weight is then carefully excised with a scalpel and purified using either a glass wool column [15] or a commercially available purification system (e.g., the QIAEX II gene extraction kit).
7. Purified DNA can then be used directly with PCR cloning vectors or can be treated with restriction endonucleases to create 'sticky-ended' DNA for cloning [31]. We have used both approaches and find that the PCR cloning vectors are generally less troublesome in that clones have been readily isolated. Digestion of amplified DNAs with restriction enzymes has worked for us, but required at least one and often two digestions of recovered DNA with phenolic extractions between each. The loss of DNA during the extractions, coupled with the likelihood of incomplete digestion of recovered DNAs, has in our opinion resulted in reduced recovery of clones.
8. Cloned PCR fragments must be carried in an appropriate *E. coli* strain. In our hands, the strain has been DH10B. DH10B exhibits high transformation efficiencies by electroporation using 0.2 cm cuvettes in the BioRad Gene Pulser system. Cells transformed with cloned

DNAs are plated on LB agar plates containing appropriate antibiotics and, if desirable, the indicator dye X-Gal.

G. Assembly of novel gene fusions using splice-over-lap extension PCR (SOE-PCR)

1. The technique of SOE-PCR, or genetic knitting, has been used in the past to combine two distinct regions of DNA, resulting in new and novel sequences [16, 17]. We have used the method to link the promoter region of the *S. mutans* ATPase operon to the chloramphenicol acetyltransferase (CAT) gene. The promoter and upstream regions of the *S. mutans* ATPase possess 2 putative stem-loop structures immediately upstream of the –35 box of the promoter [40], see Figure 2A. The goal was to connect in-frame, in varying increments, the stem-loop structures with the CAT gene. The resulting constructs will permit the examination of the role of stem-and-loops in the regulation of the *S. mutans* ATPase operon.
2. The first step in SOE-PCR was to design primers for the construction of the chimeric proteins. The procedure is performed in two steps. In the first, two fragments of DNA are amplified independently, but with a region of common DNA incorporated into one primer for each fragment to be amplified. The incorporated homologous region, represented in one end of each amplified DNA product, is then used to guide annealing of the two products during amplification in the second step. The annealing of the two DNAs formed from the reactions in the first step leads to amplification of a novel, hybrid DNA during the second phase of the process.
3. The four 'forward' or 'upstream' *S. mutans* primers were designed to create the desired portion of stem-loop region (see Figure 2A). The 30 residue complementary strand primer (Primer 5, Figure 2A) was synthesized with homology to the 15 base pair sequence that is immediately upstream of ATPase c subunit gene, followed by the first 15 base pairs of the CAT gene. The pairing of one of the primers (1–4) with primer 5 was used to produce the promoter region of the ATPase.
4. Primers 6 and 7 (shown in Figure 2A) were used to amplify the CAT gene fragments. The 5′ CAT primer, primer 6, was synthesized to be homologous to the upstream 15 base pairs of the ATPase gene and the first 15 base pairs of the CAT gene, inversely complementary to primer 5. The 3′ primer for the CAT gene, primer 7, was located downstream of the gene and was 24 base pairs long with a BamHI site engineered into the 5′ end of the primer. Taq polymerase and accompanying buffer (Life Technologies, Inc.) was used to obtain the desired products. Note that the reaction buffer concentrations included 2 mmol/l $MgCl_2$. Efficiency of amplification in each reaction was measured by examining the products of the reactions on a 1.4% agarose gel in TAE buffer. Product formation was observed in each of the 4 *S. mutans* promoter region reactions as well as the reaction designed to amplify the CAT gene.
5. For the second set of reactions, a series of dilutions of the step 1 PCR products were used as template, as concentrations of products can vary. Serial dilutions of the *S. mutans* promoter fragments and CAT gene reactions were constructed in sterile water. Generally, we have found that dilutions on the order of 1:10–1:1000 are effective.
6. The reactions used were 50 μl in size and consisted of the following: 5 μl 10× BRL Taq DNA polymerase buffer; 0.8 μl 25 mmol/l dNTP's; 2 μl 50 mmol/l $MgCl_2$; 1 μl of a 50 pmol/l stock solution of each primer, 1 μl of each template at corresponding dilutions (i.e., 1:10 dilution of *S. mutans* promoter piece with 1:10 dilution of CAT gene) and 0.5 μl of Taq polymerase. The cycling conditions for these reactions were: 94 °C, 2 min for one cycle; 94 °C 1 min, 50 °C 1 min, 72 °C 4 mins for 35 cycles; and 72 degrees for 10 min for 1 cycle.
7. Verification of the correct alignment in the fusions is essential and can only be accomplished by nucleotide sequence determination. Typically, we have cloned the PCR products by excising the desired band from a gel and purifying the DNA using the QIAEX II kit, as described above. In the example of SOE-PCR gene knitting shown here, the products were first cloned with the pGEM-T kit. *E. coli* transformants exhibiting a white phenotype on LB medium containing X-Gal (40 mg/l, final concentration of 40 μg/ml) and ampicillin (100 μg/ml) were screened for the presence of an insert. Clones that appeared to have the correct insert were screened further by nucleotide sequencing.

H. Growth of *S. mutans*

1. *S. mutans* strains UA159 [25] and UR100 [30] were grown at steady-state at pH values of 5 and 7 using methods described in the paper by Chen and Burne [5]. We have also grown cells in batch culture using methods described in the paper by Spatafora [41].

I. Preparation of *S. mutans* cell-free extracts

1. Cells are taken from their growth vessel, either flasks in the case of batch growth or pumped from the chemostat in the case of steady-state growth situations, and collected by centrifugation in pre-weighed Oakridge tubes at 8000 rpm (7,650 rcf) for 10 min at 4 °C.

2. Supernatant solutions are gently poured off and discarded.
3. Cell pellets are suspended at 1 g/ml in breaking buffer: 10 mmol/l Tris, pH 7.5; 50 mmol/l NaCl; 1 mmol/l EDTA.
4. One ml of suspension is transferred to a pre-chilled 2 ml cryovial equipped with a screw-cap and O-ring and containing 1 g glass beads.
5. The chilled vials are placed in the holding compartment of a mini-Bead Beater. The timer is set to 'timed' for 2 min and the instrument is turned on to initiate breakage of the cells by the shearing action of the glass beads. After the cycle is complete, the vials are placed on ice for 1 min, and then re-cycled in the mini-Bead Beater for an additional 2 min. Following the second cycle, the cell extract is clarified by centrifugation of the vials in an Eppendorf centrifuge at 8000 rpm (7300 rcf) for 15 min at 4 °C.
6. The supernatant solution is transferred by micro-pipette to a fresh vial and the centrifugation is repeated for 10 min. The centrifugation steps function to remove glass beads, unbroken cells, and cell membranes.
7. The extracts are dialyzed overnight in 2 litres of TE buffer at 4 °C with one complete buffer change after four hours. In the morning, the protein preparations are transferred from dialysis tubing into sterile 15 ml Corning tubes to estimate the volume. Using the stock solution, PMSF is added to a final concentration of 100 μg/ml.
8. Aliquots of dialyzed protein are withdrawn for the estimation of protein using the Bio-Rad dye-binding system or the Quanti-Gold kit.

J. Two-dimensional gel preparation and staining
   1. We have conducted two-dimensional gel separations in our laboratory using the Investigator 2-D system with the instructions provided by the manufacturer. An alternative to the purchase of sophisticated equipment is the use of commercial vendors for the preparation and execution of two-dimensional gel separations. The procedure can be performed rapidly and cost effectively on a custom-gel basis. We have had separations performed commercially by Kendrick Labs, Inc. (Madison, WI) according to the method of O'Farrell [26, 27] as follows: isoelectric focusing (IEF) was carried out in glass tubes of inner diameter 2.0 mm using 2.0% pH 4–8 ampholines for 9600 volt-hour. Fifty ng of an IEF internal standard, tropomyosin protein, with lower spot of mw 33,000 and pI of 5.2 was added to the samples. The tropomysin spot is indicated with an arrow on the 2-D gel pattern. After equilibration for 10 min. in buffer '0' (10% glycerol, 50 mmol/l dithiothreitol, 2.3% SDS and 0.0625 mol/l Tris, pH 6.8) the tube gel was sealed to the top of a stacking gel which is on top of a 10% acrylamide slab gel (0.75 mm thick). SDS slab gel electrophoresis was then carried out for about 4 hours at 12.5 mA/gel. The slab gels were fixed in a solution of 10% acetic acid/50% methanol overnight. The following proteins (Sigma Chemical Co., St Louis, MO) were added as mw standards to the agarose which sealed the tube gel to the slab gel: myosin (220,000), phosphorylase A (94,000), catalase (60,000), actin (43,000), carbonic anhydrase (29,000) and lysozyme (14,000). These standards appear as horizontal lines on the silver-stained [28] 10% acrylamide slab gel. The gels were dried between sheets of cellophane paper with the acid edge to the left.

K. Analysis of the two-dimensional gels
   1. Digital images of the 2-D gels are created using the Apple Color One Scanner, or similar device, to scan the dried gels. Scans are conducted in the 8-bit mode and at a resolution of 150 dpi. The images can be saved as either PICT or TIFF files.
   2. The image files are analyzed with NIH Image version 1.6 for the Macintosh. Using tools available in the software, high background staining and non-protein streaks can be subtracted from the image. The position and intensity of each protein, seen as a spot on the gel, can be quantitated and catalogued for comparison with other gels.

4. Results and discussion

Amplification of genetic fragments from S. mutans using degenerate PCR. The availability of nucleotide sequences, from a large and growing number of bacterial genomes, permits the rapid Clustal alignment of deduced amino acid sequences, such as that for bacterial ATPase β-subunits shown in Figure 1, Panel a. Such alignments provide the background information necessary to begin designing nucleotide primers for use in PCR.

The design of primers can be automated using computer alogrithms to determine those regions of DNA whose sequence would result in similar melting temperatures for the forward and reverse primers. Of the commercially available packages, we most often have used MacVector to search for appropriate primer sets and for estimation of primer melting temperatures. The algorithm used for computation of T_m includes nearest neighbor analysis because of the effects of base composition on the overall melting or dissociation kinetics of probes. The methods have been well described in the literature and we have found the temperatures estimated by the program to

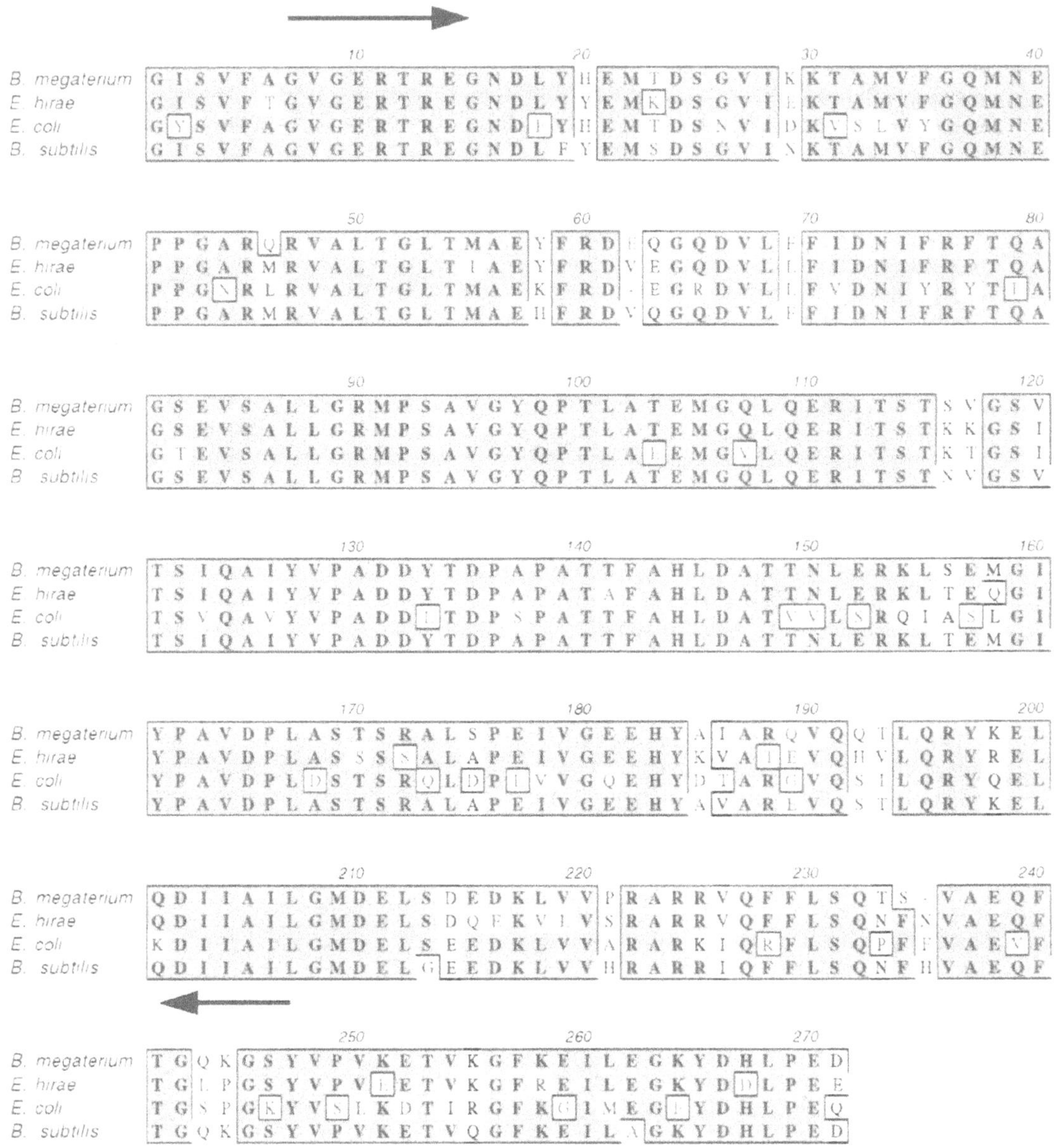

Figure 1. Panel a. Computer generated Clustal alignment of deduced amino acid sequences from bacterial ATPase β-subunits. Arrows indicate the position of degenerate oligodeoxynucleotide primers.

be useful starting points in our hands [33, 34]. We have also found that highly degenerate primers can be used successfully to amplify gene fragments using fragments encoding portions of the highly conserved, catalytic subunit of the proton-translocating ATPase (Figure 1, Panel b). Following the amplification of the fragments (Figure 1, Panel b), cloning and nucleotide sequence determination confirmed the identities of the DNAs [31]. The fragments were then used as probes to screen for clones of the intact ATPase operon from *S. mutans* [40]. In subsequent studies, we substituted inosine at the wobble base positions of codons. The effect was to greatly relax the requirement for including every possible combination of bases in our primer sequences. The practical result of using inosine was a large increase in the amount of primer able to anneal to our template, leading to the formation of clear bands at the end of the first round of amplification, eliminating the need for re-amplification and thereby simplifying cloning of our reaction products.

The reaction conditions that we have described here were successful for amplification of a specific streptococcal DNA fragment. It must be pointed out, that the results of a given amplification procedure are potentially effected by a significant number of variables, including the following: annealing temperature, magnesium concentration, primer: template

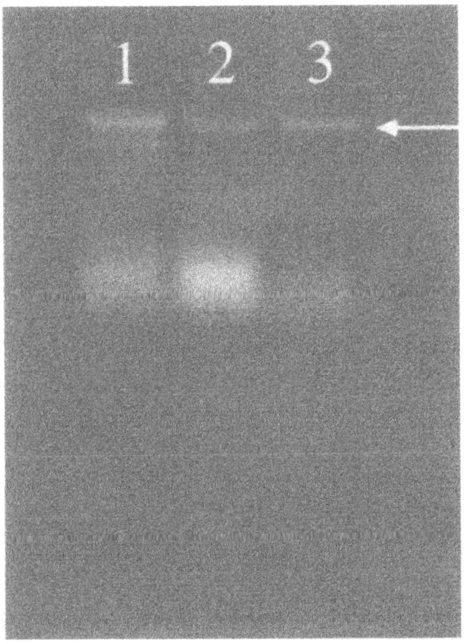

Figure 1. Panel b. Amplification of a DNA fragment encoding the beta-subunit of the proton-translocating ATPase. Lane contents as follows: Lane 1, *S. mutans* DNA amplified 30 cycles with annealing at 37 °C for 30 cycles; followed by 30 additional cycles with annealing at 50 °C; Lane 2, pWSB100 plasmid DNA containing the *Bacillus megaterium* ATPase operon amplified 30 cycles with primer annealing at 50 °C; *S. mutans* DNA, boiled for 5 min, followed by 30 cycles of PCR with primer annealing at 50 °C. The arrow at right indicates the bands which were subsequently cloned and their nucleotide sequences verified.

ratio, and ramp times (the period of time that a thermal cycler needs to achieve a given temperature). The investigator faces a possibly bewildering number of variables in experimental conditions, particularly when using degenerate primers. Our approach to reducing the number of conditions examined has been to establish a set of reagent concentrations and then to vary the annealing temperature, the second temperature in the cycle. Nevertheless, it is usually the case that differences exist in predicted annealing temperature between the two primers for a given reaction. The result is that some time may be spent in establishing optimal conditions for PCR using any primer, but especially those containing non-homologous bases (degenerate mixtures of primers) or bases with non-Watson-Crick base-pairing properties (such as inosine). For example, we examined 11 sets of conditions, differing in the annealing temperatures, before successfully amplifying a clearly defined band in the ATPase-specific reactions. Because the thermal cycler we used has three independently programmable heating blocks, the time spent on completion of all 11 reactions was not great. Reproducibility of the degenerate technique for ATPase was good, once appropriate conditions were established, though we have not quantitated the amounts of material generated in all of our experiments. However, the results shown in Figure 1b clearly indicate the general applicability of the approach, with the same fragment amplified from three organisms using the same primers, albeit with different annealing temperatures. Nucleotide sequencing of the fragments confirmed the identity of the products.

Several approaches have been developed to reduce the amount of time spent searching for temperature optima. One such method is the 'touchdown' approach to PCR [9, 14, 32], wherein the thermal cycler is programmed to begin the cycling temperatures well above the predicted T_M's for primers and to reduce the annealing temperature during the course of the reaction sequence such that the program finishes with annealing temperatures below the optimum T_m. The method relies on the trade-off between specific amplification of fragments at high temperature and the higher yields, in general, at lower annealing temperatures. The drawback to using the touchdown approach is that the thermal cycler must be capable of running a cycling program with a large number of changes in annealing temperature. Such instruments are now available and should in the future facilitate more timely exploration of optimal conditions for degenerate PCR. Many machines now are capable of a modified touchdown approach, referred to as step-down PCR, where the annealing temperature is gradually ramped down in increments of a few degrees. The result is similar to the touchdown procedure and far fewer programming steps are required, putting the demand more within range of older thermal cycler's abilities. The results of touchdown PCR in our hands show that the advantage to the technique is a more rapid establishment of optimum primer annealing temperatures and far fewer products amplified during initial experiments.

Amplification experiments in the future will undoubtedly be based on established sequences from various genomic sequencing. Thus, the need for highly degenerate PCR primers may become diminished over time. It seems likely that more physiologically-driven approaches, using completely homologous sequences to construct specific genetic fusions (as described in the following section), will be applied to streptococcal pathogenesis. However, it is possible that degenerate PCR will be used extensively for some time since, for example, the number of streptococcal strains capable of pathogenesis is unclear (see for example, Sansone et al. [37]).

Construction of gene fusions using PCR. The ability to create specifically designed DNA sequences is one of the most powerful uses of the PCR technique. Frequently, useful restriction endonuclease sites are unavailable for the in vitro juxtaposition of putative regulatory elements of DNA and reporter gene frag-

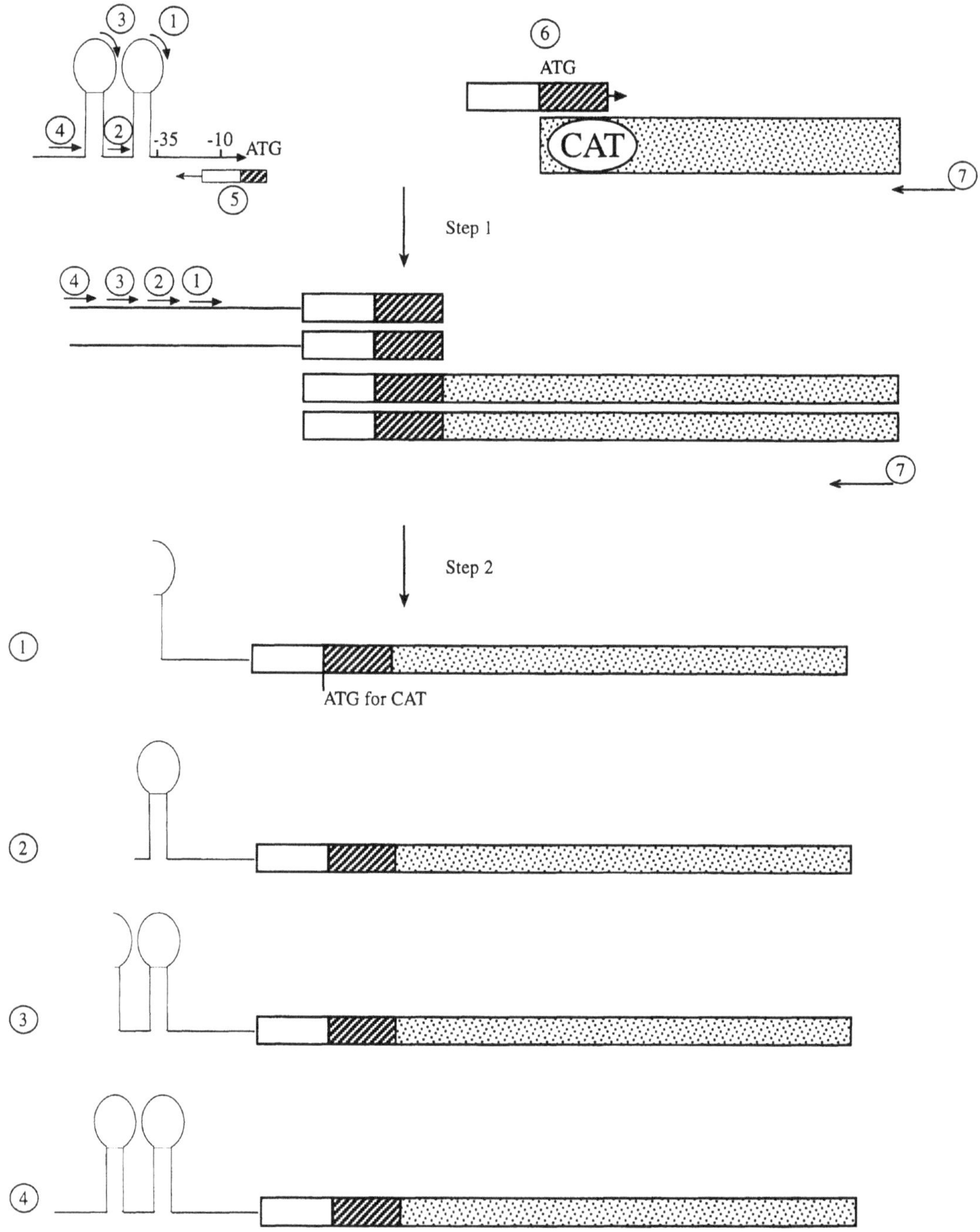

Figure 2. Panel a. Splice-overlap extension PCR to create a genetic fusion of the ATPase operon promoter with a chloramphenicol acetyl transferase gene as the reporter system. In Step 1 of the procedure, the two regions of DNA that will eventually become fused are amplified independently. In this case, the promoter region of the ATPase operon was amplified using one each of the primer labeled 1–4 and primer 5. Primer 5 contained 15 bases of sequence that were precisely homologous to the last 15 bases of the presumptive ATPase promoter region (shown as the open box), up to the ATG initiation codon for the c subunit in the operon. Primer 5 also contained the first fifteen bases encoding the CAT gene (shown as the hatched box). The CAT gene was amplified using Primer 6, which contained the same sequence as Primer 5 but in the reverse orientation, and primer 7. In Step 2 of the procedure, the products from Step 1 are mixed and used as templates for amplification with one 'forward' primer, either 1, 2, 3 or 4 and Primer 7.

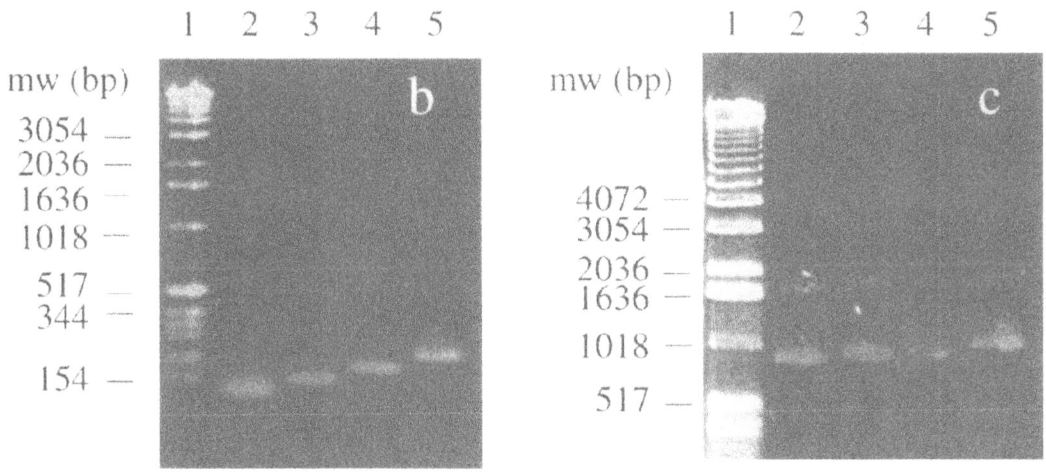

Figure 2. Panel b. Electropherogram of PCR products resulting from the Step 1 reactions described in the legend for Panel 2a. Lane contents were as follows: Lane 1, 1 kbp DNA molecular weight markers; Lane 2, DNA amplified with primer 1 and 5; Lane 3, DNA amplified with primer 2 and 5; Lane 4, DNA amplified with primer 3 and 5; Lane 4, DNA amplified with primer 4 and 5. **Panel c.** Electropherogram of PCR products resulting from the Step 2 reactions, described in the legend for Panel 2a. Lane contents were as follows: Lane 1, 1 kbp DNA molecular weight markers; Lane 2, DNA amplified with primer 1 and 7; Lane 3, DNA amplified with primer 2 and 7; Lane 4, DNA amplified with primer 3 and 7; Lane 4, DNA amplified with primer 4 and 7.

ments. Methodologies, other than PCR, are available in other bacterial systems for the construction of genetic fusions; for example, the use of mini-transposable elements to create in-frame fusions. Until similar approaches are well-established in the oral streptococci, and in the absence of useful restriction sites, knitting together diverse DNAs by PCR will be an alternative. The scheme shown in Figure 2a illustrates our approach to linking varying portions of the putative ATPase operon promoter region to a CAT reporter gene using splice-overlap extension PCR. The result of those procedures are shown in Figure 2b, where the predicted laddering effect during amplification of the four desired fusions is shown. The figure shows clearly the expected increase in the molecular weights of the amplification products, as predicted from the scheme in Figure 2a. The results of knitting the operon fragments to the CAT gene are shown in Figure 2c. Subsequent cloning of the amplicons verified the successful alignment of the CAT gene with the ATPase promoter fragments. Comment should be made here that we did observe single base changes in some of the operon clones. For example, multiple isolates for each of the fusions were examined by nucleotide sequencing. The goal was to verify a single correct fusion for each of the four constructs that we were seeking. Our sequencing results showed the following: for the shortest fusion, the first clone sequenced was a faithful and precise copy of the desired product, completely lacking any changes; of the one and one-and-a-half stem-loop fusions, only one of three isolates sequenced was unchanged in any base. For the longest fusion, containing both stem-loops, of two sequenced clones, one was a precise product, the other contained a single base change. Thus, it can be seen that while we had no trouble finding a completely authentic clone for each of the four fusions, nucleotide sequence characterization of multiple isolates was necessary to establish the desired fusion constructs. In terms of the CAT portion of the fusions, we observed CAT gene activity in *E. coli* for all of the fusions examined, whether or not they contained a base change in the ATPase promoter region or not (data not shown here). We concluded from those observations that we had either fortuitously isolated CAT fusions without changes in the CAT gene or that changes were silent. Future experiments will provide insights into the ability of these promoter fragments to drive CAT synthesis in *S. mutans* as a function of external pH value.

Development of additional targets for PCR amplification using proteomic databases. It has become apparent from genomic sequencing studies, that bacterial chromosomes can be catalogued for open-reading frames, but that significant portions of those reading frames can not be aligned with proteins in existing databases, thus obscuring their function [10, 12]. Further, the repertoire of proteins expressed in varying environmental conditions is not yet understandable from primary nucleotide sequence information. As part of the solution to the problem, two-dimensional gel electrophoresis of proteins

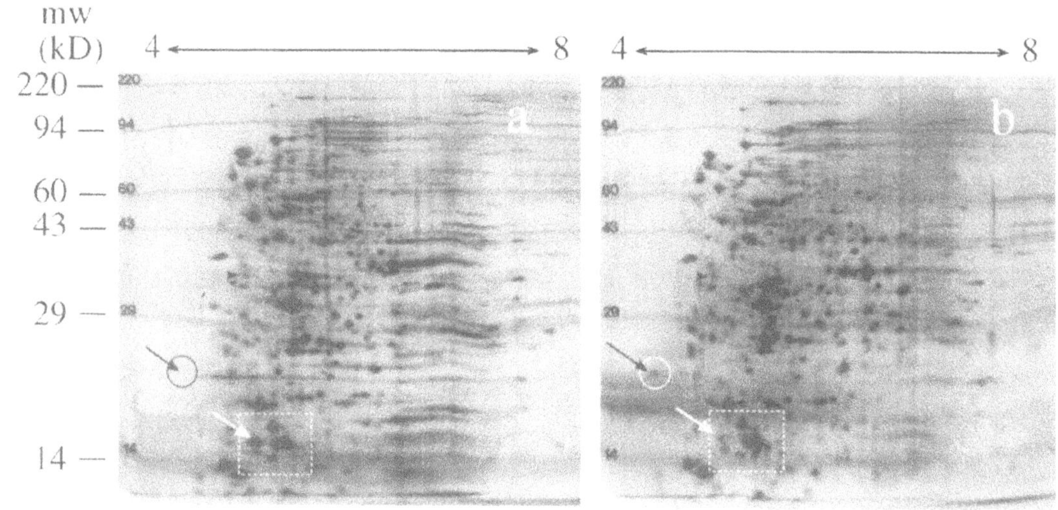

Figure 3. Panel a. Two-dimensional gel electrophoretic separation of protein extracts prepared from *S. mutans* strain UR100 grown at steady-state on BHI medium containing glucose at a pH value of 5. Approximately 100 μg of protein was used in the separation. Gels were silver-stained as described in the text. **Panel b.** Two-dimensional gel electrophoretic separation of protein extracts prepared from *S. mutans* strain UR100 grown at steady-state on BHI medium containing glucose at a pH value of 7. Approximately 100 μg of protein was used in the separation. Gels were silver-stained as described in the text.

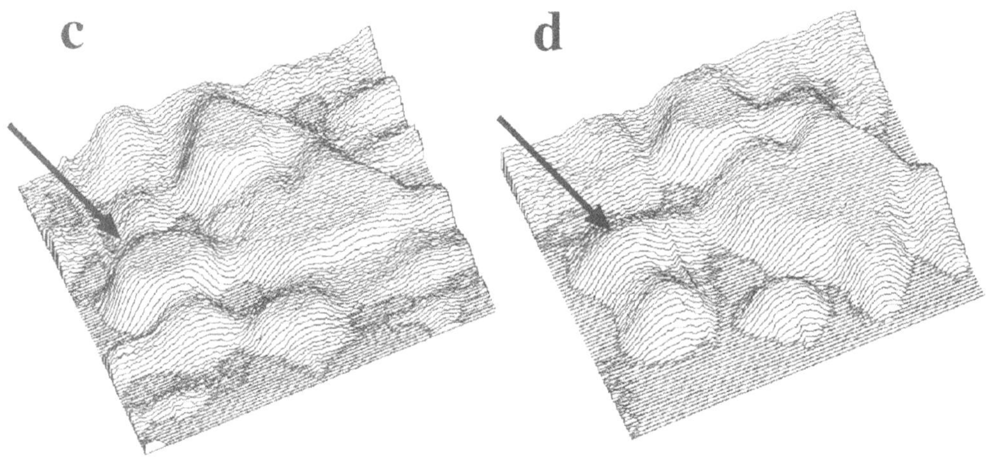

Figure 3. Panel c. Surface plots of the enclosed square shown in Panel a. **Panel d.** Surface plots of the enclosed square shown in Panel b.

prepared from cells grown under specific conditions can be used to establish a given microbe's proteome, or set of expressed proteins, in a defined environment. Proteomic maps have now been established for several organisms, grown in varying conditions. The most well-defined maps to date are those for *E. coli* [43, 44] and for the yeast, *Saccharomyces cerevisiae* [29]. The impact of changes in environmental conditions can easily be seen in two-dimensional gel separations of the proteins prepared in extracts from *E. coli* [42].

As genome nucleotide sequences become available for *S. mutans*, it will become possible, using two-dimensional gels, to map the effect of mutations on the microbe's expressed protein repertoire. The amino acid sequences of proteins can be determined directly following transfer of the gel-separated spots to membrane supports. The resulting 'blots' of proteins can be processed for sequence determination using, in some cases, as little as a femtomole of material [19, 38, 39, 46]. The amino acid sequences resulting from those analyses will be used as the basis for either direct identification by comparison with existing databases, or the proteins may be novel observations. If the genome of the organism has been determined, then the protein sequence can be aligned with the contigs derived from sequencing projects. Matches of empirically determined protein sequence with genomic nucleotide data will permit the design of homologous primers for use in amplifying the corresponding gene of interest. If the organism's genome has not been determined, then the amino acid sequence can be used as the basis for synthesizing degenerate oligonucleotides as we have described above. The degenerate primers can then be used to amplify all or part of the gene encoding the target protein.

Our goal has been to understand the changes in protein expression deployed by *S. mutans* during adaptation to acidic environmental conditions. We have included in this article examples of two-dimensional gels prepared with extracts of cells grown in two different conditions. The resolution of the gels potentially permits the identification of proteins whose expression is shown to be dependent on an environmental variable, in this case culture pH value. Proteins that are shown to be present in varying proportion, depending on culture conditions, then become candidates for micro-sequencing methods to establish their amino acid sequences.

Comparisons between gels can be readily accomplished with the unaided eye and illumination of gels on a light box. However, quantitation of the gel spots, representing individual proteins, can be more easily achieved using analytical software. Extraction of quantitative data from images is dependent on the type of staining used for the gels and on the sensitivity of the scanner used to input images of gels into a computer. For example, gel fixation with glutaraldehyde results in non-arithmetic deposition of silver grains on proteins owing to the non-stoichiometric interaction of glutaraldehyde with proteins [8]. Thus, quantitation of silver-stained gels represents an approximation of the amount of a given protein present in the gel. On the other hand, gels stained with Coomassie blue can be quantitated more readily because of the established stoichiometry of the dye with protein [4].

Once digitized images of gels have been stored in the computer, the intensity of a particular protein in the image can be normalized by comparing its integrated density to the total intensity on a gel. By establishing normalized values for proteins that may be of interest, comparisons between gels become more robust and permit the generation of a statistical set of values. A number of comparative studies showing the profile of proteins prepared from cells grown in differing conditions are now in the literature. Regarding the reproducibility of 2-D separations, we have taken the approach of requiring at least three separations of the differing preparations of cell extracts to establish a given protein as being statistically different under two different growth conditions. Our working guideline is to use the Student's t-test to compare the normalized intensities of a protein in question on two sets of three gels each [45]. Our arbitrarily chosen criteria for selecting proteins for further evaluation is a 2-fold or greater difference between protein samples prepared from cells grown under differing conditions. Clearly, proteins which appear in one set of extracts and not another would be among the first priorities.

The results of two-dimensional gel separations of proteins prepared from *S. mutans* grown at steady-state at pH values of 5 and 7 are shown in Figure 3, Panels a and b. Regions with examples of proteins obviously differing between the two gels are indicated by the boxes. The differences are more clearly seen when surface plots of the regions are made, showing the intensities of the indicated spots in comparison to the same area in the companion gel (Figure 3, Panels c and d). The arrows call the reader's attention to a protein whose relative abundance is increased at pH 5 relative to the gel prepared with pH 7 extracts. By way of example, we have identified a protein of less than 29 kD that is present in the pH 7 grown organisms and a difference in the profile of pH 5 proteins, in the region of 14 kD, when compared to the pH 7 grown cells. The proteins so indicated are examples of potential targets for future investigations, using protein micro-sequencing methods (see, for example, Matsudaira [22, 23]). Amino acid sequences developed from these types of gels will permit the development of primers to be used in PCR experiments. The goal of those experiments is to amplify part or all of the genes encoding pH-regulated proteins from the oral pathogen *S. mutans*.

Ackowledgments

We thank Nancy Kendrick for helpful discussion. This work was supported by DE-10174, PO1-DE11549 and the Rochester Cariology Training Program T32-DE07165 (W.L.K).

Notes on suppliers

1. AAPER Alcohol and Chemical Company, DSP-KY-417, Shelbyville, KY 40065, USA
2. Ambion Inc., 2130 Woodward St. #200, Austin, TX 78744, USA
3. American Biorganics, Inc., 2236 Liberty Drive, Niagara Falls, NY 14304-3796, USA
4. Bellco Glass, Inc. 340 Edrudo Rd., P.O. Box B, Vineland, NJ 08360-0017, USA
5. BioRad Laboratories, 2000 Alfred Nobel Drive, Hercules, CA 94547, USA
6. BioSpec Products, Bartlesville, OK 74005-0722, USA
7. Boehringer Mannheim Biochemicals, P.O. Box 50414, Indianapolis, IN 46250, USA
8. Brinkmann Instruments Inc., One Cantiague Road, P.O. Box 1019, Westbury, NY 11590-0207, USA
9. Corning Inc., Science Products Division, Corning, NY 14831, USA

10. Difco Corporation, P.O. Box 331058, Detroit, MI 48232, USA
11. Diversified Biotech, 1208 VFW Parkway, Boston, MA 02132, USA
12. ESA, 22 Alpha Rd., Chelmsford, MA 01824-4171, USA
13. Squibb and Sons Inc., Princeton, NJ 08540, USA
14. Hybaid, 8 East Forge Parkway, Franklin, MA 02038, USA
15. Invitrogen, 1600 Faraday Avenue, Carlsbad, CA 92008, USA
16. JT Baker Inc., 222 Red School Lane, Phillipsburg, NJ 08865, USA
17. Laboratory Product Sales, 1665 Buffalo Road, Rochester, NY 14624, USA
18. Life Technologies Inc., P.O. Box 6009, 8717 Grovemont Circle, Gaithersburg, MD 20877, USA
19. Molecular Probes, Inc., 4849 Pitchford Ave., Eugene, OR 97402, USA
20. Nalgene, 75 Panorama Creek Dr., Rochester, NY 14625, USA
21. New Brunswick Scientific, P.O. Box 4005, 44 Talmadge Road, Edison, NJ 08818-4005, USA
22. Oxford Molecular Group, 2105 South Bascom Ave., Suite 200, Campbell, CA 95000, USA
23. Perkin Elmer, 850 Lincoln Centre Drive, Foster City, CA 94404, USA
24. Promega Corporation, 2800 Woods Hollow Road, Madison, WI 53711, USA
25. Qiagen Inc., 28159 Avenue Stanford, Santa Clarita, CA 91355, USA
26. Rainin Instrument Company, Mack Road, Woburn, MA 01801, USA
27. Sigma Chemical Company, P.O. Box 14508, St. Louis, MO 63178, USA
28. Sorvall, Inc., 31 Pecks Lane, Newtown, CT 06470-2337, USA
29. Stratagene, 11011 North Torrey Pines Road, La Jolla, CA 92037, USA
30. VWR, P.O. Box 626, 200 Center Sq. Rd., Bridgeport, NJ 08014, USA
31. Pharmacia Biotech, 800 Centennial Ave., P.O. Box 1327, Piscataway, NJ 08855-1327, USA
32. Fotodyne, Inc. 950 Walnut Ridge Drive, Hartland, WI 53029, USA

References

1. Ausubel FM, Brent R, Kingston RE, et al. (1987). Current protocols in molecular biology. New York: John Wiley & Sons.
2. Belli WA, Marquis RE (1991). Adaptation of *Streptococcus mutans* and *Enterococcus hirae* to acid stress in continuous culture. Appl Environ Microbiol 57: 1134–1138.
3. Bender GR, Sutton SVW, Marquis RE (1988). Acid tolerance, proton permeabilities, and membrane ATPases of oral streptococci. Infect Immun 53: 331–338.
4. Burgess-Cassler A, Johanson JJ, Santek DA, Ide JR, Kendrick NC (1989). Computerized quantitative analysis of coomassie-blue-stained serum proteins separated by two-dimensional electrophoresis. Clin Chem 35: 2297–2304.
5. Burne RA, Chen Y-YM (1998). The use of continuous flow bioreactors to explore gene expression and physiology of suspended and adherent populations of oral streptococci. Methods Cell Sci 20: 181–190 (this issue).
6. Chassy BM (1976). A gentle method for the lysis of oral streptococci. Biochem Biophys Res Comm 68: 603–608.
7. Daspher SG, Reynolds EC (1992). pH regulation by *Streptococcus mutans*. J Dent Res 71: 1159–1165.
8. Dion AS, Pomenti AA (1983). Ammoniacal silver staining of proteins: mechanism of glutaraldehyde enhancement. Anal Biochem 129: 490–496.
9. Don RH, Cox PT, Wainwright BJ, Baker K, Mattick JS (1991). 'Touchdown' PCR to circumvent spurious priming during gene amplification. Nucl Acids Res 19: 4008.
10. Fleischmann RD, Adams MD, White O, et al. (1995). Whole-genome random sequencing and assembly of *Haemophilus influenzae* Rd. Science 269: 496–512.
11. Foster JW, Hall HK (1990). Adaptive acidification tolerance response of *Salmonella typhimurium*. J Bacteriol 172: 771–778.
12. Fraser CM, Gocayne JD, White O, et al. (1995). The minimal gene complement of *Mycoplasma genitalium*. Science 270: 397–403.
13. Hamilton IR, Buckley ND (1991). Adapation by *Streptococcus mutans* to acid tolerance. Oral Microbiol Immun 6: 65–71.
14. Hecker KH, Roux KH (1996). High and low annealing temperatures increase both specificity and yield in touchdown and stepdown PCR. BioTechniques 20: 478–485.
15. Heery DM, Gannon F, Powell R (1990). A simple method for subcloning DNA fragments from gel slices. Trends Genet 6: 173.
16. Ho SN, Hunt HD, Horton RM, Pullen JK, Pease LR (1989). Site-directed mutagenesis by overlap extension using the polymerase chain reaction. Gene 77: 51–59.
17. Horton RM, Hunt HD, Ho SN, Pullen JK, Pease LR (1989). Engineering hybrid genes without the use of restriction enzymes: gene splicing by overlap extension. Gene 77: 61–68.
18. Hotchkiss RD (1957). Methods for characterization of nucleic acid. In: Colowick SP, Kaplan NO (eds), Methods Enzymol III: 708–715.
19. James P (1997). Of genomes and proteomes. Biochem Biophys Res Comm 231: 1–6.
20. Kobayashi H, Suzuki T, Kinoshita N, Unemoto T (1984). Amplification of the *Streptococcus faecalis* proton-translocating ATPase by a decrease in cytoplasmic pH. J Bacteriol 158: 1157–1160.
21. Kramer MF, Coen DM, Dorit RC, et al. (1987). The polymerase chain reaction. In: Ausubel FM, Brent R, Kingston RE (eds), Current protocols in molecular biology, vol 2, pp 15.0.1–15.8.8. New York: John Wiley & Sons.
22. Matsudaira PT (1990). Limited N-terminal sequence analysis. Methods Enzymol 182: 602–613.
23. Matsudaira PT (1989). A practical guide to protein

and peptide purification for microsequencing. New York: Academic Press.
24. Miwa T. Esaki H, Umemori J, Hino T (1997). Activity of H^+-ATPase in ruminal bacteria with special reference to acid tolerance. J Bacteriol 63: 2155–2158.
25. Murchison JJ, Barrett JF, Cardineau GA, Curtiss III R (1986). Transformation of *Streptococcus mutans* with chromosomal and shuttle plasmid (pYA629) DNAs. Infect Immun 54: 273–282.
26. O'Farrell PH (1975). High resolution two-dimensional electrophoresis of proteins. J Biol Chem 250: 4007–4021.
27. O'Farrell PZ, Goodman HM, O'Farrell PH (1977). High resolution two-dimensional electrophoresis of basic as well as acidic proteins. Cell 12: 1133–1141.
28. Oakley BR, Kirsch DR, Morris NR (1980). A simplified ultrasensitive silver stain for detecing proteins in polyacrylamide gels. Anal Biochem 105: 361–363.
29. Payne WE, Garrels JI (1997). Yeast protein database (YPD): a database for the complete proteome of *Saccharomyces cerevisiae*. Nucl Acids Res 25: 57–62.
30. Quivey RG, Faustoferri RC (1992). In vivo inactivation of the *Streptococcus mutans recA* gene mediated by PCR amplification and cloning of a *recA* fragment. Gene 116: 35–42.
31. Quivey RG, Faustoferri RC, Belli WA, Flores JS (1991). Polymerase chain reaction amplification, cloning, sequence determination and homologies of streptococcal ATPase-encoding DNAs. Gene 97: 63–68.
32. Roux KH (1994). Using mismatched primer-template pairs in touchdown PCR. BioTechniques 16: 812–814.
33. Rychlik W, Rhoads RE (1989). A computer program for choosing optimal oligonucleotides for filter hybridization, sequencing and in vitro amplification of DNA. Nucl Acids Res 17: 8543–8551.
34. Rychlik W, Spencer WJ, Rhoads RE (1990). Optimization of the annealing temperature for DNA amplification in vitro. Nucl Acids Res 18: 6409–6412.
35. Saiki RK, Scharf S, Faloona F, et al. (1985) Enzymatic amplification of β-globin genomic sequence and restriction site analysis for diagnosis of sickle cell anemia. Science 230: 1350–1354.
36. Sambrook J, Fritsch EF, Maniatis T, eds (1989). Molecular cloning, a laboratory manual, 2nd edn. Cold Spring Harbor, NY: Cold Spring Harbor Press.
37. Sansone C, Van Houte J, Joshipura K, Kent R, Margolis HC (1993). The association of mutans streptococci and non-mutans streptococci capable of acidogenesis at a low pH with dental caries on enamel and root surfaces. J Dent Res 72: 508–516.
38. Shevchenko A, Jensen ON, Podtelejnikov AV, et al. (1996). Linking genome and proteome by mass spectrometry: large-scale identification of yeast proteins from two dimensional gels. Proc Natl Acad Sci USA 93: 14440–14445.
39. Shevchenko A, Wilm M, Vorm O, Mann M (1996). Mass spectrometric sequencing of proteins from silver-stained polyacrylamide gels. Anal Chem 68: 850–858.
40. Smith AJ, Quivey Jr. RG, Faustoferri RC (1996). Cloning and nucleotide sequence analysis of the *Streptococcus mutans* membrane-bound, proton-translocating ATPase operon. Gene 183: 87–96.
41. Spatafora GA, Moore MW (1998). Growth of *Streptococcus mutans* in an iron-limiting medium. Methods Cell Sci 20: 217–221 (this issue).
42. VanBogelen RA, Kelley PM, Neidhardt FC (1987). Differential induction of heat shock, SOS, and oxidation stress regulons and accumulation of nucleotides in *Escherichia coli*. J Bacteriol 169: 26–32.
43. VanBogelen RA, Neidhardt FC (1991). The gene-protein database of *Escherichia coli*, 4th ed. Electrophoresis 12: 955–994.
44. VanBogelen RA, Sankar P, Clark RL, Bogan JA, Neidhardt FC (1992). The gene-protein database of *Escherichia coli*, 5th ed. Electrophoresis 13: 1014–1054.
45. Walpole RE, Myers RH (1972). Probability and statistics for engineers and scientists. New York: Macmillan.
46. Wasinger VC, Cordwell SJ, Cerpa-Poljak A, et al. (1995). Progress with gene-product mapping of the Mollicutes: *Mycoplasma genitalium*. Electrophoresis 16: 1090–1094.

Address for correspondence: Dr Robert G. Quivey, Jr, Department of Dental Research, Box 611, 601 Elmwood Ave., University of Rochester, School of Medicine and Dentistry, Rochester, NY 14642, USA
Phone: (716)-275-0382; Fax: (716)-473-2679
E-mail: robert_quivey@urmc.rochester.edu

Methods in Cell Science **20**: 181–190 (1998)

The use of continuous flow bioreactors to explore gene expression and physiology of suspended and adherent populations of oral streptococci

Robert A. Burne & Yi-Ywan M. Chen
Center for Oral Biology and Department of Microbiology and Immunology, University of Rochester School of Medicine and Dentistry, 601 Elmwood Avenue, Rochester, New York, USA

Abstract. Most pathogenic bacteria elicit diseases as populations colonizing surfaces of a susceptible host. In the biofilms that form on host tissues, pathogenic bacteria can encounter huge variations in environmental conditions that can critically affect persistence and expression of virulence. Gaining a comprehensive understanding of molecular aspects of pathogenesis by streptococci and enterococci will require a detailed dissection of the phenotypic capacities of these organisms under conditions that may be experienced in vivo. Described herein are technologies that i) merge the study of bacterial gene expression with continuous chemostat culture, allowing for the strict modulation of key physiologic parameters, and ii) for the cultivation of oral streptococci as adherent populations which are suitable for analysis using reporter gene fusions and advanced microscopic techniques. These techniques extend the capabilities of investigators to control key physiologic parameters and to cultivate the organisms in environments that may more closely mimic conditions in vivo. Thus, these approaches should be useful in enhancing our understanding of bacterial pathogenesis.

1. Introduction

The streptococci, enterococci, and lactococci have a somewhat limited repertoire of pathways to generate energy equivalents and many, particularly the host-associated streptococci, are able to grow only within a comparatively narrow range of environmental conditions. Yet, these bacteria display a remarkable degree of phenotypic plasticity. Substrate availability, growth rate, pH and oxygen concentration are among a wide variety of environmental parameters which have been shown to radically alter biochemical and physiologic capabilities of these organisms, either at the genetic level or through allosteric or covalent modification of enzymes and cell constituents. In the cases of the streptococci and enterococci, many of which are capable of infecting and causing diseases in mammalian hosts, the environment can play a critical role in modulation of presentation and activities of attributes critical to virulence (for example see [8, 13]).

This chapter will focus primarily on techniques that we and others have employed to explore physiology, gene expression and virulence in suspended and adherent populations of oral streptococci. The work is built largely on two foundations. The first is that virtually all bacterial diseases are elicited by microorganisms that are colonizing surfaces. Overt or opportunistic pathogens exist as adherent populations that coat the various tissues or cells of the body and thrive in a matrix rich in complex polymers, in biofilms. Importantly, it is well established that the phenotypic characteristics of biofilm, or sessile, populations can dramatically differ from those of their suspended, or planktonic, counterparts [7]. Respiratory and growth rates, display of virulence determinants, and catabolism of substrates are among those critical determinants of pathogenic bacteria that can be influenced by growth on surfaces. The second consideration underpinning the approaches we have employed is that bacteria in the biofilm environment are subjected to environmental pressures which can dramatically impact phenotypic characteristics. These include pH, carbohydrate source and availability, and oxygen tension and reducing environment, along with many other chemical and physical forces. Notably, it is well established that there is a great deal of overlap in regulatory systems and cascades in bacteria in response to these particular external forces. Consequently, in order to dissect how these various influences impact gene expression and virulence in oral streptococci, it is critical to utilize systems which allow for study of responses to environmental stimuli in a setting that severely restricts variations in other environmental parameters so as not to confound interpretation of the outcomes. Continuous chemostat cultivation of lactic acid bacteria, although more technically cumbersome than batch culture, is among the best technologies to study particular environmental influences. This chapter will focus first on methods for the employment of continuous culture to study virulence gene expression in oral streptococci, and then later on continuous flow systems to analyze various properties of adherent populations of oral bacteria.

2. Materials

A. Equipment
1. Homogenizer, model mini-Bead Beater, Bio-Spec Products.[3]
2. Micro-centrifuge, refrigerated, Eppendorf, VWR.[14]
3. Adjustable pipettes: nos. P-20; P-200, P-1000, Rainin.[10]
4. Centrifuge, refrigerated-preparative, Sorvall RC5C, Sorvall, Inc.[12]
5. SS-34 rotor for RC5C centrifuge.[12]
6. Pipette-Aid, Drummond, VWR. no. 53498-001.[14]
7. Beckman DU640 Spectrophotometer, Beckman Instruments.[1]
8. New Brunswick Chemostat, New Brunswick Scientific.[8]
9. Modified Rototorque, Center for Biofilm Engineering.[4]

B. Culture medium
- Base Medium: Tryptone 30 g; yeast extract, 5 g (TY).[6]

E. Solutions
- 10 mM Trizma base (Tris), no. T-153027, adjusted to a pH value of 7.8 with HCl.[10]
- Bio-Rad dye-binding reagent, 500-0006.[1]

H. Glassware and plastics
- Oakridge tubes, no. 3119-0050.[13]
- 50 ml conical Corning tube, no. 25330-50.[7]
- 2 ml screw cap microfuge tubes
- glass beads, no. 34007.[13]
- cryovials, with O-rings and screw-caps, no. 4204-S.[7]
- 0.45 µM autoclavable filters.[8]

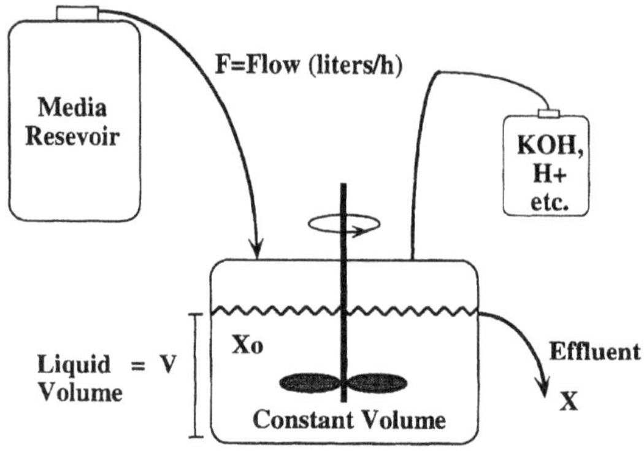

F = **Flow rate of medium (l/h)**
V = **Liquid Volume in Vessel (Constant)**
Xo = **Cell mass/liter in vessel (g/l)**
X = **Cell mass in effluent (g/l)**

Figure 1. A schematic representation of a single-stage chemostat for continuous cultivation of microorganisms. A media reservoir delivers sterile medium into the culture vessel at a fixed rate. The volume of the culture in the vessel is kept constant either by pumping out fluid or by letting it trickle out a port in the side of the vessel. Particular variables (F, V, X_0, and X) are defined and treated in Equations 4–6.

3. Methods, results and discussion

A. *Use of continuous culture to modulate physiology of oral streptococci: General Considerations*

The basic workings of the continuous chemostat are depicted in Figure 1. In reality, this is a fairly simple, albeit not inexpensive instrument. The chemostat consists of a vessel in which bacteria can be cultured in a fixed volume. With existing instrumentation, one can control dissolved oxygen, pH, and rate of input of nutrient sources, among other chemical and physical parameters. Operation of the vessel in batch mode is adequate for examining effects of certain environmental influences, but the cells grow as if in batch culture; passing through lag, exponential and into stationary phase. This means that influences on gene expression as a result of varying growth rates and growth phase phenomena cannot be excluded. A more appropriate use of the chemostat for the types of studies described above is to employ the instrumentation for continuous cultivation. In this mode, one can achieve 'steady-state' conditions, and this has important implications as detailed below. First, though, a brief overview of some principles of continuous chemostat culture are necessary or useful.

One runs a continuous chemostat by introducing fresh medium into the vessel at a fixed rate while maintaining a constant culture volume. In some cases, a reflux system that reintroduces expelled bacteria is also employed, but this is unusual in most applications of the technology to basic research questions. The rate of introduction of the medium, or dilution rate, *D*, is expressed as the amount of medium introduced per hour as a fraction of the total culture volume. For example, for a 1 liter culture, introduction of 100 ml per hour would correspond to a dilution rate, *D*, of 0.1 h^{-1}. After a number of generations, 10 appear to be sufficient, the culture can achieve steady state. At steady state, the optical density of the culture should not change and therefore, the yield, Y, remains constant under steady state conditions. This occurs because the rate at which the bacteria divide is directly related to the rate of introduction of fresh medium at steady state. Another key point is that some particular nutrient, depending on medium composition, will always be limiting under true steady state conditions. The final yield that is observed will be correlated with the yield achievable with the limiting substrate (S). Yield is also related to the maintenance coefficient for that particular substrate (Ks), which is usually an empirically determined mathematical expression of the material and energy spent by the organisms not for growth, but to maintain physiologic homeostasis. By way of example, for the oral streptococci, yields will be lower at pH 5.0 compared with pH 7.0 because the cells will be spending more energy to pump protons and maintain a comparatively alkaline cytoplasm; higher maintenance. The yields will also vary depending on whether the cells are grown on sorbitol or glucose.

Working from established equations, one can demonstrate that at steady state the specific growth rate, μ (sometimes μ_g), is directly related to the dilution rate D. First, using the standard equation to calculate cell number after a certain period of exponential growth,

$$X_t = X_0 \cdot e^{\mu t} \quad (1)$$

where X_t is the final number of cells present after a given time, X_0 is the number of cells at time 0, μ the specific growth rate, and t, the time, usually in hours. Taking the natural log of Equation 1,

$$\log_e X_t = \log_e X_0 + \mu t \quad (2)$$

then taking the derivative of the equation and allowing for one doubling ($X_t = 2$, $X_0 = 1$)

$$\frac{0.693}{\mu} = t \quad (3)$$

Equation 3 relates the specific growth rate to the generation time, t_g.

Now, mathematically considering what is happening with the population under steady state conditions in a typical chemostat, the following equation is applicable where the variables are defined in Figure 1:

$$\frac{FX_o}{V} - \frac{FX}{V} + \mu X - \alpha X = \frac{dX}{dt} \quad (4)$$

Cells in – Cells out + Growth – Death = Change in cell # with time; at steady state = 0. The letter α is a coefficient related to the rate of death of cells in the population.

In a 'single stage' chemostat, the flow of cells in is 0 because the feed stream is sterile, so that portion of the equation cancels. At steady state, $dX/dt = 0$, and if one assumes the contribution to the equation of cell death to be essentially = 0, which is reasonable for these exponentially growing populations, then,

$$-\frac{FX}{V} + \mu X = \frac{dX}{dt} = 0 \quad (5)$$

$$\mu = \frac{F}{V}, \text{ and } D = \frac{F}{V}, \text{ therefore, } \mu = D \quad (6)$$

The really useful aspect of steady state cultures is that one can modulate a single environmental parameter, for example pH, carbohydrate source, carbohydrate availability, or oxygen concentration, among others, while maintaining the same specific growth rate and without necessarily changing the limiting nutrient. Conversely, the growth rate can be altered while keeping constant the aforementioned variables. This is particularly easy with the lactic acid bacteria because respiration is not a factor. With organisms that can grow on a truly defined medium, it is possible to define the limiting nutrient. For example, one can simply obtain a sample of the spent medium and measure residual glucose to insure carbohydrate limitation, or measure ammonia, if that is the sole nitrogen source, to insure that cultures were nitrogen limited. Practically speaking though, with the oral streptococci, studies have been restricted to examining cultures growing in the presence of limiting or excess carbohydrate. Carbohydrate excess cultures have been referred to as nitrogen limiting in some past publications, but it is not clear whether the carbohydrate excess conditions used in those studies really resulted in limitation for a nitrogen source or for some other essential nutrient, e.g., a vitamin. Nonetheless, since the Gram-positive cocci that are focused on in these volumes rely almost solely on carbohydrate for growth, and have no functional respiratory cascades, the modulation of carbohydrate availability is a central physiologic control point and thus highly relevant in many cases. Notably, some studies we have conducted indicate a readily metabolizable nitrogen source, probably glutamine, may become limiting under carbohydrate excess conditions with the particular medium we use.

B. *Key tips in maintaining 'long-term' continuous culture*

If one is conducting experiments at low dilution rates, e.g., $D = 0.05\ h^{-1}$, it takes roughly 5.8 days to achieve steady state. So even an experiment designed to take measurements at three different pH values under glucose limiting and glucose excess conditions can easily extend for 6 weeks. High dilution rates mandate that media reservoirs and other additives be changed frequently. As might be predicted then, one of the most common problems encountered in continuous culture studies is contamination. This can occur through introduction of more competitive species during the initial set up of the chemostat or with changeover of input vessels, from sampling, or as a result of failure of some component of the systems, generally pump tubing. Some fairly simple modifications to the vessel and common aseptic techniques, as well as some preventive measures can essentially eliminate contamination problems. Most importantly, thanks largely to techniques developed in the laboratory of R. E. Marquis (University of

Rochester Medical Center), we now routinely employ 1/4 inch stainless steel tubing and Swagelock™ brand, stainless steel compression fittings on all of the connection points for media and additive feed lines. This allows for the connection points of input lines to be flamed prior to fastening, and to be heated directly over a flame for several seconds afterward to decontaminate the connection point. We have found this far superior to the 'poke and hope' method of connecting tubing using plastic fittings that have been covered with aluminum foil and autoclaved. Use of stainless steel fittings allows for employment of rigorous aseptic techniques, a major advantage for long term continuous culture. The only other suggestion is that tubing, particular any that passes through pump heads, should be changed on a regular basis. The cost of tubing is minor compared to the cost of media and time devoted to setting up and maintaining a chemostat.

C. *The use of continuous culture to modulate physiology of oral streptococci: Growth limitations*

The following sections detail the media and specifics of setting up a chemostat for oral streptococci, in particular, we have used these conditions for *S. mutans* and *S. salivarius*. We, and others who have used similar conditions have found that the prescribed base medium will support the growth of oral streptococci using sugar concentrations between 10 and 200 mM. Dilution rates can vary widely, but at too slow a rate, the organisms tend to colonize the surfaces of the vessel (wall growth), whereas at too high a dilution rate, the organisms cannot divide fast enough and the population steadily declines (wash out). D values of 0.05 h^{-1} to 0.5 h^{-1}, or even greater, can be accomplished without unacceptable wall growth or wash-out, respectively. The lower limits of pH for growth while still maintaining steady state is usually around 5 for *Streptococcus mutans* and about 5.5 for *Streptococcus salivarius*. Variation in the growth characteristics occurs in a species and strain dependent fashion, so it is usually necessary to empirically determine what are acceptable growth conditions for a given strain when operating the chemostat at predicted extremes.

D. *Medium preparation*

The base medium consists of 30 g of tryptone and 5 g of yeast extract per liter. In general, material for 11 liters of base medium is dissolved in 10.8 liters of H_2O when supplementing with low concentrations of carbohydrates (e.g., 10 or 25 mM final concentrations) or 10 liters of H_2O when supplementing excess concentration of carbohydrates (e.g., 200 mM). The carbohydrates are dissolved in 200 ml or 1 liter of H_2O for low or high concentrations of sugars, respectively, and autoclaved separately. For oral streptococci, the medium should be completely free of sucrose to prevent wall growth or fouling of the pH probe. To remove trace amounts of sucrose present in the base medium and the sugar solutions, about 1 mg (approx. 1000 U/mg, Sigma) of invertase is added to both the base medium and sugar additives prior to autoclaving. The base medium is autoclaved for 60 min on a liquid cycle. It is strongly recommended that the medium be allowed to cool for one hour with the autoclave door open before removing the vessel. The vessels can burst, or jarring the hot container of medium can cause the rapid release of steam. After the medium cools, additives, e.g., carbohydrates, antibiotics, etc., can be introduced. This is done as shown in Figure 2. The two vessels, with the tubing clamped, are joined using the stainless steel fittings, wrenches are used to tighten the connection, the tube clamps are removed and a flame is applied for about 10 to 15 s. After the connection has cooled, a hand held pump with tubing that will fit the exposed end of the 0.45 μm filter on the base medium vessel is attached and a mild vacuum is applied. This will cause the sugar solution in the reservoir to flow into the vessel. If it is desired, antibiotics can be added to the sugar solution or directly to the base medium. We introduce the heat labile compounds by fitting the vessel with a curved stainless steel tube that can be covered with aluminum foil. We dissolve the additives and filter sterilize them directly into the tube, flaming afterward.

The preparation of the chemostat culture vessel is fairly straightforward. The culture vessel is fitted

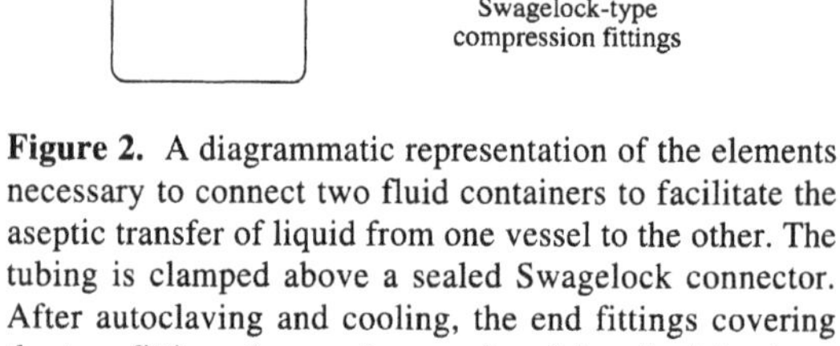

Figure 2. A diagrammatic representation of the elements necessary to connect two fluid containers to facilitate the aseptic transfer of liquid from one vessel to the other. The tubing is clamped above a sealed Swagelock connector. After autoclaving and cooling, the end fittings covering the two fittings (one male, one female) to be joined are removed near a flame. The ends are passed briefly through the flame and the ends are connected finger tight. Using open ended wrenches, the fittings are securely tightened, and then the connection is heated directly in a flame for about 10 s. The clamps to the tubing are then removed and the fittings are heated for an addition 20–30 s. See text for additional details regarding the appropriate treatment of vessels before and after autoclaving.

with Swagelock connections that are compatible with those on the medium reservoir and the alkali source. We generally place about one half of the final volume of medium into the vessel before autoclaving. Follow the suppliers recommendation for calibrating any electrodes, pH, dO_2, etc. prior to autoclaving. We autoclave our 1 liter capacity vessels for 30 min. Be certain to place one, and preferably two 0.45 μm exhaust filters on the culture vessel to release pressure built up with heating of the liquid and do not over tighten the top stainless steel portion of the vessel onto the glass bottom. After autoclaving and cooling, the chemostat is set up per manufacturers directions, and the media and alkali reservoirs are joined as above. One useful tip is to use a T-connection for the culture vessel medium input. This will allow for the connection of two media reservoirs, which can be useful for runs that might exhaust one vessel in the middle of the night. Both the alkali solution and medium source are connected to the culture vessel through Swagelok tube fittings.

E. *Running the chemostat*

After inoculation, cultures are stirred at 250 rpm and the pH is controlled by addition of 2 M KOH. When the culture reaches an $O.D._{600}$ of approximately 0.6–0.8, the medium pump is turned on. From this point, steady state, where μ, the specific growth rate, is equal to D, the dilution rate of the vessel, is assumed to be established after at least 10 generations at any single set of growth parameters. Some debate exists over whether 10 generations is sufficient time for steady state to be established. Studies that we have conducted, using urease expression in *S. salivarius*, or with chloramphenicol acetyltransferase gene fusions to promoters in *S. mutans*, support that 10 generations are probably sufficient to establish steady state. For example, using a *cat* gene fusion to the *ftf.* gene promoter, we have shown that, following transition from one growth condition to another, reporter gene enzyme activities reached the new steady state values within hours [13]. Likewise, when cultures of *S. salivarius* are shifted from pH 7.0 to pH 5.5, the acid inducible urease genes are rapidly activated, and the new steady state levels of urease are reached after only 5–6 generations. Shift back to pH 7.0, which shuts down transcription of the urease genes results in loss of roughly 1/2 of this very stable enzyme per generation, until the new steady state level is reached, and activities remains constant after 7–8 generations (Figure 3) [5]. Extension of the run at a given steady state condition did not result in any additional changes in activities. Obviously, if samples are removed or environmental conditions are altered, one must wait an additional 10 generations for a new steady state to become established.

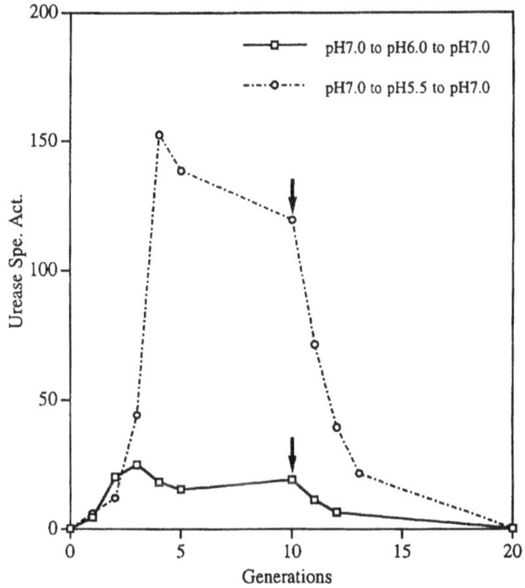

Figure 3. Time course of induction of urease enzyme activity. Steady state cultures of *Streptococcus salivarius* 57.I, growing at $D = 0.3\ h^{-1}$ were shifted (time 0) from pH 7.0 to pH 5.5. Urease activity increased rapidly and stabilized after 5 generations. After the cells equilibrated to steady state at pH 5.5, the culture pH was adjusted to 7.0 (arrow) and urease activity was measured. Apporoximately half of cellular urease activity was lost per generation, indicating that de nove synthesis of this very stable enzyme was rapidly arrested. Urease specific activity is expressed as the nanomoles of urea hydrolyzed per minute per mg cell dry weight.

F. *Harvesting cells and preparation for enzyme assays*

1. Cultures are harvested by aspiration.
2. Cells are recovered by centrifugation at 5000 rpm, 4 °C for 15 min in a Sorvall SS34 rotor.
3. The residual medium is removed by washing the cell pellets once with 10 mM sodium phosphate buffer, pH 7.0, 1 mM $MgCl_2$.
4. Cells are then resuspended in 10% of the original volume in the same buffer.
5. Cell dry weights are determined from 10 ml of concentrated cell suspensions that have been baked overnight at 80 °C, and equilibrated to room temperature in a desiccator before weighing.

The dry weight of the same amount of buffer alone is also determined in parallel, and the difference between these two values represents the net weight of the cell suspension. Alternatively, if the cells are stable in dH_2O, like many oral streptococci, cells can be washed, resuspended, dried and weighed in the same fashion without need of taking into consideration the modest contribution from buffers and other salts.

G. *Preparation of cell-free lysates for assaying of chloramphenicol acetyltransferase (CAT) activity*

Other chapters in these volumes deal more directly with the use of reporter gene fusions. However, two useful observations about reporter proteins have been gleaned from our studies with *cat* and *lacZ* fusions in continuous culture and biofilms. CAT is not stable for any great period in oral streptococci. For example, if *cat* is induced from the *gtfBC* promoter, it rises rapidly, but falls within 1 hour to the previous steady state level [13]. This is unlike many reporter enzymes, the levels of which remain high following induction even though this may not represent the actual level of transcription initiation. LacZ appears somewhat more stable in streptococci. These observations should be taken into consideration when evaluating data generated using these constructs in oral streptococci.

1. Cells are harvested by centrifugation at 5,000 rpm in a Sorvall centrifuge equipped with an SS34 rotor, at 4 °C, for 15 min.
2. Cells are washed once with an equal volume of 10 mM Tris-HCl, pH 7.8, and then concentrated to 2.5% of the original volume in the same buffer.
3. The concentrated cell suspension (1 ml) is transferred to a 2-ml screw cap microfuge tubes containing 0.75 g (roughly 1/3 volume) of glass beads (0.1 mm avg. dia., Sigma, St. Louis, MO).
4. Cell suspensions are homogenized in a Bead Beater for 1 min, at 4 °C, followed by chilling the tubes on ice for 2 min.
5. Repeat step 4 one more time.
6. Cleared lysates are generated by centrifugation at 10,000 rpm for 10 min at 4 °C in an Eppendorf Microcentrifuge, and the protein content of each lysate is determined by using the Bio-Rad Protein Assay, based on the method of Bradford [1]. Bovine serum albumin serves as the standard. Cell extracts retain CAT activity when stored at –80 °C for at least 1–2 months. Generally, we use the cell extracts immediately.
7. There are a number of methods for measuring chloramphenicol acetyltransferase activity. We use the following kinetic assay as detailed by Shaw [11]. The assay is best carried out with a recording spectrophotometer equipped with a temperature controlled cuvette chamber. We employ a Beckman DU640 equipped with a Peltier effect temperature-controlled system with 6 cuvette holders and utilize the kinetics software package. With the 6 cuvette holder, we load three identical test samples and one blank. The instrument plots rates and automatically subtracts any background detectable in the absence of chloramphenicol.
   i. Prepare the reaction mixture freshly by dissolving 4 mg 5,5′-Dithiobis-2-nitrobenzoic acid (DTNB) in 9.8 ml Tris-HCl, pH 7.8. To each 9.8 ml mixture, 0.2 ml of 5 mM acetyl-CoA solution is added. Equilibrate the mixture to 37 °C prior to use.
   ii. To each 1 cm light path cuvette, add pre-warmed reaction mixture and pre-warmed cell lysate to a final volume of 955 μl ml. All assays are carried out at 37 °C, in triplicate, with one blank (no chloramphenicol added).
   iii. The reactions are initiated by the addition of chloramphenicol (Cm) to a final concentration of 0.1 mM.
   iv. The rate of increase in absorption at 412 nm is monitored. The difference between the samples and the blank is recorded. The rates of Cm acetylation of each protein lysate are calculated as described by Shaw [11]. The net change in extinction per minute is divided by 13.6 to obtain μmoles per min of Cm-dependent DTNB reduction, and the specific activities are expressed as μmoles of Cm acetylated/min/mg protein.

H. *Quantitation of LacZ activity*

This method is simply an adaptation of the protocol of Miller [10] for assay of β-galactosidase in permeablized cells of *E. coli*.

1. Cells are harvested by centrifugation as in section F2.
2. Cells are washed twice with equal volume of an 10 mM $NaPO_4$, pH 7.0, 1 mM $MgCl_2$, and resuspended to 10% of the original culture volume in the same buffer.
3. Aliquots of the cultures are then added to the assay medium (Z buffer: 0.06 M $Na_2HPO_4 \cdot 7H_2O$, 0.04 M $NaH_2PO_4 \cdot H_2O$, 0.01 M KCl, 0.001 M $MgSO_4 \cdot 7H_2O$, 0.05 M β-mercaptoethanol, pH 7.0) to final volume of 1 ml. After chilling in an ice-water bath briefly, 50 μl of toluene is added to the cell suspensions followed by vortexing at high speed for 5 min.
4. All cell suspensions are then incubated at 37 °C for 40 min with the top open to evaporate toluene prior to assay.
5. Equilibrate all samples to 28 °C for a minimum of 5 min. Reactions are initiated by addition of 0.2 ml of *o*-nitrophenyl-β-D-galactoside (ONPG, 4 mg/ml in 0.1 M phosphate buffer, pH7.0) at 28 °C, with gentle shaking for a few seconds.
6. Record the time of the reaction. The reactions are terminated when sufficient yellow color has developed by addition of 0.5 ml of 1 M Na_2CO_3.
7. Record the optical density at both 420 nm and 550 nm. Units are calculated as follow:

$$Units = 1000 \times \frac{O.D._{420} - 1.75 \times O.D._{550}}{t \times v \times O.D._{600}}$$

where t = time of the reaction in minutes and v =

volume of the culture used in the assay. Cell dry weights can also be utilized for normalization, or cell lysates prepared and protein concentration used to normalize the data. However, the ease of permeablization of the cells and the stability of LacZ to this treatment make it ideal, particularly for a large number of samples or kinetic studies of gene induction.

I. *Biofilm model systems: Considerations*

There are a wide variety of biofilm model systems currently in use to study adherent populations of oral bacteria from the standpoint of ecology, biological activities, physiologic concerns and monitoring of gene expression. These encompass fairly simple systems in which oral streptococci are cultivated on wires or glass rods in batch type systems. These simpler models usually are composed of a single species and can yield high quality repeatable results for certain applications. Like other systems, when using monospecies cultures of oral streptococci, efficient formation of biofilms that are greater than a few microns in depth usually requires that sucrose is present in the medium as a substrate for glucosyltransferases, which, in conjunction with other glucan binding proteins promote tenaciously adhering populations. The next level in complexity of biofilm model systems contain a single species but employ continuous flow. These come in a variety of shapes and sizes, including micro flow cells, with working volumes of 50 to 100 μl and surface areas of about 1 cm^2 that are amenable to microscopic evaluation, as well as fairly large (1 L) reactor vessels with more elaborate systems to control environmental conditions and fluid flow. The choice of system is mainly dictated by the research question, or by preferences among investigators. For example, studies on dynamics of biofilm formation and structure can be accomplished on a small scale in microflow cells. Expectedly, microcells are often the best suited for these types of studies especially if the substrates to be provided are available in limited quantities, e.g., salivary secretions. Alternatively, systems which can produce substantial amounts of cellular material usually utilize media components that are neither limiting nor prohibitively expensive. These latter systems are useful to study effects of biofilm heterogeneity on physiologic parameters or on the expression of particular genes, as the larger models provide sufficient biomass and one can obtain multiple samples to obtain statistically significant results. Importantly, the continuous flow systems incorporate some key features that static, batch systems do not, including laminar flow, which can somewhat mimic salivary fluids passing over the plaques, and more flexibility for controlling critical environmental parameters, such as pH, dissolved oxygen, and carbohydrate source and concentration in the planktonic phase. Some systems also employ material in the substratum that is more consistent with surfaces found in the oral cavity, such as hydroxylapatite. Although this may be a more relevant material, there are pros and cons to using these types of material. Consideration for the specific research question must dictate whether one selects a leaching or non-leaching surface. Various continuous flow systems have been used to examine mixed population biofilms; composed either of defined multispecies consortia or used to cultivate populations following inoculation with plaque samples obtained from humans. The former has been elegantly employed to examine the role of various environmental parameters, e.g., pH, oxygen, and the presence of particular species on the persistence of organisms in mixed populations [2]. Nevertheless, the purpose of this chapter is not to review available biofilm models, which has been done quite well in several other publications [4, 6, 12]. Instead, we will detail methods which we have utilized to examine monospecies cultures of oral streptococci for the purposes of examining gene regulation and expression in these adherent populations.

J. *Cultivation of strains of oral streptococci in a commercially available biofilm reactor*

We began our studies using a commercially available, continuous flow biofilm reactor, known as a Rototorque [4]. The vessel was originally designed for studying biofilms formed by aquatic microorganisms, and included a reflux loop, which could recycle sloughed and shed microorganisms, and planktonically growing cells back through the systems. The Rototorque was designed with the idea that significant shear forces could be generated, similar to conditions in water lines, streams, etc. However, high shear is neither pertinent or favorable to biofilm formation by oral bacteria. Consequently, we had the vessel modified by the supplier to remove the reflux loop and we operate the system at very low shear forces. This allows for the formation of biofilms that can be several hundred microns in depth and still has the advantage that fairly large quantities of material can be generated for analyses.

Configuration of the Rototorque is as detailed in Figure 4. Specifically, the modified Rototorque is roughly 5 cm shorter than the standard instrument and the port in the bottom of the vessel was sealed, yielding a vessel with a 'working volume' of 0.6 L. The average surface area of the exposed face of one of the 12 polystyrene slides in the vessel is approx. 31 cm^2. Basically, we operate the instrument much like a chemostat. An input pump delivers fresh medium at a fixed dilution rate, D, expressed as a proportion of the total liquid volume in the vessel. A constant volume is maintained by pumping liquid

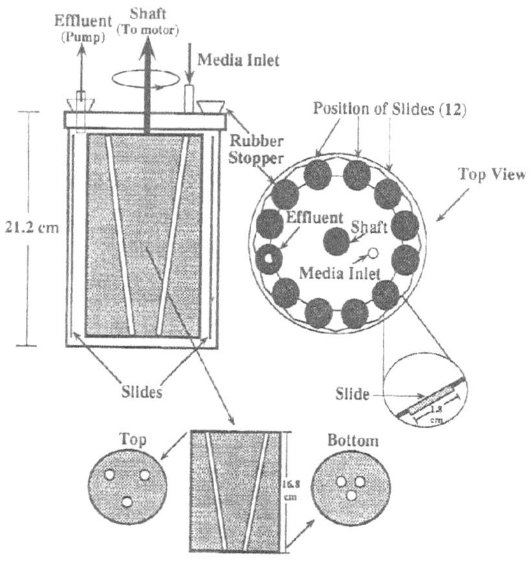

Figure 4. A schematic diagram of the modified Rototorque. Medium is pumped into the vessel and effluent was pumped from the top of the vessel as shown. The speed of the rotating inner drum was kept constant at 75 RPM. The temperature of the vessel was maintained by immersion in a 37 °C, circulating water bath. Some dimensions of the vessel, the inner drum, and the width of the slides are shown for reference.

from the vessel. As with the chemostat, we use Swagelock type fittings to connect the instrument to media feeds. The control of pH in the vessel, unlike a chemostat, is far more technically challenging. Since the biofilms are formed under conditions which promote the colonization of surfaces, use of a pH probe in the vessel is impractical because the probe will quickly become fouled. The use of buffers is a practical, although less than ideal approach. However, it must be kept in mind that this does not really allow for control of the pH in the biofilms, and that gradients will certainly develop within the surface-growing organisms. The other major consideration is that attainment of true steady state conditions, as detailed above, may not be possible in biofilms, particularly thicker films. The films thus formed are heterogeneous and access of all cells in the population to a uniform pool of nutrients is impossible. Moreover, the growth phase of bacteria within the films can differ dramatically and biofilm populations can contain considerable proportions of apparently dead cells. As a result of these confounding factors, investigators frequently report their data at 'quasi-steady state', where D has been constant for the equivalent of 10 generations, as calculated above.

K. *Protocols specific for growth of oral streptococci in the Rototorque*

Overnight cultures of oral streptococci used for inoculation of the biofilm reactor vessel are grown at 37 °C in a 5% CO_2, aerobic atmosphere in TY with 10 mM glucose, and with antibiotic if necessary. For cultivation of cells in biofilms, TY medium supplemented with 10 mM sucrose (TYS) is used. The Rototorque reactor is filled with about 1/2 the working volume of TYS, autoclaved for 30 min, equilibrated at 37 °C, and inoculated with 10 ml of an overnight culture of the strain to be studied. Fresh medium is introduced immediately after inoculation. The instrument can be operated with the motor connected directly to the shaft of the rotating inner drum. Rotation of the inner drum is controlled at a speed determined by a rheostat setting of '20' on the instrument (approximately 75 RPM). In the past, we have conducted analyses of biofilm cells after 48 h or for 7 days post inoculation, and visible biofilms begin to form in as little as 18 to 24 h [3].

L. *Sample preparation and assays of reporter gene product activity*

The thickness of the biofilms formed is dependent largely on sucrose concentration, the strain of bacteria being used, and the length of time the biofilms are allowed to form. In the past, we have found that 48 h biofilms are generally about 20 μm in depth, and frequently have properties similar to their planktonic counterparts. Under the growth conditions we utilize, biofilms of *S. mutans* formed in 7 days achieve thicknesses of of greater than 100 μm. It is with these thicker films that we have found that the gene expression patterns of the exopolysaccharide synthases, glucosyltransferase and fructosyltransferase, are altered as compared with their planktonic counterparts [3]. Parenthetically, strains of *S. sobrinus* can form biofilms that are much thicker under these conditions, while the soft tissue microorganism, *S. salivarius*, forms thinner films that are more easily removed from the polystyrene slides.

In the past, we have examined 'quasi-steady state' gene expression and induction in a kinetic fashion. The ability to do this is one major advantage to the Rototorque. Specifically, one or more of the twelve slides can be removed aseptically over a time course. For example, we have looked at induction of the *gtf* and *ftf* genes in cells in 48 h or 7 day cultures and then induced with addition of 25 mM sucrose [3]. An alternative application might include monitoring biomass, polysaccharide content and biochemical activities over a 12 day course. Slides are removed from the vessel, cells are mechanically dissociated from the slides into a solution of ice cold 10 mM/L Tris-HCl (pH 7.0) containing 10 μg of rifampicin and tetracycline per ml, to arrest transcription and translation, respectively. One can use a rubber policeman or glass microscope slide to assist in removal of the cells. The cells are immediately centrifuged at 8000× *g* for 10 min at 4 °C. Cells are washed in ice cold 10 mM Tris-HCl (pH 7.8) and the cell pellets kept on ice until all samples are ready for processing.

For assays of activities of reporter gene products, we essentially utilize the protocols described above. The only key considerations to take into account are that biofilm cells occur in an aggregated state and the biofilms themselves can be particularly rich in polysaccharides. For these reasons, it is not recommended to utilize dry weights for normalization of activities. Rather, quantification of protein is the preferred approach. In the case of *cat* gene fusions, we homogenize the material in the films as above and simply measure protein by the Bradford assay. We have confirmed that this assay is suitable for this purpose by completely hydrolyzing the biofilms, and conducting amino acid analysis and quantification of total polysaccharide. The results we have obtained using analytical chemistry correlate well with estimates of protein from the Bradford assay. For measuring LacZ activity, we have normalized the assay to $O.D._{600}$. In these cases, the cells are dispersed by sonication at 350 W with a microprobe for 30 s, twice, and then assays are performed and optical density readings are obtained.

M. *Preparation of cells for microscopy*

For imaging of biofilms, cells are cultured as above. Slides are removed from the vessel and briefly dipped in room temperature dH_2O to remove adventitiously bound cells. The polystyrene slides with the biofilms can then be cut into sections of about 5 cm and mounted on a glass microscope slide. A #1 glass slide is then gently placed over the biofilm populations and the adherent populations can be examined by phase contrast microscopy.

We have also effectively used fluorescence microscopy and scanning confocal laser microscopy to optically section biofilms that have been stained with live:dead stains, fluorescein- or rhodamine-labeled antibodies, or by providing 4-methylumbelliferyl or other fluorescent substrates for LacZ to cells. A detailed description of the methods for SCLM will not be presented because more detailed articles exist [4, 9]. More importantly, it is best to obtain assistance from individuals who have expertise in the areas of microscopy for these studies.

4. Conclusions

Batch cultivation of pure cultures of planktonically growing bacteria is the cornerstone of virtually all of our current understanding of the physiology, biochemistry and genetics of prokaryotes. Nonetheless, there are valid scientific reasons to believe that batch cultivation and the study of planktonic cells may constrain the phenotypic capacities of microorganisms. By using alternative methods to alter cellular physiology, such as tighter control in chemostats or growth of organisms in biofilms, additional key biological activities or behaviors may be revealed that will be central to understanding initiation or progression of disease. Moreover, the potential biological activities of consortia of microorganisms should not be overlooked. For example, it is established that degradative communities, consisting of populations of spatially arranged microorganisms, can completely catabolize compounds that any single species in the consortia cannot. There is ample reason, but comparatively little solid evidence, to believe that such communities form and play an active role in diseases progression. Future studies coupling the use of physiologic tools and studies of adherent populations with sophisticated and sensitive techniques in molecular biology and non-invasive microscopic technologies promise to open a new era of understanding of the pathogenic properties of individual and populations of bacteria.

Acknowledgments

This work was supported by PHS grants DE10362 and DE12236.

Notes on suppliers

1. Beckman Instruments Inc., 2500 Harbor Blvd. Fullerton, CA 92634, USA
2. Bio-Rad Laboratories, 2000 Alfred Nobel Drive, Hercules, CA 94547, USA
3. BioSpec Products, Bartlesville, OK 74005-0722, USA
4. Brinkmann Instruments Inc., One Cantiague Road, P.O. Box 1019, Westbury, NY 11590-0207, USA
5. Center for Biofilm Engineering, Montana State University, Bozeman, MT 59717, USA
6. Difco Corporation, P.O. Box 331058, Detroit, MI 48232, USA
7. Laboratory Product Sales, 1665 Buffalo Road, Rochester, NY 14624, USA
8. Nalgene, 75 Panorama Creek Dr., Rochester, NY 14625, USA
9. New Brunswick Scientific, P.O. Box 4005, 44 Talmadge Rd., Edison, NJ, USA
10. Rainin Instrument Company, Mack Road, Woburn, MA 01801, USA
11. Sigma Chemical Company P.O. Box 14508, St. Louis, MO 63178, USA
12. Sorvall, Inc., 31 Pecks Lane, Newtown, CT 06470-2337, USA
13. Swagelok Co., Solon, OH, USA
14. VWR, P.O. Box 626, 200 Center Sq. Rd., Bridgeport, NJ 08014, USA

References

1. Bradford MM (1976). A rapid and sensitive method for the quantitation of microgram quantities of protein utilizing the principle of protein-dye binding. Analyt Biochem 72: 248–254.
2. Bradshaw DJ, Marsh PD, Schilling KM, Cummins D (1996). A modified chemostat system to study the ecology of oral biofilms. J Appl Bacteriol 80: 124–130.
3. Burne RA, Chen Y-Y M, Penders JEC (1997). Analysis of gene expression in *Streptococcus mutans* in biofilms in vitro. Advances Dent Res 11: 100–109.
4. Characklis WG (1990). Laboratory biofilm reactors. In: Characklis WG, Marshall KC (eds), Biofilms, pp 55–89. New York: John Wiley and Sons.
5. Chen YM, Burne RA (1996). Analysis of *Streptococcus salivarius* urease expression using continous chemostat culture. FEMS Microbiol Letters 135: 223–229.
6. Costerton JW, Cheng K-J, Geesey GG, Ladd TI, Nickel JC, Dasgupta M, Marrie TJ (1987). Bacterial biofilms in nature and disease. Annu Revs Microbiol 41: 435–464.
7. Costerton JW, Lewandowski Z, Caldwell DE, Korber DR, Lappin-Scott HM (1995). Microbial biofilms. Annu Revs Microbiol 49: 711–745.
8. Gibson CM, Caparon MG (1996). Insertional inactivation of *Streptococcus pyogenes sod* suggests that *prtF* is regulated in response to a superoxide signal. J Bacteriol 178: 4688–4695.
9. Lawrence JR, Korber DR, Hoyle BD, Costerton JW, Caldwell DE (1991). Optical sectioning of microbial biofilms. J Bacteriol 173: 6558–6567.
10. Miller JH (1972). Experiments in molecular genetics. Cold Spring Harbor, NY: Cold Spring Harbor Laboratories.
11. Shaw WV (1979). Chloramphenicol acetyltransferase activity from chloramphenicol-resistant bacteria. Methods Enzymol 43: 737–755.
12. Sissons CH (1997). Artificial dental plaque biofilm model systems. Advances Dent Res 11: 110–126.
13. Wexler DL, Hudson MC, Burne RA (1993). *Streptococcus mutans* fructosyltransferase (*ftf*) and glucosyltransferase (*gtfBC*) operon fusion strains in continuous culture. Infect Immun 61: 1259–1267.

Address for correspondence: Robert A. Burne, Center for Oral Biology and Department of Microbiology and Immunology, University of Rochester School of Medicine and Dentistry, 601 Elmwood Avenue, Rochester, NY 14642, USA

Phone: (716) 275-0381; Fax: (716) 473-2679

E-mail: robert_burne@urmc.rochester.edu

Methods in Cell Science **20**: 191–201 (1998).

In vitro systems for investigating group B streptococcal: host cell and extracellular matrix interactions

Scott B. Winram, Glen S. Tamura & Craig E. Rubens
Children's Hospital and Regional Medical Center, Division of Pediatrics, Department of Infectious Diseases, Box 5371/CH-32, Seattle, Washington, USA

Abstract. *Streptococcus agalactiae* or group B streptococci (GBS) are gram-positive diplococci and are the leading bacterial cause of pneumoniae, sepsis, and meningitis in neonates. Neonatal GBS infections may occur prior to or during birth. GBS have been cultured from the chorioamnionic membrane of pregnant women and have therefore been associated with chorioamnionitis and premature labor. A potential route for GBS to establish infection of a neonate would be to penetrate the placental membrane of colonized pregnant women. In our laboratory, we have constructed in vitro systems to emulate certain events during the colonization and invasion of host epithelial cell tissues by GBS. By utilizing techniques to grow primary cultures of both chorion cells and amnion cells isolated from human C-section placentas, we have established a relevant model to investigate certain aspects of GBS adherence and invasion into the placental membrane. To identify relevant molecules required for GBS to colonize the multiple tissues it encounters during an infection, we have applied a variety of biochemical approaches with host cell membrane preparations as well as purified extracellular matrix proteins. These techniques are enabling us to further characterize the pathogenic mechanisms utilized by GBS.

Key words: Adherence, Chorioamnion, Extracellular matrix, Group B streptococci, Invasion fibronectin

1. Introduction

Group B streptococci (GBS) are the leading cause of bacterial pneumoniae, sepsis, and meningitis in neonates and are a major cause of bacteremia in immunocompromised adults [2]. GBS frequently colonize the human recto-vaginal tract and colonization of the placental membrane by GBS may lead to chorioamnionitis or premature rupture of the placental membrane. GBS have been hypothesized to actively invade through the placental membrane resulting in growth of the organism in the amnionic fluid [12]. Neonatal exposure to GBS during or prior to birth can lead to colonization of the lung. However, aside from the capsular polysaccharide, little is known about the bacterium's pathogenesis and only a few virulence factors have been identified to date. Furthermore, there are currently few biologically relevant model systems available to study the various adherence and invasive properties which GBS require to interact successfully with many different host tissues.

Recently, our laboratory has applied tissue culture techniques to grow primary chorion and amnion cells from the human placental membrane. This has enabled us to study the adherence and invasive properties of GBS in the laboratory using biologically relevant cells rather than transformed cell lines. Tumor cell lines may have unknown altered characteristics in comparison to freshly isolated primary cells. Using these techniques we have demonstrated that GBS adhere to the chorion cell surface, then invade into the chorion cell. This process could potentially allow the bacteria access to deeper sites of the placental membrane in vivo; and correlates with epidemiological studies where GBS has been cultured within the chorioamnion [1].

GBS adherence to and invasion into epithelial cells may occur via interactions between specific ligands on GBS and specific receptors on the surface of epithelial cells. To identify these molecules, we have also developed biochemical techniques to assess the binding of GBS to purified extracellular matrix proteins (ECM), both in solution and immobilized on a solid phase surface. Additionally, ligand blot assays have been developed which demonstrate adherence of GBS to epithelial cell membrane proteins as well as to purified ECM proteins. This approach has allowed us to identify proteins to which GBS specifically adhere. These interactions may play an important role in the pathogenesis of GBS disease.

The procedures outlined below have expanded the available methods to study GBS pathogenesis in vitro.

2. Materials

A. Bacterial growth media
   1. Todd-Hewitt (TH) broth- DIFCO (Cat. #DF0492-07-8, Fischer Scientific).[1]

2. Luria Bertani (LB) medium (Cat. #L3152, Sigma Chemical Co.).[2]
3. Agar select (Cat. #A5054, Sigma Chemical Co.).[2]

B. Primary cell isolation
  1. Reagents
    - Normal term human placentas of C-section procedures.
    - Phosphate buffered saline (PBS), pH 7.5
      - Sodium phosphate, monobasic, anhydrous (Cat. #S8282, Sigma Chemical Co.).[2]
      - Sodium chloride (Cat. #S7653, Sigma Chemical Co.).[2]
    - Penicillin G (Benzylpenicillin, Cat. #P3032, Sigma Chemical Co.).[2]
    - Streptomycin sulfate (Cat. #S9137, Sigma Chemical Co.).[2]
    - Vancomycin (Cat. #V1130, Sigma Chemical Co.).[2]
    - Amphotericin B-solubilized (Cat. #A9528, Sigma Chemical Co.).[2]
    - Dulbecco's modified Eagle's medium nutrient mixture F-12 Ham (DME/ F-12 1:1 mixture) (Cat. #D6421, Sigma Chemical Co.).[2]
    - Fetal bovine serum (Cat. #F-4135, Sigma Chemical Co.).[2]
    - L-glutamine-penicillin-streptomycin solution (Cat. #G1146, Sigma Chemical Co.).[2]
    - ITS liquid media supplement, 100× (Cat. #I3146, Sigma Chemical Co.).[2]
    - Trypsin 1:250 (Cat. #T4799, Sigma Chemical Co.).[2]
    - Trypsin-EDTA solution, 1× (Cat. #T3924, Sigma Chemical Co.).[2]
    - Trypan blue solution (0.4%) (Cat. #T8154, Sigma Chemical Co.).[2]
  2. Miscellaneous supplies
    - Sterile surgical gloves.
    - Forceps.
    - Scissors.
    - 50 ml volume disposable conical centrifuge tubes.
    - Plastic 25 ml volume pipettes (Cat. #431032, Fisher Scientific).[1]
    - Falcon large (150 mm × 15 mm) sterile petri dishes (Cat. #1058, Fisher Scientific).[1]
    - Corning 24-well tissue culture plates (Cat. #25820, Fisher Scientific).[1]
    - Falcon 75 cm^2 flat bottom flasks (Cat. #3013, Fisher Scientific).[1]
    - Hemacytometer (Clay Adams, New York).
    - 0.22 μm pore size 500 ml volume filter unit (Cat. #450-0020, Nalgene).[3]
    - Tissue culture incubator.
    - Jouan GR412 swinging bucket centrifuge.

C. Radiolabeling GBS and cellular adherence assays
  1. Reagents
    - Bovine serum albumin (Cat. #A7030, Sigma Chemical Co.).[2]
    - PBS containing 5.0% wt/vol bovine serum albumin (BSA/PBS) bovine serum albumin (Cat. #A7030, Sigma Chemical Co.).[2]
    - S^{35} Translabel (Cat. #51006, ICN)[4] or ^3H leucine (Cat. #TRI510, Amersham).[5]
    - RPMI without leucine, methionine, glutamine, lysine, or sodium bicarbonate (Cat. #R7130, Sigma Chemical Co.).[2]
    - Sodium bicarbonate (Cat. #S5761, Sigma Chemical Co.).[2]
    - Lysine (Cat. #L5501, Sigma Chemical Co.).[2]
    - Leucine (Cat. #L5652, Sigma Chemical Co.).[2]
    - Methionine (Cat. #M6039, Sigma Chemical Co.).[2]
    - Glutamine (Cat. #G7029, Sigma Chemical Co.).[2]
    - Sodium hydroxide (Cat. #VW3247-1, VWR-SP).[6]
  2. Miscellaneous supplies
    - 24-well tissue culture plates.
    - Beckman Ready Caps (Cat. #534181, Beckman Instruments).[7]
    - Beckman Scintillation Counter (Model #LS6001C, Beckman Instruments).[7]
  3. Invasion assays
    - 24-well tissue culture plates.
    - Penicillin G.
    - Gentamicin sulfate (Cat. #G1264, Sigma Chemical Co.).[2]
    - Trypsin-EDTA solution, 1×.
    - Triton X-100 (Cat. #X100, Sigma Chemical Co.).[2]

D. Solid phase adherence assays
  1. Reagents
    - Radiolabeled bacteria (see above supplies).
    - Bovine serum albumin.
    - Non-fat dry milk.
    - PBS.
    - Sodium hydroxide.
    - Fibronectin (Cat. #F0895, Sigma Chemical Co.).[2]
    - SuperSignal Chemiluminescence Substrate (Cat. #34080, Pierce).[8]
    - FN-15 monoclonal antibody to human fibronectin (Cat. #F7387, Sigma Chemical Co.).[2]
    - Anti-mouse immunoglobulin-horseradish peroxidase conjugate (Cat. #A9917, Sigma Chemical Co.).[2]
    - Tween-20 (polyoxyethylene monolaurate, Cat. #P7949, Sigma Chemical Co.).[2]
  2. Miscellaneous supplies
    - Beckman Ready Caps® (Cat. #534181, Beckman Instruments).[7]

- Beckman Scintillation Counter (Model #LS6001C, Beckman Instruments).[7]
- Falcon round bottom ELISA/EIA plates (Cat. #08-772-5, Fischer Scientific).[1]

E. Isolation of eukaryotic cell membranes and bacterial adherence assays to isolated membranes

1. Reagents
   - A549 cells (Cat. #CCL185, American Type Culture Collection).[9]
   - RPMI-1640 (R5886, Sigma Chemical Co.).[2]
   - Fetal bovine serum.
   - N-[2-hydroxyethyl]piperazine-N′[2-ethane sulfonic acid] (HEPES, Cat. #H3375, Sigma Chemical Co.).[2]
   - Mannitol (Cat. #M9647, Sigma Chemical Co.).[2]
   - Trypan blue solution, 0.4%.
   - Phenyl-methyl-sulfonyl-fluoride (PMSF, Cat. #P7626, Sigma Chemical Co.).[2]
   - Dimethyl sulfoxide (DMSO, Cat. #D8779, Sigma Chemical Co.).[2]
   - Acrylamide/Methylene-bis-acrylamide 29:1 (Cat. #161-0124, Bio-Rad Laboratories).[10]
   - Tris-hydroxymethyl-amino-methane (Tris base, Cat. #T1378, Sigma Chemical Co.).[2]
   - Hydrochloric acid (Cat. #H7020, Sigma Chemical Co.).[2]
   - Sodium dodecyl sulfate (Cat. #161-0302, Bio-Rad Laboratories).[10]
   - Tetra-methyl-ethylene-diamine (Cat. #161-0800, Bio-Rad Laboratories).[10]
   - Ammonium persulfate (Cat. #161-0700, Bio-Rad Laboratories).[10]
   - Methanol (Cat. #M1775, Sigma Chemical Co.).[2]
   - Glycine (Cat. #G7403, Sigma Chemical Co.).[2]
   - Bovine serum albumin.
   - Nonfat dry milk.
   - ^{35}S-Trans label.
   - Todd-Hewitt Broth.
   - RPMI-1640 without methionine.
2. Miscellaneous supplies
   - Mini-Protean II Electrophoresis System (Cat. #165-2940, Bio-Rad Laboratories).[10]
   - Corning tissue culture flasks (Cat. #10-126-29, Fischer Scientific).[1]
   - Dounce homogenizer (Cat. #1984-70007, Bellco Glass Co.).[11]
   - Immobilon-P (Cat. #IPVH00010, Millipore Corp.).[12]
   - Mini Trans-Blot Cell (Cat. #170-3930, Bio-Rad Laboratories).[10]
   - Whatman 3 mm thickness paper (Cat. #28458-005, VWR-SP).[5]
   - Plate-spinning centrifuge (Jouan).
   - Ultracentrifuge (Model #L350, Beckman Instruments).[7]
   - SW41Ti Rotor (Beckman Instruments).[7]
   - SW41 Polyallomer centrifuge tubes, 14 × 89 mm (Cat. #331372, BeckmanInstruments).[7]
   - Glad-wrap (Union Carbide).

3. Procedures

A. Bacterial stains and growth conditions

GBS strain A909 is a type Ia clinical isolate; GBS strain COH1 is a type III clinical isolate [7]. Group B streptococcal cultures were aerated in Todd-Hewitt (TH) medium at 37 °C. *E. coli* DH5α (F-ϕ80*dlac*ZΔM15 *recA1 endA1 gyrA96 thi-1 hsdR17* (r_k^- m_k^-) *supE44 relA1 deoR* Δ*(lacZTA argF) U169*) was grown in aerated Luria Bertani (LB) medium.

B. Chorion and amnion cell isolation and growth conditions

Normal, term placentas of C-section procedures are obtained immediately following delivery, placed into a sterile container, and set on ice. Primary cell isolation should be performed as soon as possible after obtaining the placenta. Gloves, eye protection, and protective clothing should be worn to protect against any potential blood borne human pathogens. The placenta is transferred to a laminar flow tissue culture hood to minimize subsequent bacterial contamination. It is helpful to use sterile surgical gloves to minimize contamination of the placental membrane since handling it is unavoidable.

The chorioamnionic membrane is cut from the placenta and placed into a beaker of 200 ml PBS containing penicillin (10 μg/ml), streptomycin (100 μg/ml), vancomycin (40 μg/ml), and amphotericin B (2.5 μg/ml). The beaker is gently swirled to remove red blood cells which are loosely attached to the membrane. The membrane is then asepticly transferred with forceps to a new beaker with fresh PBS/antibiotics-fungicide and incubated for 20 min at room temperature (RT).

The chorionic membrane is then peeled away by hand from the amnionic membrane. The respective layers can be separated by rolling an edge of the membrane between the thumb and index finger (similar to opening a plastic produce bag in the grocery store). Once started, the layers should peel apart fairly easily. The layers may come apart in one or more pieces, so care should be taken to make sure that the two layers are completely separated to avoid contaminating one cell type with the other. The chorion layer is the maternal side in vivo and may still contain small exposed venous and arterial vessels. This layer is thicker than the amnion layer and is yellowish in color. The thinner amnion layer appears more transparent. It is important that the layers are completely separated so that one cell type will not contaminate the other in subsequent culture.

The respective layers are cut into 1 cm × 5 cm strips using surgical scissors and forceps, and then placed into separate large sterile petri dishes containing 50 ml DME/Ham's F-12 medium with 0.5% wt/vol trypsin (Sigma 1:250 trypsin) and 100 μm EDTA. Petri dishes are then incubated for 45 min at 37 °C, 5% CO_2 in a tissue culture incubator. Amnion membrane strips are then removed with forceps and placed into a petri dish containing fresh DME/Ham's F-12/trypsin solution and incubation continued for an additional 1.5 h. Chorion strips are similarly transferred to fresh trypsin solution and incubation continued for an additional 2.5 h.

Membrane tissues are then asepticly transferred to 50 ml conical centrifuge tubes containing 20 ml DME/Ham's F-12 medium with 5% fetal bovine serum (FBS) which has been heat treated for 10 min at 65 °C (HI). Total volume should not exceed approximately 30 ml. Tubes are vortexed thoroughly to release cell monolayers from the basement membranes. Chorion cells require thorough vortexing (4–6, 30-sec bursts) at the highest setting to be released from the extracellular matrix (ECM). A major cause of poor cell recovery is insufficient vortexing. In contrast, amnion cells should be released by vortexing for 3, 20-sec bursts. Amnion cell isolation is otherwise identical to that of the chorion cell procedure. The ECM proteins will clump together, while the cells remain in the supernatant. The sample is poured into a large petri dish and the ECM clumps removed with forceps. Liquid containing the cells is transferred to a new 50 ml conical tube and centrifuged at 1500 ×*g* for 10 min at 6 °C in a Jouan swinging bucket centrifuge. The supernatant is decanted and both amnion and chorion cell pellets are individually suspended in fresh cell growth medium. Cell growth medium is DME/Ham's F-12 medium with 5% vol/vol FBS (HI), penicillin (10 U/ml), streptomycin (100 μg/ml), amphotericin B (2.5 μg/ml), 1× ITS (Sigma mixture of 5 μg/ml insulin, 5 μg/ml transferrin, and 5 ng/ml selenium), and 2 mM L-glutamine. If bacterial contamination becomes a problem, vancomycin (40 μg/ml) can be added as an additional antibiotic as Staphylococci are often the culprit. There will be a significant number of red blood cells in the chorion cell pellet, but they do not seem to interfere with the initial adherence and growth characteristics of the chorion cells. Far fewer red blood cells are usually present in the amnion preparation. There will be some small clumps of cells following centrifugation which can be dispersed by drawing the sample up and down in a 25 ml pipette.

Cell densities are then determined using a standard hemacytometer and staining the cells with trypan blue. Cells are seeded into 75-cm^2 tissue culture treated flasks or 24-well tissue culture treated plates at a density of 1 × 10^5 cells/cm^2. Allow 24–48 h incubation time at 37 °C, 5% CO_2 for initial cell adherence and growth. An inverted microscope is necessary to assess cell growth. Cell layers are then rinsed gently with PBS using a pipette to remove residual red blood cells and fresh medium is then added. Cell growth medium is subsequently changed every three days. Cell monolayers should reach confluence within about 5 days. The purity of the cell cultures is generally determined by morphological observation. Chorion cells are oblong whereas amnion cells more cuboidal. The purity of the cell cultures has also been confirmed on several experiments by transmission electron microscopy.

Growth rate will vary among placental preparations. Cells are not usually passaged more than one time. Alternatively, cells grown in 75-cm^2 flasks may be frozen in liquid nitrogen for later use. Cells are removed from the flask by incubating in Hank's Balanced Salt's Solution containing 0.5 g/l porcine trypsin and 0.2 g/l EDTA (trypsin/EDTA solution). Cells are transferred to a conical centrifuge tube and pelleted by centrifugation at 1500 ×*g* for 10 min at 6 °C. Pellet is suspended in 5 ml of growth medium containing 40% FCS (HI) and 10% DMSO and aliquoted in cryogenic vials. Cells are frozen at –70 °C overnight before storing in liquid nitrogen. Frozen cells are rapidly thawed in a 37 °C water bath, and then immediately added to fresh chorion cell growth medium. The isolation of chorion and amnion cells was initially based on modifications of previously described protocols [3, 4, 6].

C. Radiolabeling of GBS

Deficient RPMI 1640 is made by reconstituting one bottle of medium with 1 liter of distilled water and adding 300 mg of glutamine, 50 mg of leucine, 40 mg of lysine, 2 g of sodium bicarbonate, and either 50 mg of leucine (for methionine-deficient media) or 15 mg of methionine (for leucine-deficient media). The medium is then filter-sterilized using a 0.2 μm pore size filter flask. The pH of the RPMI should be adjusted to approximately 7.0, as a more alkaline pH may result in reduced labeling efficiency.

Four hundred microliters of a stationary-phase bacterial culture is grown to log phase (OD at 600 nm = 0.5–0.6) in 10 ml of THB (streptococci) or LB (*E. coli*) by incubation at 37 °C with aeration for approximately 2 h. A 1 ml aliquot of bacteria is washed twice by centrifugation in a Hermle Z230M microcentrifuge, and suspended in 2 ml of RPMI 1640 without L-leucine or L-methionine, depending on the radiolabel to be used. One to two hundred μCi of ^{3}H leucine or ^{35}S-Trans label are then added, and GBS are incubated for 30 to 60 min at 37 °C in a shaking incubator. Bacteria are pelleted by centrifugation, suspended in 4 ml of TH or LB broth, and incubated for 1 h to an OD of 0.5–0.6 at 600 nm. Bacteria are then washed once with PBS and suspended in 4 ml of TH or LB broth with 30% glycerol. Radiolabeled bacterial preparations are assessed for both incorporated activity (cpm) by using Beckman Ready Caps (a solid-phase scintillant) and a Beckman LS6001C Scintillation Counter, and for colony forming units (CFU) per ml by plating dilutions of 10^{-5}, 10^{-6} and 10^{-7} onto TH or LB agar plates. A CFU:cpm ratio is derived for each preparation. Radiolabeled bacteria are then frozen down in 0.5 ml aliquots at −70 °C. Bacterial viability and bacterial adherence are minimally affected by freezing in TH or LB broth with 30% glycerol.

Care should be taken to avoid inhalation of volatile compounds in ^{35}S-Trans label. All procedures involving uncapped ^{35}S-Trans label samples should be performed in a certified fume hood until unincorporated label has been removed.

D. Bacterial adherence assays

Assessment of bacterial attachment to cell monolayers is performed using a modification of a previously described protocol [11]. Radiolabeled bacteria are used (see above procedure) to determine the adherent characteristics of GBS. Aliquots of radiolabeled bacteria are thawed by the addition of 1 ml THB and then incubated at 37 °C for 15 min. Bacteria are pelleted by centrifugation in microtubes and washed 1× in PBS. The cell pellets are then suspended in PBS containing 1.0% wt/vol bovine serum albumin (BSA/PBS) and placed on ice for 2 h.

Chorion or amnion cell monolayers are grown to confluence in 24-well tissue culture plates (as described above) and washed 3× with PBS. The PBS is removed by aspiration. Five hundred microliters of BSA/PBS is added per well and plates are incubated at 4 °C for 2 h. Incubation of both the bacteria and tissue culture wells with 1.0% BSA is necessary to reduce non-specific interactions of bacteria with both the cell monolayer as well as the plastic plate. The BSA/PBS is aspirated off of cell monolayers and dilutions of bacteria in 200 μl volumes of BSA/PBS are added. Bacteria are allowed to settle onto the monolayer for 6 h at 4 °C or alternatively, bacteria can be brought into contact with the cell surface by centrifugation at 800 ×*g* prior to a 2 h incubation at 4 °C. Incubation at 4 °C will inhibit potential bacterial invasion into the cells and prevent bacterial growth during the assay. Plates are then washed 6× with PBS. It is essential that the plates are vortexed between each wash to eliminate high background. The PBS is aspirated at each wash steps. Two hundred microliters of 2 M NaOH are then added to each well and incubated for 20 min at 65 °C to lyse cells and adherent bacteria. Fifty microliters of the lysate suspension are added to Ready-Caps (any method for scintillation counting may be used but we have found the Ready-Caps to be very convenient) for determination of cell associated cpm of radioactivity.

The number of CFU associated per well is determined by converting the total cell associated cpm per well into CFUs as described in the above radiolabeling bacteria section. This number is an approximation since it assumes that all of the adherent counts are from viable bacteria. All samples are performed in triplicate.

E. Invasion assays

Qualitative determination of intracellular bacteria has been performed by a modification of the procedure described [9]. Primary cell cultures are grown to confluence in 24-well tissue culture treated plates as described above. Cell monolayers are washed 3 times in PBS followed by the addition of 340 μl cell growth medium per well. Stationary phase cultures of bacteria are washed one time with sterile PBS and suspended in chorion cell growth medium. Bacteria are then diluted to 10^{-2} so that approximately 5×10^5 CFU in a 60 μl vol are added per well. Initial contact of the bacteria with the cell monolayer is aided by centrifugation at 800 ×*g* for 10 min, at 6 °C in a Jouan swinging bucket centrifuge. Plates are then incubated at 37 °C, in 5% CO_2, for 2 h.

Non-adherent, extracellular bacteria are removed by washing cell monolayers four times with sterile PBS. The plates are gently swirled between washes and the PBS is then removed by aspiration. Five hundred μl of cell growth medium containing 10 μg/ml penicillin and 100 μg/ml gentamicin are added to each well to kill adherent extracellular bacteria. Plates are again incubated at 37 °C, in 5% CO_2, for 2 h. Following the incubation period, an aliquot of the medium can be spread on TH agar plates to verify that the extracellular bacteria have been eliminated. Growth medium is aspirated off the cell monolayer and the cells are subsequently washed five times with sterile PBS as described above.

Two hundred microliters of trypsin/EDTA solution are added per well to release the adherent cell layer. The plates are incubated at 37 °C for 30 min or until the cell layers are no longer visibly adherent (an inverted microscope may aid in assessing the degree of cells still attached). This step is followed by the addition of 800 μl of 0.025% vol/vol Triton X-100 to lyse the eukaryotic cell membranes. Lysis may take up to 30 min at 37 °C. Samples are then drawn up and down in a 1 ml pipette to break up small clumps of cellular debris, added to individual microfuge tubes, and vortexed thoroughly for at least 12 sec each. Fifty microliter aliquots of the lysis mixture are plated on TH or LB agar plates to determine the number of intracellular CFU. Agar plates are incubated at 37 °C overnight and individual colonies counted.

The CFU count is multiplied by 20 to determine the total number of intracellular bacteria per well. This number is divided by the number of CFU initially added to the cell monolayer and multiplied by 100 to determine the percent invasion of the inoculum. All samples are performed in triplicate and several dilutions of each inoculum can be used to determine at what concentration of bacteria saturates the system.

F. Detection of liquid-phase fibronectin binding to GBS by immunoblot analysis
One ml of a log-phase culture (generated as described above) is pelleted by centrifugation in a microcentrifuge, washed twice with 1 ml of PBS, and suspended in 1 ml of 5% BSA/PBS. Ten micrograms of fibronectin is added to the tube containing bacteria and the mixture is incubated on an Orbitron rotating platform for 1 h at 4 °C. Bacteria are pelleted, washed twice with 1 ml of 5% BSA/PBS, and three times with 1 ml of PBS. Bacterial pellets are then boiled for 5 min with 100 μl of SDS-PAGE sample buffer. Bacteria are pelleted, and proteins contained in the supernatants are separated by electrophoresis on a 5% SDS-polyacrylamide gel using a BioRad Mini-Protean II electrophoresis cell following the manufacturer's instructions. One hundred nanograms of fibronectin is loaded onto the gel as a positive control.

Proteins are then transferred to a PVDF membrane (Immobilon-P) using a Bio-Rad Laboratories Mini Trans-Blot Cell and the protocol established by the manufacturer. Nonspecific binding to the membrane is prevented by soaking the membrane in PBS containing 5% nonfat dry milk and 0.1% Tween-20 (Blotto). The membrane is then reacted with anti-fibronectin FN-15 ascites diluted 1:1000 in Blotto and washed three times for 1 min in Blotto. The membrane is then incubated with rabbit-anti-mouse IgG-horseradish peroxidase conjugate diluted 1:1000 in Blotto for 1 h. The blot is then washed once for 20 min, twice for 5 min with Blotto, and three times for 5 min with PBS containing 0.1% Tween-20. Fibronectin reacting bands were visualized with an enhanced chemiluminescence (ECL) system and exposed to XAR-2 film for 0.2 sec to 1 min.

G. Solid-phase ECM binding assays
Fibronectin or other ECM proteins are diluted with PBS for a range of concentrations in 50 μl volumes and incubated in round bottom microtiter wells at 4 °C for 18 h to allow proteins to adhere. Wells are then incubated for 2 h with 5% BSA/PBS to reduce non-specific binding of bacteria. Fifty microliter aliquots of radiolabeled GBS ($1–3 \times 10^{-6}$ CFU $well^{-1}$) are added. The plate is centrifuged at 800 $\times g$ for 10 min at 4 °C and incubated for 1 h at 4 °C. Supernatant is then aspirated from the wells. Wells are washed three times with 5% non-fat dry milk in PBS, and then three times with PBS. Following each addition of wash buffer, GBS bound by low avidity interactions are removed from the plate by shaking the plate on a Vortex Genie 2 (VWR Scientific) at level 2 for 10 sec. To determine the extent of high avidity bacterial binding, 50 μl of 2 N sodium hydroxide is added to each well and the plate is incubated at 65 °C for 1 h.

The sodium hydroxide lysates are then dispensed onto solid scintillant (Ready Caps), dried at RT, and well-associated cpm are quantified with a scintillation counter. Initial input cpm are also quantified. Background counts from scintillant without lysate added are also quantified. All experiments are performed in duplicate. The percent adherence is calculated as follows: (cpm bound – cpm background)/cpm added × 100. The error bars shown are 95% confidence intervals for the mean; 95% confidence interval = (standard deviation/square root (N)) × 1.962, where N is the number of replicates.

GBS adherence to eukaryotic cell membrane proteins

Isolation of A549 membranes
ATCC A549 cells are grown in one to three 150 mm^2 tissue culture flasks using RPMI-1640 supplemented with 10% fetal bovine serum and 2 mM L-glutamine as growth medium. Generally, cells will divide to form confluent monolayers in 3–5 days. Adherent cells are removed from the flask by adding 10 ml of 1 mM EDTA in PBS and incubating the flask for 15 min at 37 °C. The flask is then struck vigorously onto a countertop to loosen the cells. If cells remain adherent, they are incubated for additional 5 min intervals and checked for up to 45 min until cells can be loosened. EDTA treatment may dissociate an indeterminate proportion of fibronectin and other ECM proteins from cell surface receptors during this cell removal process. Cells are then removed with a Pasteur pipette and pelleted by centrifugation at 200 $\times g$ for 10 min at 4 °C. The supernatant is aspirated and the cells are suspended in 10 ml of ice-cold lysis buffer (10 mM HEPES, pH 7.4, 50 mM

mannitol). For the remainder of the procedure all solutions are kept on ice to prevent proteolytic degradation of membrane proteins.

Cells are homogenized in a Dounce homogenizer and with 30 strokes of the plunger. The proportion of disrupted cells is assessed by trypan blue staining. Fifty microliters of the cell suspension is mixed with 50 µl of trypan blue solution, and 10 µl of this mixture is placed in a hemacytometer. Intact cells (which exclude trypan blue) and disrupted cells are counted separately. If the proportion of lysed cells is less than 90%, the homogenization procedure is repeated until greater than 90% lysis is achieved. Forty microliters of 0.5 M Phenyl-methyl-sulfonyl fluoride (PMSF) in DMSO is then added (for a final concentration of 2 mM PMSF) to inhibit serine proteases.

The homogenate is centrifuged at 200 ×*g* for 10 min at 4 °C to pellet the nuclei and unlysed cells. The supernatant, which contains the majority of membranes is removed. Some membrane material will remain in the pellet. The pellet fraction is suspended in 10 ml of lysis buffer, and the centrifugation step repeated. The supernatant is again saved and combined with the initial supernatant. The pellet is then discarded.

To separate the membranes from soluble cellular components, the mixture is centrifuged in a swinging bucket rotor (SW40) at 70,000 ×*g* for 1 h. The supernatant is discarded and the pellet, containing membrane material, is suspended in 0.5 ml of lysis buffer. The membranes are stored in aliquots of 25–100 µl at –70 °C, and are stable for at least 2 years. Repeated freeze-thaw cycles can result in degradation of membrane proteins and should be avoided.

Bacterial blots

Purified extracellular matrix proteins and eukaryotic cell membrane proteins are separated by electrophoresis in a 10% SDS-polyacrylamide gel as described by Laemmeli. A Mini-Protean II Electrophoresis System from Bio-Rad Laboratories and protocols provided by the manufacturer are used in this experiment. Proteins are then transferred to a PVDF membrane (Immobilon-P), as described by Towbin, using a Bio-Rad Laboratories Mini Trans-Blot Cell following manufacturer's protocols.

Ten milliliters of 5% BSA/PBS are placed in a plastic heat sealable bag, which should then be rotated to coat the entire interior surface of the bag. Allowing the PVDF membrane to touch an uncoated portion of the bag may result in excessive background binding of GBS to the membrane. After electrotransfer of proteins is completed, the portion of the membrane containing the standards should be cut off, rinsed briefly with water, and saved. The lower right hand corner of the membrane should then be cut off with a razor blade to allow orientation of the membrane during later steps. The side of the membrane which faced the gel will be referred to as the 'top-side'.

The membrane is placed into the bag containing 5% BSA/PBS with forceps. Care should be taken to avoid allowing any portion of the membrane except the edges to touch the forceps, and to avoid touching the membrane with gloves or fingers, as this may result in excessive non-specific binding. The bag is then sealed and incubated for 2–24 h at 4 °C. Longer or shorter incubation times may result in excessive background or decreased specific binding, respectively.

An aliquot of ^{35}S-labeled GBS is thawed (see radiolabeling of bacteria) and bacteria are pelleted by centrifugation in a microcentrifuge for 1 min. The GBS are suspended in 1 ml THB and incubated for 30 min at 37 °C. Bacteria are again pelleted by centrifugation, washed with 1 ml of 5% BSA/PBS, and then suspended in 1 ml of 5% BSA/PBS. Bacteria are then incubated for 2–4 h on ice. Shorter incubation times result in increased background binding.

The following steps should be carried out on a bench dedicated to radioactive materials, which has been covered with at least two layers of absorbent bench-top paper with a water-tight backing. A corner of the sealed bag containing the blotting membrane is cut open, and the radiolabeled GBS are added to the bag. The bag is then sealed with care taken to insure that no bubbles remain inside the bag. The edges of the bag should then all be heat-sealed adjacent to the membrane. This insures that all bacteria will be brought into contact with the membrane, and that leakage of the bag during subsequent centrifugation will be avoided. The bag is then placed in the lid of a 96-well microtiter plate which has been lined with 3 mm Whatman filter paper. The membrane must be placed with the top-side up. If the membrane is placed top-side down, no signal will be seen as minimal protein passes completely through the membrane during electrotransfer of proteins. The GBS are brought into contact with the membrane by centrifugation at 800 ×*g* for 10 min at 4 °C. The blotting membrane is then incubated for 1 h at 4 °C.

One corner of the sealed bag is carefully opened and the liquid decanted into a radioactive waste container. The bag is then placed on several layers of paper towel and cut on all four edges with a razor blade. The upper portion of the bag is then removed.

Fifty milliliters of ice-cold 5% non-fat dry milk in PBS is placed in a container large enough to hold the blot. The membrane is then placed in the container top-side up. The container is placed on ice and gently shaken on an orbital platform at approximately 60 rpm for 5 min. The blot should move freely in the container. Placing the blot upside-down during washing or allowing the blot to stick to the container during washing may result in higher backgrounds or lower signals. The washing step is repeated twice with 5% non-fat dry milk in PBS, and then three times with ice-cold PBS alone. Initial wash solutions should be discarded as radioactive waste.

A sheet of Glad-Wrap is placed on the bench top. The blotting membrane is then placed top-side down on the Glad-Wrap and the edges folded. ^{35}S is a low-energy beta-emitter and therefore signals cannot be detected through the blotting membrane or through multiple layers of Glad-Wrap. Care should therefore be taken to insure that the top-side of the membrane faces the X-ray film, and that only a single layer of Glad-Wrap is between the membrane and the film. The membrane should not be placed directly on the film as it may adhere tightly to the film, making film development impossible. Thicker food-wraps, such as Saran Wrap, partially block the signal and result in slightly longer exposure times. The film should be developed after exposure times of 1–24 h, depending on the strength of the signal.

Results and discussion

GBS adherence and invasion of chorion cells

Colonization of host tissues is the initial phase of bacterial and host interaction. GBS have been cultured from infected placental membranes in vivo [1] and GBS have been shown to adhere to intact chorioamniotic membranes in vitro by electron microscopy [5]. The adherence assay performed here demonstrates that primary human chorion cells grown in vitro also provide an analogous colonizable surface as shown in Figure 1. The two clinical isolates of GBS tested, strain A909 and strain COH1, bound to the chorion cell monolayer in a saturable manner. Similar results were observed with amnion cells grown in vitro (data not shown). Allowing bacteria to settle onto the cell monolayer as opposed to centrifugation of the bacteria gives consistently 2-fold less adherent bacteria over the range of inoculums. Background radioactivity is generally at or near undetectable levels if the wash steps are followed as described in *Procedures*.

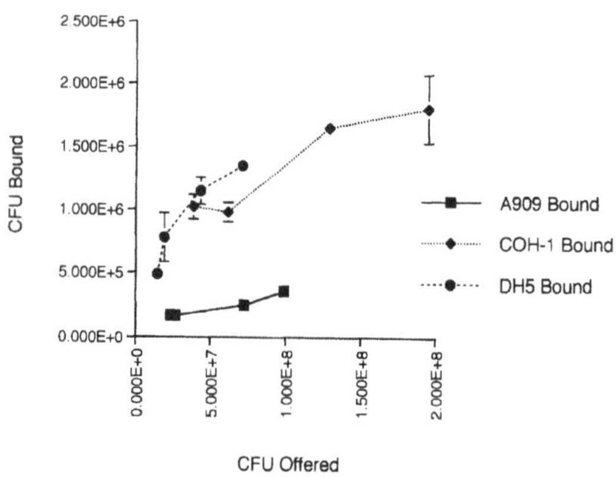

Figure 1. GBS and *E. coli* adherence to primary human chorion cells. Primary chorion cells were isolated and grown to confluence in 24-well tissue culture plates. Dilutions of radiolabeled bacteria were added to monolayers and assessed for adherence as described in *Procedures*. Error bars indicate standard deviations. $N = 3$.

Subsequent to bacterial attachment, invasion into the intracellular environment of chorion cells could allow the bacteria to penetrate the first protective layer of the chorioamnion and enable GBS to establish an infection at a deeper site. Following GBS inoculation onto chorion cell monolayers, viable GBS are recovered from the intracellular environment, indicating that these bacteria have the ability to enter chorion cells (Figure 2). Up to 6.0% of the GBS inoculum can be recovered. The degree of invasion is variable from strain to strain, but is a shared GBS phenotype. Interestingly, although *E. coli* DH5α are adherent in this system (see Figure 1), they are relatively non-invasive, suggesting that there may be specific factors expressed by GBS for adherence and invasion which enable the bacterium to enter the intracellular chorion cell environment.

We routinely see a minimum of one to four percent GBS invasion using the inoculums indicated in Figure 2. The numbers of intracellular CFU recovered among triplicate samples are very similar (see error bars in Figure 2); however, results do vary among experiments. Additionally, certain batches of chorion cells are more susceptible to bacterial invasion than are other preparations. Therefore, *E.*

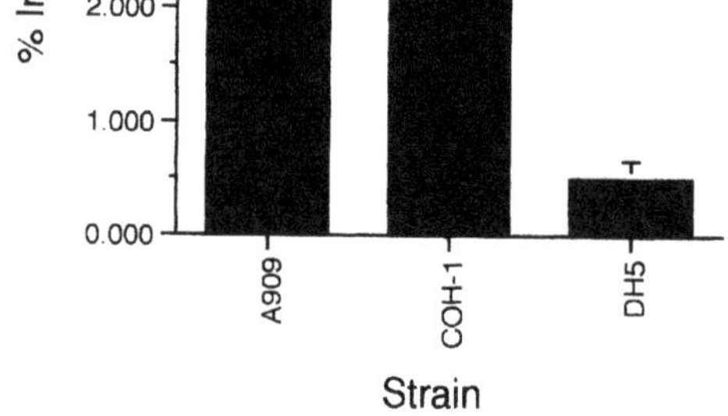

Figure 2. GBS invasion into primary human chorion cells. GBS strain A909 and strain COH1, and *E coli* DH5-alpha were inoculated onto confluent chorion cell monolayers at 5.0×10^5, 4.1×10^5, 5.8×10^5 CFU per well, respectively. *E coli* DH5-alpha is a non-invasive bacterium in this system. The percent invasion is defined as the number of CFU recovered from cellular lysates divided by the initial extracellular inoculum multiplied by 100. Error bars indicate standard deviations. $N = 3$.

coli DH5α should be included as a non-invasive control in every experiment. *Streptococcus gordonii*, an oral streptococci, has also been used reproducibly as a gram-positive non-invasive control (data not shown).

Similarly, adherence assay results will vary somewhat from experiment to experiment and among various placenta cell preparations; however, relative binding trends among both GBS strains and other bacterial species are qualitatively consistent.

GBS have previously been shown to invade lung epithelial carcinoma cells [9], and have been observed in lung tissue sections of infected *Macaca nememstrina* primates [8]. This intracellular invasion has been postulated to enable GBS to reach deeper sites which could result in bacteremia. The chorion and amnion cells form the outermost and innermost protective barriers, respectively, of a neonate from bacterial infection. It is possible that the invasive interaction described here may enable GBS to reach deeper sites of the human placental membrane leading to premature rupture of the membrane or infection of the amnionic fluid.

Adherence of GBS to fibronectin in solid phase and in solution

Bacterial pathogens often bind to extracellular matrix proteins in order to colonize a host or to coat itself with a host molecule to evade host immune surveillance. We examined the ability of GBS to bind to purified ECM proteins using the protocols described above and determined that GBS do bind to immobilized fibronectin [10]. GBS adherence to laminin, tenascin, vitronectin, or fibrinogen was not observed. The results of a typical fibronectin binding experiment are shown in Figure 3. The degree of binding is concentration-dependent, as expected. Significant adherence is seen even at very low concentrations. As little as 100 ng of fibronectin added per well shows GBS adherence significantly above background wells that were coated with BSA alone.

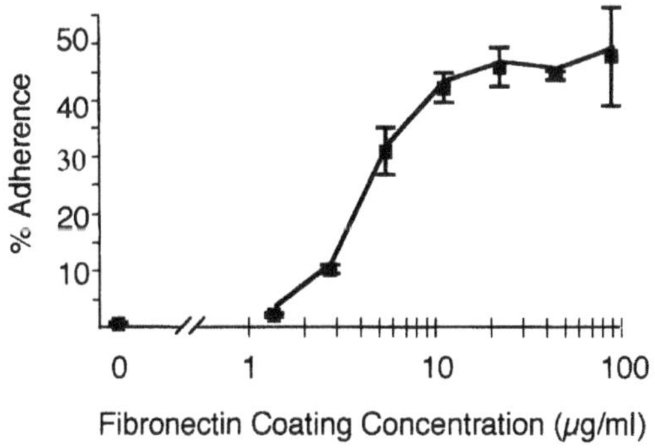

Figure 3. Adherence of radiolabeled GBS strain COH1 to immobilized fibronectin. Polystyrene microtiter plates were coated with varying concentrations of fibronectin and incubated with log-phase ^{3}H-labeled strain COH1. Results are percentages of input adherent inocula. Error bars represent 95% confidence intervals for the mean.

GBS binding is very reproducible from well to well (see error bars, Figure 3). Furthermore, binding from experiment to experiment also shows minimal variation if the same strain of GBS and the same manufacturer's lot of fibronectin is used. However, adherence can vary greatly depending on both the manufacturer and the lot of fibronectin used. GBS adherence to fibronectin from Sigma Chemical Co. has ranged from 5% to 45%. Fibronectin from GIBCO/BRL gave approximately 30% adherence, while fibronectin obtained from Calbiochem resulted in no GBS adherence (< 1%). Background levels can also vary greatly depending on a number of factors, including the manufacturer of BSA and of microtiter plates. Pilot experiments should be performed to identify optimal reagents. Non-specific binding of GBS to microtiter wells should be less than 0.2%. It is important that blocking of both plates and bacteria with 5% BSA be performed for at least 2 h prior to addition of the bacteria to the microtiter wells.

The results of a typical experiment assaying GBS binding of soluble fibronectin using strain COH1 is shown in Figure 4. Group A streptococcal (GAS) strain STRA-B498 was included as a positive control. GAS bound to material which reacted with the anti-fibronectin FN-15 ascites and was of the predicted molecular weight of fibronectin (~220 kDa). No reactive protein was eluted from GBS after incubation with fibronectin, suggesting that GBS do not adhere to fibronectin in solution. These results have been very reproducible. For complete elution of bound proteins, the bacterial samples should be boiled for at least 5 min in SDS-PAGE sample buffer.

Ligand blot analysis of GBS adherence to fibronectin and to A549 cell membrane proteins

Our laboratory has previously demonstrated that GBS adhere to cultured epithelial cells. In lane A of the ligand blot shown in Figure 5 we show that radio-labeled GBS adhere to two specific A549 cell membrane proteins. GBS adherence to fibronectin was included as a positive control. Non-reducing sample buffer (without 2-mercaptol-ethanol) was used in this experiment. GBS predominately bound two proteins bands in lane B containing fibronectin, which correspond to variants of fibronectin which are migrating at a slightly higher predicted molecular weight than that of full length fibronectin. No binding is observed when reducing sample buffer is used (data not shown).

Binding to specific ECM proteins is very reproducible from blot-to-blot when using the same lots of membranes and the same GBS strain. However, significant variability in binding can be seen among membrane preparations and among different GBS strains. It should also be noted that these membrane protein labeling experiments can and have been

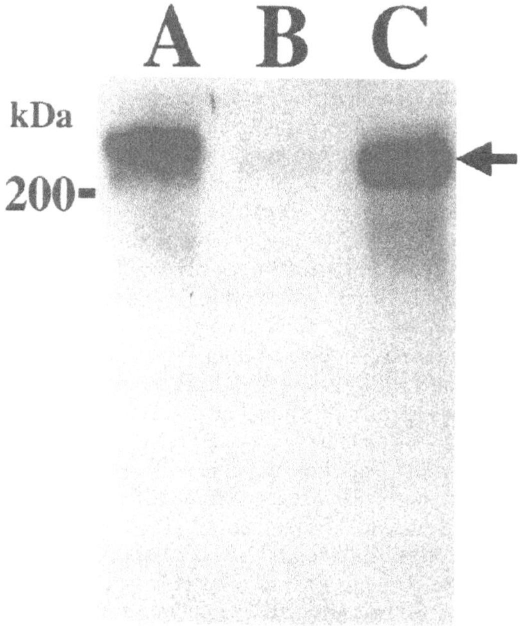

Figure 4. Binding of soluble fibronectin to GBS strain COH1 and GAS strain STRA-B498. Fibronectin was incubated with log-phase COH1 and STRA-B498. Bacteria were washed and bound material was eluted with SDS-PAGE sample buffer. Bound material was separated by electrophoresis on a 5% SDS-polyacrylamide gel. Proteins were analyzed by immunoblot using anti-fibronectin FN-15 ascites as a probe and detection as described in *Procedures*. Purified fibronectin was used as a positive control. The arrow represents the migration of fibronectin. Lane A, fibronectin bound to GAS strain STRA-B498; lane B, fibronectin bound to GBS strain COH1; C, fibronectin-only control.

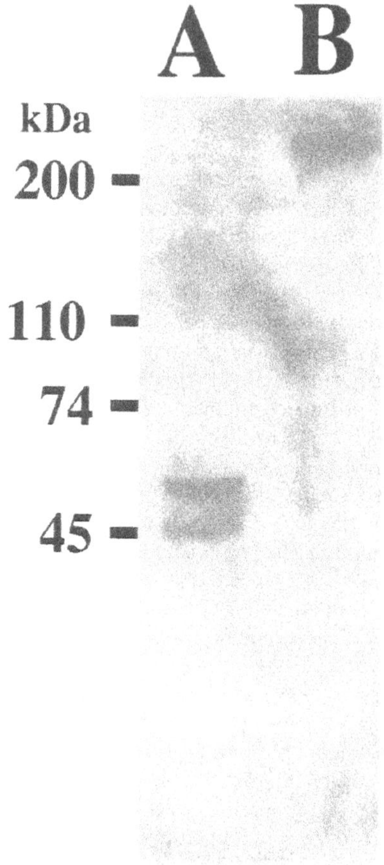

Figure 5. Ligand blot assay of GBS strain COH1 adherence to A549 cell membrane components and fibronectin. Two hundred micrograms of A549 cell membrane proteins and 2 μg of fibronectin were separated by electrophoresis under non-reducing conditions on a 10% SDS-polyacrylamide gel and electrotransferred to a nylon membrane. The membrane was then reacted with log-phase ^{35}S-labeled strain COH1 with detection by autoradiography. Lane A, A549 cell membrane proteins; lane B, purified fibronectin.

performed using primary cells (data not shown) in addition to immortalized cell lines.

Background binding is minimized if care is taken with blocking reagents and washing steps. The following steps are extremely important: (1) Do NOT allow the membrane to touch gloves, hands, or sections of the plastic bag which are not coated with 5% BSA; (2) Do NOT allow bubbles in the bag during the blocking step or the binding step; (3) Block both GBS and the blotting membrane for at least 2 h.

Conclusions

In our laboratory, we have constructed in vitro systems to emulate certain events during the colonization and invasion of host epithelial cells by GBS. The findings described in this report using primary human placental cells may have a greater relevance to human host-pathogen interactions than transformed cell lines derived from a variety of hosts which are biologically altered. The adherent and invasive properties of GBS with both chorion and amnion cells continues to be characterized in our laboratory. Additionally, the biochemical methods employed here have allowed identification of a specific ECM protein and a putative host cell receptor protein which may play a direct role in adherence of GBS with host cells.

Acknowledgments

This investigation was supported by AI30068 (SBW and CER) and AI22498 (GST).

Notes on suppliers

1. Fisher Scientific, 2170 Martin Ave, Santa Clara, CA 95050, USA
2. Sigma Chemical Company, PO Box 14508, Saint Louis, MO 63178, USA
3. Nalge/Nunc Inc., 2000 N. Aurora Rd., Naperville, IL 60563-1796, USA
4. ICN Pharmaceuticals, ICN Plaza, 3300 Hyland Ave, Costa Mesa, CA 92626, USA
5. Amersham Life Sciences Inc., Life Science Div., 2636 S. Clearbrook Dr., Arlington Height, IL 60005-4626, USA
6. VWR-SP, 1430 Waukegan Road, McGaw Park, IL 60085, USA
7. Beckman Instruments, Inc., Bioanalytical Systems Group, 2500 Harbor Blvd, Fullerton, CA 92634-3100, USA
8. Pierce, PO Box 117, Rockford, IL 61105-9976, USA
9. American Type Culture Collection, 12301 Parklawn Dr, Rockville, MD 20852-1776, USA
10. Bio-Rad Laboratories, Life Science Group/Molecular Bioscience Group. 2000, Alfred Nobel Dr, Hercules, CA 94547, USA
11. Bellco Glass, Inc., PO Box B 340 Edrudo Rd., Vineland, NJ 08360, USA
12. Millipore Corporation, 80 Ashby Road, Bedford, MA 01730, USA

References

1. Boggess KA, Watts DH, Hillier SL, Krohn MA, Benedetti TJ, Eschenbach DA (1996). Bacteremia shortly after placental separation during cesarean delivery. Obstet Gynecol 87: 779–784.
2. Centers for Disease Control and Prevention (1996). Prevention of perinatal group B streptococcal disease: A public health perspective. MMWR RR-7: 1–24.
3. Dell'aquila ML, Gaffney EV (1986). Growth of normal human amnion epithelial cells in serum-free medium. Exp Cell Res 137: 441–445.
4. Ferguson KA, Mitchell BF, Tanswell AK (1986). Human chorion cells respond to growth factors but lose steroidogenic capacity in primary monolayer cell culture. In Vitro Cell Dev Bio 22: 321–324.
5. Galask RP, Varner MW, Rosemarie PC, Wilbur SL (1984). Bacterial attachment to the chorioamniotic membranes. Am J Obster Gynecol 148: 915–926.
6. Lamont RF, Elder MG, Rose M (1985). Effect of bacterial products on prostaglandin E production by amnion cells. Lancet 2: 1331–1333.
7. Martin TR, Rubens CE, Wilson CB (1988). Lung antibacterial defense mechanisms in infant and adult rats: Implications for the pathogenesis group B streptococcal infections in the neonatal lung. J Infect Dis 157: 91–100.
8. Rubens CE, Raff HV, Jackson CJ, Chi EY, Bielitzki JT, Hillier SL (1991). Pathophysiology and histopathology of group B streptococcal sepsis in *Macaca nememstrina* primates induced after intraamniotic inoculation: Evidence for bacterial cellular invasion. J Infect Dis 164: 320–330.
9. Rubens CE, Smith S, Hulse M, Chi EY, van Belle G (1992). Respiratory epithelial cell invasion by group B streptococci. Infect Immun 60: 5157–5163.
10. Tamura GS, Kuypers JM, Smith S, Raff H, Rubens CE (1993). Adherence of group B streptococci to cultured epithelial cells: Role of environmental factors and bacterial surface components. Infect Immun 62: 2450–2458.
11. Tamura GS, Rubens CE (1995). Group B streptococci adhere to a variant of fibronectin attached to a solid phase. Mol Microbiol 15: 581–589.
12. Varner MW, Truner JW, Petzold CR (1985). Ultrastructural alterations of term human amniotic epithelium following incubation with group B betahemolytic streptococci. Am J Reprod Immun Microbiol 9: 27–32.
13. Westerlund B, Korhonen TK (1993). Bacterial proteins binding to the mammalian extracellular matrix. Mol Microbiol 9: 687–694.

Address for correspondence: Craig E. Rubens, Children's Hospital and Regional Medical Center, Division of Pediatrics, Department of Infectious Diseases, Box 5371/CH-32, Seattle, Washington 98105, USA
Phone: (206) 527-3890; Fax: (206) 527-2116
E-mail: cruben@chmc.org

Methods in Cell Science **20**: 203–207 (1998).

The rat model of endocarditis

Cindy L. Munro
Virginia Commonwealth University, School of Nursing, Richmond, Virginia, USA

Abstract. The rat model of endocarditis is a well established experimental protocol which closely approximates human native valve endocarditis. The rat model of endocarditis has been used to examine the role of particular streptococcal virulence factors, to assess immunoprotective strategies, and to evaluate the efficacy of selected antibiotic treatment regimens for streptococcal endocarditis. Like humans, rats are generally susceptible to endocarditis only if the cardiac valves have been damaged. In the rat model of endocarditis, damage to the aortic valve and sterile vegetation formation is accomplished by insertion of a polyethylene catheter through the carotid artery into the left ventricle. Following catheter insertion, an inoculum of streptococci are injected intravenously. Vegetations removed from the heart valves during thoracotomy of euthanized animals are qualitatively cultured for streptococcal infection. The method, including investigator safety considerations, is described in detail.

Key words: Endocarditis, Rat, Streptococci

1. Introduction

The viridans streptococci are part of the normal oral flora of humans, but can cause serious infections when introduced into the bloodstream. One of the most serious infections caused by the viridans streptococci is native valve endocarditis. The rat model of endocarditis is a well established experimental protocol which closely approximates the infection in humans [1, 4].

Roughening of the heart valve, such as is caused by the presence of a catheter in the heart, facilitates the establishment of infection. Like humans, rats are generally susceptible to endocarditis only if the cardiac valves have been damaged. In the rat model of endocarditis, a closed catheter tube (made of polyethylene tubing sealed with silicone) is inserted through the carotid artery into the left ventricle, where it remains for the duration of the experiment and induces damage to the aortic valve and sterile vegetation formation. Aseptic technique and general anesthesia are used for the procedure of catheter insertion. Following catheter insertion, an inoculum of streptococci are injected intravenously. Vegetations removed from the heart valves during thoracotomy of euthanized animals are qualitatively cultured for streptococcal infection.

The rat model of endocarditis was first described by Santoro and Levison [11]. The model has been used to examine the role of particular streptococcal virulence factors [3, 10, 17], to assess immunoprotective strategies [16], and to evaluate the efficacy of selected antibiotic treatment regimens for streptococcal endocarditis.

2. Materials

A. Anesthesia and procedure
- Animal scale
- Ketamine hydrochloride, provided through institutional animal resources department.
- Methoxyflurane, provided through institutional animal resources department.
- Conical centrifuge tube, 50 ml, disposable, Fisherbrand #05-539-6, Fisher.[1]
- Syringes, B-D tuberculin, 1 ml with 25 G $^{1}/_{2}$ inch needles, #BD309626, VWR.[2]
- Rat restrainer, Model 700R, Roboz Surgical Instrument Company, Inc.[3]

B. Procedure
- Electric clippers with #40 blade, such as Oster animal clippers, #10749-020, VWR.[2]
- Delta phase ™ Operating Board, Model 39OP, Braintree Scientific.[4]
- Deltaphase isothermal pads, warmed (37 °C), Model 39DP, Braintree Scientific.[4]
- Instrument soak, such as Lysol IC-instrument disinfectant presoak solution, #C6320-16, VWR.[2]
- Ethanol for use as skin cleansing solution.
- Scalpel, sterile disposable with #10 surgical blade, Fisherbrand #08-927-5A, Fisher.[1]
- 2 curved toothed forceps, such as Micro Dissecting Forceps, 4 inch, 1 × 2 teeth, #RS-5153, Roboz Surgical Instrument Company, Inc.[3]
- 2 curved serrated forceps, such as Micro Dissecting Forceps, 4 inch, #RS-4136, Roboz Surgical Instrument Company, Inc.[3]
- Iris scissors, such as Vannas Ultra Micro

Scissors, very delicate, 3 inch, curved, #RS-5611, Roboz Surgical Instrument Company, Inc.[3]
- Baby Diffenbach serrafine clamps, $1^1/_8$ inch, curved, very delicate, #RS-7421, Roboz Surgical Instrument Company, Inc.[3]
- Suture, 3-0 silk, cut in 7 mm lengths, Braintree Scientific.[4]
- Catheters, 5 cm, cut from PE10 Clay Adams polyethylene tubing, #63019-003, VWR.[2]
- Sterile 4 × 4 gauzes, such as Johnson and Johnson #B3063-9, VWR.[2]
- Michel skin clips, 7 mm, #RS-9270, Roboz Surgical Instrument Company, Inc.[3]
- Michel skin clip applying forceps, #RS-9292, Roboz Surgical Instrument Company, Inc.[3]
- Surgical gloves.
- Sterile PBS.
- Bench paper for surgical field.

C. Necropsy supplies
- Ethanol.
- Sterile 4 × 4 gauze.
- Tissue grinders, sterile disposable, one per animal, # 3505, Sage Products.[5]
- Sterile PBS.
- Syringe, B-D tuberculin, 1 ml with 27 gauge 2 inch needle, #BD309623, VWR.[2]
- Mayo scissors, $6^3/_4$ inch, straight, #RS-6872, Roboz Surgical Instrument Company, Inc.[3]
- Operating scissors, $4^3/_4$ inch, curved, sharp, #RS 6703, Roboz Surgical Instrument Company, Inc.[3]
- Hemostatic forceps, such as Rankin-Kelly forceps, $6^1/_4$ inch, straight, #RS-7140, Roboz Surgical Instrument Company, Inc.3
- Toothed forceps (see 2b above).
- Serrated forceps (see 2b above).
- Surgical gloves.
- Bench paper for necropsy field.

3. Procedures

A. Preoperative Preparation and Anesthesia

Animal investigations must comply with all applicable legal and institutional requirements. All animal experiments must be reviewed and approved by the institutional animal use committee prior to initiation of the work. The institutional committee and animal resource personnel can be valuable sources of information and assistance to the investigator.

Order male Sprague-Dawley rats, weighing 225–250 grams, so that they will arrive at least one week prior to the procedure; this enables observation of the animals for pre-existing illness and minimizes pre-procedure stress.

Prepare catheters and assemble supplies prior to the procedure. For rats weighing 225–250 grams, catheters should be 5 cm long. To prepare catheters, cut PE10 Clay Adams polyethylene tubing to 5 cm lengths and melt one end by pressing the tip with a pair of hemostatic forceps which have been flamed. Verify that the end of the catheter is sealed by placing the open end in instrument disinfectant solution; if the solution rises in the catheter by capillary action, the seal is inadequate. Cut suture material to 7 cm lengths. Soak all instruments, catheters, and suture in instrument disinfectant for the time period recommended by the manufacturer.

On the evening before catheterization, remove food from cages. On the morning of the procedure, water should be withheld. Each animal should be weighed to determined correct anesthesia dose. Calculate the dosage for induction of anesthesia with ketamine HCl (100 milligrams ketamine/kg weight). Pick the rat up by the base of his tail and place his front feet on the edge of the rat restraint; the animal should enter the restraint willingly. Secure the restraint. Following cleansing of the animal's abdomen with an alcohol swab, inject correct dose of ketamine intraperitoneally using a 1 cc syringe with $^1/_2$ inch 25 gauge needle. Return the animal to his cage, with supervision, until the anesthetic takes effect. When the animal has become very groggy, shave the neck area from chin to just below the sternum using an electric clipper with a close blade (for example, a #40 blade).

Monitor anesthesia level throughout the procedure by checking the pedal reflex; anesthesia is adequate when the animal does not respond to brisk foot rubbing. If a deeper level of anesthesia is required, supplement with whiffs of methoxyflurane. Deliver the methoxyflurane by placing a few drops of methoxyflurane on a gauze pad placed inside a 50 ml conical tube; the conical tube will fit snugly over the rat's nose. Control administration of inhalation anesthetic by replacing or removing the conical tube containing methoxyflurane from the animal's nose. Always prepare and administer methoxyflurane under a hood. Anesthesia should not be so deep as to diminish respiratory effort. If respiratory difficulties occur or if respiratory effort diminishes, discontinue methoxyflurane immediately; blow gently on the rat's nose to encourage respiratory effort and dissipate methoxyflurane. If respiratory arrest occurs, attempt resuscitation by holding the rat by the thorax and shaking gently. If gentle shaking does not elicit respiratory effort, pick the rat up and drop him onto the operating board from a height of about 3–4 inches.

B. Surgical procedure

Use sterile technique throughout the surgery; instruments should be thoroughly cleaned and disinfected before use on another animal. Wear gloves throughout the procedure. When the rat is well under the influence of anesthesia, lay him on his back on an operating board (with thermal pad in place in board and clean bench paper covering) with the animal's head toward you. Cleanse the neck area with 70% isopropanol applied to a 4 × 4 gauze pad. Using a sterile disposable scalpel with a #10 blade, make an incision vertically through only the skin layer of the neck. Begin just above the level of the sternum and end below the rat's chin. The incision should be $^1/_2$ to 1 inch in length. Gently pull the fascia of the neck apart. The carotid artery lies underneath and lateral to the sternohyoideus muscle and medial to the rectus capitus muscle. Locate the right carotid artery. Gently dissect tissues away from the artery with 2 pairs of curved toothed forceps. Slide one pair of curved toothed forceps under the artery. Gently pull the artery up out of neck cavity, stretching gently to free it from associated tissues. Throughout the remainder of the procedure, keep the artery elevated out of the neck cavity by supporting it on the open blades of one pair of forceps; the pressure provided by the blades of the forceps will keep the artery occluded and prevent unnecessary bleeding.

Tie off the right carotid artery at the exposed cephalic end with a 7 cm length of 3-0 silk suture. Place two 7 cm long pieces of suture under the artery at the exposed caudal end. Make a small slit (less than one third the radius of the vessel) in the top half of the artery with iris scissors. This slit will provide an entry hole for insertion of the catheter. A correct cut will form small drop of blood on the top of the slit. Be careful not to cut too deeply, or the artery will tear and the animal will exsanguinate. If the artery becomes dry during manipulations, rewet with sterile saline using a sterile transfer pipette. Grasp a catheter with serrated forceps in your dominant hand and thread the catheter (unsealed end first) through the slit in the artery, steadying the artery with a second pair of curved forceps in your nondominant hand. Push down on the toothed forceps supporting the artery to open the vessel. Advance the catheter toward the heart until resistance is met. The artery will bleed while the catheter is being advanced; work quickly to minimize blood loss. When the catheter is in place (as determined by advancement of a majority of its length, resistance to further advancement, and pulsation of the catheter in synchrony with the heartbeat), place a baby Diffenbach serrafine clamp on the artery above the two lengths of sutures placed at the caudal end of the artery. Tie each of the two loose sutures around the caudal end of artery, securing the catheter in place and preventing further blood loss from the artery. Remove the clamp and check for hemostasis. If bleeding occurs, tie an additional suture at the caudal end of the artery. Remove toothed forceps supporting the artery, allowing it to return to the neck cavity, and check again for hemostasis. Clip excess suture ends. Tuck the loose end of catheter under skin of the neck. Close the skin with skin clips.

C. Postprocedure

Place the rat in his cage on a warmed isothermal pad; the presence of the pad during surgery and recovery counteracts anesthesia induced hypothermia. Be certain water is available to replace volume lost during surgery and to prevent dehydration. Food may be replaced at this time. Keep handling to a minimum to avoid distress to the animal. The animal must not be left unattended until recovered from the anesthesia. Check the rat frequently during recovery from anesthesia; observe for return of the righting reflex, intake of oral fluid, and movement about cage. Observe for hematoma or bleeding at site. Assess post surgical animals at least daily. Observations should include the skin wound, activity level, intake of food and water, and behavior. Deaths beyond the immediate postoperative period may be related to ventricular arrhythmias secondary to catheter irritation of the ventricular muscle, or to cerebrovascular accidents from emboli originating from the sterile vegetations on the aortic valve.

D. Inoculation

Inoculation can be performed between 1 and 5 days post operatively, but consistency within experiments is critical. Prepare an inoculum of organisms at the desired cell count in 0.5 ml volume; draw the inoculum into a 1 cc syringe with 27 gauge $^1/_2$ inch needle. Pick the rat up by the base of his tail and place his front feet on the edge of the rat restraint. The animal should enter the restraint willingly. Secure the restraint. Cleanse the tail with 70% isopropanol; if needed, place the tail between warmed (37 °C) isothermal pads to dilate the tail veins. Steady the tail with your nondominant hand, being careful not to induce a sharp bend in the tail. Locate the lateral tail vein, just below the skin surface. Inject the inoculum intravenously. Proper injection will not raise a bleb and minimal resistance to fluid injection will be encountered. Hold pressure at the site until hemostasis occurs before returning the rat to his cage.

E. Necropsy

Sacrifice rats by carbon dioxide inhalation. Place the rat on his back with feet toward you. Wipe down the chest with 70% ethanol. Make a V-shaped incision in the chest below sternum. Undermine skin from the chest muscle using Mayo scissors. Extend the skin incision to the level of the neck on each side with Mayo scissors. Make a horizontal incision below the sternum through the peritoneum. Clip the diaphragm away. Cut the cartilage of ribs on each side of sternum, from base to apex. With a pair of hemostatic forceps, pull the flap of sternum back to expose the heart. Position the hemostatic forceps above the animal's head so that the weight of the forceps attached to the tip of the sternum keeps the chest cavity exposed. Expose the vena cava by lifting the heart gently with a pair of toothed forceps. If desired, take a blood sample from the inferior vena cava using a 1 cc syringe with 27 gauge $^1/_2$ inch needle. Pull the heart up gently (to avoid displacement of the catheter). Using curved scissors, clip through the heart tissue close to the aorta and dissect up to free the heart. Check visually for catheter placement. Cut off the apex of the heart. Locate the right ventricle. The right and left ventricles can be distinguished by wall thickness, with the right ventricle having a thinner wall. Place one blade of the curved scissors into the left ventricle through the cut at the apex, exiting through the aorta. Make a lateral cut through the left ventricular inner wall, the septum, and the right ventricle to expose the aortic valve. Place the open heart on a sterile 4 × 4 gauze and examine valve leaflets for vegetation. Check catheter placement over the valve. If desired, place the catheter tip in appropriate medium for qualitative culture. Remove vegetations from the valve with curved scissors and serrated forceps. Place the vegetations in a sterile disposable homogenizer tube with 2 ml sterile saline. Homogenize by rotating pestle in homogenizer tube. The sample is now ready for quantitative culture on appropriate media. Identity of organisms recovered should be determined, and should be the same as the inoculating strain.

4. Results and discussion

Representative results which can be obtained using the rat model of endocarditis are presented in Table 1. In the example provided (which is taken from previously published data [10]) , virulence of a wild type organism, S. mutans V403, is compared to virulence of an allelic exchange mutant defective in exopolysaccharide synthesis (V1996). Infectivity is strikingly different between wild type and mutant strains. At the highest inoculum (107 organisms per animal), 57% of animals inoculated with the wild type organism develop endocarditis, while only 17% of animals inoculated with the mutant strain developed endocarditis.

The rat model of endocarditis is an excellent method for in vivo evaluation of virulence of the viridans streptococci. As with any animal use, every effort should be made to use in vitro methods if acceptable, and to carefully design the experiments to use the minimum number of animals necessary to generate statistically valid results. Several in vitro methods to examine particular aspects of streptococcal virulence relevant to endocarditis have been described [5–7, 10, 13–15, 17, 19]. These in vitro methods include assays to measure binding to components commonly found in sterile vegetations, including fibrin and platelets [8–10, 13, 18], or to assess avoidance of host immunodefenses [10, 19]. However, the interaction of microorganisms with living animals is so complex that it is impossible to adequately model the entire system in vitro, and it is often essential to study the streptococci in live vertebrate animals in order to understand the importance of particular virulence factors in human infection. Rats are a preferred species for the endocarditis model because of similarities between human and rat endocardial infections, small size, surgical hardiness, and relatively low purchase and housing costs.

Table 1. Endocarditis cases in rats inoculated with wild type and mutant strains of S. mutans

Log_{10} inoculum	Wild type V403		Mutant V1996	
	Cases[a]	Total[b]	Cases	Total
7	11	19	4	24
6	4	6	0	6
5	4	8	0	3
4	0	3	0	2

[a] Number of animals with streptococci recovered from vegetations.

[b] Surgical survivors with correct catheter placement at necropsy.

The combination of injected ketamine for induction of anesthesia and inhalation of methoxyflurane for anesthesia maintenance is appropriate for survival surgery in rodents when the procedure is of short duration. The effects of ketamine last 30–40 minutes, while the procedure generally takes less than 20 minutes; methoxyflurane has potent analgesic effects which extend into the immediate post operative period. However, attention to investigator safety is an important aspect in the use of inhalation anesthetics. Methoxyflurane has been associated with hepatic and renal toxicity in humans [12], and the National Institute for Occupational Safety and Health recommends limiting exposure to halogenated anesthetics such as methoxyflurane to 2 ppm during anesthetic administration [2]. In addition, increased levels of methoxyflurane in the investigator's immediate environment can lead acutely to hypoxia which may be experienced as dizziness. It is imperative, therefore, that rat catheterizations be performed under a functioning fume hood. The 50 ml conical tube used to deliver whiffs of methoxyflurane to the rat should be stored inverted and toward the back of the hood to prevent release of anesthetic vapors into the investigator's work space.

The rat model of endocarditis requires specific psychomotor skills that are perfected over time. Surgical survival rates are highly dependent upon the technical expertise of the investigator. Success in the technique can be measured by the number of animals that survive the surgery, have correct catheter placement at time of necropsy, and have vegetations present on cardiac valves. Additionally, the organisms recovered from vegetations should be the same as those used in the inoculum. Assuming technical competence in the rat surgery, results generated with the rat model of endocarditis are highly reproducible.

Notes on suppliers

1. Fisher, 711 Forbes Ave., Pittsburgh, PA 15219, USA
2. VWR, P.O. Box 483, Bridgeport, NJ 08014, USA
3. Roboz Surgical Instrument Company, Inc., 1000 Connecticut Avenue, NW, P.O. Box 19148, Washington, DC 20036, USA. Telephone 202-393-1234
4. Braintree Scientific, Inc., P.O. Box 850929, Braintree, MA 02185-0929, USA. Telephone 617-843-7932
5. Sage Products, Cary, IL, USA

References

1. Baddour LM, Christensen GD, Lowrance JH, Simpson WA (1989). Pathogenesis of experimental endocarditis. Rev Infect Dis 11(3): 452–463.
2. Burkhart JE, Stobbe TJ (1990). Real-time measurement and control of waste anesthetic gases during veterinary surgery. Am Ind Hyg Assoc J 51: 640–645.
3. Burnette-Curley D, Wells V, Viscount H, Munro CL, Fenno JC, Fives-Taylor P, Macrina FL (1995). FimA, a major virulence factor associated with Streptococcus parasanguis endocarditis. Infect Immun 63: 4669–4674.
4. Carbon C (1993). Experimental endocarditis: A review of its relevance to human endocarditis. J Antimicrob Chemother 31 (Suppl D): 71–85.
5. Douglas CW, Brown PR, Preston FE (1990). Platelet aggregation by oral streptococci. FEMS Microbiol Lett 60: 63–67.
6. Ford I, Douglas CW, Preston FE, Lawless A, Hampton KK (1993). Mechanisms of platelet aggregation by Streptococcus sanguis, a causative organism in infective endocarditis. Br J Haematol 84: 95–100.
7. Herzberg MC, Gong K, MacFarlane GD, Erickson PR, Soberay AH, Krebsbach PH, Manjula G, Schilling K, Bowen WH (1990). Phenotypic characterization of Streptococcus sanguis virulence factors associated with bacterial endocarditis. Infect Immun 58: 515–522.
8. Manning JE, Geyelin AJ, Ansmits LM, Oakey HJ, Knox KW (1994). A comparative study of the aggregation of human, rat and rabbit platelets by members of the Streptococcus sanguis group. J Med Microbiol 41: 10–13.
9. Manning JE, Hume EB, Hunter N, Knox KW (1994). An appraisal of the virulence factors associated with streptococcal endocarditis. J Med Microbiol 40: 110–4.
10. Munro CL, Macrina FL (1993). Sucrose-derived exopolysaccharides of Streptococcus mutans V403 contribute to infectivity in endocarditis. Mol Microbiol 8: 133–142.
11. Santoro J, Levinson ME (1978). Rat model of experimental endocarditis. Infect Immun 19: 915–918.
12. Stimpel TM, Gershey EL (1991). Selecting anesthetic agents for human safety and animals recovery surgery. FASEB J 5: 2099–2104.
13. Sullam PM, Costerton JW, Yamasaki R, Dazin PF, Mills J (1993). Inhibition of platelet binding and aggregation by streptococcal exopolysaccharide. J Infect Dis 167: 1123–1130.
14. Sullam PM, Frank U, Yeaman MR, Tauber MG, Bayer AS, Chambers HF (1993). Effect of thrombocytopenia on the early course of streptococcal endocarditis. J Infect Dis 168: 910–914.
15. Tart RC, van de Rijn I (1991). Analysis of adherence of Streptococcus defectivus and endocarditis-associated streptococci to extracellular matrix. Infect Immun 59: 857–862.
16. Viscount HB, Munro CL, Burnette-Curley D, Peterson DL, Macrina FL (1997). Immunization with FimA protects against Streptococcus parasanguis endocarditis in rats. Infect Immun 65: 994–1002.
17. Wells VD, Munro CL, Sulavik MC, Clewell DB, Macrina FL (1993). Infectivity of a glucan synthesis-defective mutant of Streptococcus gordonii (Challis) in a rat endocarditis model. FEMS Microbiol Lett 112: 301–305.
18. Willcox MD (1995). Potential pathogenic properties of members of the 'Streptococcus milleri' group in relation to the production of endocarditis and abscesses. J Med Microbiol 43: 405–410.
19. Wu T, Yeaman MR, Bayer AS (1994). In vitro resistance to platelet microbicidal protein correlates with endocarditis source among bacteremic staphylococcal and streptococcal isolates. Antimicrob Agents Chemother 38: 729–732.

Address for correspondence: Cindy L. Munro, RN, PhD, Virginia Commonwealth University, Box 980567, Richmond VA 23298-0567, USA
Phone: 804-828-3410; Fax: 804-828-7743
E-mail: cmunro@vcu.edu

Methods in Cell Science **20**: 209–216 (1998).

Lipoproteins and other cell-surface associated proteins in streptococci

Roderick McNab* & Howard F. Jenkinson
Department of Oral and Dental Science, University of Bristol, Dental Hospital and School, Bristol, UK
*(*Present address: Department of Microbiology, Eastman Dental Institute, University of London, London, UK)*

Abstract. Methods are described for the identification and extraction of lipoproteins and other cell-surface proteins from streptococci. Polypeptides loosely-associated with the cell surface may be extracted from intact cells with detergents, while lipid-linked and other membrane-integral proteins are efficiently solubilized only from disrupted cells or isolated membranes. Incubation of intact cells with wall-hydrolyzing enzymes effects release of proteins linked covalently to cell-wall peptidoglycan. Surface proteins have been shown to be important for the colonization and virulence properties of streptococci and thus are potential targets for new anti-microbials and components of novel vaccines.

Key words: Anchorage of polypeptides, Lipoproteins, Polypeptide release, *Streptococcus* surface proteins, Wall-associated protein extraction

Abbreviations: aa = amino acid; LTA = lipoteichoic acid; PMSF = phenylmethylsulfonyl fluoride; SDS = sodium dodecyl sulfate; SDS-PAGE = sodium dodecyl sulfate-polyacrylamide gel electrophoresis; SLS = sodium lauroylsarcosinate

1. Introduction

Streptococci express a large repertoire of cell surface proteins that are involved in mediating bacterial cell adhesion, protection from host immune defences, nutrient uptake and environmental sensing [11]. It has long been appreciated that a number of these proteins are found intimately associated with the cell walls of streptococci [5, 27]. While incubation of intact bacterial cells or isolated cell walls with solutions of salts, detergents or chaotrophic agents will extract many surface-associated proteins, some polypeptides are released only following proteolytic or hydrolytic enzyme treatments suggesting their covalent linkage to cell-wall peptidoglycan [24]. Protease treatment was utilized in early purification of the major anti-phagocytic M protein from group A streptococcal cell walls [1].

In recent years, it has become established that there are two major mechanisms by which polypeptides are anchored at the streptococcal cell surface. Both of these mechanisms involve post-translational covalent modifications of polypeptide precursors that carry leader peptides directing their secretion via the general export pathway [28]. In the first, the precursors of lipoproteins, which carry the consensus sequence Leu-(Ser/Ala)-(Ala/Gly)-Cys-(Ser,Gly) at the carboxy (C)-terminal end of the leader peptide, are cleaved by signal peptidase II immediately before the Cys residue. The amino(N)-terminal Cys then becomes lipid-modified which serves to tether the polypeptide to the outer face of the cytoplasmic membrane [3]. Lipoproteins first to be documented in gram-positive bacteria were the β-lactamases of bacilli and staphylococci [25]. Virtually all of the streptococcal lipopolypeptides characterized to date are substrate-binding proteins serving ATP-binding cassette (ABC) transport systems [30].

A second mechanism of anchorage involves the covalent linkage of protein via a specialized C-terminal sequence to cell wall peptidoglycan. More than 40 streptococcal cell-surface polypeptides have now been identified that carry a C-terminal anchorage region comprising a Leu-Pro-Xaa-Thr-Gly-Xaa sequence motif followed by a hydrophobic sequence and a short charged cytoplasmic tail [4]. These sequences together serve as a 'stop-transfer' signal to protein secretion. A hypothetical enzyme termed sortase is then proposed to cleave between the Thr and Gly residues within the LPxTGx motif and the carboxyl group of threonine becomes amide-linked to a free amino group present within the peptidoglycan cross-bridge [29].

The colonization and virulence properties of many gram-positive bacteria are determined by the activities of cell-surface polypeptides. However, extraction and purification of surface-associated proteins for analysis, especially in intact form, has long been problematical. Only relatively recently have the necessary biochemical and genetic technologies been developed for more detailed structural and functional studies of streptococcal cell-surface polypeptides. This article describes methods for the extraction and identification of cell-surface polypeptides from streptococci that may be applicable also to other gram-positive bacterial systems.

2. Materials

A. Equipment

1. Anaerobic Jar, GasPak Anaerobic System, BBL.[1]
2. Spectrophotometer, model UV 120-02, Shimadzu.[2]
3. Centrifuge, model KR 20000T, Kuboto.[3]
4. Centrifuge, model 5402, Eppendorf.[4]
5. Vacuum Pump model VP190, Condensation Trap model RT400A, Speed Vac Concentrator model SVC 200H, Rotor model RH24-18, Savant.[5]

B. Growth media and reagents

1. BHY medium
   - Brain Heart Infusion, No. 0037-17-8, Difco.[6]
   - Yeast Extract, No. 0127-17-9.[6]
   - Agar, No. L11, Oxoid.[7]
   - Glycerol, No. 10118, BDH.[8]
2. TY-glucose medium
   - Tryptone, No. 0123-17-3.[6]
   - Yeast Extract, No. 0127-17-9.[6]
   - Dextrose, No. 28450.[8]
   - Dipotassium hydrogen orthophosphate, No. 10349.[8]
3. Spheroplasting buffer
   - Raffinose, No. R0250, Sigma.[9]
   - Tris, No. 15504-038, Life Technologies.[10]
   - Magnesium chloride, No. 296, Ajax.[11]
   - Mutanolysin, No. M9901.[9]
4. TEP buffer
   - Tris, No. 15504-038.[10]
   - Hydrochloric acid, No. 30721, Riedel-de Haen.[12]
   - EDTA, No. E5134.[9]
   - PMSF, No. P7626.[9]
   - Isopropanol (propan-2-ol), No. 10224.[8]
5. Protein extraction agents
   - Sodium dodecyl sulfate, No. 15535-017.[10]
   - Lauroylsarcosine, No. L9150.[9]
   - Acetone, No. 10003.[8]
   - Glass beads, 0.10 mm diam., No. 854 140/0, Braun.[13]
   - Bromophenol blue, No. 20015.[8]
   - 2 Mercaptoethanol, No. M6250.[9]
   - Sodium hydroxide, No. 10252.[8]
   - Sodium chloride, No. 10241.[8]
   - Glycine, No. 10119.[8]
   - Methanol, No. 10158.[8]

C. Lipoprotein labeling and detection

- [9,10-^{3}H]Palmitate, No. TRK 909, Amersham Life Science.[14]
- Tween 80, No. 37475, Serva.[15]
- Amplify, No. NAMP 100.[14]
- Kodak EKtamat X-ray film, No. 1884220, Eastman-Kodak.[16]

D. Glassware and plastics

- McCartney glass bottles, 28 ml with metal caps, No. 215/0054.[8]
- Duran glass bottles and caps, No. 215/0150.[8]
- Culture tubes, 12 ml PC capped, No. 1209, Labserv.[17]
- Corex glass 15 ml centrifuge tubes, No. 03286, du Pont.[18]
- Plastic 3 ml pipettes, No. 225, Samco.[19]

3. Procedures

A. Preparation of materials

1. Use standard sterilization techniques for all glassware, microbiological culture media, and buffers.
2. Preparation of culture media:
   a) BHY medium
   Dissolve 37 g Brain Heart Infusion and 5 g Yeast Extract to 1 L in distilled water, distribute 20 ml portions into glass bottles and sterilize by autoclaving.
   b) TY-glucose medium
   Dissolve 5 g Tryptone, 5 g Yeast Extract and 4 g dipotassium hydrogen orthophosphate in distilled water to 1 L (adjust to pH 7.5 if necessary), distribute into 20 ml portions and sterilize by autoclaving. Before using, add 0.4 ml of 40% (wt/vol) dextrose solution in water, sterilized by autoclaving, to each 20 ml of medium.
3. Preparation of extraction media and buffers:
   c) Spheroplasting buffer
   To a 100 ml-bottle containing 0.24 gm Tris, 0.2 g magnesium chloride hexahydrate, and 26 g raffinose, add distilled water to approximately 90 ml and dissolve ingredients by incubation at 40 °C with stirring. Adjust to pH 6.8 with 1N HCl, make volume to 99 ml, distribute into 9.9 ml portions and store at –20 °C. Prior to using, thaw a portion at 40 °C until completely dissolved and add 0.1 ml freshly-prepared solution of serine protease inhibitor PMSF (0.174 g dissolved in 10 ml anhydrous isopropanol). PMSF is highly toxic and decomposes rapidly in aqueous solution.
   d) TEP buffer
   Dissolve 1.21 g Tris and 0.372 g EDTA disodium salt in 200 ml deionized water, adjust to pH 7.5 with HCl and make to 990 ml with water. Just before using add 10 ml stock PMSF solution (see above).
   e) Sodium lauroylsarcosinate (SLS)
   Mix 10 g lauroylsarcosine with 90 ml water, gradually add sodium hydroxide pellets with stirring to effect dissolution and to achieve pH 8.5, adjust to 100 ml with water and store at room temperature. Dilute 1:10 with deionized water before use.

g) Saline
Dissolve 9 g sodium chloride in 1 L deionized water, sterilize by autoclaving and store at 4 °C.

4. SDS-PAGE gels and buffers, protein stains and electrotransfer buffer
   a) Acrylamide gels and buffers
   These methods are described in detail in reference [15] and in many laboratory manuals.
   b) Sample buffer
   To 1 g SDS in a 28-ml glass bottle add 2.5 ml 0.5 M Tris-HCl buffer, pH 6.8, 7.0 ml deionized water, dissolve by incubating at 40 °C with stirring, add 0.2 ml 2-mercaptoethanol, and store at –20 °C.
   c) Sample loading dye
   Dissolve 0.5 mg bromophenol blue in 3 ml alkalized deionized water in a glass bottle, add 7 ml glycerol, mix well and store at room temperature.
   d) Silver-staining of proteins
   This is described in detail in reference [21]. It is essential to use ultra-pure reagents silver nitrate and anhydrous sodium carbonate.
   e) Protein electrotransfer buffer
   Dissolve 3.0 g Tris and 14.4 g glycine in 800 ml distilled water and add 200 ml methanol.

B. Preparation of bacterial stocks
1. Pure cultures of streptococcal bacteria should be obtained on the surface of agar (BHY medium containing 1.5% wt/vol agar) following incubation of plates for 24 h at 37 °C in a GasPak anaerobic jar.
2. Inoculate 10 ml BHY medium, pre-warmed within a 12 ml-screw-capped polycarbonate tube at 37 °C, with several freshly-grown bacterial colonies and incubate for 16 h at 37 °C without shaking (achieves a bacterial cell density of about 2×10^9 cells ml^{-1}).
3. Harvest cells by direct centrifugation of culture tube (5000 g, 4 °C, 10 min), discard the supernatant, and suspend the pellet in 2 ml BHY medium containing 10% (vol/vol) glycerol. Distribute 0.2 ml quantities into sterile microfuge tubes and store at –80 °C.

C. Extraction of proteins from streptococcal cells with SLS
1. Inoculate 10 ml BHY medium, pre-warmed within a 12 ml-screw-capped tube at 37 °C, with 0.05 ml recently-thawed stock cell suspension (see Procedure B). Incubate without shaking for 5 h at 37 °C, or until the OD_{600} = 1.0 (1.5×10^9 cells ml^{-1}).
2. Harvest cells by centrifugation (5000 g, 4 °C, 10 min), discard the supernatant and suspend cells in 5 ml saline using a Samco plastic transfer pipette.
3. Harvest cells by centrifugation as before, discard the supernatant and suspend the pellet in 0.2 ml 1% (wt/vol) SLS solution in water by vortex-mixing. Transfer the cell suspension with pipette to a 1.5-ml microfuge tube, and incubate the suspension for 20 min at 20 °C, vortex-mixing for 10 s after 10 min.
4. Pellet the bacterial cells by centrifugation (10,000 g, 4 °C, 15 min) and pipette carefully 0.18 ml of the supernatant into a clean 1.5-ml tube. The SLS-extract may be stored for up to 24 h at –20 °C without incurring significant protein degradation.
5. In preparation for SDS-PAGE, add to the 0.18 ml extract 20 µl sample buffer followed by 20 µl sample loading dye, mix and heat for 10 min at 70 °C. Subject the sample to SDS-PAGE gel within 30 min. Otherwise, store the sample for up to 24 h at –20 °C, but heat again for 10 min at 70 °C prior to SDS-PAGE.

D. Labeling and extraction of lipoproteins
1. Inoculate 10 ml BHY medium, pre-warmed in a 12 ml-screw-capped tube at 37 °C, with 0.05 ml recently-thawed stock cell suspension. Incubate at 37 °C without shaking for 1–2 h until the OD_{600} = 0.15–0.2 (approximately 1×10^8 cells ml^{-1}). During this time, vacuum dry 0.15 ml [^{3}H]palmitate (150 µCi, 50 Ci $mmol^{-1}$) in a sterile 1.5 ml-microfuge tube and redissolve the palmitate in 0.05 ml 0.1% (vol/vol) Tween 80 in sterile distilled water at 37 °C.
2. Transfer 1.4 ml of exponential-phase bacterial culture into the pre-warmed tube containing [^{3}H]palmitate and incubate for 4 h at 37 °C without shaking.
3. Collect the bacterial cells by centrifugation (10,000 g, 4 °C, 10 min), aspirate off the supernatant and suspend the bacteria in 0.1 ml TEP buffer. Add an equal volume of glass beads (0.1 mm diam.) and vortex mix the suspension continuously for 2 min.
4. Add 0.1 ml of sample buffer to suspension, heat for 10 min at 70 °C, then centrifuge (5000 g, 5 min) to sediment glass beads and cell debris.
5. Transfer carefully 0.18 ml of the supernatant into a clean tube, add 20 µl sample loading dye and immediately subject the sample to SDS-PAGE.

E. Extraction of cell wall-associated proteins
1. Prepare late-exponential phase bacterial culture as described in Procedure C, 1.
2. Harvest cells by centrifugation (5000 g, 4 °C, 10 min), suspend in 1 ml TEP buffer and transfer the suspension to a 1.5 ml-microfuge tube. Centrifuge to collect cells (5000 g, 2 min)

aspirate off the supernatant and wash the cells by suspension and centrifugation as before.

3. To the bacterial cell pellet add 0.12 ml spheroplasting buffer, suspend cells by vortex-mixing and add 6 μl of 10000 units mutanolysin ml^{-1}. Mix gently and incubate the suspension for 30 min at 37 °C with occasional gentle mixing. Collect the spheroplasts by centrifugation (5000 g, 4 °C, 10 min), transfer carefully 0.1 ml supernatant to a clean tube, add 0.1 ml sample buffer and heat for 10 min at 70 °C in preparation for SDS-PAGE. To extract wall-associated proteins from *Enterococcus faecalis* cells, lysozyme (final concn. 0.2 mg ml^{-1}) must be added together with mutanolysin.

F. Preparation of culture supernatant proteins

1. Prepare late-exponential phase culture as described in Procedure C, 1, except in TY-glucose medium. Pellet cells by centrifugation (10,000 g, 4 °C, 10 min), collect the supernatant, transfer to a clean polycarbonate culture tube and re-centrifuge.
2. Remove 2 ml clear supernatant into a 15 ml glass Corex tube. To precipitate proteins, add 8 ml acetone at –20 °C, mix by inversion, and incubate tube without shaking for 16 h at –20 °C.
3. Centrifuge the suspension (10,000 g, 4 °C, 10 min), drain the tube carefully so as not to disturb the loose sediment, then place the tube under vacuum for 10 min to dry the precipitate. Add 0.2 ml sample buffer, vortex mix to suspend, transfer to a 1.5-ml microfuge tube and heat the suspension for 10 min at 70 °C. In preparation for SDS-PAGE, centrifuge for 5 min at 10,000 g, collect the supernatant and add 20 μl sample loading dye.

G. Protein separation and detection

1. Separate proteins by SDS-PAGE [15]. The sensitive method of staining proteins with silver [21] is necessary to visualize protein profiles of SLS extracts (Procedure C), mutanolysin extracts (Procedure E), and culture supernatant fluid (Procedure F). Care should be taken to not overload gels with Procedure F samples since these contain elevated salt levels that may cause distortion of protein banding patterns.
2. To detect [^{3}H]palmitate-labeled lipoproteins by fluorography following SDS-PAGE, first fix the gel for 30 min in 10 volumes of 25% (vol/vol) methanol, then transfer to 10 volumes of Amplify and shake gently for 30 min. Vacuum-dry the fluor-impregnated gel onto Whatman 3 mm paper at 60 °C and then expose the dried gel to X-ray film for up to 7 d at –80 °C.
3. Electroblot proteins onto nitrocellulose by using the method of Towbin et al. [32]. For efficient transfer of high molecular mass proteins > 180 kDa blotting times of 90 min at 20 V cm^{-1} are recommended. Transfer efficiency may be improved further by reducing the methanol concentration to 5% (vol/vol) in the transfer buffer. Immunoblots are developed as described in detail elsewhere [10].

4. Results and discussion

A. *Lipoproteins*

The presence of a L-(S,A)-(A,G)-C-(S,G) consensus sequence within the signal peptide of a polypeptide precursor is the 'signature' for lipid-modification of the mature protein. The simplest and most widely-utilized means of confirming that a mature protein is lipid-modified is to demonstrate that following growth of bacterial cells in radioactively-labeled fatty acid (usually palmitate), the polypeptide carries covalently-bound radiolabel. Lipoproteins are thought to be held at the cell surface through intercalation of their lipid moiety into the cell membrane. In support of this notion, lipoproteins are effectively released following incubation of cells with ionic detergents, by not by treatment of cells with solutions of salts or chaotrophes. The nature of the detergent utilized affects significantly the efficiency of extraction. Thus incubation of intact cells with SLS at ambient temperature (Procedure C) extracts upwards of 20 polypeptides from the *S. gordonii* cell surface [7]. A major component of this extract is a phosphocarrier protein (HPr) which is poorly extracted by incubating cells with SDS at elevated temperatures [7]. By contrast, SLS treatment of intact cells does not effectively remove lipoproteins, and is 50% less efficient than SDS for solubilizing [^{3}H]palmitate-labeled lipoproteins from membranes of disrupted cells [8]. Following metabolic labeling of *S. gordonii* DL1 (Challis) cells with [^{3}H]palmitate and extraction of proteins with SDS (Procedure D), 13 lipopolypeptide bands may be discerned by fluorography of SDS-PAGE gels (Figure 1). Roughly similar numbers of radioactively-labeled protein bands are detected in extracts of *Streptococcus sanguis*, *Streptococcus parasanguis* and *Streptococcus mutans* cells grown in the presence of radioactively-labeled palmitate [8, 31]. The three slowest-migrating lipopolypeptide bands in Figure 1 of apparent molecular masses 78 kDa, 78 kDa (co-migrate) and 76 kDa, which are expressed during growth of *S. gordonii* cells in BHY medium, are oligopeptide-binding proteins serving the Hpp oligopeptide uptake system [12]. Insertional inactivation of the *hppA* gene in *S. gordonii* OB285 abolishes production of the 76-kDa labeled band (Figure 1, lane 2), confirming that HppA is a lipoprotein.

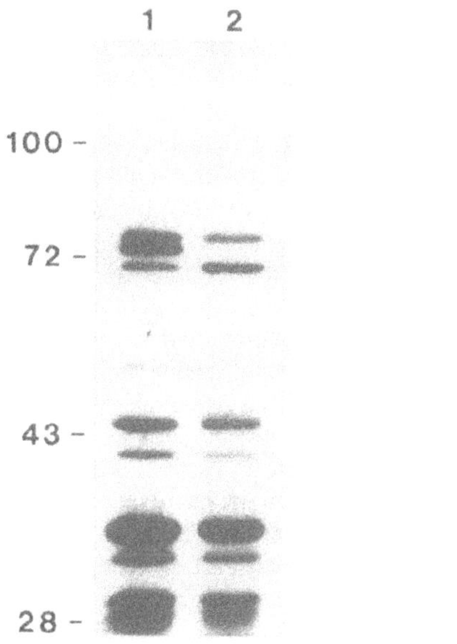

Figure 1. Fluorogram of [^{3}H]palmitate-labeled proteins extracted from *S. gordonii* DL1 wild-type (lane 1) or OB285 *hppA1* mutant (lane 2) cells. Positions of molecular mass markers (in kDa) are indicated on the left.

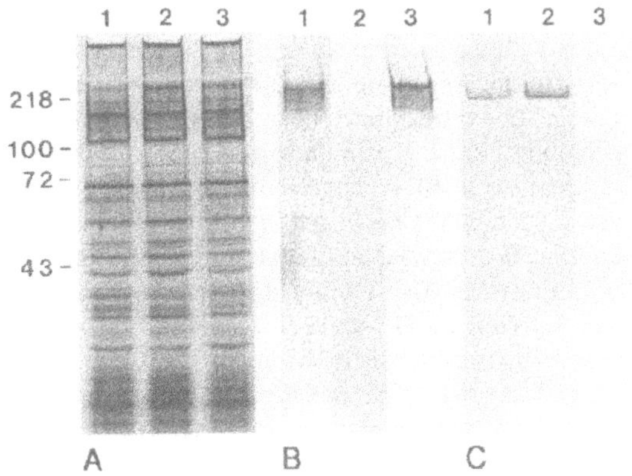

Figure 2. SDS-PAGE patterns and immunoblot analysis of proteins released following mutanolysin treatment of *S. gordonii* wild-type (lane 1), OB277 *cshA31 cshB2* (lane 2), or OB219 *sspA sspB* (lane 3). A, profile of silver-stained proteins; B, Western blot of panel A reacted with rabbit antiserum to CshA poylpeptide diluted 1:500; C, Western blot reacted with antibodies to antigen I/II polypeptide (SpaP) of *S. mutans* diluted 1:200. Primary antibody binding was detected with peroxidase-linked swine antibodies to rabbit immunoglobulin G and developed with chloronapthol [10]. Positions of molecular mass marker proteins (kDa) are indicated.

Although streptococcal lipoproteins are firmly anchored at the cell surface and fractionate along with isolated cell membranes [8] there is evidence that some release may occur, perhaps as a result of lipid turnover [30]. The β-lactamases of bacilli are in fact cleaved from the cell surface late in the growth cycle and released into the culture fluid as hydrophilic products lacking the lipid modification [25]. Functional studies of streptococcal lipoproteins have been hampered by findings, from a number of laboratories, that expression of even relatively small coding segments of the genes on multi-copy plasmids can be deleterious to growth of *Escherichia coli* cloning hosts. By contrast, expression of *Bacillus licheniformis* β-lactamase is well tolerated by *E. coli*, and this has permitted detailed analyses of the structural requirements for lipoprotein processing and membrane localization [6].

B. *Wall-associated proteins*

Proteins that are covalently-linked to cell wall peptidoglycan in oral streptococci are released from cells following treatment with the hydrolytic enzyme mutanolysin in the presence of an osmotic stabilizer to minimize cell lysis (Procedure E). Fifty or more polypeptide bands ranging in molecular mass from > 250 kDa to < 30 kDa are detected upon silver-staining of SDS-PAGE gels of mutanolysin extracts from *S. gordonii* DL1 cells (Figure 2). Lipoproteins do not appear to be released in significant amounts by this treatment. The SDS-PAGE patterns are highly reproducible for a given strain and its isogenic mutants (Figure 2). The CshA (259 kDa) and CshB (245 kDa) oligospecific adhesins [19] are recovered more or less intact from the surface of *S. gordonii* wild-type cells and are detected by reacting Western blots with CshA antibodies (Figure 2B, lane 1). Also released intact from *S. gordonii* cells are the SspA (171 kDa) and SspB (160 kDa) salivary glycoprotein adhesins of the antigen I/II family [9], detected by reacting Western blots with antigen I/II family polypeptide antibodies (Figure 2C, lane 1). This method is generally applicable to extracting these kinds of polypeptides from most other species of oral streptococci. Isogenic *S. gordonii* mutants in which *cshA* and *cshB* genes are inactivated do not produce Csh-antibody reactive bands (Figure 2B, lane 2), but still retain the SspA and SspB adhesins (Figure 2C, lane 2). Isogenic *sspA sspB* double mutants of *S. gordonii* that do not express antigen I/II-antibody reactive bands (Figure 2C, lane 3) are unaffected in expression of CshA and CshB proteins (Figure 2B, lane 3). These data demonstrate clearly that Csh and Ssp adhesins are independently expressed and presented on the streptococcal cell surface.

Sequence analyses have revealed that the aforementioned *S. gordonii* adhesins all possess a C-terminal anchorage region, viz. the LPxTGx motif, hydrophobic sequence and charged tail. However the precise structural and sequence requirements necessary for determining anchorage are still being elucidated, and the enzyme(s) involved in cleavage and transpeptidation is/are currently uncharacterized. It appears that the exact location of the anchorage region within a polypeptide may not be critical, since when the anchorage region of *Staphylococcus aureus* protein A is translocated by using recombinant technology to a central position within a hybrid polypeptide, it still becomes covalently linked to the wall [23]. On the other hand, evidence suggests that the LPxTGx motif alone might not be sufficient for recognition by the sorting machinery. This was demonstrated by analyzing the cell-surface versus culture supernatant distributions of CshA protein in *S. gordonii* wild-type or isogenic mutant cells in which various C-terminally truncated forms were generated by gene disruption. The extracellular fluid of stationary phase cultures of *S. gordonii* contains in excess of 70 polypeptides as detected by SDS-PAGE of proteins collected by precipitation (Procedure F) (Figure 3A, lane 1). Small quantities of intact CshA protein are released into the culture supernatant during growth of wild-type cells as shown by immunoblot analysis (Figure 3B, lane 1). Isogenic mutant *S. gordonii* OB186 carries an insertion within the *cshA* gene that C-terminally truncates the CshA protein by 250 aa residues [18]. Ensuing loss of the anchorage region results in release of the truncated protein from the cell surface into the culture extracellular fluid (Figures 3A and 3B, lane 2). It is of note that anchorage failure occurs despite there being present, internally within the truncated polypeptide, four additional LPxTGx motifs [19]. A more extensive C-terminally truncated polypeptide produced by *S. gordonii* OB193, which

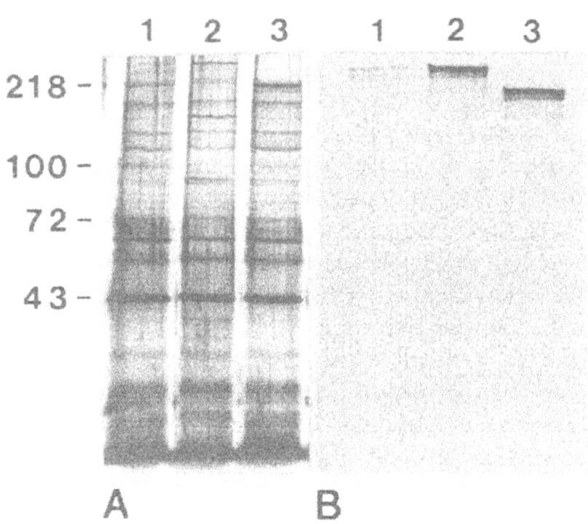

Figure 3. SDS-PAGE patterns and immunoblot analysis of culture fluid proteins present in early stationary phase cultures of *S. gordonii* wild-type or isogenic mutant strains expressing truncated forms of CshA. Lane 1, *S. gordonii* DL1 wild-type; lane 2, OB185 *cshA8*; lane 3, OB193 *cshA5*. Panel A, profile silver-stained proteins; Panel B, corresponding Western blot reacted with antibodies to CshA (see legend to Figure 2). Positions of molecular mass marker proteins (kDa) are indicated.

still carries an single LPxTGx sequence motif, is also secreted intact into the culture fluid and not anchored at the cell surface (Figures 3A and 3B, lane 3). The phenotypic effects of these mutations are to render the mutant cells deficient in CshA-mediated adhesion properties [19]. Interestingly, a human oral isolate of *S. mutans* was shown recently to release 'naturally' a truncated form of PAc adhesin (antigen I/II) protein into the culture supernatant. Sequence analysis suggests this occurs as a result of a frameshift mutation within the C-terminal coding region of the gene causing premature termination of translation [22]. Although PAc is required for adhesion of *S. mutans* cells to salivary glycoprotein pellicle in vitro [14], the consequences of protein anchorage failure on bacterial growth and survival in vivo are not yet known. It has been suggested that release of specific surface protein antigens by streptococcal cells may assist in evasion of host immune defences [17].

C. *Novel cell-surface protein anchorage mechanisms*

Currently there is much interest in determining the roles of various surface proteins in epithelial cell invasion by *Listeria monocytogenes*. While a major invasion-associated surface protein InlA carries the characteristic gram-positive bacterial LPxTGx motif and anchorage region [16], two additional invasion-associated proteins with novel cell-surface anchorage sequences have been recently described. The first of these, ActA, is required for actin polymerization following entry of *Listeria* into host cells. ActA protein carries at the C-terminus a trans-membrane helix of 22 aa residues and a short charged cytoplasmic tail, but no LPxTGx motif [13]. ActA can be extracted from intact *Listeria* cells with SDS which is consistent with it being non-covalently linked to the cell surface. A second protein, designated InlB, which is also involved in bacterial entry into the host cell [2], carries three 80 aa residue repeat blocks at the C-terminal end. These repeat block sequences, which possibly bind to cell wall polymers, are essential for surface localization of InlB, and are sufficient for allowing re-association of secreted or exogenously-added InlB protein with the *Listeria* cell surface [2]. This mode of cell-surface association of InlB is reminiscent of that of PspA which is held at the *Streptococcus pneumoniae* cell surface via interaction of C-terminal aa residue repeat blocks with the choline-containing LTA [33]. As new streptococcal surface antigens and proteins with novel biological activities are continually being discovered, it is likely that additional surface protein-anchorage sequences and mechanisms will be revealed. For example, a protein with glyceraldehyde-3-phosphate dehydrogenase activity is present on the cell surface of *Streptococcus pyogenes* [26] and is sensitive to trypsin. The primary aa sequence reveals no obvious surface-anchorage determinant, and the protein is not removed from cells with SDS, so a novel anchorage mechanism might be predicted.

D. *Wall anchor mechanisms and streptococcal diseases*

With the increasing occurrence of antibiotic resistance amongst clinical bacterial isolates there is urgent need to develop new anti-microbial agents and to identify novel bacterial targets. Since many cell-wall-associated polypeptides are intricately linked with adhesion, colonization and virulence properties of streptococci [11], developing strategies to inhibit their expression, presentation or function on the cell surface might be significant. Inhibition of the enzymic reactions involved in LPxTGx-associated cell-wall linkage could block the normal presentation of adhesins, anti-phagocytic molecules, and immunoglobulin binding proteins. Another possibility might be to inhibit the presentation of lipoproteins. Novel inhibitors of pre-lipoprotein processing might be valuable given that a number of streptococcal lipoproteins are components of nutrient uptake systems. Lastly, the wall-anchorage machinery has been made use of to direct chimeric proteins to the streptococcal cell surface. In this way it may be possible to present foreign antigens on the surfaces of commensal streptococcal or other gram-positive bacteria in order to elicit production of protective antibodies at mucosal surfaces [20].

Acknowledgments

We thank R. A. Baker and A. R. Holmes for assistance in preparation of the manuscript.

Notes on suppliers

1. BBL Microbiology Systems, Cockeysville, Md., USA
2. Shimadzu Corp., Nakagyo-ku, Kyoto 604, Japan
3. Kuboto, Bunkyo-ky, Tokoyo, Japan
4. Eppendorf, 22331 Hamburg, Germany
5. Savant Instruments Ltd., Farmingdale, NY 11735, USA
6. Difco Laboratories, Detroit, Mich., USA
7. Oxoid Ltd., Basingstoke RG24 0PW, UK
8. BDH, Lutterworth LE17 4XN, UK
9. Sigma-Aldrich Co. Ltd., Poole, UK
10. Life Technologies Ltd., Paisley PA3 4EF, UK
11. APS Chemicals, Seven Hills NSW 2147, Australia
12. Riedel-de Haen, Seelzel D-3016, Germany
13. B. Braun Melsungen D-3598, Germany
14. Amersham International plc., Little Chalfont, UK
15. Serva Products Div., Heidelberg D-69042, Germany
16. Eastman Kodak, Rochester NY, USA
17. Labserv Products Ltd., Auckland, New Zealand
18. E. I. du Pont de Nemours & Co. Inc., Wilmington DE 19898, USA
19. Samco Scientific Inc., San Fernando CA 91340, USA

References

1. Beachey EH, Stollerman GH, Chiang EY, Chiang TM, Seyer JM, Kang AH (1977). Purification and properties of M protein extracted from group A streptococci with pepsin: covalent structure of the amino terminal region of type 24 M antigen. J Exp Med 145: 1469–1483.
2. Braun L, Dramsi S, Dehoux P, Bierne H, Lindahl G, Cossart P (1997). InlB: An invasion protein of *Listeria monocytogenes* with a novel type of surface association. Mol Microbiol 25: 285–294.
3. Braun V, Wu HC (1994). Lipoproteins, structure, function, biosynthesis and model for protein export. In: Ghuysen J-M, Hakenbeck R (eds), Bacterial cell wall, pp 319–341. New York: Elsevier Science.
4. Fischetti VA (1996). Gram-positive commensal bacteria deliver antigens to elicit mucosal and systemic immunity. ASM News 62: 405–410.
5. Fischetti VA, Gotschlich EC, Siviglia G, Zabriskie JB (1976). Streptococcal M protein extracted by nonionic detergent. I. Properties of the antiphagocytic and type-specific molecules. J Exp Med 144: 32–53.
6. Hayashi S, Chang SY, Chang S, Wu HC (1984). Modification and processing of *Bacillus licheniformis* prepenicillinase in *Escherichia coli*. Fate of mutant penicillinase lacking lipoprotein modification site. J Biol Chem 259: 10448–10454.
7. Jenkinson HF (1989). Properties of a phosphocarrier protein (HPr) extracted from intact cells of *Streptococcus sanguis*. J Gen Microbiol 135: 3183–3197.
8. Jenkinson HF (1992). Adherence, coaggregation, and hydrophobicity of *Streptococcus gordonii* associated with expression of cell surface lipoproteins. Infect Immun 60: 1225–1228.
9. Jenkinson HF, Demuth DR (1997). Structure, function and immunogenicity of streptococcal antigen I/II polypeptides. Mol Microbiol 23: 183–190.
10. Jenkinson HF, Easingwood RA (1990). Insertional inactivation of the gene encoding a 76-kilodalton cell cell surface polypeptide in *Streptococcus gordonii* Challis has a pleiotropic effect on cell surface composition and properties. Infect Immun 58: 2689–2697.
11. Jenkinson HF, Lamont RJ (1997). Streptococcal adhesion and colonization. Crit Rev Oral Biol Med 8: 175–200.
12. Jenkinson HF, Baker RA, Tannock GW (1996). A binding-lipoprotein-dependent oligopeptide transport system in *Streptococcus gordonii* essential for uptake of hexa- and heptapeptides. J Bacteriol 178: 68–77.
13. Kocks C, Gouin E, Tabouret M, Berche P, Ohayon H, Cossart P (1992). *Listeria monocytogenes*-induced actin assembly requires the *actA* gene product, a surface protein. Cell 68: 521–531.
14. Koga T, Okahashi N, Takahashi I, Kanamoto T, Asakawa H, Iwaki M (1990). Surface hydrophobicity, adherence, and aggregation of cell surface protein

antigen mutants of *Streptococcus mutans* serotype c. Infect Immun 58: 289–296.
15. Laemmli UK, Favre M (1973). Maturation of the head of bacteriophage T4. I. DNA packaging events. J Mol Biol 80: 575–599.
16. Lebrun M, Mengaud J, Ohayon H, Nato F, Cossart P (1996). Internalin must be on the bacterial surface to mediate entry of *Listeria monocytogenes* into epithelial cells. Mol Microbiol 212: 579–592.
17. Lee SF (1995). Active release of bound antibody by *Streptococcus mutans*. Infect Immun 63: 1940–1946.
18. McNab R, Jenkinson HF (1992). Gene disruption identifies a 290 kDa cell-surface polypeptide conferring hydrophobicity and coaggregation properties in *Streptococcus gordonii*. Mol Microbiol 6: 2939–2949.
19. McNab R, Jenkinson HF, Loach DM, Tannock GW (1994). Cell-surface-associated polypeptides CshA and CshB of high molecular mass are colonization determinants in the oral bacterium *Streptococcus gordonii*. Mol Microbiol 14: 743–754.
20. Medaglini D, Pozzi G, King TP, Fischetti VA (1995). Mucosal and systemic immune responses to a recombinant protein expressed on the surface of the oral commensal bacterium *Streptococcus gordonii* after oral colonization. Proc Natl Acad Sci USA 92: 6868–6872.
21. Morrissey JH (1981). Silver stain for proteins in polyacrylamide gels: A modified procedure with enhanced uniform sensitivity. Anal Biochem 117: 307–310.
22. Murakami Y, Nakano Y, Yamashita Y, Koga T (1997). Identification of a frameshift mutation resulting in premature termination and loss of cell wall anchoring of the PAc antigen of *Streptococcus mutans* GS-5. Infect Immun 65: 794–797.
23. Navarre WW, Schneewind O (1994). Proteolytic cleavage and cell wall anchoring at the LPXTG motif of surface proteins in gram-positive bacteria. Mol Microbiol 14: 115–121.
24. Nesbitt WE, Staat RH, Rosan B, Taylor KG, Doyle RJ (1980). Association of protein with the cell wall of *Streptococcus mutans*. Infect Immun 28: 118–126.
25. Nielsen JBK, Lampen JO (1982). Membrane bound penicillinases in gram-positive bacteria. J Biol Chem 257: 4490–4495.
26. Pancholi V, Fischetti VA (1992). A major surface protein on group A streptococci is a glyceraldehyde-3-phosphate dehydrogenase with multiple binding activity. J Exp Med 176: 415–426.
27. Russell RRB (1979). Wall-associated protein antigens of *Streptococcus mutans*. J Gen Microbiol 114: 109–115.
28. Schatz G, Dobberstein B (1996). Common principles of protein translocation across membranes. Science 271: 1519–1526.
29. Schneewind O, Fowler A, Faull KF (1995). Structure of the cell wall anchor of surface proteins in *Staphylococcus aureus*. Science 268: 103–106.
30. Sutcliffe IC, Russell RRB (1995). Lipoproteins of gram-positive bacteria. J Bacteriol 177: 1123–1128.
31. Sutcliffe IC, Tao L, Ferretti JJ, Russell RRB (1993). MsmE, a lipoprotein involved in sugar transport in *Streptococcus mutans*. J Bacteriol 175: 1853–1855.
32. Towbin H, Staehelin T, Gordon J (1979). Electrophoretic transfer of proteins from polyacrylamide gels to nitrocellulose sheets: Procedure and some applications. Proc Natl Acad Sci USA 76: 4350–4354.
33. Yother J, White JM (1994). Novel surface attachment mechanism of the *Streptococcus pneumoniae* protein PspA. J Bacteriol 176: 2976–2985.

Address for correspondence: H. F. Jenkinson, Department of Oral and Dental Science, University of Bristol, Dental Hospital and School, Lower Maudlin Street, Bristol BS1 2LY, UK
Phone: 0044 117 928 4304; Fax: 0044 117 928 4428
E-mail: howard.jenkinson@bristol.ac.uk

Methods in Cell Science **20**: 217–221 (1998).

Growth of *Streptococcus mutans* in an iron-limiting medium

Grace A. Spatafora & Meagan W. Moore
Middlebury College, Department of Biology, Middlebury, Vermont, USA

Abstract. Microbial pathogens require iron to survive and promote disease. Reports in the literature indicate that iron can potentiate infection in the human host by triggering the expression of bacterial virulence determinants [8]. *Streptococcus mutans* is a prevalent microorganism in dental plaque and the principal causative agent of human dental caries. This oral pathogen may respond to the changing availability of iron in the plaque environment brought about by conditions of feast or famine. Our understanding of a role for iron in *S. mutans*-induced caries formation is limited, however, by the lack of an appropriate test medium. In this study we applied batch chromatography to remove iron from a chemically-defined medium [7] which we subsequently reconstituted with manganese and/or magnesium. The final metal ion concentrations in this medium were confirmed by inductively coupled argon plasma (ICAP) analysis. Importantly, the iron-depleted medium supported the growth of *S. mutans* only when supplemented with 0.01 to 10 μM iron, concentrations which are consistent with those found in human saliva. The practical applications of this medium are far-reaching and include the identification of iron sources, and iron-responsive genes and their products which may promote streptococcal pathogenesis. Other trace metals may be altered in this medium to promote investigations of their putative effect(s) on streptococcal growth and expression.

Key words: Growth, Iron, Medium, Metals, Streptococci

Abbreviations: FMC = a conventional chemically-defined medium; FMC^- = FMC prepared without the iron component; $mdFMC_{cx}$ = metal-depleted FMC, FMC prepared without the metal components followed by treatment with Chelex-100; FMC_{cx}^- = iron-free $mdFMC_{cx}$ supplemented with manganese and/or magnesium FMC_{cx}^+ = FMC_{cx}^- supplemented with exogenous ferric chloride; ICAP = inductively coupled argon plasma analysis

1. Introduction

Elemental iron is an essential micronutrient for nearly all microbial pathogens [5]. Upon colonizing a mammalian host, bacteria are challenged with a shift from high to low iron availability, and therefore must evolve sophisticated mechanisms for obtaining iron from host proteins [9]. Despite a well-documented role for iron in promoting the pathogenesis of gram-negative microorganisms, relatively little is known of its putative involvement in the etiology of gram-positive disease [3]. For instance, iron uptake and utilization is virtually unexplored in the disease-causing streptococci, with only a single report in the literature which addresses iron uptake in *Streptococcus mutans*, the principal causative agent of human dental caries [4]. Preliminary studies suggest that certain trace metals in food and drinking water, including iron, are associated with the development of dental caries [1, 2]. However, a specific role for iron in *S. mutans* growth and pathogenesis has not been investigated, owing to the lack of an appropriate test medium.

The preparation of an iron-limiting medium for *S. mutans* has proven challenging since only trace amounts of iron are required for growth. In the present study, we used batch chromatography to remove all traces of iron from a chemically-defined medium (FMC) [7]. This medium was then repleted with various concentrations of iron (0.01–300 μM) so that the effect(s) of iron availability on streptococcal growth, metabolism, and pathogenesis might be analyzed. Specifically, we used this medium to define sources of iron suitable for *S. mutans* growth as well as to identify the differential expression of iron-responsive proteins on SDS-PAGE gels. In addition, FMC agarose plates can be used to identify streptococcal mutants altered in iron uptake and/or storage. The modification of this medium with respect to other metals, such as manganese and magnesium, is also possible so that their putative role(s) in streptococcal growth and disease may be examined.

2. Materials

A. Equipment

- CO_2 incubator (NAPCO model 5300, E series).[1]

- Inductively Coupled Argon Plasma Analyzer (Thermo Jarell Ash, ICAP 61).[2]
- Sorvall RC-5B Centrifuge.[3]
- SS34 rotor.
- Spectrophotometer (visible range).
- Orbit Environ Shaker (Labline) (4 °C).[4]
- Milli-Q water purification system (Millipore No. ZD52115 84).[5]

B. Glassware
- Pyrex Laboratory Bottles (1-liter), acid-cleaned and sterilized by autoclaving (Fisher No. 06-414-1D).[6]
- Qorpak Amber Boston Round Bottles (1-liter), sterile (Fisher No. 03-326-2R).[6]

C. Supplies
- 15 ml polypropylene centrifuge tubes (Fisher No. 05-539-11).[6]
- 50 ml polypropylene centrifuge tubes (Fisher 05-539-8).[6]
- sterile Nalgene 0.2 or 0.45 μm bottle top filters (Fisher No. 09-740-37V).[6]
- sterile 10 ml serological pipets (Fisher No. 13-676-10J).[6]
- sterile 25 ml serological pipets (Fisher No. 13-676-10K).[6]
- Supelco Diaion CR11 (Chelex 100) (Sigma No. 1-3545).[7]

D. Media and media components
- Bacto-tryptone (Fisher No. BP1421-500).[6]
- glucose (Fisher No. D16-1).[6]
- Todd Hewitt broth (DIFCO No. 0492-17-6).[8]
- Ultra pure agarose (BIO RAD No. 162-0125).[9]
- ddH_2O.

E. Chemicals
- concentrated HCl (12M) (Sigma No. H-7020).[7]
- KH_2PO_4 (monobasic) (Sigma No. P-5379).[7]
- K_2HPO_4 (dibasic) (Sigma No. P-3786).[7]
- $(NH_4)_2SO_4$ (Sigma No. A-2939).[7]
- Na_2HPO_4 anhydrous (dibasic) (Sigma No. S-7907).[7]
- $NaH_2PO_4 \cdot H_2O$ (monobasic) (Fisher No. S-369).[6]
- Na acetate$\cdot 3H_2O$ (Fisher No. BP334-500).[6]
- NaCl (Fisher No. S640-3).[6]
- Na_2CO_3 (Fisher No. S-263).[6]
- $MnSO_4 \cdot H_2O$ (Fisher No. M-113, 75755).[6]
- $MgSO_4 \cdot 7H_2O$ (Sigma No. M-9272).[7]
- *$FeCl_3 \cdot 6H_2O$ (Sigma No. F-2877).[7]
- L-cysteine (Sigma No. C-7880).[7]
- DL-tyrosine (Sigma No. T-3379).[7]
- L-glutamine (Sigma No. G-3126).[7]
- adenine (Sigma No. A-8751).[7]
- guanine (Sigma No. G-6377).[7]
- uracil (Sigma No. U-0750).[7]
- thiamine·HCl (Sigma No. T-4625).[7]
- D-biotin (Sigma No. B-4501).[7]
- riboflavin (Sigma No. R-4500).[7]
- folic acid (Sigma No. F-7876).[7]
- nicotinic acid (Sigma No. N-4126).[7]
- pantothenate hemicalcium (Sigma No. P-6045).[7]
- p-amino benzoic acid (Sigma No. A-1174).[7]
- pyridoxine monohydrochloride (Vitamin B_6) (Sigma No. P-9755).[7]

F. *Possible substitutions for $FeCl_3 \cdot 6H_2O$
- $FeCl_2 \cdot 4H_2O$ (Sigma No. F-2130).[7]
- $Fe(NO_3)_3 \cdot 9H_2O$ (Sigma No. F-3002).[7]
- $Fe_2(SO_4)_3$ (Sigma No. F-1135).[7]
- $FeSO_4 \cdot 7H_2O$ (Sigma No. F-7002).[7]
- $Fe(NH_4)_2(SO_4)_2 \cdot 6H_2O$ (Sigma No. F-3754).[7]
- Ferric citrate ($FeC_6H_5O_7$) (Sigma No. F-6129).[7]
- Ferric ammonium citrate ($NH_4FeC_6H_5O_7$) (Sigma No. F-5879).[7]
- Ferrous fumarate ($C_4H_2FeO_4$) (Sigma No. F-5381).[7]
- apo-lactoferrin (Sigma No. L-0520).[7]
- holo-lactoferrin (Sigma No. L-3770).[7]
- apo-transferrin (Sigma No. T-4515).[7]
- holo-transferrin (Sigma No. T-3400).[7]
- hemoglobin (Sigma No. H-7379).[7]

3. Procedures

A. Preparation of media components

Dissolve the components of Solution A (see Table 1 below) in 830 ml of ddH_2O, adjust pH to 6.9 with concentrated HCl and autoclave.

In each of 2 separate flasks, autoclave 100 ml 5% tryptone and 50 ml 20% glucose. These volumes are sufficient for the preparation of 1 L of medium.

In a 2 L flask, autoclave 30 g Chelex-100 (per liter of medium).

Dissolve the components of solution B (see Table 1) in 1 L of 30 mM HCl and filter sterilize through a 0.2 μm filter.

Dissolve the components of solution C (see Table 1) in 1 L of ddH_2O and filter sterilize as described above. Store at 4 °C in an amber bottle.

Dissolve the components of solution D (see Table 1) in 1 L of 30 mM HCl and filter sterilize. Store at 4 °C.

Prepare and sterilize the following stock solutions:

1) 100 mM $MgSO_4 \cdot 7H_2O$
2) 10 mM $MnSO_4 \cdot H_2O$

Determine the magnesium, manganese and iron concentrations of these stock solutions by inductively coupled argon plasma (ICAP) analysis (see below). These solutions should be virtually free of iron.

Prepare 1 mM FeCl (or another iron source) and sterilize by autoclaving. Confirm iron concentration by ICAP analysis (see below).

B. To prepare conventional FMC medium

To solution A, aseptically add 100 ml 5% tryptone, 50 ml 20% glucose, 5 ml solution B, 5

Table 1. Media components (g/L)

Solution A (1×):	
KH_2PO_4	0.44
K_2HPO_4	0.30
Na_2HPO_4 anhydrous (dibasic)	3.15
$NaH_2PO_4 \cdot H_2O$ (monobasic)	2.36
Na acetate$\cdot 3H_2O$	9.95
$(NH_4)_2SO_4$	0.60
NaCl	0.01
L-cysteine	0.10
DL-tyrosine	0.20
L-glutamine	0.005
Na_2CO_3	2.10
Solution B (200×):	
$MgSO_4 \cdot 7H_2O$	4.0
$MnSO_4 \cdot H_2O$	0.15
$FeCl_3 \cdot 6H_2O$	0.04
Solution C (200×):	
pantothenate hemicalcium	0.16
p-amino benzoic acid	0.02
thiamine$\cdot$HCl	0.08
nicotinic acid	0.40
pyridoxine monohydrochloride	0.16
D-biotin	0.002
riboflavin	0.08
folic acid	0.02
Solution D (100×):	
adenine	1.0
guanine	1.0
uracil	1.0

ml solution C and 10 ml solution D. Mix gently and store at 4 °C in a 1 L Pyrex laboratory bottle. Under these conditions, the medium is stable for up to 2 weeks.

C. To Prepare Modified FMC Medium

To solution A, aseptically add 100 ml 5% tryptone, 50 ml 20% glucose, 5 ml solution C, and 10 ml solution D. Transfer this medium to the sterilized Chelex-100 resin and shake vigorously on an orbital shaker (350–400 rpm) at 4 °C overnight. Pass the medium through a 0.2 μm bottle top filter to remove the resin and perform ICAP analysis. Reserve 50 ml of this metal-depleted FMC (mdFMC$_{cx}$) for subsequent washes. Reconstitute the remaining medium with $MgSO_4 \cdot 7H_2O$ and $MnSO_4 \cdot H_2O$ to a final concentration of 20 μM and 1 μM, respectively (FMC$_{cx}^-$). Confirm the final metal ion content of this iron-depleted medium by ICAP analysis.

Supplement FMC$_{cx}^-$ with various concentrations of an appropriate iron source (e.g., 0.01–10 μM $FeCl_3$). Specifically, transfer 50 ml aliquots of the FMC$_{cx}^-$ to separate sterile 50 ml polypropylene tubes and supplement each with iron. Aliquots should be derived from a single batch of medium to ensure that any differences observed are due to variations in iron concentration. Use this iron-repleted medium (FMC$_{cx}^+$) for subsequent growth experiments.

D. To prepare modified FMC agarose plates

Agar can carry with it residual iron which cannot be monitored by conventional ICAP analysis. Thus, we recommend that ultrapure agarose be used as the solidifying agent when preparing plates, and that the following protocol be applied to minimize iron carry-over.

Dissolve the components of Solution A (Table 1) in 830 ml of ddH_2O and adjust pH to 6.9 with concentrated HCl. To this solution add 100 ml 5% tryptone and 50 ml 20% glucose.

Transfer this solution to a 2 L flask containing 30 g Chelex-100 and shake vigorously on an orbital shaker (350–400 rpm) at 4 °C overnight. Pass the medium through a 0.2 μm filter to remove the resin.

Reconstitute the medium with $MgSO_4 \cdot 7H_2O$ and $MnSO_4 \cdot H_2O$ to a final concentration of 20 μM and 1 μM, respectively. Supplement, if desired, with exogenous iron and confirm the metal ion concentration by ICAP. Add 15 g of ultrapure agarose and autoclave.

Cool the medium to 45–50 °C and add 5 ml solution C and 10 ml solution D (Table 1). Stir gently and pour plates.

E. Inductively coupled argon plasma (ICAP) analysis

Perform the general operation of the ICAP analyzer according to the recommendations of the manufacturer. Standardize the analyzer for magnesium, manganese and iron using stock solutions containing these metals at 10 ppm. After a 15 minute equilibration period, adjust the sample flow rate into the plasma using 1000 ppm Yttrium to visualize the argon flame. Program the machine (ThermoSpec software) to perform four readings per sample and to report the data as an average. Analyze the media samples in 5 ml aliquots at t = 30 seconds post-uptake. Rinse the uptake tubing thoroughly with 5% HCl and then with ddH_2O between readings. Convert metal ion concentrations provided in ppm to μM units.

F. Bacterial growth

Inoculate 14 ml of Todd Hewitt broth contained in a 15 ml polypropylene tube with 150 μl of a *S. mutans* cell concentrate and incubate at 37 °C and 5% CO_2 for 16 hours. Harvest the cells by centrifugation at 7000 rpm for 10 minutes and pour off the supernatant. Wash the cell pellet with 1 ml of mdFMC$_{cx}$ and repeat two more times to remove all traces of iron. Resuspend the pellet in 1 ml mdFMC$_{cx}$. Use 100–150 μl of this cell concentrate to inoculate each of five sterile 15 ml polypropylene tubes containing 14 ml of iron-repleted FMC (FMC$_{cx}^+$). Incubate all tubes standing at 37 °C and 5% CO_2. To monitor growth, obtain OD_{600nm} values for individual tubes at t = 8, 10, 12, 16 or 24 hours post-inoculation (i.e., use one culture per timepoint). These early, mid-, late-log and stationary phase cultures may then be applied to nucleic acid and/or protein isolation protocols to characterize putative iron-regulated expression.

Furthermore, a *S. mutans* transposon library may be replica-plated onto FMC^- and FMC_{cx}^+ agarose plates to identify mutants altered in iron uptake and/or storage.

4. Results and discussion

This work describes the generation of an iron-depleted, chemically-defined medium which supports the growth of streptococci upon supplementing with exogenous $FeCl_3$. The final metal ion concentrations in this medium were confirmed by ICAP analysis. Alternatively, atomic absorption spectroscopy or inductively coupled plasma mass spectroscopy (ICP-MS) may be applied to confirm metal concentrations.

As shown in Figure 1a, the treatment of FMC^- with Chelex (FMC_{cx}^-) markedly reduced the residual iron content of the medium from 1.6 to 0.005 μM. The growth of *S. mutans* was also restricted in this medium as measured after a 24 hour incubation at 37 °C and 5% CO_2 (Figure 1b). Growth was restored to this medium when 0.01 to 10 μM iron was added (FMC_{cx}^+). Importantly, these iron concentrations are comparable to those present in human saliva (unpublished observations). As the concentration of iron in FMC_{cx}^+ was increased from 0.01 to 10 μM, *S. mutans* growth increased in parallel. This growth pattern leveled off, however, when iron was provided at concentrations greater than 10 μM (data not shown). *S. mutans* did not grow in any of the media described herein unless it was supplemented with exogenous magnesium (20 μM). Therefore, we conclude that *S. mutans* has nutritional requirements for iron as well as for magnesium.

Preliminary experiments in our laboratory suggest that manganese can support the growth of *S. mutans* in iron-free FMC_{cx}^-. Since *S. mutans* can grow in the presence of manganese or iron, it appears that either metal can satisfy the nutritional requirements for this oral microbe. Consistent with this finding are reports in the literature suggesting an interchangeability of manganese and iron in promoting the growth of *S. mutans* [6]. Finally, surface protein profiles for *S. mutans* cultures grown in FMC_{cx}^- supplemented with manganese were different from those derived from cultures grown in FMC_{cx}^- supplemented with iron (data not shown). This indicates that manganese has a unique effect on *S. mutans* expression.

The reproducibility of this protocol is sound but can vary with water quality. For instance, trace amounts of iron can persist in ddH_2O as well as on glassware. A well-maintained water still and purification system is required, and the use of disposable plasticware is recommended whenever possible. Moreover, the preparation of iron-free magnesium and manganese stock solutions is particularly tedious. Thus, we recommend that only ultra-pure reagents be used and that highly concentrated metal solutions (at least 100 mM) be prepared to limit the carry-over of iron into the supplemented medium. Since we found the removal of iron from manganese stock solutions to be especially challenging, we often omitted manganese from the preparation of FMC_{cx}^+ so that the sole effect(s) of iron on *S. mutans* growth and expression could be assured.

In conclusion, the medium described in the present study can be used to investigate the effect(s) of multiple iron sources on the expression of strepto-

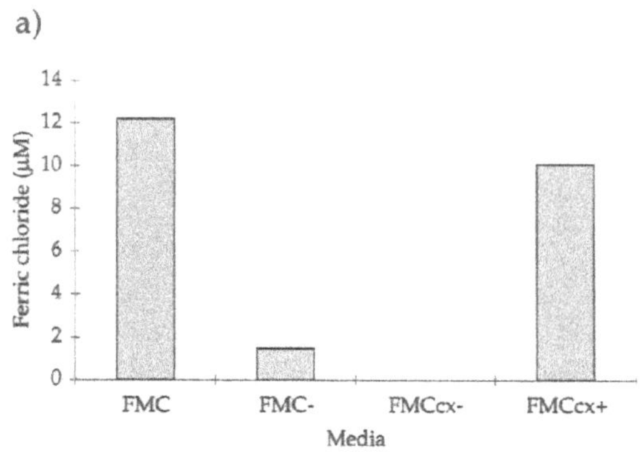

Figure 1a. Iron concentrations in FMC, FMC without iron (FMC^-), Chelex-treated FMC^- (FMC_{cx}^-), and FMC_{cx}^- repleted with 10 μM $FeCl_3$ (FMC_{cx}^+) as determined by ICAP analysis. While each of these media conditions was tested in triplicate, the results of a single representative experiment are shown. All media were supplemented with manganese and/or magnesium as described in the text.

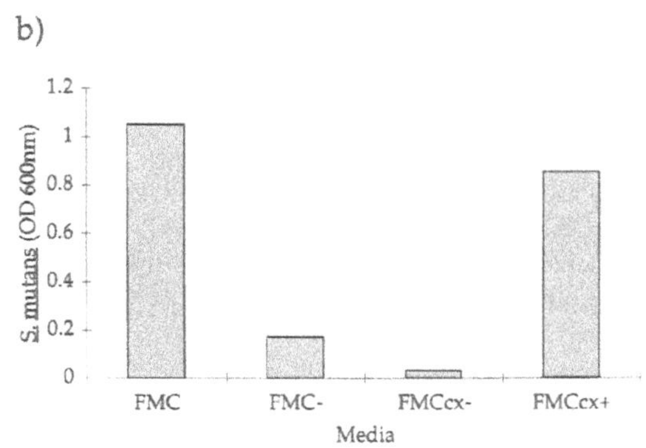

Figure 1b. Effect of iron on *S. mutans* growth. *S. mutans* was grown to stationary phase (t = 24 h) in the media described in Figure 1a. While each of these media conditions was tested in triplicate, the results of a single representative experiment are shown. All media were supplemented with manganese and/or magnesium as described in the text.

coccal virulence determinants. The medium can also be modified by varying the concentration of other trace metals so that their putative effect(s) on streptococcal infections may be analyzed. Preliminary evidence presented in this work suggest that magnesium and/or manganese can influence streptococcal growth and expression. Indeed, this medium can be employed to elucidate the involvement of these metals in streptococcal metabolism and pathogenesis.

Acknowledgments

This work was supported by the American Society for Microbiology as an Undergraduate Research Fellowship to M.W.M. We also acknowledge the Middlebury College Department of Biology for supporting this work.

Notes to suppliers

1. NAPCO Scientific Co., 20210 S.W. Teton, P.O. Box 1000, Tualatin, OR 97062, USA
2. Thermo Jarell Ash Corp., 8E Forge Pkwy, P.O. Box 9101, Franklin, MA 02038, USA
3. Sorvall, Dupont Biomedical Dept., Wilmington, DE 19898, USA
4. Labline Instruments, 15th and Bloomingdale Ave., Melrose Park, IL 60160, USA
5. Millipore Corp., 80 Ashby Rd., Bedford, MA 01730, USA
6. Fisher Scientific, 52 Fadem Rd., Springfield, NJ 07081, USA
7. Sigma Chemical Company, P.O. Box 14508, St. Louis, MO 63178, USA
8. DIFCO Laboratories, Inc., Detroit, MI 48201, USA
9. BIO RAD, 2000 Alfred Nobel Dr., Hercules, CA 94547, USA

References

1. Adkins BL, Losee FL (1970). A study of covariation of dental caries prevalence and multiple trace element content of water supplies. NY State Dent J 36: 618–622.
2. Aranha H, Strachan RC, Arceneaux J, Byers B (1982). Effect of trace metals on growth of *Streptococcus mutans* in a Teflon Chemostat. Infect Immun 35: 456–460.
3. Bullen JJ, Rogers HJ, Griffiths E (1978). Role of iron in bacterial infections. Curr Top Micriobiol Immunol 80: 1–35.
4. Evans S, Arceneaux J, Byers B, Martin M, Aranha H (1986). Ferrous iron transport in *Streptococcus mutans*. J Bacteriol 168: 1096–1099.
5. Finkelstein R, Sciortino C, McIntosh M (1983). Role of iron in microbe-host interactions. Rev Infect Dis 5: S759–S777.
6. Martin ME, Byers BR, Olson M, Salin ML, Arceneaux J, Tolbert C (1986). A *Streptococcus mutans* Superoxide dismutase that is active with either manganese or iron as a cofactor. 261: 9361–9367.
7. Terleckyj B, Willett NP, Shockman GD (1975). Growth of several cariogenic strains of oral Streptococci in a chemically defined medium. Infect Immun 11: 649–655.
8. Weinberg ED (1978). Iron and infection. Microbiol Revs 42: 45–66.
9. Wooldridge, Williams P (1993). Iron uptake mechanisms of pathogenic bacteria. FEMS Microbiol Rev 12: 325–348.

Address for correspondence: Dr G. A. Spatafora, Department of Biology, Middlebury College, Middlebury, Vermont 05753, USA
Phone: (802) 443-5431; Fax: (802) 443-2072
E-mail: spatafor@panther.middlebury.edu

Methods in Cell Science **20**: 223–231 (1998).

Identification of oral streptococci using PCR-based, reverse-capture, checkerboard hybridization

Bruce J. Paster, Irena M. Bartoszyk & Floyd E. Dewhirst
Department of Molecular Genetics, Forsyth Dental Center, Boston, Massachusetts, USA

Abstract. A PCR based, reverse capture checkerboard hybridization methodology was designed to rapidly detect and enumerate oral species of *Streptococcus* and other genera in plaque samples to assess the bacterial etiology of root surface caries or other oral diseases. The procedure circumvents the need for in vitro bacterial cultivation. Up to 30 reverse capture probes that target regions of 16S rRNA genes are deposited on a nylon membrane in separate horizontal lanes using a MiniSlot apparatus. 16S rRNA genes of DNA from plaque or pure cultures are PCR amplified using a digoxigenin-labeled primer. Hybridizations are performed in vertical channels in a Miniblotter apparatus with digoxigenin-labeled amplicons from up to 45 samples. Consequently, 1350 hybridizations are performed simultaneously using a single membrane. Hybridization signals are detected using chemifluorescence procedures. Bacterial numbers are determined by analyzing hybridization signals using a Storm system. A genus-specific probe is used to assess the total number of streptococci and universal probes are used to assess total bacterial counts. Use of this procedure enables laboratories to analyze thousands of clinical or environmental samples with many probes in a relatively short period. This methodology has broad range applications in microbiology to monitor population distributions in a given environment.

Key words: 16S rRNA, Checkerboard hybridization, DNA probes, Phylogeny, *Streptococcus*

1. Introduction

While the development of DNA probes has provided investigators with a major breakthrough for bacterial detection and identification, hybridization assays have been limited by the number of probes that can be tested simultaneously with large numbers of bacterial or clinical samples. These limitations are now circumvented by the use of the 'checkerboard' hybridization methodology as developed in our laboratories [19]. The 'checkerboard' format offers a number of advantages for DNA-DNA hybridization techniques. The main advantage is that large numbers of bacterial or clinical samples can be hybridized with multiple probes (i.e., 1350 hybridizations) at the same time using a single solid support membrane about 14 cm square. Thus, the technique lends itself readily to any of a number of rapid screening procedures. The technique is relatively inexpensive in that the small volume of the hybridization chambers (70 to 130 μl depending on the apparatus) provides an enormous saving in reagents and probes. Costs of solid support membranes are decreased because large numbers of reactions may be performed on individual membranes. In addition, the use of non-isotopic techniques in this format eliminates the problem of radioactive disposal and provides probes that are stable for prolonged periods of time.

Whole genomic DNA probes have been the primary type used in the checkerboard hybridization assays reported to date, but methods have also been developed to employ rRNA-based DNA probes or oligonucleotides [8, 19]. In contrast to whole genomic probes, oligonucleotide probes can differentiate between closely related species or even subspecies [6]. This is especially important for the identification of species of oral streptococci, which are very closely related phylogenetically [2, 10, Figure 1]. Species-specific, oligonucleotide probes for oral and nonoral species of *Streptococcus* have been previously described [8, 1, 3–4, 13]. Recently, other investigators [8] used the Miniblotter apparatus for assays, which they termed 'line blot' hybridization, to screen 399 isolates presumptively identified as strains of '*Streptococcus milleri*'. PCR amplified 16S rRNA amplicons were hybridized with species-specific, biotinylated oligonucleotide probes to identify target oral streptococcal species.

One of the major advantages of the checkerboard hybridization methodology is that clinical samples can be analyzed directly without the need of bacterial cultivation. In order to use oligonucleotide probes for this direct analysis and to maintain the sensitivity of whole genomic probes, we developed a PCR-based, reverse capture, checkerboard hybridization protocol based on a modification of the procedures of Saiki et al. [17] and Morrissey and Collins [12]. We utilized this modified procedure to detect and enumerate species of oral *Streptococcus* and other oral genera directly from plaque samples in order to

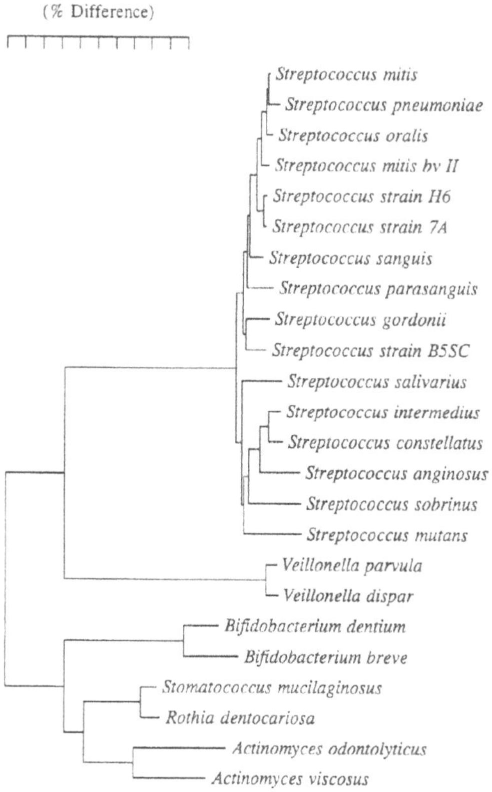

Figure 1. Phylogenetic tree of oral streptococci and other oral genera based on 16S rRNA gene sequence comparisons [4, 5, unpublished data]. Similarity matrices were constructed from aligned sequences by using only those sequence positions for which 90% of strains had data. Similarity matrices were corrected for multiple base changes by the method of Jukes and Cantor [9]. The neighbor-joining method of Saitou and Nei [18] was used for phylogenetic tree construction. The scale bar represents a 10% difference in nucleotide sequence as determined by measuring the lengths of horizontal lines connecting two species.

assess their role in the etiology of root surface caries. Briefly, the procedure is as follows. DNA probes are synthesized with multiple thymidines (T) at the 5′ end of the oligonucleotide. The poly-T tails are crosslinked to the membrane support via UV irradiation, which leaves the specific probe available for hybridization. The 16S rRNA genes of DNA isolated from bacterial plaque are amplified using universal primers (the forward primer is labeled with digoxigenin). The digoxigenin-labeled amplicon is hybridized to the capture probes bound to the membrane. The digoxigenin residues are complexed to anti-digoxigenin antibody covalently bound to alkaline phosphatase. Standard chemifluorescence (Storm system) or chemiluminescence (X-ray film) procedures are used for detection. Figure 2 represents a schematic representation of the procedure.

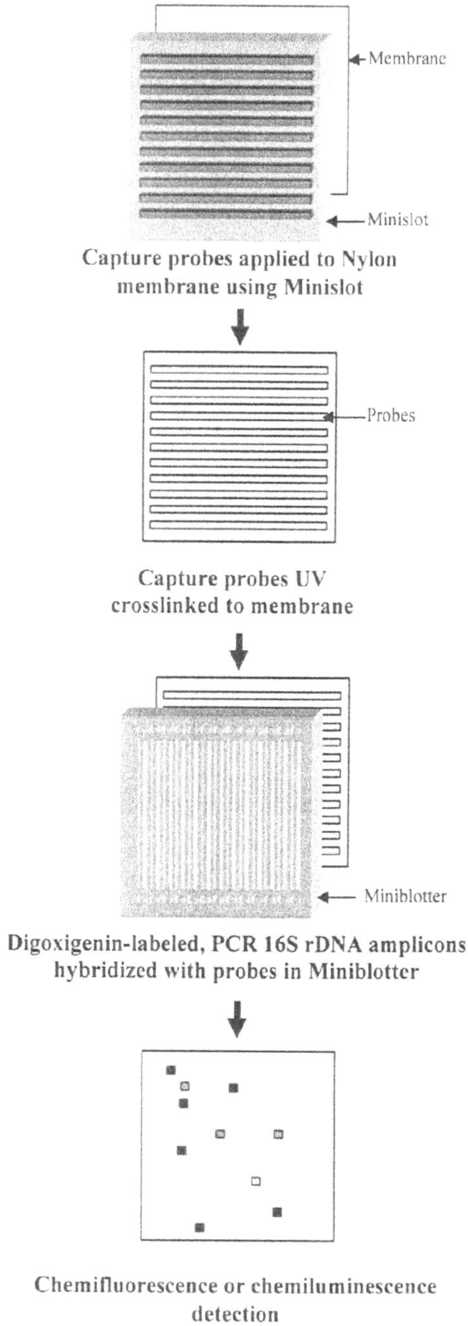

Figure 2. Schematic representation of the PCR-based, reverse-capture, checkerboard hybridization methodology. The Minislot is made of aluminum, whereas the Miniblotter is made of plastic. Note that the bottom bases of the Minislot and Miniblotter are not shown. Actual number of slots on the Minislot is 30, and number of hybridization channels on the Miniblotter is 45 (thus, a potential of 1350 simultaneous hybridizations on a single solid support).

2. Materials

A. Equipment
1. Anaerobic Chamber.[7]
2. Dry Bath Incubators, 11-718-2.[8]
3. Eppendorf centrifuge 5415C, 2236280-1.[5]
4. pH meter model 340, 13-641-299.[8]
5. Micro combination pH electrode, 13-644-6.[8]
6. Ultrasonic processor (2 mm probe), L-04710-00.[6]
7. DNA thermal cycler Model 480, N801-0100.[18]
8. Power supply model 200/2.0, 165-4761.[3]
9. Horizontal gel system.[17]
10. Transilluminaor, FB-T1-88.[8]
11. Incubator Model 310, 11-702-16.[8]
12. Checkerboard System 2—Minislot™ 30 and Miniblotter® MN100-45, SB-02.[10]
13. Platform rocker, 55S.[13]
14. Fisher Clinical rotator, 14-251-200.[8]
15. Storm system model 840.[14]
16. Stratalinker UV crosslinker 1800, 400071.[22]
17. Votex-Genie 2 mixer, 58815-178.[24]
18. Hot plate model 320, 33918-830.[24]
19. Pipetman (P-10, P-100, P-200).[20]

B. Medium and chemicals
1. Trypticase Soy Agar, 4311043.[2]
2. Defibrinated Sheep Blood, SB100.[15]
3. Tris HCl, 4103-01.[11]
4. EDTA, Disodium salt:dihydrate, E4884.[21]
5. Proteinase K (10 mg/ml), V3021.[19]
6. Polyoxyethylene-sorbitan monolaurate (Tween 20), P1379.[21]
7. NaOH, S-0899.[21]
8. HCl (1N), SA48-500.[8]
9. Gene Amp PCR Core Reagents, N808-0009.[18]
10. AmpliTaq Gold, N808-0241.[18]
11. Sterile distilled deionized water (ddH_2O)
12. Light mineral oil (Paraffin Oil), 0121-1.[8]
13. Ethidium bromide (10 mg/ml), E2515.[21]
14. SeaKem Le Agarose, 50000.[9]
15. Tris (base), 4099-06.[11]
16. Boric acid, B0252.[21]
17. Oligonucleotides (primers/probes).[16]
18. NaCl, 5671-3.[8]
19. $NaH_2PO_4 \cdot H_2O$, 3818-01.[11]
20. SDS, Sodium salt, L-5750.[21]
21. Ficoll, Type 400, F-4375.[21]
22. Polyvinylpyrrolidone, PVP-360.[21]
23. Bovine serum albumin Fraction V, A2934.[21]
24. DNA, Herring sperm (10 mg/ml), 223646.[4]
25. Maleic acid, M-0375.[21]
26. Casein (technical, Bovine milk), C-7078.[21]
27. Anti-digoxigenin-AP Fab fragments, 1093274.[4]
28. Vistra ECF substrate (Attophos), RPN 5785.[1]
29. CDP-Star, 1685627.[4]
30. 1 kB DNA ladder, 15615-016.[12]
31. Bromophenol Blue, B6131.[21]
32. Xylene cyanole, X4126.[21]
33. Glycerol, G6279.[21]
34. REACT 2 buffer, 16302-010.[12]

C. Supplies
1. Nalgene filters (0.22 μm), 09-740-3A.[8]
2. Petri dishes, 08-757-12.[8]
3. Eppendorf tubes, 05-407-10.[8]
4. Microcentrifuge tubes (0.5 ml thin wall), 05-408-1A.[8]
5. Blotting paper grade 238, 28303-102.[24]
6. Nylon Membrane, positively charged, 1417240.[4]
7. Tupperware™ containers.
8. Saran wrap.
9. Plastic bags.
10. Plastic cushions, PC2.[10]
11. Forceps, 10-300.[8]
12. Pipet tips for general use
    - 1000 μl, 21-197-8F.[8]
    - 200 μl, 21-278-51.[8]
    - 10 μl, 21-197-2F.[8]
13. Pipet tips for PCR
    - 0.5–10 μl, 1018-3810.[23]
    - 0–200 μl, 1018-8810.[23]
    - 0–155 μl, 1024-5810.[23]

3. Procedure

A. Preparation of solutions
1. Media for stock cultures
   Blood plates: 40 g Trypticase Soy Agar, 950 ml ddH_2O. Stir and let come to boil for 1 min. Autoclave. Cool to 50 °C in water bath. Add 50 ml of sheep blood. Pour about 30–40 ml per petri dish. Let sit overnight on bench. Store at 4 °C.
2. Prehybridization solution
   Boil sheared herring sperm DNA (2 ml of 10 mg/ml stock) for 10 min; put on ice. Warm the following solution (at 55 °C):
   - 50× Denhardt's Solution 10 ml
   - 20× SSPE 25 ml
   - 10% SDS 5 ml
   - ddH_2O 58 ml

   Add the iced sperm DNA to the rest of the warmed solution. Prepare prehybridization solution *fresh* every time.
3. 50× Denhardt's solution
   - Ficoll (Type 400, Pharmacia) 5 g
   - Polyvinylpyrrolidone 5 g
   - Bovine Serum Albumin (Pentex Fraction V) 5 g
   - ddH_2O 500 ml

   Low heat is needed for dissolving BSA. Filter solution (0.2 μm filter). Aliquot and store at –20 °C.

4. 20× SSPE
   - NaCl 175.3 g
   - $NaH_2PO_4 \cdot H_2O$ 27.6 g
   - EDTA (0.5 M) 40 ml
   - dd H_2O 800 ml

   Adjust pH to 7.4 with NaOH. Adjust volume to 1 liter. Filter solution (0.2 μm filter). Autoclave.
5. 1.25× Hybridization buffer
   - 50× Denhardt's Solution 400 μl
   - 20× SSPE 1.25 ml
   - Herring Sperm DNA (10 mg/ml) 400 μl

   Bring to 20 ml with sterile dd H_2O and aliquot. Store at –20 °C.
6. Denaturing solution
   - 400 mM NaOH
   - 10 mM EDTA

   Filter solution (0.2 μm filter).
7. Neutralizing solution
   - 0.5N HCl
   - 500 mM Tris HCl, pH 8.3

   Filter solution (0.2 μm filter).
8. Blocking buffer
   - Maleic acid 5.80 g
   - NaCl 4.35 g
   - NaOH 3.75 g
   - Distilled H_2O 500 ml

   Do not autoclave.

   Add 50 g casein, and dissolve by mixing at about 60 °C on a hot plate. Autoclave. Store 10 ml aliquots at –20 °C.
9. Buffer 1 (Conjugate buffer)
   - 150 mM NaCl
   - 100 mM TrisHCl, pH 7.5

   Filter solution (0.2 μm filter). Autoclave.
10. Buffer 2 (Substrate buffer)
    - 100 mM Tris HCl
    - 100 mM NaCl

    pH to 9.5 with NaOH. Filter solution (0.2 μm filter). Autoclave.
11. TE buffer
    - 50 mM Tris HCl
    - 1 mM EDTA

    pH to 7.6 with NaOH. Filter solution (0.2 μm filter). Autoclave.

B. Cell lysis, Proteinase K treatment
1. In a 1.5 ml microcentrifuge tube, suspend cells in 39.6 μl TE buffer.
2. Add 4.5 μl 5% Tween 20 and 0.9 μl Proteinase K (10 mg/ml stock).
3. Incubate at 55 °C for 2 h.
4. Incubate at 95 °C for 10 min to inactivate the enzyme.
5. Spin tubes down very briefly.
6. Add 5.0 μl 5N NaOH.
7. Samples are used immediately for PCR or are stored –70 °C until needed.

C. Polymerase chain reaction
1. To the Proteinase K lysed sample, add neutralizing solution until sample reaches a pH between 7.3–7.5. (If this neutralized solution was sonicated for 3, 10 sec bursts, additional DNA was made available for PCR amplification, especially for the 'hard to lyse' species, such as *Actinomyces* sp., *S. mutans*, and Bifidobacterium sp. However, sonication is not presently part of the protocol. In addition, dialysis methods or spin columns are presently being evaluated for pH adjustment to avoid the cumbersome task of measuring the pH of each sample.)
2. Reagents for PCR:

Reagent	Final concentration
– 10× PCR Buffer II	1×
– 25 mM $MgCl_2$	2.5 mM
– dNTPs	200 μM of each
– Forward digoxigenin-labeled primer (E02)	0.5 μM
– Reverse primer (E08)	0.5 μM
– Taq polymerase	2.5 Units/reaction
– DNA template	2.1 μl*
– Water	to 100 μl

* Amount of template may be increased.

3. PCR conditions

   Perform PCR reactions in thin-walled tubes using a Perkin-Elmer 480 thermal cycler (or other manufacturer). Use the following conditions for amplification with primers E02 and E08 (Table 2); denaturation at 94 °C for 45 sec, annealing at 50 °C for 45 sec, and elongation at 72 °C for 90 sec with 5 additional sec added for each cycle. Run for 30 cycles followed by a final elongation step at 72 °C for 15 min.
4. Agarose gel electrophoresis

   Add 0.75 g of SeaKem Le Agarose and 75 ml of 1× TBE buffer. Melt agarose in microwave, cool for 15 minutes and add 3 μl ethidium bromide stock (10mg/ml). Pour into tray with the combs set in. Let solidify. Load wells with samples [5 μl of sample + 2.0 μl 6× dye (0.25% Bromophenol blue, 0.25% Xylene cyanole, 30.0% Glycerol)]. Load 4 μl of 1 Kb ladder stock (25 μl, 1 Kb ladder (1 μg/ul); 50 ul, 6× dye; 150 μl, ddH_2O; 25 μl, REACT 2 buffer). Run at 95 mV for 1 h. Visualize DNA bands with UV transilluminator.

D. Probe design

Target regions suitable for probe targeting are identified by examining aligned 16S rRNA sequences. A computer alignment program (developed by F. Dewhirst [15]) which displays sequences in a difference format relative to the target organism is particularly useful for locating hypervariable regions. Regions of 17–22 bases with 1 or more base differences and a melting temperature (T_m) of approximately 51 °C is selected for probe targeting. The target sequence is then verified for uniqueness against all sequences in the sequence databases. Sequences are checked for internal hairpins or primer dimers using a commercially available computer program (Oligo 5.0). DNA probes with 19 to 22 thymidine residues added at the 5′ end are synthesized by Operon (or other comparable company), with no additional purification steps.

E. Checkerboard hybridization procedure (see Figure 2 for schematic representation)
  1. Capture probes
     Probes are complementary to the 16S rRNA gene sequence with 19 to 22 T′s on the 5′ end (Table 1). T_m's of the probe (not including the T′s) are about 51 to 52 °C.
  2. Set up Minislot
     a) Prepare a stack of 18 sheets of blotting paper cut to size.
     b) Cut nylon membrane to size (DO NOT TOUCH MEMBRANE, even with gloved hands).
     c) Place the 18 sheets of blotting paper on top of the bottom plate of the Minislot-30 apparatus (this apparatus is made of aluminum – smaller units are made of plastic). Using forceps, place nylon membrane on top of the 18 sheets of blotting paper. Mark top lightly with pencil and cut left top edge of membrane off to ensure proper orientation.
     d) Place top of Minislot-30 on top of nylon membrane and clamp the unit tightly (treat apparatus with care; wash after use by gentle rinsing with distilled water. Do not wipe dry – it may scratch the surface).
  3. Preparing and applying probes
     a) Put 58 pmole of capture probe (5.8 µl of 10 pmole/µl stock) into 994.2 µl of 10 mM Tris HCl, 1 mM EDTA, pH 8.0. Vortex and spin down briefly.
     b) Apply all (1 ml) into wells. Gently rock apparatus to ensure that probe is distributed along the entire length of the well. Let soak in for 10 min (use vacuum aspirator, if necessary).
     c) Put the membrane on TE soaked blotting paper (probe side up). The probes are cross-linked to the membrane by UV irradiation

Table 1. Reverse capture probes and PCR primers

ID	Specificity	Position	Sequence (5′–3′)
D43	*Streptococcus anginosus*	187-208	TTTTTTTTTTTTTTTTTTTTCTTTCAAGCATCTAACATGTG
D45	*Streptococcus anginosus/ S. gordonii*	90-111	TTTTTTTTTTTTTTTTTTTTCAACTCACAGTYTATGGTGTAG
E61	*Streptococcus anginosus/ S. intermedius*	444-465	TTTTTTTTTTTTTTTTTTTATGGATTCTCACACTTGTTCTTCCT
D46	*Streptococcus constellatus/ S. intermedius*	172-192	TTTTTTTTTTTTTTTTTTTTCAGTAAATGTTCTTATGCGGTA
D38	*Streptococcus mitis/ S. oralis/S. pneumoniae*	187-208	TTTTTTTTTTTTTTTTTTTTCCTTTTAAGYAAATGTCATGC
E78	*Streptococcus mitis* biovar II	175-187	TTTTTTTTTTTTTTTTTTTTCAATAACTGCTATTATGCGG
E37	*Streptococcus mutans*	587-608	TTTTTTTTTTTTTTTTTTTTTACTCCAGACTTTCCTGAC
E33	*Streptococcus parasanguis*	169-190	TTTTTTTTTTTTTTTTTTTTGTCGACTTTTATGCGGTATTA
E34	*Streptococcus salivarius*	175-194	TTTTTTTTTTTTTTTTTTTTGTCATCCATTGTTATGCGG
E54	*Streptococcus salivarius/ S. sanguis*/H6	89-108	TTTTTTTTTTTTTTTTTTTATGCAACTCATCCAAGAAGAG
E35	*Streptococcus sobrinus*	175-194	TTTTTTTTTTTTTTTTTTTTGTTAACTCCTCTTATGCGG
D37	*Streptococcus* sp. (strains 7A, 1F, H6)	175-196	TTTTTTTTTTTTTTTTTTTTATGCGATAATCCATTTTATGCG
E57	*Streptococcus* sp. strain B5SC	178-194	TTTTTTTTTTTTTTTTTTTTCATGCAATAGTCAATGTTATG
E72	All *Streptococcus*	491-509	TTTTTTTTTTTTTTTTTTTTTAGCCGTCCCTTTCTGGT
E76	*Actinomyces odontolyticus*	998-1010	TTTTTTTTTTTTTTTTTTTTCAGTGCCGCCGTGC
E45	*Actinomyces viscosus*	87-104	TTTTTTTTTTTTTTTTTTTTCACTCATCCAGAACCAG
E52	*Rothia/Stomatococcus*	179-189	TTTTTTTTTTTTTTTTTTTTGCGGAGATTGGTCGTAT
E49	All *Bifidobacterium*	228-241	TTTTTTTTTTTTTTTTTTTGGACGCGACCCCAT
E79	All *Veillonella*	217-235	TTTTTTTTTTTTTTTTTTTTAATCCCCTCCTTCAGTGA
E40	Universal 1089	1089-1107	TTTTTTTTTTTTTTTTTTTTCTCGTTGCGGGACTTAAC
E29	Universal 341	341-357	TTTTTTTTTTTTTTTTTTTTCTGCTGCCTCCCGTAGG
E02	Universal forward primer (digoxigenin labeled)	8-23	Digoxigenin-AGAGTTTGATYMTGGC
E08	Universal reverse primer	1492-1509	GYTACCTTGTTACGACTT

using a Stratalinker 1800 (Stratagene) on autocrosslink position. The thymidine tails are preferentially UV crosslinked to the nylon which leaves the specific probe available for hybridization.
   d) Place membrane in container and wash the membrane in 5× SSPE, 0.5% SDS for 30 min at 55 °C on shaker (gentle shaking).
   e) Rinse in distilled water and proceed to the prehybridization step or dry the membrane (probe side up on scrap membrane). Wrap in Saran Wrap for later use. Store dried membranes at room temperature in a dark, dry environment.
4. Prehybridization
   a) Prepare prehybridization solution (must be made fresh daily). Place membrane in 60 ml of prehybridization solution in a Tupperware container.
   b) Close container's lid tightly and place container on a shaker (gently shaking) at 55 °C for 1 h. Prehybridization can go overnight, but ensure that the seal is secured tightly.
   c) Pre-warm Miniblotter apparatus at 55 °C.
5. Setting up Miniblotter
   a) Position one foam pad on top of the bottom plate of the Miniblotter-45.
   b) Lift membrane (using forceps) from prehybridization solution and drip off excess prehybridization solution.
   c) Place the membrane face up on top of the foam pad (with cut corner facing upper left) and mark with a marker on the edge of the foam where the first well and last well begin and end.
   d) Place the top of the Miniblotter apparatus on the top of the membrane making sure that the top part of the Miniblotter channels are about 3–4 mm above the first Minislot well and that the bottom part of the Miniblotter channels are below the last Minislot well. Clamp the whole apparatus tightly.
6. Preparing and applying samples
   a) Heat 5 ml of 1.25× hybridization buffer to 55 °C. Add 9.6 μl denaturing solution to 9.6 μl DNA sample (PCR product), and then add 108.1 μl of the heated hybridization buffer.
   b) Heat sample to 55 °C. Just prior to loading Miniblotter, add 7.7 μl neutralizing solution. Mix. Load 135 μl of each sample into a hybridization channel on the Miniblotter via the small holes at the end of each channel (avoid bubbles). For empty lanes, load 135 μl of 1.25× hybridization buffer. With practice, 45 samples can be loaded in 15 min.
   c) Wrap apparatus in Saran wrap and place in ziplock bag with two wet paper towels. Seal bag well and place on tilter (slow speed) at 55 °C for 2 to 4 h.
   d) Heat 200 ml of 1× SSPE, 0.1% SDS at 55 °C, and take out blocking buffer to thaw, then put on ice.
7. Washing and detection
   a) Carefully remove Miniblotter from the bag by keeping it level. Position the washing manifold (that comes with the apparatus) in the slots where the loading holes are and press in firmly. Using appropriate-sized tubing, connect one side of the manifold to a vacuum source and the other side into the 200ml of 1× SSPE, 0.1% SDS and then aspirate the samples (it takes approximately 1 min).
   b) Disassemble the Miniblotter. Remove membrane with a forceps and place the membrane face down in buffer 1 (conjugate buffer) for 1 min.
   c) Pour off buffer 1 and put in 10 ml of blocking buffer and 50 ml of buffer 1. Block the membrane for 1 h with gentle shaking at room temperature.
   d) Pour off the blocking solution and put in 60 ml of buffer 1 + 6 μl of anti-digoxigenin-AP Fab fragments (antibody). Gently shake for 30 min at room temperature.
   e) Pour off the antibody solution and wash with 60 ml of buffer 1 for 10 min (gently shaking at room temperature). Repeat this step once.
   f) Pour off second wash and put in buffer 2 (substrate buffer). Pour in just enough of buffer 2 so as to cover membrane and let sit for 2 min without shaking. Keep probe-side down.
   g) Lift membrane from buffer 2 and drip excess buffer off. Place membrane (probe side up) on top of a piece of Saran wrap. Pour about 3 ml of Attophos on membrane. Cover the top of membrane with Saran wrap. Spread the Attophos evenly for 1 minute using a 10 ml pipette (roll pipette back and forth). Drip excess Attophos off membrane and place membrane between two mylar sheets (Saran wrap may be used if wrinkles are avoided).
   h) Expose membrane for 2–4 h away from direct light (in a drawer between two sheets of paper). May have to expose for longer time – as much as overnight (longer exposures often results in higher background).
   i) Scan the membrane using the Storm system. Scanning time is about 5 min. If a Storm system (or equivalent system which

measures chemifluorescence) is not available, then replace Attophos with any of the commercially available chemiluminescence substrates (e.g., CDP Star) and expose blot to X-ray film for 15 to 30 min (the optimal time of exposure should be tested empirically). Comparable qualitative results are obtained using chemiluminescence detection.

j) Quantitation of the hybridization signals can be obtained on the Storm system using the provided software, ImageQuant. We are presently optimizing our protocol to obtain quantitative data using ImageQuant. Quantitation of the data can be obtained from chemiluminescence data using laser densitometry; however the level of sensitivity is only about 2 orders of magnitude. ImageQuant can achieve about 5 levels of magnitude.

4. Results and discussion

The PCR based, reverse capture checkerboard hybridization methodology described in this study can be used to rapidly detect and enumerate oral species of *Streptococcus* and other genera (Figure 3). The assay works well for the identification of pure cultures and for the detection and enumeration of specific species or higher taxa directly in clinical samples, without the need for in vitro bacterial cultivation.

In the pilot study shown in Figure 3, the standards represent 5×10^6 cells of each specific target and the clinical samples represent bacterial dental plaque from 7 healthy volunteers. One sample (BK-4) was a saliva specimen from one of the patients. The limit of detection for the assay was about 1×10^3 cells. Many of the sites had high levels of *S. sanguis*, the *S. oralis* cluster (*S. oralis*, *S. pneumoniae* or *S. mitis*), species of *Rothia* or *Stomatococcus*, a potential new species B5SC, *S. mitis* biovar II, and species of *Veillonella*. Only some of the sites had *A. viscosus*, *S. gordonii*, and *S intermedius*. No or very few sites had detectable levels of *S. mutans*, *S. sobrinus*, *S. salivarius*, *Bifidobacterium* sp. or *A. odontolyticus*. In the analysis of the BK-4 saliva sample, the relatively high level of hybridization with the all streptococci probe and the relatively low levels of hybridization with specific streptococcal probes indicated that an unknown species of *Streptococcus* (for which there was no probe) was present.

Some species could not be differentiated adequately on the basis of 16S rRNA sequence base comparisons (Figure 1). For example, *S. oralis*, *S.*

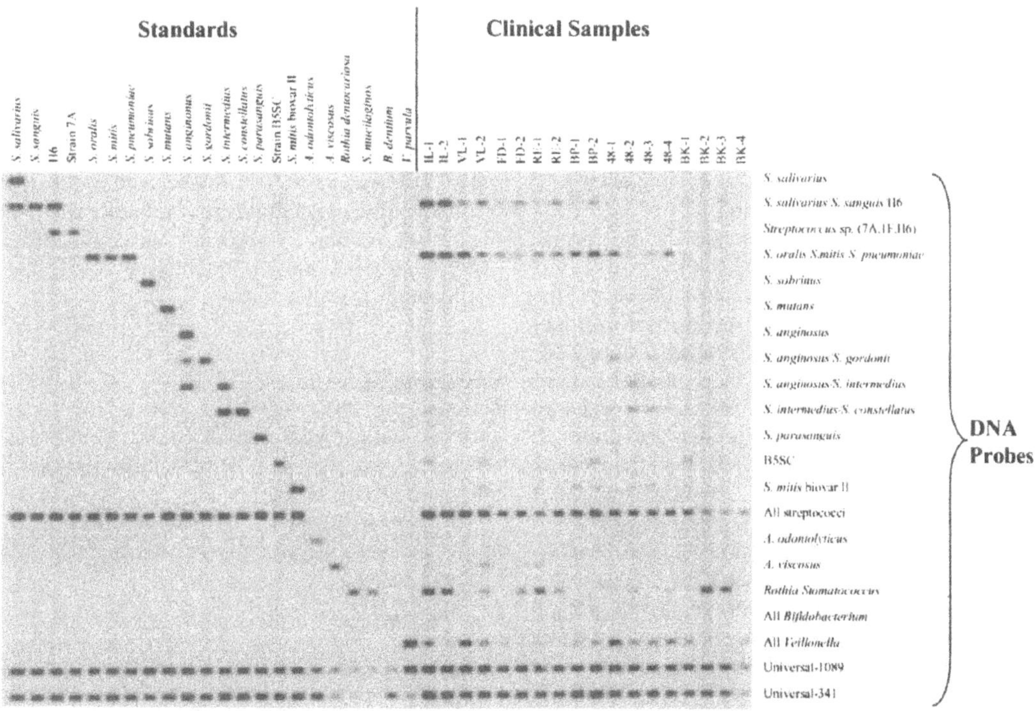

Figure 3. Blot demonstrating hybridization of digoxigenin-labeled, PCR products of standards and clinical samples with reverse-capture, oligonucleotide probes. This figure illustrates the population distribution of streptococcal species and other genera in plaque or saliva samples from 7 patients.

pneumoniae, and *S. mitis* are essentially indistinguishable, and thus a single probe (D38) was designed to target all three 'species' – the *S. oralis* cluster (Table 1; Figure 3). Other species required more than one probe for identification. For example, no single probe for *S. sanguis* could be designed. Consequently, probe D48 that targeted *S. sanguis*, *S. salivarius*, and a potentially new species (designated as H6) was used to determine the levels of all three species. Levels of hybridization with individual probes to *S. salivarius* and H6 can then be subtracted from the levels obtained with probe D48. Thus, as described above, little or no *S. salivarius* or H6 were present in any of the clinical samples, but *S. sanguis* was present in high numbers.

Although not presented here, we have shown in preliminary experiments that quantitative data can be obtained from hybridization blots. The Storm system allows for the direct quantitation of chemifluorescence samples without the need of X-ray film. Signal intensities obtained from the Storm system for each sample can be expressed as microbial counts by reference to standards run at known concentrations. In addition, percent of total streptococci or percent of total bacteria can be obtained by dividing the signal intensities from the individual probes by the signal intensities obtained from the all streptococci probe or universal probes, respectively.

Another advantage of the reverse capture format is that the crosslinked probes can be preserved indefinitely (i.e., years) on dry membranes stored in the dark at room temperature. Consequently, 'stock' membranes with probes can be made prior to processing of the clinical or bacterial samples. Another advantage is that short hybridization times (from 2 to 4 h) of the oligonucleotide probes (compared to overnight with whole genomic probes) allow for same day hybridization.

The reproducibility and accuracy of the reverse capture, checkerboard hybridization methodology is dependent upon the extreme care by the researcher with regard to the following precautions. Gloves must be used for all steps of the hybridization. The nylon membrane must not be touched with bare or gloved hands. The membranes should be picked up with clean forceps by grabbing only at its corners. It is important to filter all solutions since particulate matter (e.g., dust) may appear as aberrant spots after development. We have recently detected different background levels of hybridization with different lots of Nylon membrane – some lots have no detectable background, whereas others have some background. Nevertheless, these background signals, as shown in Figure 3, can be subtracted out during quantitative analysis. If validated probes are used and the precautions are heeded, good results can be achieved the first time through. With practice, hybridization blots such as shown in Figure 3 are achieved routinely in about 8 hours (not including time for PCR). If new probes are to be developed, significant time must be spent in optimizing probes for standard hybridization conditions.

One major advantage of utilizing probes that target 16S rRNA genes is that they can be used to monitor species that can not be presently cultivated. The phylogenetic identity of these 'uncultivable' organisms can be determined by sequencing cloned 16S rRNA genes that are amplified directly from environmental samples [7, 14]. These procedures have been used to deduce the identity of many unknown cultivable and not yet cultivable bacterial species from free-living and host-associated sources [5, 11, 16, 20, 23]. Probes to these not yet cultivable species can be designed just easily as they can be for cultivable species.

One of our initial goals in developing this assay was to measure bacterial population densities of the predominant microflora on dental root surfaces. Although it is widely accepted that mutans streptococci play a major role in caries production, there is growing evidence indicating that 'low pH' nonmutans streptococci and other genera, such as *Actinomyces*, may also play a role in the disease [21, 22]. With the use of validated probes, it is now possible to analyze a statistically sufficient number of sites to assess the role of the low pH, nonmutans streptococci, mutans streptococci, and other bacteria in root surface caries. Such a comprehensive analysis of the bacterial flora on root surfaces has not been previously performed. We are also utilizing checkerboard methodology to determine the bacterial population densities of predominant microorganisms involved in oral health, periodontitis, acute necrotizing ulcerative gingivitis, and refractory periodontal disease. This methodology has obvious utility beyond that of the oral cavity – it can be used for ecological studies of many environments that harbor a diverse, cultivable and not-yet-cultivable bacterial population, such as extraoral areas of the body, animals, insects, soil, and deep sea sediments.

Acknowledgments

These studies were supported by Public Health Service grant P50-DE-070009 from the National Institute of Dental Research.

We thank Dr W. G. Weisburg and Dr D. J. Lane for their helpful discussions on DNA-DNA hybridization.

Notes on suppliers

1. Amersham Corp., Arlington Heights, IL, USA
2. Benton Dickinson and Co., Cockeysville, MD, USA
3. Bio-Rad Laboratories, Hercules, CA, USA
4. Boehringer Mannheim Corp., Indianapolis, IN, USA

5. Brinkmann, Westbury, NY, USA
6. Cole Palmer Instrument Co., Chicago, IL, USA
7. Coy Laboratory Products, GrassLake, MI, USA
8. Fisher Scientific, Pittsburgh, PA, USA
9. FMC BioProducts, Rockland, ME, USA
10. Immunetics, Inc., Cambridge, MA, USA
11. J.T. Baker, Phillipsburg, NJ, USA
12. Life Technologies, Gaithersburg, MD, USA
13. Marsh Biomedical Products, Inc., Rochester, NY, USA
14. Molecular Dynamics, Sunnyvale, CA, USA
15. Northeast Labs, Waterville, ME, USA
16. Operon Technologies, Alameda, CA, USA
17. Owl Scientific, Inc., Woburn, MA, USA
18. Perkin Elmer, Foster City, CA, USA
19. Promega, Madison, WI, USA
20. Rainin, Emeryville, CA, USA
21. Sigma Chemical Co., St. Louis, MO, USA
22. Stratagene, LaJolla, CA, USA
23. USA/Scientific Plastics, Ocala, FL, USA
24. VWR Scientific Products, S. Plainfield, NJ, USA

References

1. Bentley RW, Leigh JA (1995). Development of PCR-Based hybridization protocol for identification of streptococcal species. J Clin Microbiol 33: 1296–1301.
2. Bentley RW, Leigh JA, Collins MD (1991). Intrageneric structure of *Streptococcus* based on comparative analysis of small-subunit rRNA sequences. Intern J System Bacteriol 41: 487–494.
3. Bentley RW, Leigh JA, Collins MD (1993). Development and use of species-specific oligonucleotide probes for differentiation of *Streptococcus uberis* and *Streptococcus parauberis*. J Clin Microbiol 31: 57–60.
4. Cangelosi GA, Iversen JM, Zuo Y, Oswald TK, Lamont RJ (1994). Oligonucleotide probes for mutans streptococci. Mol Cell Probes 8: 73–80.
5. Choi BK, Paster BJ, Dewhirst FE, Göbel UB (1994). Diversity of cultivable and uncultivable oral spirochetes from a patient with severe destructive periodontitis. Infect Immun 62: 1889–1895.
6. Dewhirst FE, Paster BJ (1991). DNA probe analyses for the detection of periodontopathic bacteria in clinical samples. In: Hamada S, Holt SC, McGhee JR (eds), Periodontal disease: Pathogens and Host Immune Responses (pp 367–377). London: Quintessence Publishing Co., Ltd.
7. Giovannoni S (1991). The polymerase chain reaction, p. In: Stackebrandt E, Goodfellow M (eds), Nucleic Acid Techniques in Bacterial Systematics (pp 177–203). John Wiley & Sons, New York.
8. Jacobs JA, Schot CS, Bunschoten AE, Schouls LM (1996). Rapid species identification of '*Streptococcus milleri*' strains by line blot hybridization: Identification of a distinct 16S rRNA population closely related to *Streptococcus constellatus*. J Clin Microbiol 34: 1717–1721.
9. Jukes TH, Cantor CR (1969). Evolution of protein molecules. In: Munro HN (ed), Mammalian Protein Metabolism, vol. 3 (pp 21–132).Academic Press, Inc., New York.
10. Kawamura Y, Hou XG, Sultana F, Miura H, Ezaki T (1995). Determination of 16S rRNA sequences of *Streptococcus mitis* and *Streptococcus gordonii* and phylogenetic relationships among members of the genus Streptococcus. Intern J System Bacteriol 45: 406–408.
11. Liesack W, Stackebrandt E (1992). Unculturable microbes detected by molecular sequences and probes. Biodiversity Conserv 1: 250–262.
12. Morrissey DV, Collins ML (1989). Nucleic acid hybridization assays employing dA-tailed capture probes. Single capture methods. Molec Cell Probes 3: 189–207.
13. Nelms LF, Odelson DA, Whitehead TR, Hespell RB (1995). Differentiation of ruminal and human *Streptococcus bovis* strains by DNA homology and 16S rRNA probes. Current Microbiol 31: 294–300.
14. Pace NR, Stahl DA, Lane DJ, Olsen GJ (1986). The analysis of natural microbial populations by ribosomal RNA sequences. Adv Microb Ecol 9: 1–55.
15. Paster BJ, Dewhirst FE (1988). Phylogeny of campylobacters, wolinellas, *Bacteroides gracilis*, and *Bacteroides ureolyticus* by 16S ribosomal ribonucleic acid sequencing. Inter J System Bacteriol 38: 56–62.
16. Paster BJ, Dewhirst FE, Cooke SM, Fussing V, Poulsen LK, Breznak JA (1996). Phylogeny of not-yet-cultured spirochetes from termite guts. Appl Environ Microbiol 62: 347–352.
17. Saiki RK, Walsh PS, Levenson CH, Erlich HA (1989). Genetic analysis of amplified DNA with immobilized sequence-specific oligonucleotide probes. Proc Natl Acad Sci 86: 6230–6234.
18. Saitou N, Nei M (1987). The neighbor-joining method: A new method for reconstructing phylogenetic trees. Mol Biol Evol 4 :406–425.
19. Socransky SS, Smith C, Martin L, Paster BJ, Dewhirst FE, Levin AE (1994). 'Checkerboard' DNA-DNA hybridization. Biotechniques 17: 788–792.
20. Stackebrandt E, Liesack W, Goebel BM (1993). Bacterial diversity in a soil sample from a subtropical Australian environment as determined by 16S rDNA analysis. FASEB J 7: 232–236.
21. van Houte J, Lopman J, Kent R (1994). The predominant cultivable flora of sound and carious human root surfaces. J Dent Res 73: 1727–1734.
22. van Houte J, Lopman J, Kent R (1996). The final pH of bacteria comprising the predominant flora on sound and carious human root and enamel surfaces. J Dent Res 75: 1008–1014.
23. Ward DM, Weller R, Bateson MM (1990). 16S rRNA sequences reveal numerous uncultured microorganisms in a natural community. Nature 345: 63–65.

Author for correspondence: Bruce Jay Paster, Forsyth Dental Center, 140 Fenway, Boston, MA 02115, USA
Phone: (617) 262-5200 x288; Fax: (617) 262-4021
E-mail: bpaster@forsyth.org

Methods in Cell Science **20**: 233–239 (1998).

Pulsed-field gel electrophoresis as an epidemiologic tool for enterococci and streptococci

Jan E. Patterson[1,2] & Cindy C. Kelly[1]

[1] *Departments of Medicine (Infectious Diseases) and* [2] *Departments of Pathology, University of Texas Health Science Center at San Antonio, San Antonio, Texas, USA*

Abstract. Enterococci (*Enterococcus faecium* and *Enterococcus faecalis*) and streptococci such as *Streptococcus pyogenes* (Group A streptococcus), *Streptococcus agalactiae* (Group B streptococcus), and *Streptococcus pneumoniae* are increasing in importance as both hospital-acquired and community pathogens. Emerging resistance and increasing incidence of these organisms has necessitated the analysis of their epidemiologic mechanisms of spread. Pulsed-field gel electrophoresis (PFGE) has emerged as the one of the most widely applicable, reproducible, and stable methods to examine strain identity in bacterial organisms. The procedure used in our laboratory for PFGE typing of whole cell DNA digested with *Sma*I for enterococci, *S. pneumoniae*, *S. pyogenes*, and *S. agalacatiae* is presented. Issues regarding interpretation are also reviewed and discussed.

Key words: Enterococci, Epidemiologic typing, Pulsed-field gel electrophoresis, Streptococci

Abbreviations: PFGE = pulsed-field gel electrophoresis; CHEF = contour-clamped homogenous electric fields

1. Introduction

Enterococci and streptococci are important causes of infection in hospitalized patients. Enterococci have become a leading cause of hospital-acquired infection and are now known to be cross-transmitted in the hospital setting [11]. Multiple new mechanisms of antibiotic resistance have emerged in *Enterococcus* spp. in the last decade, including vancomycin resistance [11]. Streptococci and enterococci continue to cause serious community-acquired infections which require hospitalization [13, 14, 15]. Streptococci such as *S. pneumoniae*, *S. agalactiae* (Group B), and *S. pyogenes* (Group A) continue to be important pathogens and are of epidemiologic importance. Group B streptococcus has long been recognized as important in the neonate and peripartum female; results of more recent studies have documented the increasing incidence of this pathogen in non-pregnant adults [4]. Invasive Group A streptococcal disease appears to be on the increase and determination of strain identity has been important in cluster investigations [14, 15]. Drug-resistant *S. pneumoniae* isolates continue to increase in the US and globally, and the epidemiologic mechanisms of emergence of these strains are of interest [8, 10].

To study the evolving epidemiology of enterococci and these streptococci, epidemiologic typing methods which are widely applicable, reproducible, and stable over time are needed to determine strain identity. Molecular typing approaches in recent years have included plasmid typing, restriction endonuclease analysis (REA) of whole cell DNA using conventional agarose gel electrophoresis, pulsed-field gel electrophoresis (PFGE) of whole cell DNA, and polymerase chain reaction assays such as randomly amplified polymorphic DNA (RAPD). A disadvantage of REA is the large number of small fragments of DNA to be resolved by conventional agarose electrophoresis. REA patterns are often difficult to discern and compare. The disadvantage of plasmid typing is that this procedure does not examine the chromosome, a more stable genetic marker over time. Polymerase chain amplification procedures are advantageous due to the rapidity with which results can be obtained, but they examine only a portion of the chromosome. The advantages of PFGE typing methods include the applicability to a broad spectrum of organisms, reproducibility, stability over time, examination of the entire chromosomal DNA, and standard criteria for interpretation [16, 17]. PFGE typing has been used successfully to type enterococcal and streptococcal species [1, 2, 8, 9, 12, 14].

In the PFGE procedure, the organism of interest is embedded in agarose and then lysed. The whole cell DNA is then digested by a restriction endonuclease for which there are infrequent recognition sites on the bacterial DNA. This procedure generates large fragments of DNA which cannot be resolved by conventional agarose gel electrophoresis, but are resolved using PFGE. A commonly used PFGE technology for clinical and infection control applications is contour-clamped homogenous electric fields (CHEF) [7].

This review will focus on PFGE typing procedures for the above-mentioned organisms: enterococci (including *Enterococcus faecium* and *E. faecalis*), Group B streptococci, Group A streptococci, and *S. pneumoniae*.

2. Materials

A. Equipment and supplies
- Sterile inoculating loop (Secure Medical Products, Inc., Cat. #58335).[1]
- Test tube rack (Bel Art Products, Cat. #F18746-0002).[2]
- Pipetter (Ulster Scientific, Inc., Mfr. #V200TE).[3]
- Tabletop centrifuge (IEC, Cat. #2438).[4]
- Sterile transfer pipet (Samco Scientific, Order #202-15).[5]
- 5.0 McFarland Equivalence Turbidity Standard (Remel, Cat. #20-415).[6]
- Sterile cotton-tipped applicator (Citmed, Order #22-9589).[7]
- Microwave oven (Sears, Roebuck & Company, Kenmore Model #721.89652).[8]
- 50 °C water bath (Precision Scientific, Order #66800).[9]
- Lead ring (Fisher Scientific, Order #05-869-1).[10]
- Plug molds (BioRad, Cat. #170-3622).[11]
- Kimwipes (Kimberly Clark Corp., Order #34155).[12]
- Bulb (Baxter, Cat. #P5215).[13]
- Teflon spatula (Fisher Scientific, Cat. #14-356-8).[10]
- 37 °C water bath (Precision Scientific, Cat. #66800).[9]
- Rack for microcentrifuge tubes (Fisher Scientific, Cat. #03-213-2).[10]
- Contour-clamped homogeneous electric fields system (CHEF DRII System Cat. #170-3612 **or** CHEF DRIII System, BioRad Laboratories, Cat. #170-3695).[11]
- Gel casting stand and comb (BioRad Laboratories, Cat. #170-3620 and 170-4326).[11]
- UV transilluminator (254–360 nm) (Fotodyne, Inc., Cat. #3-3000, 120V).[14]
- Polaroid camera and photographic hood (Fotodyne, Inc., Cat. #5-5330 and 5-5343).[14]
- Polaroid film (Sigma, Cat. #F-3390).[15]

B. Optional equipment:
- Computerized scanner (HP ScanJet II cx, Hewlett Packard) **or** gel documentation system (Gel Doc 1000, Biorad Cat. #170-7266).[11]
- Molecular epidemiology analysis software program (Dendron software, Solltech, Inc.[17] **or** Molecular Analyst Fingerprinting software, Biorad Laboratories, Cat. #170-7637).[11]

C. Glassware and Plasticware
- Sterile pipet tip (USA Scientific Plastics, Cat. #1011-1006).[18]
- Graduated conical centrifuge tubes with caps, (Sarstedt, Inc., Cat. #62.553.002PS).[19]
- 125 ml Erlenmeyer flask (Baxter Diagnostics, Order #F4253-125).[13]
- Petri dish lid (Fisher Scientific, Order #8-757-12).[10]
- Pasteur pipet (WWR Scientific Products, Order #TXATI14673-0).[20]
- Microcentrifuge tube (Fisher Scientific, Cat. #05-407-10).[10]

D. Chemicals/Reagents
- Trypticase soy agar with 5% sheep blood (Becton Dickinson & Company, Order #21261).[21]
- Trypticase soy broth (TSB) (Becton Dickinson & Company, Order #4397342).[21]
- Brain heart infusion broth (Becton Dickinson & Company, Order #21812).[21]
- Todd-Hewitt broth (Difco Laboratories, Order #0492-17-6).[22]
- 5% yeast extract (ICN Biomedicals, Inc., Order #103303).[23]
- Glucose (ICN Biomedicals, Order #901521).[23]
- Choline chloride (Sigma, Order #C-1879).[15]
- NaCl (EM Science Industries, Inc., Order #SX0420-1).[24]
- Tris-base (Sigma, Cat. #T-8524).[15]
- Low melting point agarose (Boehringer Mannheim, Order #100-450).[25]
- TRIZMA hydrochloride (Tris), (Sigma, Cat. #T-7149).[15]
- Ethylenediamine tetraacetic acid (EDTA) (Sigma, Cat. #E-1644).[15]
- Polyoxyethylene 20 cetyl ether (Brij 58) (Sigma, Cat. #P-5884).[15]
- Deoxycholic acid (deoxycholate) (Sigma, Cat. #D-6750).[15]
- N-lauroyl sarcosine (sarcosyl) (Sigma, Cat. #L-9150).[15]
- Ribonuclease A (Sigma, Cat. #R-6513).[15]
- Lysozyme (Sigma, Cat. #L-6876).[15]
- Mutanolysin (Sigma, Cat. #M-9901).[15]
- Proteinase K (Sigma, Cat. #P-6556).[15]
- Pulsed-field certified agarose (BioRad Laboratories, Cat. #162-0137).[11]
- Tris/borate/EDTA buffer (ICN, Aurora, OH, Cat. #816202).[23]
- Ethidium bromide (Sigma, Cat. #E1510).[15]

3. Procedures

Using sterile technique, streak a single colony for isolation onto a trypticase soy agar plate with 5% sheep blood using a sterile inoculating loop and incubate the plate at 37 °C with 5% CO_2 overnight.

For enterococci, Group A streptococci, and Group B streptococci, transfer 2–3 isolated colonies to 5 ml of brain heart infusion broth and incubate at 37 °C overnight in a test tube rack.

For *Streptococcus pneumoniae*, transfer 2–3 isolated colonies to 5 ml TSB, add choline chloride to 1%, and 50 µl of lysed horse blood (LHB). Incubate overnight at 37 °C in the presence of 5% CO_2.

LHB preparation

Aseptically distribute 250 ml of defibrinated horse blood (Remel order #54-096)[6] to 4 or 6 250 ml capacity Nalgene centrifuge bottles. Add an equal volume of sterile distilled water to each bottle.

- Run each bottle through 5 freeze-thaw cycles as follows:

  Tilt bottles slightly and store overnight at –20 °C. Remove bottles from freezer in AM and allow to thaw at room temperature during the day. Return to freezer at the end of the day.
- Centrifuge at 12,000 ×*g*, 5 °C, 20 min.
- Decant supernatant (LHB) into appropriate containers and store at –70 or –20 °C.

Pour the overnight growths into sterile 15 ml graduated conical centrifuge tubes with caps and centrifuge at 4000 rpm for 20 minutes. Pour off the supernatants completely. Resuspend the cell pellets completely in PIV buffer. Any clumps of bacteria in the cell suspension will result in a white streak in the lane of the gel.

PIV buffer, pH 7.6

- 1.0 M NaCl
- 10 mM Tris-base

The amount of buffer added to each cell pellet will depend on the size of the pellet (typically between 0.5 ml and 1 ml). Pour a small quantity of the PIV buffer stock into a separate sterile tube and add the PIV buffer drop-wise with a sterile transfer pipet. The resulting suspension should have the opacity of a 5.0 McFarland standard. The same optical density should be used forsolates being compared to each other.

Arrange the tubes in a test tube rack in rows of three. There will be three isolates per ten-sample plug mold and three plugs (wells) per isolate.

If results are urgently needed, the cell suspensions can be made by taking colonies directly from an agar plate of the isolated bacteria using a sterile cotton-tipped applicator and suspending them in 0.5–1.0 ml of PIV buffer.

In a glass 125 ml Erlenmeyer flask, prepare a quantity of 1.6% low melting point agarose in PIV buffer equal to the volume of all cell suspensions combined plus some dead volume (plus 5–10 mls extra). A microwave oven, at 30% power for approximately 2–3 minutes, can be used to melt the agarose. The agarose should be checked at 30-second intervals, and the contents swirled to ensure even heating. The contents should also be examined for complete melting of the agarose.

Add weight to the Erlenmeyer flask with a lead ring (to prevent tipping) and place into a 50 °C water bath. Place a Petri dish lid over the opening of the flask. Also place the test tube rack with the cell suspensions into the 50 °C water bath.

Clean the plug molds with sterile dH_2O and Kimwipes and assemble.

Beginning with the first graduated tube in the front of the rack, move left to right, examining the tube to determine the volume of the cell suspension. Add an equal volume of the low melting point agarose to make a 1:2 dilution using a $5^3/_4$ inch Pasteur pipet and bulb. Take care not to contaminate the flask of agarose with any of the cell suspensions. Mix thoroughly by pulling the mixture up into the pipet and expelling it back into the tube several times. Finally, draw the cell/agarose suspension up into the pipet. Fill three wells of the plug mold, from right to left, three isolates per mold, leaving the last well blank. Fill the wells of the mold slowly to avoid bubble formation in the wells. Refrigerate at 4 °C for 20 minutes.

Prepare fresh centrifuge (or other sterile) tubes by labeling them and organizing them in the same order as those in the 50 °C water bath. Fill each with 5 ml lysis buffer.

Lysis buffer, pH 7.6

- 6 mM TRIZMA hydrochloride (Tris)
- 1.0 M NaCl
- 100 mM ethylenediamine tetraacetic acid (EDTA)
  0.5% polyoxyethylene 20 cetyl ether (Brij 58)
- 0.2% deoxycholic acid (deoxycholate)
- 0.5% N-lauroyl sarcosine (sarcosyl)
- 50 µg/ml ribonuclease A
- 1 mg/ml lysozyme

For *S. pneumoniae* and Group B streptococcus, add 200 U mutanolysin per ml of lysis buffer.

When the plugs are cool, remove them from the molds when cooled with the curved end of a Teflon spatula. Move from left to right, keeping the blank space next to third isolate of each group, and insert in tubes containing lysis buffer.

Incubate overnight in a 37 °C water bath with gentle shaking at approximately 65 rpm (add weight to the test tube rack with a lead ring).

Pour off the lysis buffer and replace with 5 ml of ESP buffer.

ESP buffer, pH 9.0–9.5

- 0.4M EDTA
- 1% sarcosine
- 0.5 mg/ml proteinase K

Incubate overnight in a 50 °C water bath with gentle shaking (65 rpm).

Pour off the ESP buffer completely and replace with 5 ml of TE buffer.

TE buffer, pH 7.5
- 20 mM Tris
- 5 mM EDTA

Incubate at 37 °C for 30 minutes with gentle shaking. Repeat 5×.

Place plugs in fresh TE buffer and store at 4 °C until used.

Digestion of the agarose plugs prior to electrophoresis. Create a legend of the samples including an appropriate DNA size standard. This should indicate which position each plug will have in the gel. It should also contain data regarding the parameters at which the gel will run (i.e., pulse interval, voltage, duration of run.)

Label a 1.5 ml snap cap microcentrifuge tube for each plug on your proposed gel with a permanent marker. Place the tubes in the same order as they will appear on the gel in a rack for microcentrifuge tubes. Add 270 μl of sterile dH_2O to each of the microcentrifuge tubes.

Then add 30 μl of the 10× reaction/incubation buffer intended for use with the restriction endonuclease selected for digestion of the DNA to each of the microcentrifuge tubes. For enterococci, *S. pneumoniae*, Group B streptococci and Group A streptococci, *Sma*I, which recognizes the sequence CCCGGG can be used. These recognition sites are infrequent in most Gram-positive organisms which have A-T rich genomes [6].

Next, cut a third of a single plug for each isolate to be run on the gel with the Teflon spatula by placing the blade perpendicular to the plug and pressing down in a single motion. The plugs are semi-soft and should be cut easily. Then, using the spatula, lift the piece of plug, wick the excess TE buffer with a Kimwipe from the plug and place it into the labeled microcentrifuge tube. The remaining two-thirds and extra two plugs are returned to a labeled tube containing fresh TE buffer and refrigerated for future use.

Add 3 μl of the restriction endonuclease chosen for digestion (*Sma*I) to each microcentrifuge tube. The enzyme must remain on ice or in a chilled container during this time, and returned to the freezer as soon as possible. Close the microcentrifuge tubes and mix the contents of each by inverting gently 4 or 5 times. Then gently tap the pointed end of the tube on the benchtop to ensure that the plug is completely covered by the enzyme/buffer mixture. Examine the tube to check that the plug is not stuck in the cap, and to make sure that it is indeed present in the tube. Return the tube to the rack and incubate the tubes overnight at the temperature recommended by the manufacturer for the restriction endonuclease (20 °C for *Sma*I).

After the incubation period is complete, pour off the enzyme/buffer mixture and keep the plug in the tube. Add 1ml of TE buffer to the tube. Incubate the tubes at 37 °C for 1 hour.

Running the gel. Make 120 ml of 0.8% pulsed-field certified agarose in 0.5× Tris/borate/EDTA buffer. Allow the agarose to cool in a 50 °C water bath. (Weight the flask and cover the opening of the flask with a Petri dish.)

Clean the gel casting stand and comb with Kimwipes and dH_2O. Attach the end plates securely with the screws. Place the casting stand on a level surface. Adjust the comb to 2 mm above the casting stand surface and 5–10 mm away from one of the end plates. Pour the agarose gently into the casting stand so that no bubbles are formed. Inspect the agarose for bubbles. If there are any, pop them with a Pasteur pipet before the agarose solidifies. Allow the agarose to set for 1 hour. Prepare the contour-clamped homogenous electric fields instrument by pumping any residual buffer from the lines and draining the instrument. Gently wipe the inside of the CHEF with Kimwipes. Be careful in the area of the electrodes; wipe them very gently in a single motion along the length of the platinum wire. Plug the drain tubing of the instrument, and adjust the CHEF so that it is level. Use a leveling instrument to ensure this is done accurately.

Add 2 liters of 0.5× TBE buffer to the CHEF, and turn on the pump and chiller at least 30 minutes prior to the run. Be sure that the buffer is circulating at one liter per minute and that the chilling device keeps the buffer at 14 °C during the electrophoresis.

Remove the comb from the solidified agarose and add some 0.5× TBE buffer to the wells. This will facilitate loading of the DNA inserts and reduce the possibility of the formation of bubbles in the wells. Using the Teflon spatula, place each DNA insert into the well designated on your template for that plug. If bubbles form in the wells, use the spatula to gently manipulate the DNA insert until the bubble is eliminated. It may be necessary to add more 0.5× TBE buffer to the well in order to accomplish this.

Move the casting stand to the CHEF instrument and loosen the screws of the casting stand. Remove the end plates and be careful that the gel does not slip out of the casting stand. Position the gel on the casting stand over the surface of the buffer in the CHEF and lower the casting stand and gel into the buffer leading with the end of the gel that contains the wells. It is necessary that it be done this way so that the plugs are not popped out of the wells by the pressure of the buffer under the gel. Gently push the gel off the casting stand into the buffer and position it behind the gel stops. Seat the gel gently by pushing down lightly on the gel. Check to be sure that the gel is sitting flat on the bottom of the gel box. Replace the lid of the CHEF and take care not to bend the contacts of the interlock. Set the pulse times (initial and final), run time and voltage.

Pulse times. See Table 1 for pulse times. Settings may need to be adjusted due to individual instrument variation and conditions. Use shorter pulse times for resolution of smaller DNA fragments. Use longer pulse times for resolution of larger DNA fragments.

After the electrophoresis is complete, stain the gel in 0.5 μg/ml ethidium bromide for 20–30 min. It may be necessary to decolorize the gel in dH_2O for 30 min to 1 hour. The bands can be visualized (using eye protection) on a UV transilluminator (254–360 nm) and photographs can be taken with a Polaroid camera and photographic hood and Polaroid film. Dispose of the used ethidium bromide as directed by your institutional safety committee.

Table 1. Pulse times applied for PFGE of total cell DNA from different species of enterococci and streptococci

Organism	Pulse time
Enterococcus faecium	1–20 seconds for 15 hours at 200 volts
Enterococcus faecalis	1–20 seconds for 15 hours at 200 volts
Group B streptococcus	1–30 seconds for 15 hours at 200 volts
Group A streptococcus	1–20 seconds for 16 hours at 200 volts
Streptococcus pneumoniae	1–30 seconds for 15 hours at 200 volts

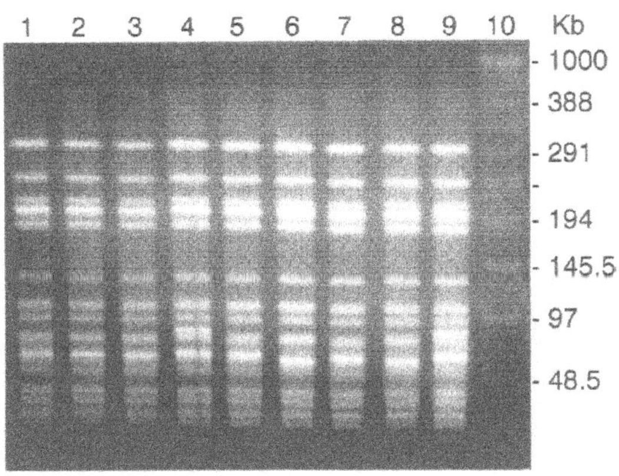

Figure 1. Pulsed-field gel electrophoresis (PFGE) of *Enterococcus faecium* whole cell DNA digested with *Sma*I. The PFGE patterns of isolates in Lanes 1–10 suggest highly related, if not identical, strains. λ size standard is shown in Lane 11.

4. Results, interpretation, and discussion

There should be at least 10 bands in the PFGE pattern for optimal evaluation and comparison to other patterns. PFGE patterns may be compared visually or with the use of a computerized scanning and software system. A computerized system for analysis is quite useful for a large number of isolates or for comparison of isolates over a long period of time. Standard guidelines exist for the interpretation of strain relatedness in PFGE [16, 17]. In general, identical PFGE patterns among isolates indicate that those isolates are the same strain. Choose the predominant pattern or the pattern from the index patient of an outbreak to be the index PFGE pattern. PFGE pattern differences of only 2 to 3 bands compared to the

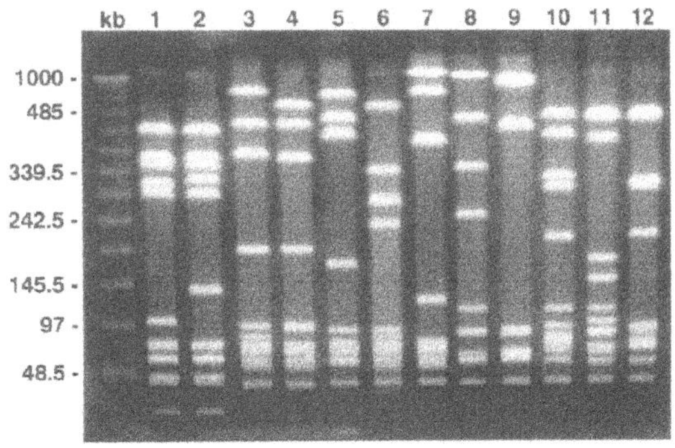

Figure 2. Pulsed-field gel electrophoresis (PFGE) of Group B streptococcus (*Streptococcus agalactiae*) whole cell DNA digested with *Sma*I. The PFGE patterns suggest a heterogeneous group of strains. λ size standard is shown at left.

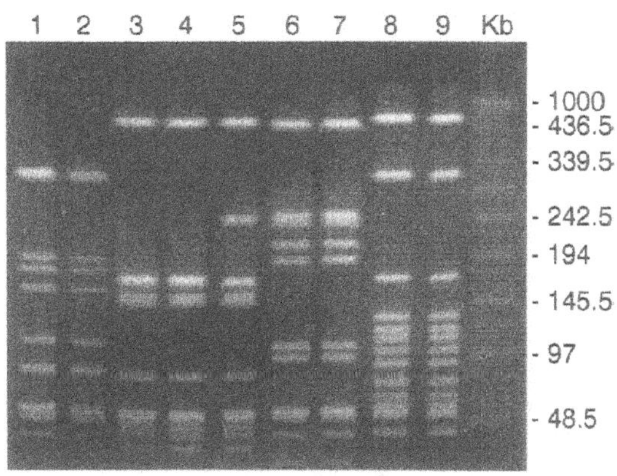

Figure 3. Pulsed-field gel electrophoresis (PFGE) of Group A streptococcus (*Streptococcus pyogenes*) whole cell DNA digested with *Sma*I. There are four very different patterns present, but the strains providing the pattern in Lanes 1 and 2 are highly related (Pattern A), as are those providing the pattern in Lanes 3, 4 and 5 (Pattern B); Lanes 6 and 7 (Pattern C); and Lanes 8 and 9 (Pattern D). λ size standard is shown in the right lane.

index PFGE pattern may indicate a single genetic event, and these isolates may be closely related to the index PFGE pattern. PFGE pattern differences of 4 to 6 bands are possibly related to the index PFGE pattern. PFGE pattern differences of ⩾ 7 bands suggest that the isolates are distinct strains. If a computerized analysis is done, computerized dendograms can be generated . PFGE patterns with coefficients of similarity ⩾ 85% are examined closely for strain relatedness [5, 8].

Since PFGE of whole cell DNA examines the chromosome, a relatively stable genetic marker, it is a reproducible and accurate method of determining bacterial strain identity. Be careful to correlate antibiotic susceptibility results, since isolates that are the same by PFGE type can occasionally vary by antibiotic susceptibility results, due to acquisition or loss of an antibiotic resistance trait encoded by an extrachromosomal element such as a plasmid. Differences based on small, usually circular, extrachromosomal genetic elements will not be detected in PFGE of whole cell DNA, which resolves larger fragments of linear double-stranded DNA (75–600 kb range). Examples include vancomycin resistance in enterococci, and macrolide-lincosamide-streptogramin, tetracycline, or clindamycin resistance in enterococci or staphylococci. In addition, correlate serotyping results if available for pneumococcal isolates, since capsular transformation has been reported to occur, and thus isolates with the same PFGE type may occasionally vary by serotype [1, 3]. Most importantly, results of strain identity should be reviewed in the context of, and not independently of, existing epidemiologic data for the isolates [16, 17]. Examples of the types of PFGE patterns produced by highly related, as well as divergent, strains of bacterial species are provided in Figures 1 (*E. faecium*), 2 (*S. agalactiae*), 3 (*S. pyogenes*), and 4 (*S. pneumoniae*).

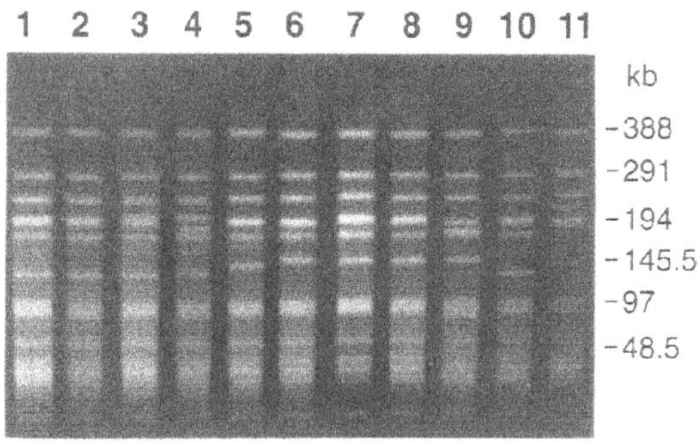

Figure 4a. Pulsed-field gel electrophoresis (PFGE) of *Streptococcus pneumoniae* whole cell DNA digested with *Sma*I. The PFGE patterns are highly related. λ size standard is shown at right.

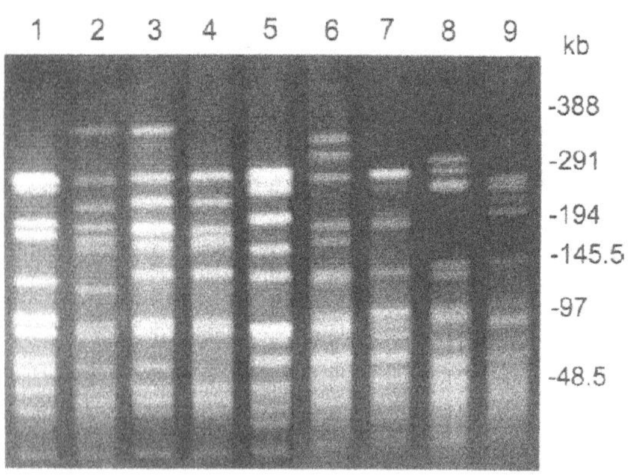

Figure 4b. Pulsed-field gel electrophoresis (PFGE) of *Streptococcus pneumoniae* whole cell DNA digested with *Sma*I. The PFGE patterns in Lanes 2, 3, and 4 represent related strains, whereas those in Lanes 1, 5, 6, 7, 8, and 9 were produced from DNA from distinct strains. λ size standard is shown at right.

Notes on suppliers

1. Secure Medical Products, Inc., 1000 Allanson Rd., Mundelein, IL, 60060, USA
2. Bel Art Products, Pequannock, NJ 07440-1992, USA
3. Ulster Scientific, Inc., PO Box 819, New Paltz, NY 12561-0819, USA
4. IEC, 300 Second Avenue, Needham Heights, MA 02194, USA
5. Samco Scientific, 1050 Arroyo Avenue, San Fernando, CA, USA
6. Remel, 12076 Sante Fe Dr., Lenexa, KS 66215, USA
7. Citmed, Cottontail Medical Products, 18601 South Main Street, Citronelle, AL, USA
8. Sears, Roebuck & Company, Hoffman Estates, IL 60179, USA
9. Precision Scientific, 3737 West Cortland Street, Chicago, IL 60647-9987, USA
10. Fisher Scientific, 711 Forbes Avenue, Pittsburgh, PA 15219, USA
11. BioRad Laboratories, 2000 Alfred Nobel Drive, Hercules, CA 94547, USA
12. Kimberly Clark Corporation, Roswell, GA 30076-2199, USA
13. Baxter Diagnostics, Inc., Scientific Products Division, 1430 Waukegan Road, McGaw Park, IL 60085, USA
14. Fotodyne, Inc., 950 Walnut Ridge Drive, Hartland, WI 53029-9910, USA
15. Sigma Chemical Company, PO Box 14508, St. Louis, MO, USA
16. Hewlett Packard
17. Solltech, Inc., Technology Innovation Center, Suite No. 9, Oakdale, IA 52319, USA
18. USA Scientific Plastics, PO Box 3565, Ocala, FL 34478, USA
19. Sarstedt, Inc., Newton, NC 28658-0468, USA
20. WWR Scientific Products, West Chester, PA 19380, USA
21. Becton Dickinson & Company, Cockeysville, MD 21030, USA
22. Difco Laboratories, Detroit, MI 48232-7058, USA
23. ICN Biomedicals, Inc., 1263 Chillicothe Road, Aurora, OH 44202, USA

24. EM Science Industries, Inc., 400 South Democrat Road, Gibbstown, NJ, USA
25. Boehringer Mannheim, 9115 Hague Road, PO Box 50414, Indianapolis, IN, USA

References

1. Barnes DM, Whittier S, Gilligan PH, Soares S, Tomasz A, Henderson FW (1995). Transmission of multidrug-resistant serotype 23F *Streptococcus* pneumoniae in group day care: Evidence suggesting capsular transformation of the resistant strain in vivo. J Infect Dis 171: 890–896.
2. Blumberg HM, Stephens DS, Modansky M, Erwin M, Elliot J, Facklam RR, Schuchat A, Baughman W, Farley MM (1996). Invasive group B streptococcal disease: The emergence of serotype V. J Infect Dis 173: 365–373.
3. Coffey TJ, Dowson CG, Daniels M (1991). Horizontal transfer of multiple penicillin-binding protein genes, and capsular biosynthetic genes in natural populations of *Streptococcus pneumoniae*. Mol Microbiol 5: 2255–2260.
4. Harrison LH, Ali A, Dwyer DM, Libonati JP, Reeves MW, et al. (1995). Relapsing invasive Group B streptococcal infection in adults. Annals Int Med 123: 421–427.
5. Hellstein J, Vawter-Hugart H, Fotos P, Schmid J, Soll DR (1993). Genetic similarity and phenotypic diversity of commensal and pathogenic strains of *Candida albicans* isolates from the oral cavity. J Clin Microbiol 31: 3190–3199.
6. Holt JG (editor-in-chief) (1984). Bergey's manual of systematic bacteriology, Volumes 1 and 2. Baltimore: Williams and Wilkins.
7. Lai E, Birren BW, Clark SM, Simon MI, Hood L (1989). Pulsed field gel electrophoresis. BioTechniques 7: 34–42.
8. Moreno F, Crisp C, Jorgensen JH, Patterson JE (1995). The clinical and molecular epidemiology of bacteremias at a university hospital caused by pneumococci not susceptible to penicillin. J Infect Dis 172: 427–432.
9. Moreno F, Grota P, Crisp C, Magnon K, Melcher GP, Jorgensen JH, Patterson JE. Clinical and molecular epidemiology of vancomycin-resistant *Enterococcus faecium* during its emergence in a city in southern Texas. Clin Infect Dis 21: 1234–1237.
10. Munoz R, Coffey TJ, Daniels M, et al. (1991). Intercontinental spread of a multiresistant clone of serotype 23F *Streptococcus pneumoniae*. J Infect Dis 164: 302–306.
11. Murray BE (1990). The life and times of the enterococcus. Clin Microbiol Rev 3: 46–65.
12. Murray BE, Singh KV, Heath JD, Sharma BR, Weinstock GM (1990). Comparison of genomic DNAs of different enterococcal isolates using restriction endonucleases with infrequent recognition sites. J Clin Microbiol 28: 2059–2063.
13. Patterson JE, Sweeney AH, Simms M, Carley N, Mangi R, Sabetta J, Lyons RW (1995). An analysis of 110 serious enterococcal infections: Epidemiology, antibiotic susceptibility, and outcome. Medicine 74: 191–200.
14. Ramage L, Green K, Pyskir D, Simor AE (1996). An outbreak of fatal nosocomial infections due to group A streptococcus on a medical ward. Infection Control Hosp Epidemiol 17: 429–431.
15. Stevens DL, et al. (1989). Invasive Group A streptococcal infections associated with a toxic shock-like syndrome and scarlet fever toxin A. New Engl J Med 321: 1–7.
16. Tenover FC, Arbeit RD, Goering RV, Mickelsen PA, Murray BE, Persing DH, Swaminathan B (1995). Interpreting chromosomal DNA restriction patterns produced by pulsed-field gel electrophoresis: Criteria bacterial strain typing. J Clin Microbiol 33: 2233–2239.
17. Tenover FC, Arbeit RD, Goering RV, and the Molecular Typing Working Group of the Society for Healthcare Epidemiology of America (1997). How to select and interpret molecular strain typing methods for epidemiological studies of bacterial infections: A review for healthcare epidemiologists. Infection Control Hosp Epidemiol 18: 426–439.

Address for correspondence: Jan E. Patterson, MD, Department of Medicine/ID, University of Texas Health Science Center at San Antonio, 7703 Floyd Curl Dr., San Antonio, TX 78284-7881, USA
Phone: (210) 358-2927; Fax: (210) 358-4702
E-mail: pattersonj@uthscsa.edu

Methods in Cell Science **20**: 241–249 (1998).

Cell-based panning as a means to isolate phage display Fabs specific for a bacterial surface protein

Aimee E. Stephenson, Paula Fives-Taylor & Robert J. Melamede
University of Vermont, Department of Microbiology and Molecular Genetics, Burlington, Vermont, USA

Abstract. Surface proteins provide a multitude of functions for the bacterial cell. Antibodies to these proteins can provide tools for tagging bacteria and characterizing protein function. Phage display technology has emerged as a powerful method for producing monoclonal Fabs in *Escherichia coli.* In an effort to study the adhesion mechanisms of *Streptococcus parasanguis* FW213, Fabs specific for the surface adhesin protein Fap1 were produced using phage display. The immune repertoire of a mouse injected with purified Fap1 was cloned into the phagemid vector pCOMB3, and a combinatorial Fab library was expressed in *E. coli.* A cell-based panning method using whole *S. parasanguis* cells was developed and has been shown to be a means for enriching for Fabs specific for the Fap1 protein.

Key words: Adhesin, Antibody, Bacteria, Fab, Phage display

Abbreviations: Fab = fragment antigen binding

1. Introduction

There are a variety of proteins found on the surface of bacteria that provide numerous functions for the cell including protection, adhesion, transport and signal transduction to name only a few. In pathogenic species, proteins involved in adhesion to host surfaces and invasion of host tissues are of particular interest. Monoclonal antibodies to bacterial surface proteins provide useful tools for tagging bacteria as well as dissecting functional epitopes. Here we present an efficient means for enriching for functional antibody fragments to bacterial surface proteins from combinatorial phage display libraries.

Although most monoclonal antibodies have historically been produced using classical hybridoma methods [15], phage display technology has become the method of choice by many investigators for the isolation of monoclonal Fabs [for reviews see 7, 11, 25]. A monovalent Fab (fragment antigen binding) consists of a heavy chain fragment (Fd) covalently linkled to a light chain. Although they are lacking an Fc (fragment constant) portion, Fabs function as whole antibodies in terms of antigen recognition. Since they are smaller and have fewer disulfide bonds, they can be expressed in and secreted from *E. coli* [4]. Fabs can be advantageous immunoreagents because they lack the Fc portion that can interact with receptors found on *S. parasanguis* as well as on other cell types.

The current technology for producing phage display Fabs originated from two breakthroughs: 1) the ability to display proteins on the surface of filamentous phage as fusions to the minor capsid protein, pIII [22] and 2) the ability to produce functional Fab molecules in *E. coli* [4]. Although for this application Fabs were generated, production of scFv (heavy and light chain variable domains joined by a flexible linker in a single chain) fragments presents an alternative technology [18]. A comparison of these different methods can be found elsewhere [13].

Production of Fabs using phage display has several unique advantages. The time required to isolate Fabs from phage libraries can take as little as several weeks compared to the several months required for conventional methodologies. Common cloning techniques and bacterial growth methods make stocks easier to establish and maintain than hybridoma cell lines. Unlike classical hybridoma technology, phage display allows for the coupling of genotype with phenotype; i.e., the phage provides a direct link between the binding capacity of a Fab on its surface with the DNA that encodes that product encapsulated inside. This coupling allows for antigen driven selection via 'panning' where several million potential binding Fabs can be functionally surveyed as contrasted to only a few thousand screened using traditional hybridoma methods [7].

The direct availability of molecular manipulations has additional advantages. Sequencing of heavy and light chain genes that encode a given Fab allows for computer modeling, which aids in visualization of the antibody/antigen interaction [17, 24]. site-directed and random mutagenesis of Fab genes provide ways for testing the importance of specific amino acid residues in this interaction and for increasing the binding capacity of a Fab [6, 16, 20].

In this paper, we describe the use of phage display isolate monoclonal Fabs to Fap1, an adhesin protein found naturally on the surface of *S. parasanguis*. Briefly, the immune repertoire of a mouse injected with purified Fap1 was cloned into the phagemid vector pCOMB3 and a combinatorial phage display library was expressed in *E. coli* [1, 2]. In the pCOMB3 system, the Fd fragment is cloned in frame with gene III which encodes the minor capsid protein, pIII [22]. The light chain is cloned into another site on pCOMB3 and synthesized from a separate promotor. The resulting Fd/pIII fusion protein and light chain are directed to the periplasm by *pelB* leader sequences where they fold into a functional Fab fragment and are assembled on the surface of the phage via the pIII capsid protein. Enrichment for Fap1 specific clones was achieved with solution-based panning against whole *S. parasanguis* cells.

The customary method of panning involves incubation of the library with antigen that is immobilized on a solid support such as a well in a microtiter plate. Bound Fabs are eluted with acid, base, or ligand, and used to re-infect *E. coli* for amplification. The process is then repeated for several rounds to attain enrichment. Our use of a solid-based method using a cell-based panning was developed. In this method, whole bacterial cells in suspension are mixed with the phage display library. Phage that bind to antigens on the surface of the cells are trapped by centrifugation and recovered by means of competition-driven elution.

In a similar cell-based panning method, an epitope from p21*ras* was expressed within the *E. coli* outer membrane protein LamB. 'Living columns' of these recombinant *E. coli* were used to separate a phage bound Fv fragment, specific for the p21*ras* eptiope, from a non-binding phage species [5]. Cell-based panning has also been used against eukaryotic cells to isolate phage bound Fabs [19, 23]. This method can have several advantages over traditional solid-based panning. First, the Fap1 protein on the surface of *S. parasanguis* maintains its native conformation and provides an unlimited supply of this protein so that no purification is necessary. Second, solution-based panning prevents exclusion of epitopes that might be hidden or denatured in solid-based panning. Third, using whole cells facilitates competition-driven elution of bound phage after each round of panning. By mixing phage bound *S. parasanguis* cells with a large excess of *E. coli* used for amplification of the panned library, phage are competed off of antigens on the surface *S. parasanguis* by a disproportionate number infection sites on the *E. coli.* This method was successful in enriching for Fabs specific for the Fab1 protein.

2. Materials

A. Biologicals
- Male BALB-c mice Charles River Laboratories, Inc.[1]
- *E. coli* XL1-Blue Cat. #200268 Stratagene.[2]
- Helper phage VCSM13 Cat. #200251 Stratagene.

B. Media/Antibiotics
- Superbroth Cat. #3010-032 BIO 101, Inc.[3]
- Tetracycline Cat. #T-3258 Sigma.[4]
- Carbenicillin Cat. #C-3416 Sigma.
- Kanamycin Cat. #S-9888 Sigma.
- IPTG (isopropyl β-D-thiogalactopyranoside) Cat. #15529-019 Gibco BRL.[5]

C. Buffers/Enzymes/Reagents
- TriReagent Cat. #TT-118 Molecular Research Center, Inc.[6]
- RIBI Adjuvant Cat. #R-700 RIBI Immunechem Research, Inc.[7]
- *XhoI* plus manufacturer's buffer Cat. #899 194 Boehringer Mannheim.[8]
- *SpeI* plus manufacturer's buffer Cat. #1 008 943 Boehringer Mannheim.
- *XbaI* plus manufacturer's buffer Cat. #674 257 Boehringer Mannheim.
- *SacI* plus manufacturer's buffer Cat. #669 792 Boehringer Mannheim.
- *SpeI* plus manufacturer's buffer Cat. #1 008 943 Boehringer Mannheim.
- *NheI* plus manufacturer's buffer Cat. #885 843 Boehringer Mannheim.
- Amplitaq polymerase Cat. #166 Gibco BRL.
- T4 Ligase plus manufacturer's buffer Cat. #15224-D17 Gibco BRL.
- Superscript II reverse. transcriptase plus manufacturer's buffer Cat. #18064-D14 Gibco BRL.
- Horse-radish peroxidase conjugated goat anti-mouse IgG Cat. #A2304 Sigma.
- Horse-radish peroxidase conjugated anti-M13 Cat. #27-9411-01 Pharmacia Biotech.[9]
- o-phenylenediamine Cat. #P-9029 Sigma.
- ECL Western blotting detection reagents Cat. #RPN 2209 Amersham.[10]
- RNase H Cat. #786 349 Boehringer Mannheim.
- Wizard plasmid kits (mini and maxi) Cat. #A7100 and A7270, respectively Promega.[11]
- pCOMB 3 phagemid vector DNA The Scripps Research Institute.[12]
- Nuclease free water Cat. #P119C Promega.
- dNTPs Cat. #1774 Idaho Technologies.[13]
- Enzyme diluent (for Amplitaq) Cat. #1773 Idaho Technologies.
- 2mM Mg^{2+}-Sucrose Buffer Cat. #1783 Idaho Technologies.
- 3mM Mg^{2+}-Surcose Buffer Cat. #1782 Idaho Technologies.
- 4mM Mg^{2+}-Surcose Buffer Cat. #1781 Idaho Technologies.

- Sephadex G-50 DNA Grade F Cat. #17-0573-02 Pharmacia Biotech.
- Spin Columns Cat. #C1281.
- Dye Terminator Cycle Sequencing Ready Reaction Kit Part #402079 Perkin Elmer/Applied Biosystems.[14]

D. Supplies and equipment
- Centricon 100 Cat. #4212 Amicon.[15]
- Gene Pulser Apparatus Cat. #165-2105 Bio-Rad.[16]
- Rapidcycler Idaho Technologies.
- 96 well tissue culture treated plates Cat. #25860 Costar.[17]
- E.I.A./R.I.A. Plate A/2 96 well microtiter plates Cat. #3690 Costar.
- Nunc-Immuno Wash 12 Cat. #62409-160 VWR Scientific Products.[18]
- Automated Microplate Reader EL311sBio-Tek Instruments, Inc.[19]

3. Procedure

A. Immunization

Mix antigen (purified Fap1, 10–20 μg per injection) with RIBI adjuvant as per manufacturer's instruction and inject into two BALB-c mice. Give each mouse two 100 μl injections, one under each arm, and boost twice at three week intervals with the same injections. Three days after the final boost, euthanize the mice by asphyxiation, remove their spleens and place immediately in TriReagent for RNA extraction. Divide spleen/TriReagent in half, place in two microfuge tubes, and homogenize with a plunger centrifuge at 10,000 rpm for 10 minutes in a microfuge to isolate serum.

B. Immune response

Assess the general serum response of each mouse to antigen (purified Fap1 protein) by ELISA assay. Suspend antigen in phosphate buffered saline (PBS) consisting of 0.14 M NaCl, 1 mM KH_2PO_4, 20 mM Na_2HPO_4, and 3 mM KCl, pH7.4 [21] and immobilize on a Costar A/2 microtiter plate (0.24 μg Fap1 protein/well) by incubating at 37 °C for one hour. Block non-specific binding sites with PBS/gelatin (0.5%). Add a 1:1000 dilution of mouse sera in PBS and detect antibody binding with 1:5000 dilution of HRP (horse radish peroxidase) conjugated anti-mouse IgG secondary antibody. Incubate each addition for one hour at 37 °C. Before each new addition, wash three times with PBS. After incubating with secondary, wash twice with PBS and once with TPBS (PBS + 0.5% Tween 20). Perform all washes using a Nunc-Immuno Wash 12. Detect a positive immune response with an OPD (o-phenylenediamine) based substrate or a similar substrate that produces a color reaction that is detectable by a spectrophotometer.

C. RNA extraction and cDNA synthesis

Extract RNA from the spleen cells with TriReagent according to the manufacturer's instructions [for explanation see 8]. Briefly, homogenize spleen cells in 5ml TriReagent and centrifuge to remove cell debris. Extract the resulting supernatant with 0.2ml chloroform/ml supernatant. Save the upper aqueous phase and use a Centricon 100 to concentrate while exchanging buffer for nuclease free water.

Reverse transcribe the first strand of cDNA with Superscript II. Heat 5 μg of total RNA and 1 μg of oligo(dT) (15-mer) primer to 70 °C for 10 minutes and then quickly cool on wet ice. Add manufacturer's buffer to a final concentration of 1× along with 500 units of Superscript II. Incubate the reaction at 40 °C for 1–1.5 hours and inactivate the enzyme by incubating at 70 °C for 10 minutes. Concentrate the reaction mix and exchange the buffer for water with a Centricon 100.

D. PCR amplification of heavy and light chain sequences

Use the following primers to PCR amplify heavy chain Fd (C_{H1} + V_H) DNA fragments: 5′ primers – Hc1-8 agg tcc a(g/a)c t(g/t)c tcg agt c(t/a)g and Hc9 agg tii aic tic tcg agt c(a/t)g g; 3′ class specific primers – IgG_1 agg ctt act agt aca atc cct ggg cac aat, IgG_{2A} ggt ctg act agt ggg cac tct ggg ctc, IgG_3 ggg ggt act agt ctt ggg tat tct agg ctc. Use the following primers to PCR amplify light chain genes: 5′ primers – Lc1 cca gtt ccg agc tcg ttg tga ctc agg aat ct, Lc2 cca gtt ccg agc tcg tgt tga cgc agc cgc cc, Lc3 cca gtt ccg agc tcg tgc tca ccc agt ctc ca, Lc4 cca gtt ccg agc tcc aga tga ccc agt ctc ca, Lc5 cca gat gtg agc tcg tga tga ccc aga ctc ca, Lc6 cca gat gtg agc tcg tca tga ccc agt ctc ca, Lc7 cca gtt ccg agc tcg tga tga cac agt ctc ca; 3′ primer – kappa gcg ccg tct aga att aac act cat tcc tgt tag a. Heavy chain extension primers are: V_{H1} Ext gag aga gag aga gag aga agg tcc arc tkc tcg a, V_{H9} Ext gag aga gag aga aga agg tii aic tic tcg a, IG_1 Ext gag aga gag aga gag aga agg ctt act agt, IG_2 Ext gag aga gag aga gag aga gtt ctg act agt, and IG_3 Ext gag aga gag aga gag aga ggg ggt act agt. Light chain extension primers are: V_L Ext gag aga gag aga gag aga cca cwt cyt agc tcg and C_L Ext gag aga gag aga gag aga gcg ccg tct aga [3].

All PCR was carried out with an Idaho Technologies Rapidcycler that works by air convection and requires significantly less time than traditional thermocyclers. Specified concentrations of each reaction component in a 1× reaction are as follows: 200 μM dNTPs, 0.5 μM each primer, 0.4 units/μl Amplitaq polymerase in enzyme diluent composed of 10 mM Tris pH8.3, 2.5 mg/ml BSA (bovine serum albumin), and a reaction buffer containing 50 mM Tris, 250 μg/ml

BSA, 2% sucrose, 0.1 mM Cresol Red and either 2 mM, 3 mM or 4 mM Mg^{2+}. PCR amplify heavy chains using a combination of the 5′ primers (Hc1-8, 9) plus each 3′ primer (IgG_1, IgG_{2a}, or IgG_3) in three separate reactions. Amplify light chains with each 5′ primer (Lcl-7) plus the 3′ primer (kappa) in seven separate reactions. Carry out PCR from the cDNA (1:10 dilution) in 10 μl reactions with the following PCR conditions: initial denaturation cycle of 95 °C for 5 s (seconds) followed by 37–40 rounds of 94 °C for 0 s, 52–54 °C for 5 s and 72 °C for 35 s. Complete with a final extension step of 72 °C for 60 s.

E. Vector preparation and restriction digests
Gel purify PCR products from 2.0% agarose gels using the following method. Resuspend cut get pieces in water (approximately 2× gel volume) and homogenize in a microfuge tube with a 1 ml syringe plunger. Centrifuge at 13,000 rpm in a microfuge to remove agarose, remove supernatant and repeat extraction. Concentrate supernatants containing DNA in a Centricon 100 while exchanging buffer for distilled water. Combine light chain cDNA PCR products while retaining heavy chains in three separate reactions for extension PCR. Carry out preparative extension PCR in 50 μl reactions to total 250–500 μl all together. Follow an initial denaturation step at 96 °C for 10 s with 25 cycles at 94 °C for 0 s, 60 °C for 5–10 s, and 72 °C for 50 s. Complete the reaction with a final extension at 72 °C for 60 s. Gel purify products as described above. Combine heavy chain products to result in a single population of heavy chain PCR products and a single population of light chain products overall.

Isolate vector pCOMB3 (carbenicillin resistance) from *E. coli* XL1-Blue ($recA^-$ recA1, lac^-, endA1, gyrA96, thi, hsdR17, supE44, relA1, {F′ proAB, $lacI^q$, lacΔM15, Tn*10*}) using Wizard plasmid kits according to the manufacturer's instructions. Carry out restriction digests with the following units (u) of enzyme/μg DNA: digest 1.7 μg of heavy chain PCR products with *SpeI* (17 u/μg) and *XhoI* (70 u/μg), digest 2 μg of light chain PCR products with *SacI* (35 u/μg) and *XbaI* (70 u/μg), digest 10 μg of pCOMB3 vector with *SpeI* (3 u/μg) and *XhoI* (10 u/μg) and digest 10 μg of pCOMB3 with *SacI* (5 u/μg) and *XbaI* (10 u/μg). Carry out *SpeI/XhoI* and *SacI/XbaI* double digests at 37 °C overnight in manufacturer's buffers H and A, respectively. Digested vector preparations were gel purified on 0.5% agarose gels as described above.

F. Library construction
Perform all ligations with T4 DNA ligase in the manufacturer's buffer at room temperature overnight. For test ligations, use 50 ng of insert (heavy or light chain PCR products) plus 250 ng of vector (pCOMB3). Transform *E. coli* XL1-Blue with 1–2 μl of the ligation reaction by electroporation. Plate dilutions on Superbroth (SB) agar plates containing carbenicillin (50 μg/ml) and tetracycline (10 μg/ml) to assess library size. Use 250 ng of insert plus 1 μg of vector in preparative library ligations. Construct redundant libraries by ligating heavy chains first followed by light chain PCR products and vice versa. Ligate the first PCR product, and transform *E. coli* XL1-Blue by electroporation with the entire ligation reaction. Isolate plasmid the following day. Insert the second PCR product, and again, transform *E. coli* XL1-Blue by electroporation with the entire ligation reaction. Synthesize a high titer phage library in the following manner. Combine electroporations in 100 ml of SB containing 10 μg/ml tetracycline that selects for the F factor in XL1-Blue cells. Harboring the F factor. Incubate at 37 °C for 30 minutes followed by addition of carbencillin to a concentration of 20 μg/ml. After an additional hour of incubation at 37 °C, add carbenicillin to a final concentration of 50 μg/ml. Carbenicillin selects for cells harboring the pCOMB3 plasmid. Incubate culture for another hour at 37 °C and add helper phage VCSM13 (10^{12} pfu) which carried a gene for kanamycin resistance. Continue to incubate for two more hours at 37 °C before adding kanamycin to a final concentration of 50 μg/ml to select for helper phage infection. At this point, move the culture to 30 °C and incubate overnight. Filamentous phage are secreted into the medium as the culture grows. In the morning, centrifuge culture to remove cells and precipitate the phage from supernatant (100 ml) with the addition of 30 ml of 20% PEG 8000/2.5 M NaCl. Recover phage by centrifugation at 10,000 rpm in a Sorvall GSA rotor for 40 minutes and resuspend in 1.5 ml PBS/gelatin (high titer Fab library).

G. Panning
Perform solid-based panning by immobilizing protein on two wells of a Costar A/2 microtiter plate and incubating at 37 °C for 1 hour (Fap 1–100 ng/well). Normal procedure calls for 1 μg protein/well, but less was used here due to limited supplies of purified Fap1. Block non-specific binding sites with PBS/gelatin before adding 50 μl (~1 × 10^{12}/ml) of the high titer Fab library. Wash with distilled water (3–4×) between each step and incubate at 37 °C for 1 hour after each addition. After incubating with phage, wash 1× with TPBS and 4× with water to remove non-specific binders. Elute phage with 50 μl 0.1 M glycine HCl (pH2.2), agitating and scraping sides of well with pipetman intermittently for 10–15 minutes. Add phage to 2 ml freshly grown XL1-Blue and dilute with 8 ml of SB containing tetracycline (10 μg/ml). Plate dilutions (usually > 1:10,000) of culture onto tetracycline (10

μg/ml) and carbenicillin (50 μg/ml) SB agar plates to assess phage yields and to provide colonies to test individual clones. Add carbenicillin (20 μg/ml) to the 10 ml culture after a half hour of incubation. After an additional hour of incubation, add 10 ml culture to 100 mls of SB containing 50 μg/ml carbenicillin and grow up to synthesize phage as described above. Use the resulting high titer phage stock from the first panning in a second round of panning and so on. Monitor enrichment by testing individual clones from each panning as described below.

Perform cell-based panning against whole *S. parasanguis* FW213 cells by centrifuging 1.2ml of overnight culture (grown at 37 °C in 5% CO_2) in a microfuge tube at 3000 rpm for 2–3 minutes. Resuspend cells in 200 μg PBS/gelatin and incubate for 20–30 minutes to block non-specific binding sites on the plastic tube before adding 50 μl (~1 × 10^{12}/ml) of the high titer Fab library. Incubate cells with phage for 2–3 hours at room temperature and centrifuge at 3000 rpm for 1 minute to trap Fabs that are bound to cells. Wash cells repeatedly by resuspending in 1 ml of PBS followed by centrifugation (approximately ten times). Resuspend the final pellet in 200 μl PBS and add phage-bound *S. parasanguis* cells to 2 ml freshly grown *E. coli* XL1-Blue. At this point, the phage are competed off of *S. parasanguis* by infection sites on the *E. coli*. Culture the *E. coli* to produce phage as described above for solid-based panning.

H. Clone characterization and soluble Fab production

Pick individual colonies from plates resulting from each round of panning and synthesize phage as described above. Test individual clones for the ability to bind wildtype FW213 cells versus an insertion mutant, VT1205, that does not express the Fap1 protein using a whole cell ELISA assay (BactELISA) [9]. Briefly, dry cells (FW213 and VT1205) suspended in 50 mM carbonate buffer (16 mM sodium carbonate/34 mM sodium bicarbonate) onto 96 well tissue culture treated plates. Carry out ELISA in the same general fashion as described above. Add a 50 μl/50 μl mixture of culture supernatant containing approximately 10^9 phage and PBS/gelatin to the well. Detect Fab binding with a mixture of HRP conjugated anti-mouse IgG and HRP conjugated anti-M13 secondary antibodies (1:5000 dilution). Although it is not necessary to use two secondary antibodies, enhanced signals can be achieved in this way because the anti-mouse IgG reacts with the antibody and the anti-M13 reacts with the Fab bearing phage. Select positive clones for sequence analysis. Heavy and light chain genes from pCOMB3 DNA were sequenced on an ABI 373 Stretch Sequencer and analyzed with Sequence Analysis Software Version 2.1.1. Perform dideoxy sequencing using a dye terminator cycle sequencing ready reaction kit, and purify products using spin columns packed with G-50 fine grade sephadex. Use the following DNA sequencing primers: SeqT3 ata acc cct cac taa ag or Cγ ggc cag tgg ata gac aga for heavy chains and KEF gaa ttc taa act agc tag tcg or Cκ cac tgg atg gtg gga aga tg for light chains.

Generate soluble Fabs by first purifying pCOMB3 DNA from individual clones using the Wizard mini prep kit. Then digest 2 μg pCOMB3 DNA with *NheI* (9 u/μg) and *SpeI* (3 u/μg) to remove the gene III DNA that fuses the heavy chain to the pIII capsid protein. Digestion with these enzymes produces compatible cohesive ends. Gel purify digested, linear vector and ligate cohesive vector ends together using ligation conditions as describe above. Transform *E. coli* XL1-Blue with 1 μl of the ligation by electroportation and plate dilutions to obtain single colonies. Inoculate 10 ml SB plus carbenicillin (50 μg/ml) with single colonies and incubate at 37 °C for six hours. Induce by adding 300 μM IPTG and moving to 30 °C to incubate overnight. Harvest soluble Fab by concentrating with ammonium sulfate (50%) directly from the growth medium or release Fabs from cells by suspending in PBS and freezing at –70 °C followed by thawing at 37 °C three times.

FW213 and VT1205 ($Fap1^{-1}$ kanamycin insertion mutant) growth culture medium was precipitated with 45% ammonium sulfate and dialyzed in 0.5 × PBS to provide samples for Western blots. Precipitates from FW213 supernatants contain the 200 kDa Fap1 protein whereas VT1205 contains all other proteins with the exception of Fap1. Run samples (25 μg/well) in SDS loading buffer containing beta-mercaptoethanol on SDS polyacrylamide gels (5%). Transfer proteins to nitrocellulose and after blocking with 5% non-fat dry milk/PBS, probe blots with Fabs. Detect Fab binding with the HRP conjugated anti-mouse IgG as described above (1:5000 dilution). React blots with a chemi-illuminescent substrate and expose to x-ray film to visualize protein bands. Two clones that were isolated by phage display were used in Western blots (Ph5 and Ph12). An ammonium sulfate concentrated stock of soluble Fab was used for clone Ph5, while the phage bound form of Ph12 was used directly from culture supernatants. Since affinity purified phage were not used, functional Fab concentrations were determined using ELISA assays (data not shown).

4. Results and discussion

The serum immune response to antigen was tested by ELISA against purified Fap1 protein compared to the reaction with blocking agent alone (PBS/gelatin). There was no apparent immune response in mouse #1. However, a ten-fold immune response over background in mouse #2 indicated the immune repertoire of this mouse contained antibodies specific for the Fap1 protein (Figure 1).

Total RNA was isolated from the spleen of mouse #2 using the method developed by Chomczynski and Sacchi [8]. cDNA was reverse transcribed from mRNA encoding all the antibodies expressed within mouse #2. Light chain genes and the Fd fragments of the heavy chain genes that encode $V_H + C_{H1}$ were amplified from the cDNA by polymerase chain reaction (PCR). A second round of extension PCR provided unique restriction sites for cloning products into the phagemid vector pCOMB3. Libraries were constructed in a redundant fashion resulting in two identical libraries differing only in the order in which the light or heavy chain PCR products were inserted. This method prevents loss of clones due to restriction of the first cloned PCR product with the enzymes used to clone the second PCR product. Ligations were transformed into *E. coli* XL1-Blue by electroporation and upon the addition of helper phage, a combinatorial Fab library was synthesized. Insertion of heavy chain PCR products first resulted in poor electroporation efficiencies. However, insertion of light chains first gave good efficiencies and upon cloning heavy chain PCR products, a library on the order of 3×10^6 clones resulted. Each clone within this library has the potential to possess a different combination of heavy and light chains encoding a unique Fab. However, it is expected that with a large enough library some of the original heavy and light chains will be recombined [12].

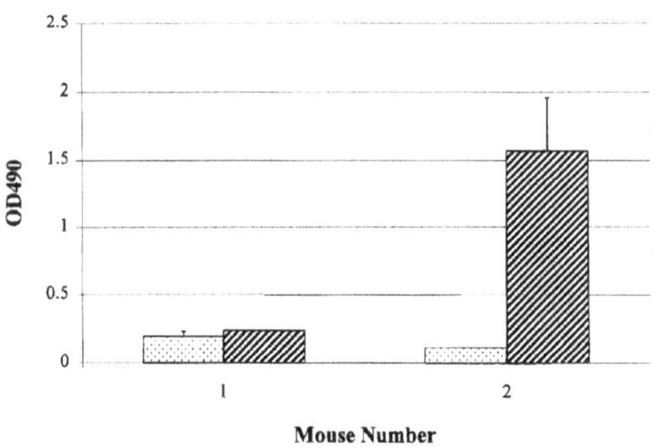

Figure 1. Serum immune response of mouse #1 and mouse #2 as tested by ELISA. Mouse sera was diluted 1:1000 in PBS and added to blocked wells on a microtiter plate containing 0.24 μg Fap1 protein/well or PBS/gelatin alone. Binding of antibodies from the serum was detected with an HRP conjugated anti-mouse IgG secondary antibody. Gelatin ▒; Fap1 protein ▨.

Panning is the process by which antigen specific Fabs are selectively sorted from phage libraries. Multiple rounds of panning are often necessary to bring enrichment to a detectable level [2]. Initially, solid-based panning against purified Fap1 immobilized on a microtiter plate was used, but five rounds of panning did not yield any positive clones. This may have been the result of using small amounts of the Fap1 protein, which was in limited supply. Alternatively, high affinity binders may not have eluted efficiently.

Fap1 is associated with fimbriae on the surface of *S. parasanguis* and is important for the adhesion of these bacteria to the tooth surface [26]. As an alternative, a cell-based panning method using whole *S. parasanguis* FW213 cells was developed. This technique has several advantages over traditional solid-based methods. Whole cells provide ample supplies of the Fap1 antigen. Additionally, native proteins are presented on cells preventing epitopes from being concealed or denatured by binding to plastic microtiter plates [10]. Thus, whole cells increase the chance of isolating antibodies that will be more relevant in analyzing a protein's biologically active state. Using whole cells also facilitated the elution of phage after panning. *S. parasanguis* cells with Fabs bound to their surface were added directly to the *E. coli* culture, which allowed for the phage to be competed from the *S. parasanguis* by infection sites on the *E. coli*. Because of the greater numbers of *E. coli* present in the mixture and the longer elution time allowed by this method, even Fabs with a high affinity for the Fap1 protein would be expected to elute. Traditional elution methods may not elute the highest affinity Fabs in the given time frame. Because *S. parasanguis* does not grow in the absence of carbon dioxide and it does not have a natural resistance to the antibiotics used to culture the *E. coli*, there is little chance of *S. parasanguis* contamination. With another bacterium where this might not be the case, it would be feasible to use heat killed cells [5].

Two populations of phage were used in cell-based panning against *S. parasanguis* cells: the original phage library (PL_1) and the library that resulted from two rounds of solid-based panning against purified Fap1 (PL_2). Selective enrichment can be monitored by following the percentage yield of phage from each panning if input phage levels and panning surfaces are approximately equivalent for each round of panning [2]. Five rounds of panning surfaces are approximately equivalent for each round of panning [2]. Five rounds of panning were carried out and enrichment was evident with increasing phage yields in each round of panning. Enrichment was also quantitated by screening ten random clones from each PL_1 panning one through four. Clones were initially screened for their ability to bind to FW213 and not to VT1205, a Fap1 insertion mutant that does not express the 200 kDa Fap1 protein [26]. A hybridoma monoclonal antibody specific for the Fap1 protein (MAb F51) was used as a positive control. An increase in the proportion of positive clones can be seen with each successive panning. The first panning contained no positive three and the fourth panning contained nine positive clones (Figure 2). A combined total of forty-nine clones from the fifth PL_1 and PL_2 pannings were screened and forty eight of these were positive (data not shown).

Six clones from each of the fifth PL_1 and PL_2 pannings were sequenced to differentiate clones. Heavy and light chain genes were sequenced from pCOMB3 DNA isolated from each clone. All six clones from PL_1 pannings proved to be identical and this Fab was designated Ph5. Three of the clones from PL_2 pannings were also identical to Ph5. However, the remaining three clones had identical sequences that were distinct from Ph5. This clone was specified Ph12. Isolation of a distinct clone from PL_2 supports our hypothesis that this library was biased as compared to the PL_1 library, which had not been panned on microtiter plates. Although no positive clones were isolated from solid-based pannings alone, this library was probably enriched to some degree.

Sequencing clones from the second, third and fourth PL_1 pannings revealed Ph5 was present after two rounds of panning and in every panning thereafter. Due to the prevailing presence of Ph5, it became apparent that further sequencing would not be a productive means of identifying new clones. The genes encoding the Ph5 Fd fragment and light chain were searched for unique restriction sites to use as a means of eliminating this clone from the library, but none were found. An attempt was also made to block enrichment of Ph5 by coating FW213 cells with soluble Ph5 Fab prior to panning. However, Ph5 clones were still isolated from these pannings and no additional clones were identified. In fact, the continuous reappearance of this clone in independent pannings not only demonstrated the strength of this clone, but was also affirmation of the reproducibility of the cell-based panning method.

Results of Western blot analyses with Ph5 and Ph12 demonstrated the specificity of these Fabs for the Fap1 protein (Figure 3). MAb F51 was used as a positive control and has been shown to react with a 200 kDa protein identified as Fap1 as well as a slightly smaller protein of approximately 180kDa [26]. Presumably this is a modification, degradation product or precursor of the 200 kDa Fap1 protein. Ph12 reacted with the same two bands corresponding to Fap1 as F51. While Ph5 also reacted with these same bands, it also reacted with an uncharacterized lower molecular weight protein that may represent a breakdown product of Fap1. Neither Ph5 nor Ph12 reacted with the $Fap1^-$ mutant, VT1205. Additionally, Ph5 reacted weakly in immunoblots as compared to Ph12. Although Ph5 and Ph12 Fab stocks used in Western blots reacted comparably in ELISA assays (data not shown), Ph5 was only minimally detectable

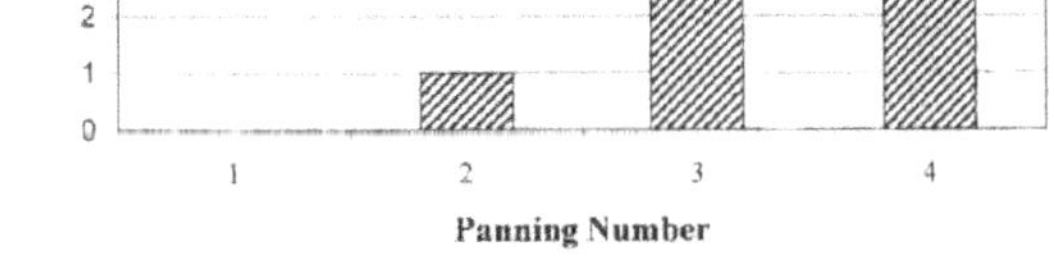

Figure 2. Selective enrichment in each round of panning quantitated by whole cell ELISA assay. Ten clones from each round of panning 1–4 were tested for their ability to bind to *S. parasanguis* FW213 and VT1205, a *fap 1* insertion mutant. Binding of phage bound Fabs directly from culture supernatants was detected with a mixture of HRP conjugated anti-mouse IgG and HRP conjugated anti-M13 secondary antibodies. Positive clones were scored as those with the ability to bind specifically to FW213 and not to VT1205.

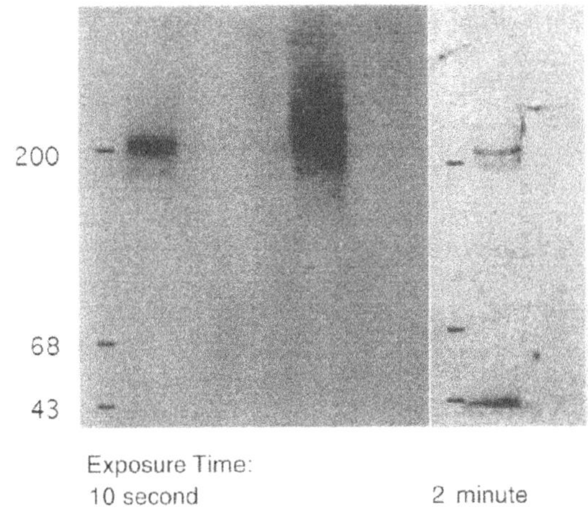

Figure 3. Western blot analyses with Ph5 and Ph12. Ammonium sulfate precipitated samples from wildtype FW213 (FW) and VT1205 *fap1* mutant (Fap-) culture supernatants were probed in Western blots with a concentrated ammonium sulfate precipitated stock of soluble Fab clone Ph5, a stock of clone Ph12 in phage bound form directly from culture supernatants or MAb F51 (positive control) antibodies. Binding was detected with HRP-conjugated anti-mouse IgG and a sensitive chemi-illuminescent substrate.

using a more concentrated ammonium sulfate stock, while Ph12 binding was readily detectable using a dilute stock of phage bound Fabs. With sensitive detection with chemi-illuminescent substrate, only faint bands were detected after an exposure time of two minutes for Ph5 compared with Ph12 that gave a strong signal within ten seconds. This indicates that Ph12 reacts well with the denatured protein while Ph5 does not, another difference that may be attributed to differential enrichment of the PL_2 library with initial solid-based pannings. Western blots of whole cell extracts of FW213 and VT1205 gave similar results further demonstrating the specificity of Ph5 and Ph12 for the Fap1 protein (data not shown).

Immunization with the purified protein allowed for the isolation of Fabs specific for the Fap1 protein and not to other *S. parasanguis* proteins. It should be possible to isolate a variety of Fabs to different surface proteins by immunizing with whole cells. Unique proteins could be identified by isolating specific Fabs and using these antibodies to immunoaffinity purify unknown antigens. Reverse genetics could ultimately be employed to clone the gene product. In a complementary approach, protein products from a genomic library were phage displayed and panned against different ligands of interest [14]. This approach was shown to be successful for cloning known receptor sequences from *Staphylococcus aureus*. In addition to the pCOMB3 system [1, 2], several kits are also available to make phage display techniques even more user friendly. These include the Recombinant Phage Antibody System made by Pharmacia Biotech, the T7 Select Phage Display System by Novagen, and the SurfZAP Vector system by Stratagene.

In conclusion, cell-based panning was successfully used to isolate phage display Fabs to the *S. parasanguis* surface protein Fap1. Panning against whole cells is an efficient means of enrichment, producing positive clones in as little as two pannings. This method would appear to be generally applicable for cloning phage display Fabs to various cell surface antigens.

Acknowledgments

This work was supported by US Department of Energy Grant DE-FG02-88ER60742, NIH-NIDR Grant R37-DE11000, the Vermont Cancer Center, and the Lake Champlain Cancer Research Organization. We would like to thank Dr Dennis Burton for the SeqT3, KEF and PCR primer sequences, and Dr Ivan Bespalov for Cκ and Cγ primer sequences. We would like to thank Dr Keith Mintz for providing the purified Fap1 protein.

Notes on suppliers

1. Charles River Laboratories, Inc., 251 Ballardvale Street, Wilmington, MA 01887-1000, USA
2. Stratagene, 11011 N. Torrey Pines Road, La Jolla, CA 92037, USA
3. BIO 101, Inc., 1070 Joshua Way, Vista CA 92083, USA
4. Sigma Chemical Company, P.O. Box 14508, St. Loius, MO 63178, USA
5. Gibco BRL, P.O. Box 68, Grand Island, NY 14072-0068, USA
6. Molecular Research Center, Inc., 5645 Montgomery Road, Cincinnati, OH 45212, USA
7. Ribi ImmuneChem Research, Inc., 553 Old Corvallis Road, Hamilton, MT 59840, USA
8. Boehringer Mannheim, 9115 Hague Road, P.O. Box 50414, Indianapolis, IN 46250-0414, USA
9. Pharmacia Biotech, Inc., 800 Centennial Ave, P.O. Box 1327, Piscataway, NJ 08855-1327, USA
10. Amersham Corporation, 2636 S. Clearbrook Drive, Arlington Heights, IL 60005, USA
11. Promega, 2800 Woods Hollow Road, Madison, WI 53711-5399, USA
12. The Scripps Research Institute, 10666 North Torey Pines Road, La Jolla, CA 92037, USA
13. Idaho Technologies, P.O. Box 50819, Idaho Falls, ID 83405, USA
14. Perkin Elmer Applied Biosystems, 850 Lincoln Centre Drive, Foster City, CA 94404, USA
15. Amicon, A GRACE Company, 72 Cherry Hill Drive, Beverly MA 01915, USA
16. Bio-Rad Laboratories, 1000 Alfred Nobel Drive, Hercules, CA 94547, USA
17. Costar, One Alewife Center, Cambridge, MA 02140, USA
18. VWR Scientific Products, 405 Heron Drive, P.O. Box 626, Bridgeport, NJ 08014, USA
19. Bio-Tek Instruments, Inc., Laboratory Division, Highland Park, Box 998, Winooski, VT 05404-0998, USA

References

1. Barbas CF III, Kang AS, Lerner RA, Benkovic SJ (1991). Assembly of combinatorial antibody libraries on phage surfaces: The gene III site. Proc Natl Acad Sci USA 88: 7978–7982.
2. Barbas CF III, Lerner RA (1991). Combinatorial immunoglobulin libraries on the surface of phage (Phabs): Rapid selection of antigen-specific Fabs. Methods: Companion Methods Enzymol 2: 119–124.
3. Bespalov I, Purmal AA, Glackin MP, Wallace SS, Melamede RJ (1996). Recombinant phabs reactive with 7,8–dihydro–8–oxoguanine, a major oxidative DNA lesion. Biochemistry 35: 2067–2078.
4. Better M, Chang CP, Robinson RR, Horwitz AH (1988). *Escherichia coli* secretion of an active chimeric antibody fragment. Science 240(4855): 1041–1043.
5. Bradbury A, Persic L, Werge T, Cattaneo A (1993). Use of living columns to select specific phage antibodies. Biotechnology (NY) 11: 1565–1569.

6. Casson LP, Manser T (1995). Random mutagenesis of two complementarity determining region amino acids yields an unexpectedly high frequency of antibodies with increased affinity for both cognate antigen and autoantigen. J Exp Med 182: 743–750.
7. Chiswell DJ, McCafferty J (1992). Phage antibodies: Will new 'coliclonal' antibodies replace monoclonal antibodies? Trends Biotechnol Mar 10: 80–84.
8. Chomczynski P, Sacchi N (1987). Single-step method of RNA isolation by acid guanidinium thiocyanate-phenol-chloroform extraction. Anal Biochem 162: 156–159.
9. Elder BL, Boraker DK, Fives-Taylor PM (1982). Whole-bacterial cell enzyme-linked immunosorbent assay for *Streptococcus sanguis* fimbrial antigens. J Clin Microbiol 16: 141–144.
10. Friguet B, Chaffotte AF, DJavadi-Ohaniance L, Goldberg ME (1985). Measurements of the true affinity constant in solution of antigen-antibody complexes by enzyme-linked immunosorbent assay. J Immunol Methods 77: 305–319.
11. Hill HR, Stockley PG (1996). Phage presentation. Mol Microbiol May 20: 685–692.
12. Huse WD, Sastry L, Iverson SA, Kang AS, Alting-Mees M, Burton DR, Benkovic SJ, Lerner RA (1989). Generation of a large combinatorial library of the immunoglobulin repertoire in phage lambda. Science 246(4935): 1275–1281.
13. Iba Y, Kurosawa Y (1997). Comparison of strategies for the construction of libraries of artificial antibodies. Immunol Cell Biol 75: 217–221.
14. Jacobsson K, Frykberg L (1996). Phage display shotgun cloning of ligand-binding domains of prokaryotic receptors approaches 100% correct clones. Biotechniques 20: 1070–1076.
15. Kohler G, Milstein C (1975). Continuous cultures of fused cells secreting antibody of predefined specificity. Nature 256: 495–497.
16. Marks JD, Hoogenboom HR, Griffiths AD, Winter G (1992). Molecular evolution of proteins on filamentous phage. Mimicking the strategy of the immune system. J Biol Chem 267: 16007–16010.
17. Martin AC, Cheetham JC, Rees AR (1991). Molecular modeling of antibody combining sites. Methods Enzymol 203: 121–153.
18. McCafferty J, Griffths AD, Winter G, Chiswell DJ (1990). Phage antibodies: filamentous phage displaying antibody variable domains. Nature 348(6301): 552–554.
19. Pereira S, Maruyama H, Siegel D, Van Belle P, Elder D, Curtis P, Herlyn D (1997). A model system for detection and isolation of a tumor cell surface antigen using antibody phage display. J Immunol Methods 203: 11–24.
20. Ruff-Jamison S, Glenney JR Jr (1993). Molecular modeling and site directed mutagenesis of an anti-phosphotyrosine antibody predicts the combining site and allows the detection of higher affinity interactions. Protein Eng 6: 661–668.
21. Sambrook J, Maniatis T, Fritsch EF (1989). Molecular cloning: a laboratory manual. New York: Cold Spring Harbor Laboratory Press.
22. Smith GP (1985). Filamentous fusion phage: Novel expression vectors that display cloned antigens on the virion surface. Science 228(4705): 1315–1317.
23. Watters JM, Telleman P, Junghans RP (1997). An optimized method for cell-based phage display panning. Immunotechnology 3: 21–29.
24. Webster DM, Pedersen J, Staunton D, Jones A, Rees AR (1994). Antibody-combining sites. Extending the natural limits. Appl Biochem Biotechnol 47: 119–132.
25. Winter G, Griffiths AD, Hawkins RE, Hoogenboom HR (1994). Making antibodies by phage display technology. Annu Rev Immunol 12: 433–455.
26. Wu H, Mintz KP, Ladha M, Fives-Taylor P (1997). Isolation and characterization of Fap1, a fimbriae associated adhesin of *Streptococcus parasanguis* FW213. Mol Micro (submitted).

Address for correspondence: Paula Fives-Taylor, Department of Microbiology and Molecular Genetics, College of Medicine and College of Agriculture and Life Sciences, Stafford Hall, University of Vermont, Burlington, Vermont 05405, USA
Phone: (802) 656-1121; Fax: (802) 656-8749
E-mail: pfivesta@zoo.uvm.edu

Index, Vol. 20/1–4

www.ingramcontent.com/pod-product-compliance
Lightning Source LLC
LaVergne TN
LVHW062121200926
848484LV00007B/1729

* 9 7 8 0 7 9 2 3 5 8 1 5 2 *